MECHANICS OF MATERIALS

South Carolina Edition

JAMES W. DALLY
ROBERT J. BONENBERGER

University of Maryland, College Park

College House Enterprises, LLC
Knoxville, Tennessee

Copyright © 2018 by College House Enterprises, LLC.

Reproduction or translation of any part of this work beyond that permitted by Sections 107 and 108 of the 1976 United States Copyright Act without permission of the copyright owner is unlawful. Requests for permission or for further information should be addressed to the Publisher at the address provided below.

This textbook is intended to provide accurate and authoritative information regarding the topics described. It is distributed and sold with the understanding that the publisher is not engaged in providing legal, accounting, engineering or other professional services. If legal advice or other expertise advice is required, the services of a recognized professional should be retained.

The front cover photograph is the new Abraham Lincoln Bridge (also known as the Downtown Crossing), which connects downtown Louisville and Jeffersonville, Ind., running parallel to the Kennedy Bridge. The purpose of the Louisville-Southern Indiana Ohio River Bridges Project is to increase cross-river mobility by improving safety, alleviating traffic congestion and connecting major highways.

The body of the manuscript was prepared with 11 point Times New Roman font using Microsoft Word. The Word.doc files were converted to pdf files using Adobe Acrobat Pro DC. The book was printed "on demand" by Ingram Spark.

College House Enterprises, LLC.
5713 Glen Cove Drive
Knoxville, TN 37919, U. S. A.
email   jwd@collegehousebooks.com
Visit our web site at http://www.collegehousebooks.com

13 Digit ISBN 978-1-935673-37-8

ABOUT THE AUTHORS

James W. Dally (Jim) obtained a Bachelor of Science and a Master of Science degree, both in Mechanical Engineering from the Carnegie Institute of Technology. He obtained a Doctoral degree in mechanics from the Illinois Institute of Technology. He has taught at Cornell University, Illinois Institute of Technology, the U. S. Air Force Academy and served as Dean of Engineering at the University of Rhode Island. He is currently a Glenn L. Martin Institute Professor of Engineering (Emeritus) at the University of Maryland, College Park.

Dr. Dally has also held positions at the Mesta Machine Co., IIT Research Institute and IBM. He is a Fellow of the American Society for Mechanical Engineers, Society for Experimental Mechanics, and the American Academy of Mechanics. He was appointed as an honorary member of the Society for Experimental Mechanics in 1983 and elected to the National Academy of Engineering in 1984. Professor Dally was selected by his peers to receive the Senior Faculty Outstanding Teaching Award in the College of Engineering and the Distinguish Scholar Teacher Award from the University. He was also a member of the University of Maryland team receiving the 1996 Outstanding Educator Award sponsored by the Boeing Co. More recently he received the Daniel C. Drucker Medal from the ASME and the Archie Higdon Distinguished Educator Award from the Mechanics Division of the ASEE.

Professor Dally has co-authored several other books: *Experimental Stress Analysis, Experimental Solid Mechanics, Photoelastic Coatings, Instrumentation for Engineering Measurements, Packaging of Electronic Systems, Mechanical Design of Electronic Systems, Production Engineering and Manufacturing, Statics, Instrumentation and Sensors for Engineering Measurements and Process Control, and Introduction to Engineering Design, Books 1, 2, 3, 4, 5, 6, 7, 8, 9, 10 and 11*. He has authored or coauthored about 200 scientific papers and holds five patents.

Robert J. Bonenberger, Jr. (Bob) obtained B.S.E., M.S. and Ph.D. degrees in Mechanical Engineering from the University of Maryland, Baltimore County and College Park campuses. He has taught undergraduate and graduate students in mechanics, strength of materials, and experimental stress analysis. Currently, he is the Coordinator for the Materials Instructional Laboratory and a Keystone Instructor in the Clark School of Engineering at the University of Maryland, College Park.

Previously, Dr. Bonenberger worked in the Fracture Mechanics Section at the Naval Research Laboratory, both as a postdoctoral fellow and as a contract employee. He is a member of the American Society for Mechanical Engineers, Society for Experimental Mechanics, American Society for Materials, and American Society for Engineering Education. His research interests include material behavior at high strain rates, experimental stress analysis, and fracture mechanics. He has authored or co-authored 25 scientific papers.

PREFACE

This textbook is one of the many titles that exist for a course in Mechanics of Materials. The content in the Mechanics of Materials text is a revision of a part of the content in a book titled Design Analysis of Structural Elements, which dates back to 2002. The textbook "Design Analysis of Structural Elements" was an integrated treatment of both Statics and Mechanics of Materials. However, it was too long and the students complained that it was too heavy. It also required a supplement for the homework exercises. Separate textbooks for Statics and Mechanics of Materials published in 2106, and were based largely on the content in Design Analysis of Structural Elements. Over the past decade several editions of both Statics and Mechanics of Materials have been published. The revisions and reorganization of the content was based on reactions of instructors and students. The organization of this edition of Mechanics of Materials was directed by Professor Michael Sutton, for his classes at the University of South Carolina in Columbia, SC.

We began developing notes for the first edition of Design Analysis of Structural Elements with a pilot offering in the spring semester of 1998. Many revisions were made before a limited first edition of the textbook was published in the summer of 1999. The second edition was published in 2000 and was used by about a thousand students. The third edition, written in 2001, included six new chapters to expand the coverage necessary for a complete Mechanics of Materials course. The fourth edition, published in 2004, was similar to the third edition except for relatively small changes. Since that time, more than 10,000 students have used the various editions of the Statics and Mechanics of Materials textbooks.

Errors discovered during this extensive usage have been corrected; however, errors always occur even with careful proof reading by many diligent people. We would greatly appreciate students and instructors calling errors to our attention. The e-mail address of one of the authors is given on the copyright page.

ACKNOWLEDGEMENTS

As is always the case when major changes are made to the curriculum, the College administrators must lead the way. We are indeed fortunate to have several administrators who not only supported this effort, but also insisted that the mechanics offerings be markedly improved. They seek not only a much more favorable educational experience for the students, but also much better understanding and retention of the course content. We thank Dr. William L. Fourney, Chair of the Keystone Program, Dr. Balakumar Balachandran, Chair of the Mechanical Engineering Department, and Dr. Ali Haghani, Chair of the Civil Engineering Department for their support. Particular thanks are due to Dr. Nariam Farvardin who established the Keystone Program, when he was Dean of the College of Engineering and to Darryll Pines, the current Dean, who has continued the strong emphasis on continuously improving the educational experience for our engineering students.

Special thanks are due to several individuals for their significant contributions to the development of this book. Professor Michael Sutton from USC outlined the reorganization of the content and recommended the addition of separate chapters on Stress and Strain at a point. Dr. Bill Fourney led our efforts to change the curriculum to provide a more effective approach for presenting Statics and Mechanics of Materials so students would better understand the material and retain this knowledge. Bill has stayed with the project for more than 19 years, taught sections of students every semester, and provided the leadership needed to keep others involved and interested.

Many instructors teaching the course made valuable suggestions for improvements to the textbook. These include: Dr. Michael Sutton, Dr. Mary Bowden, Dr. Thomas Brodrick, Dr. Hugh Bruck, Dr. James Duncan, Dr. Bongtae Han, Dr. Kwan-Nan Yeh, Dr. Peter Sandborn, Dr. Charles Schwartz, Mr. Christopher Baldwin, and Mr. Thomas Beigel of the University of Maryland, College Park.

James W. Dally
College Park, Maryland
Summer 2017

ABOUT THE AUTHORS
PREFACE

CONTENTS

LIST OF SYMBOLS   xii

CHAPTER 1 REVIEW OF STATICS

1.1	Introduction	1
1.2	Statics and Mechanics of Materials	2
1.3	Newton's Laws	3
1.4	Forces	4
1.5	Basic Quantities, Units, Prefixes and Conversions	5
1.6	Significant Figures	7
1.7	Scalars, Vectors and Tensors	7
1.8	Internal and External Forces	8
1.9	Moments	10
1.10	Free Body Diagrams and Equilibrium	11
1.11	Free Body Diagrams (FBDs)	12
1.12	Centroid of an Area	17
1.13	Summary	22
	Problems	22

CHAPTER 2 STRESS AT A POINT

2.1	Introduction	23
2.2	Three Dimensional Stresses at a Point	23
2.3	Stress Equations of Transformation	26
2.4	Principal Stresses	30
2.5	Mohr's Circle	36
2.6	Mohr's Circle in Three Dimensions	42
2.7	Failure Theories	42
2.8	Summary	46
	Problems	48
	References	49

CHAPTER 3 STRAIN AND THE STRESS STRAIN RELATIONS

3.1	Introduction	50
3.2	Definitions of Displacement and Strain	50
3.3	Strain Equations of Transformation	54
3.4	Stress Strain Relations	59
3.5	Determining Principal Stresses with Strain Gages	62
3.6	Thermal Strains and Thermal Stresses	66
	Problems	70

CHAPTER 4 MATERIAL PROPERTIES

4.1 Introduction	73
4.2 The Tensile Test	73
4.3 Material Properties	77
4.4 Fatigue Strength	84
4.5 Summary	95
Problems	98
Reference	98

CHAPTER 5 AXIALLY LOADED STRUCTURAL MEMBERS

5.1 Introduction	99
5.2 Characteristics of Cable, Rods and Bars	99
5.3 Design analysis Methods	100
5.4 Shear Stresses	116
5.5 Stresses on Oblique Planes	117
5.6 Axial Loading of a Tapered Bar	120
5.7 Axial Loading of a Stepped Bar	123
5.8 Stress Concentrations	125
5.9 Scale Models	129
5.10 Statically Indeterminate Axial Members	132
5.11 Summary	136
Problems	139
References	148

CHAPTER 6 TORSION OF STRUCTURAL ELEMENTS

6.1 Torsion Loading	149
6.2 Deformation of a Shaft Due to Torsion	151
6.3 Stresses Produced by Torsion	152
6.4 Shear Stresses on Different Planes	160
6.5 Principal Stresses in Shafts	163
6.6 Angle of Twist	164
6.7 Design of Power Transmission Shafting	168
6.8 Torsion of Non-Circular Shafts	171
6.9 Shear Center	174
6.10 Thin Walled Tube with Arbitrary Shape	178
6.11 Stress Concentrations in Circular Shafts Subject to Torsion	184
6.12 Summary	187
Problems	189

CHAPTER 7 STRESSES IN BEAMS

7.1 Pure Bending of Symmetric Beams	196
7.2 Deformation of a Beam in Pure Bending	199
7.3 Properties of Cross Sections	205
7.4 Bending of Beams with Transverse Forces	206
7.5 Bending of Composite Beams	222
7.6 Plastic Bending	230
7.7 Stress Concentrations in Beams	234
7.8 Summary	241
Problems	243

CHAPTER 8 DEFLECTION OF BEAMS

8.1	Introduction	254
8.2	The Elastic Curve for a Beam	254
8.3	Establishing the Elastic Curve by Integration	255
8.4	Deflections and Slopes with Singularity Functions	269
8.5	Superposition Concepts	281
8.6	Statically Indeterminate Beams	288
8.7	Summary	293
	Problems	295

CHAPTER 9 STRESSES IN THIN WALLED PRESURE VESSELS

9.1	Introduction	305
9.2	Spherical Pressure Vessels	305
9.3	Cylindrical Pressure Vessels	306
9.4	Summary	309
	Problems	310

CHAPTER 10 COMBINED LOADING

10.1 Introduction	313
10.2 Summary	329
Problems	329

CHAPTER 11 BUCKLING OF COLUMNS

11.1	Introduction	333
11.2	Buckling of Columns with Both Ends Pinned	334
11.3	Influence of End Conditions	337
11.4	Column Stresses and Limitations of Euler's Theory	344
11.5	Eccentrically Loaded Columns	347
11.6	Stresses in Columns with Eccentric Loading	349
11.7	Design Codes	353
11.8	Summary	356
	Problems	358

CHAPTER 12 ENERGY METHODS

12.1	Introduction	364
12.2	The Energy Theorems	365
12.3	Strain Energy Density	366
12.4	Strain Energy in Structural Elements	368
12.5	Dynamic (Impact) Loading	376
12.6	Castigliano's Theorem	383
12.7	Summary	398
	Problems	400

APPENDICES — 411

APPENDIX A Wire and Sheet Metal Gages — 412

APPENDIX B1 Physical Properties of Common Structural Materials — 413
APPENDIX B2 Tensile Properties of Common Structural Materials — 414
APPENDIX B3 Tensile Properties of Non Metallic Materials — 415

APPENDIX C Properties of Areas

	C.1	Area	416
	C.2	First Moment of Area	417
	C.3	Centroid of the Area A	417
	C.4	Locating the Centroid of a Composite Area	422
	C.5	Second Moment of the Area	423
	C.6	The Parallel Axis Theorem	426
	C.7	Moments of Inertia of Composite Areas	427
	C.8	Summary	431

APPENDIX D GEOMETRIC PROPERTIES OF ROLLED STEEL SHAPES

	D.1	Wide Flange Shapes	434
	D.2	American Standard Beams	436
	D.3	Structural Tees	438

APPENDIX E EQUATIONS FOR DEFLECTION OF BEAMS

	E.1	Simply Supported Beams	440
	E.2	Cantilever Beams	441

INDEX   442

LIST OF SYMBOLS

A	area	s	distance or dimension
%A	percent reduction in area	S_{design}	design strength
a	acceleration vector	S_f	fatigue strength
a, b, c	dimensions	S_y	yield strength
a_{CR}	critical crack length	S_{ys}	yield strength in shear
C	constant, center dimension	S_u	ultimate tensile strength
D	diameter	S_{us}	ultimate tensile strength in shear
d	diameter or distance	SF	safety factor
%e	percent elongation	SE	structural efficiency
E	elastic modulus or Young's modulus	SR	slenderness ratio
E	strain energy	T	temperature, torque
E	modulus scale factor	t	time
e	eccentricity	**u**	unit vector
e	strain energy density	U	Distortional energy
F	force magnitude	V	shear force
F	force as a vector	**v**	velocity vector
F_f	friction force or failure force	V	volume
FBD	free body diagram	W	weight, watt
g	gravitational constant	W	work
G	shear modulus	w	width dimension
GF	gage factor	x, y z	Cartesian coordinates
H	horizontal shear force	Z	section modulus
HP	horsepower		
h	height	α	temperature coefficient of expansion
I	moment of inertia	α, β, γ	direction cosines
J	polar moment of inertia	Δ	delta
K	stress concentration factor	$\Delta\varepsilon$	strain range
K_I	stress intensity factor	$\Delta\sigma$	stress range
K_{IC}	plane strain fracture toughness	δ	deflection or displacement
K_c	fracture toughness	ε	strain
k	number, spring rate	ε_p	plastic strain
L	length dimension	ε_T	true strain
L	load scale factor	ϕ	angle of friction, angle of twist
ln	natural logarithm	γ	shear strain, density
M	moment magnitude	κ	curvature
M_Y	yield moment	π	3.1416 radians
M_L	limit moment	μ	coefficient of friction
M	moment as a vector	ν	Poisson's ratio
M	multiplier, momentum	ρ	radius of curvature
MA	mechanical advantage	Σ	Summation sign
MOS	margin of safety	σ	stress
m	mass, subscript for model	σ_a	alternating cyclic stress
N	normal force, number, Newton, RPM	σ_B	Bearing stress
n	number	σ_{design}	design stress
P	internal force	σ_f	failure stress
P_{CR}	critical buckling force	σ_m	mean cyclic stress
p	pressure, subscript for prototype	σ_T	true stress
Q	first moment of the area, flow rate	σ_{fail}	stress to produce failure
q	distributed loading	σ_{yield}	stress to produce yielding
r	radius, radius of gyration, distance	θ	angle
r	position vector	θ_s	angle of repose
R	reaction force, radius, resistance	τ	shear stress
S	geometric scale factor	ω	angular velocity

CHAPTER 1

REVIEW OF STATICS

1.1 INTRODUCTION

The subject of mechanics is usually divided into four different courses, which include:

Statics, Dynamics, Mechanics of Materials and Fluid Mechanics

Statics and dynamics both deal with rigid bodies that are subjected to a system of forces. The study of statics, involves the determination of either internal and/or external forces acting on a structural element, which is in a state of equilibrium (usually at rest). In dynamics, the forces acting on the body produce motion and the body either accelerates or decelerates. The analysis in dynamics deals with determining position, velocity (angular or linear) and acceleration as some function of time.

Newton's laws guide studies of all branches of mechanics[1]. Consider Newton's second law:

$$\sum F = \frac{d}{dt}(mv) \qquad (1.1)$$

where $\Sigma \mathbf{F}$ is the sum of all of the forces acting on the rigid body, m is the mass of the body, **v** is the velocity and d/dt is the derivative operator[2].

When dividing the study of mechanics into four separate subjects, educators considered a special situation, where the body's velocity is either constant or zero. The subject called **statics** is based on this simplification. In this special situation, Eq. (1.1) reduces to:

$$\sum \mathbf{F} = 0 \qquad (1.2)$$

In the study of both statics and dynamics, the material from which the body is fabricated is not a concern, providing the body remains essentially rigid when the forces are applied. However, in the study of mechanics of materials, the deformation of the body is an essential consideration. Two assumptions are made in studying mechanics of materials:

1. The deformations of the body are small
2. Plane sections in the body remain plane during and after deformation.

These assumptions enable us to determine the distribution of internal forces and stresses in a body. Note that the material from which the body is fabricated is of critical importance in determining the distribution

---

[1] Sir Isaac Newton (1642-1727) formulated three laws of motion and the law of universal gravitational attraction.
[2] Symbols in bold font indicate vector quantities.

of internal forces and stresses of materials. Also, the deformations of the body due to the application of external forces are markedly affected by the rigidity of its material. Whether a body fails or not, depends on the strength of its material. Hence, you should recognize that the rigidity and the strength of the material from which the body is fabricated are important parameters.

1.2 STATICS AND MECHANICS OF MATERIALS

This textbook integrates two closely related subjects in mechanics, namely **statics** and **mechanics of materials**. In studying **statics**, it is assumed that the body under consideration is perfectly rigid, and does not deform when forces are applied. Solutions to statics problems involve determining **external forces** acting on some structures and the **internal forces** developed in others by using only **free body diagrams** and the appropriate **equilibrium equations**. Solutions for forces acting on bodies in equilibrium involve only three basic steps:

- **Construct a complete set of free body diagrams.**
- **Apply the appropriate equations of equilibrium.**
- **Execute the mathematics required to solve one or more of the equilibrium equations.**

Mechanics of materials is an extension of statics. In the study of Mechanics of Materials we account for the effect of **material deformations**[3] on the **internal stresses** generated in a body. Solutions to problems that arise in mechanics of materials involve the five steps. These include the same three basic steps, and two additional steps. These additional steps are to accommodate for the effect of the deformations on the distribution of the internal stresses:

- **Construct a complete set of free body diagrams.**
- **Apply the appropriate equations of equilibrium.**
- **Assume the geometry of the deformations (plane sections remain plane).**
- **Employ the appropriate relations between stress and strain.**
- **Execute the mathematics required in solving the equations.**

It is evident that statics and mechanics of materials are closely related. The equilibrium equations from statics must be used to determine the internal forces and moments, **P**, **V** and **M** before beginning to solve problems in mechanics of materials. In studying mechanics of materials, the concepts of stress and strain are quickly encountered. Mathematically these concepts are somewhat complex, because they are tensor quantities, but physically they are simple and easily understood. **Stress is a concept based on the equilibrium** of a portion of a body, which is produced by sectioning the body. The internal force acting on a section cut is developed by stresses distributed over the area exposed by that cut.

Strain, on the other hand, is a **geometric concept**. Strain is determined by the change in geometry that occurs when a body deforms under load. When considered individually, both stress and strain are independent of the material from which the body is fabricated. It is only when we attempt to write a relation connecting stress and strain that the material properties (modulus of elasticity and Poisson's ratio) must be considered.

---

[3] In mechanics of materials, we assume the deformations are so small that they do not significantly affect the magnitude or the direction of the internal and external forces acting on the body.

Materials used in constructing most structures are usually ignored in the study of statics. The equilibrium equations are identical for all materials, and the internal and external forces for statically determinant structures[4] do not depend upon the structure's material. Materials are much more important in studies of the mechanics of materials. Stress and strain are usually related in solutions, and the elastic constants that describe the rigidity of the materials must be employed. In addition, the margin of safety for a structure subjected to specific loading is often predicted. In these predictions of safety or structural failure, the appropriate **"strength"** of the material is required.

In this textbook, the physical aspects of problems solved using principles of both statics and mechanics of materials are stressed. It is essential that complete free body diagrams be constructed to model the structure and to define the unknown forces. Understanding the equations of equilibrium is as important as the use of **vector algebra** to solve them. Of course, it is **essential** that the equations resulting from the application of the principle of equilibrium be solved correctly.

1.3 NEWTON'S LAWS

The first of Newton's laws is derived from Eq. (1.2) and may be written as:

$$\mathbf{F}_1 + \mathbf{F}_2 + \ldots\ldots + \mathbf{F}_n = 0 \tag{1.2a}$$

where n forces are acting on the body.

Equation (1.2a), in one form or the other, is used extensively in both statics and mechanics of materials, when the **equilibrium equations** are written. The vector representation of Eq. (1.2a) may be recast by writing the equivalent **scalar equations** as:

$$\Sigma F_x = 0 \qquad \Sigma F_y = 0 \qquad \text{and} \qquad \Sigma F_z = 0 \tag{1.2b}$$

$$F_{1x} + F_{2x} + \; + F_{nx} = 0 \quad F_{1y} + F_{2y} + \; + F_{ny} = 0 \quad \text{and} \quad F_{1x} + F_{2x} + \ldots + F_{nx} = 0 \tag{1.2c}$$

The direction of the forces in this equation have been taken into account by considering only those forces in either the x, y or z directions. When the forces are constrained to the three Cartesian directions, the **scalar form of the equilibrium equations** is employed. However, when the forces act in directions other than the Cartesian directions, the force vectors must first be decomposed into their **components** acting in the x, y and directions, before they can be used in the scalar equations.

Newton's second law $\Sigma \mathbf{F} = (d/dt)(m\mathbf{v}) = m\mathbf{a}$ is the equation used most frequently in the study of **dynamics**. When the sum of the forces is not zero, a body of mass m is subjected to an acceleration $\mathbf{a}$. The study of **dynamics** in most engineering programs is scheduled later in the curriculum.

Newton's third law is called the law of action and reaction. The concept of active and reactive forces is illustrated in Fig. 1.1. In this illustration, the sphere is in equilibrium under the action of two forces — the first is the weight W due to gravity that acts downward and the second is the reaction force acting upward at its contact point. The reaction force R is equal in magnitude to the weight W, but opposite in direction. When any two bodies are in contact, the two forces that develop at the contact point are equal in magnitude and opposite in direction.

---

[4] Structures are classified as statically determinant if the reactions at their supports may be determined using only the equations of equilibrium.

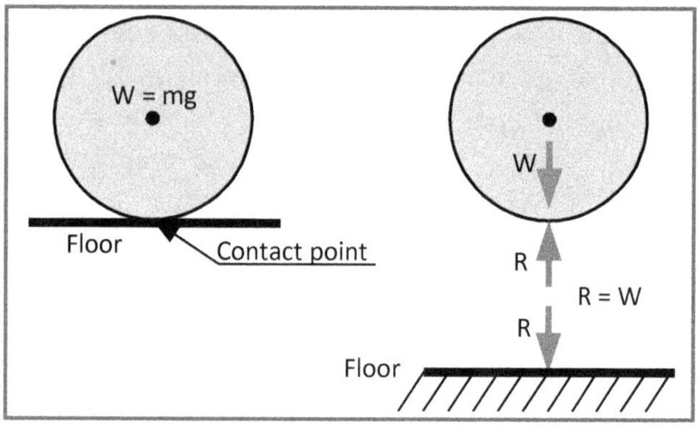

Fig. 1.1 The active force W due to gravity produces an equal and opposite reaction force R at the contact point.

1.4 FORCES

Statics involves a study of external forces that act on a body and internal forces that act within the body. In studying statics the external forces acting on a body, and the internal forces developed by stresses within the body, are determined. Some forces occurring under static (steady state) conditions include:

1. Gravitational
2. Pressure acting over a defined area
3. Friction
4. Magnetic
5. Electrostatic.

In addition, forces developed under dynamic conditions are referred to as inertial forces and include:

1. Centrifugal
2. Centripetal
3. Coriolis.

Let's examine the forces due to gravity, because they are usually the most important forces in engineering applications. We continuously work and expend huge amounts of energy to overcome gravitational forces. Gravitational forces are the primary concern when designing building and bridge structures against failure by collapse or rupture. Even in vehicle design, where dynamic forces are significant, gravitational forces are critical in the design of the vehicle's structure and its power train.

Weight is a force produced by the Earth's gravitational pull on the mass of our body, as illustrated in Fig. 1.2. The force on a body due to gravity is given by:

$$F = G\, m_e\, m_b\, /R_e^2 \qquad (1.3)$$

where m_e is the mass of the Earth, m_b is the mass of a body, R_e is the radius of the Earth, and G is the universal gravitational constant [6.673×10^{-11} m^3/ (kg-s^2)].

In writing Eq. (1.3) it was assumed that Earth bound bodies, either those of people or objects at or near the Earth's surface, are very small compared to the radius of the Earth. This is a reasonable assumption because radius of the Earth is 3,960 mi. or 6.37×10^6 m. Collecting together the quantities in Eq. (1.3) that are related to the Earth and setting them equal to g_e gives:

$$g_e = G m_e /R_e^2 \qquad (1.4)$$

Note that g_e is the **gravitational constant** equal to 32.17 ft/s² or 9.807 m/s². The subscript will be dropped in subsequent discussions of the gravitational constant, with the understanding that g is to be applied to Earth bound bodies.

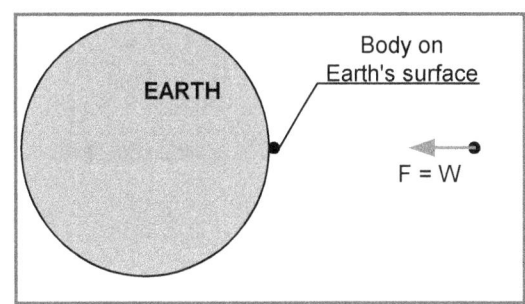

Fig. 1.2 Bodies on Earth's surface are small relative to the Earth's radius.

Combining Eq. (1.3) and Eq. (1.4) gives:

$$F = m_b g = W \qquad (1.5)$$

From these definitions, it is clear that the units of g are ft/s² or m/s². When the weight of an object is measured, its weight is essentially constant regardless of location if the object remains Earth bound. However, the constant quantity in Eq. (1.5) is not the weight W, but the mass m_b. To prove this statement, consider the weight of a known mass on the moon. It is known that a person weighs much less on the moon — about one sixth as much as here on Earth. Because the mass m_b is constant, the object weighs less because the gravitational constant for the moon is only about $g_e/6$. The smaller gravitational constant for the moon is due to its smaller mass when compared to the Earth's mass. The mass of any body is the same whether it is on the moon, Mars, Earth or anywhere in space.

1.5 BASIC QUANTITIES, UNITS, PREFIXES AND CONVERSIONS

In the study of mechanics, four basic quantities — length, time, force and mass — are encountered. These quantities are shown with their respective units for the SI and the U. S. Customary systems of units in Table 1.1.

The basic units are not independent, because they must be **dimensionally homogenous** in any equation. For example, the dimension homogeneity of Eq. (1.1) is maintained by defining the units for length, time and mass in the **SI** system, and then deriving the remaining basic unit for the force, the **newton**, in terms of length, time and mass. For the **U. S. Customary** system, the basic unit for mass, the **slug**, is derived in terms of the units for length, time, and force.

In the International System of Units (SI), length is given in meters (m), time in seconds (s), and the mass in kilograms (kg). The unit for force is called a newton (N) in honor of Sir Isaac Newton. The newton is derived from Eq. (1.1) so that a force of 1 N will impart an acceleration of 1 m/s² to a mass of 1 kg [1 N = (1 kg) (1 m/s²)]. For dimension homogeneity, it is clear that N is equivalent to (kg-m)/s². In the SI system the gravitational constant g = 9.807 m/s². With this value of the acceleration due to gravity on Earth, the weight of a mass of 1 kg is:

$$W = mg = (1\text{kg})(9.807 \text{ m/s}^2) = 9.807 \text{ N} \qquad (a)$$

Table 1.1 Unit Conversion Factors

Quantity	U. S. Customary	SI Equivalent
Acceleration	ft/s^2	0.3048 m/s^2
	in/s^2	0.0254 m/s^2
Area	ft^2	0.0929 m^2
	in^2	645.2 mm^2
Distributed Load	lb/ft	14.59 N/m
	lb/in.	0.1751 N/mm
Energy	ft-lb	1.356 J
Force	kip = 1000 lb	4.448 kN
	lb	4.448 N
Impulse	lb-s	4.448 N-s
Length	ft	0.3048 m
	in	25.40 mm
	mi	1.609 km
Mass	lb mass	0.4536 kg
	slug	14.59 kg
	ton mass	907.2 kg
Moments or Torque	ft-lb	1.356 N-m
	in-lb	0.1130 N-m
Area Moment of Inertia	in^4	0.4162 x 10^6 mm^4
Power	ft-lb/s	1.356 W
	hp	745.7 W
Stress and Pressure	lb/ft^2	47.88 Pa
	lb/in^2 (psi)	6.895 kPa
	ksi = 1000 psi	6.895 MPa
Velocity	ft/s	0.3048 m/s
	in/s	0.0254 m/s
	mi/h (mph)	0.4470 m/s
Volume	ft^3	0.02832 m^3
	in^3	16.39 cm^3
	gal	3.785 L
Work or Energy	ft-lb	1.356 J

In the U. S. Customary System, length is given in feet (ft), force in pounds (lb), and the time in seconds (s). The unit for mass is called a slug, which is derived from Eq. (1.1) so that a force of 1 lb will impart an acceleration of 1 ft/s^2 to a mass of 1 slug [1 lb = (1 slug)(1 ft/s^2)]. For dimension homogeneity, it is clear that a slug is equivalent to (lb-s^2)/ft. In the U. S. Customary System the gravitational constant g = 32.17 ft/s^2. With this value of the acceleration, the mass of a body weighing 1 lb is:

$$m = W/g = 1 \text{ lb}/ (32.17 \text{ ft/s}^2) = 0.03108 \text{ slug} \qquad (b)$$

In addition to the basic quantities, several other quantities encountered in studies of statics and mechanics of materials are presented in another table shown on the page prior to the back cover.

Often the metric prefixes are employed in expressing numerical results. For example, a metric prefix M with the unit for stress (Pa) is often employed. Metric prefixes are useful in dealing with either very large or very small numbers. Accordingly, it is common practice to employ them together with the symbol for the units (e.g. MPa). A table of the metric prefixes is provided in another table shown on the page prior to the back cover.

The quantities in these tables are presented in terms of the basic units for length, time and force. However, in practice other units are often employed. For instance, moments may be expressed as in-lb instead of ft-lb, and a stress expressed as MPa instead of Pa or psi (lb/in^2) instead of (lb/ft^2). It is fortunate that strain is a dimensionless quantity and it is not necessary to convert from one system of units to another, or to convert from one unit to another within the same system. Table 1.1 shows the factors used to convert from U S Customary to SI units.

1.6 SIGNIFICANT FIGURES

With a hand-held calculator, numerical results with ten or more figures are common. These results are misleading, because the values used for material properties are rarely accurate to more than 0.2% or one part in 500. To avoid the implication of fictitious accuracies, it is recommended that the results from a calculator be written with four significant figures. This practice yields a computational accuracy of 1/1000 or 0.1% and is consistent with the accuracy of the physical data and the analytical model upon which the formula is based.

1.7 SCALARS, VECTORS AND TENSORS

Quantities of three types — **scalars, vectors** and **tensors** — arise in the study of Mechanics. In ordinary life it is not essential to recognize the difference among them. A person is comfortable if he or she has friends, food, shelter and some money. Incidentally, friends, food, shelter and money are all scalar quantities, because it is only necessary to count them. Other quantities encountered in engineering that are scalar include: area, energy, length, mass, moment of inertia, power, volume, work, etc.

Scalar quantities are the easiest of the three types with which to work. Simple arithmetic operations (add, subtract, multiply and divide) are sufficient in manipulating scalar quantities. However, arithmetic operations are not sufficient for manipulating vectors and tensors, because additional mathematical operations and descriptions must be introduced to manipulate equations involving these quantities.

While both vector and tensor quantities are found in mechanics, they are not **difficult to manipulate.** However, non-scalar quantities must be recognized and treated appropriately. Mathematics beyond arithmetic is required when dealing with vectors and tensor quantities. For example, vector quantities require two descriptors — **magnitude and direction**. The magnitude indicates the size of the quantity, and the direction gives its orientation. Quantities that require vector representation for their complete description include: acceleration, force, impulse, momentum, moments, velocity, etc.

A typical vector quantity, such as a force, is represented with an arrow, as shown in Fig. 1.3. The length of the arrow is proportional to the magnitude of the force, and its inclination relative to an x-y coordinate system gives its direction. Vectors are used extensively in statics, and it is necessary to accommodate both their magnitude and directions in the solution of many different types of equilibrium problems. Vectors are **independent** of the orientation chosen for the x-y coordinate system. However, the scalar value of the vector's direction and the magnitude of its components relative to the x and y directions are **dependent** on the orientation of the reference coordinate system. This property is important when coordinate systems are selected to solve equilibrium problems, because choices can be made that simplify a solution without changing the physical description of the problem.

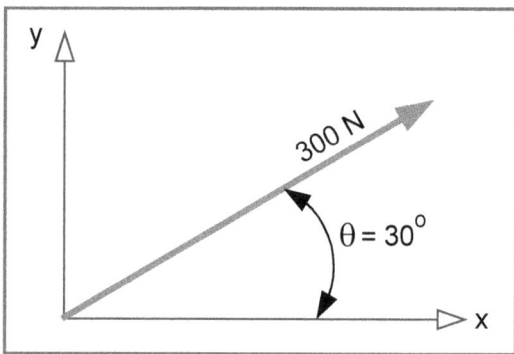

Fig. 1.3 A force vector with a magnitude of 300 N and direction of 30° relative to the x-axis.

Tensor quantities are more difficult to describe than vector quantities, because their complete description requires information pertaining to three characteristics. Like vectors, the magnitude and direction of a tensor quantity must be specified. In addition, the orientation of the plane upon which a tensor quantity acts must be specified. For example, stresses σ are tensor quantities. The three characteristics — **magnitude, direction, and the orientation of the plane** upon which the stresses act are illustrated in Fig. 1.4. The round bar is loaded with forces in the axial direction. A section cut exposes an internal surface (plane) of the bar normal to its axis, where stresses act in the axial direction with a magnitude of 1,000 kPa.

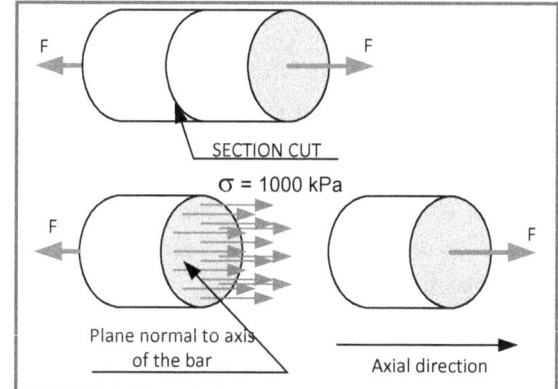

Fig. 1.4 Stresses of 1,000 kPa act in the direction of the axis of the bar. The stresses are acting on the plane normal to the bar's axis.

1.8 INTERNAL AND EXTERNAL FORCES

In dealing with forces, it is important to distinguish between those applied to a structure **(external)**, and those that develop within a structural element **(internal)**. external forces include the active loads applied to the structure, such as those shown in Fig. 1.5a-c, and the reaction forces, shown in Fig. 1.5d, that develop at the supports to maintain the structure in equilibrium.

In Fig.1.5a, a simply supported beam is loaded with a **concentrated force F** at a point near its center. The concentrated force, applied at a point, is an idealization. Forces are always distributed over some area; however, with concentrated forces, we assume that the area is so small that it approaches a point. In Fig. 1.5b, the beam is loaded with a **uniformly distributed load q** that is applied over most of its length. Uniformly distributed loads along beams are specified in terms of force/unit length (i.e. lb/ft or N/m). A distributed load that is increasing from the left end of the beam to its right end is illustrated in Fig. 1.5c. Again, the symbol **q** is used to designate the magnitude of the distributed forces; however, in this case q is a function of the position x along the length of the beam, and is designated by **q(x)**.

The final example of external forces is shown in Fig. 1.5d. The beam and its loading are identical with that shown in Fig. 1.5b; however, the supports have been removed. Reaction forces **R**, developed by the supports to maintain the beam in equilibrium, are shown as concentrated forces.

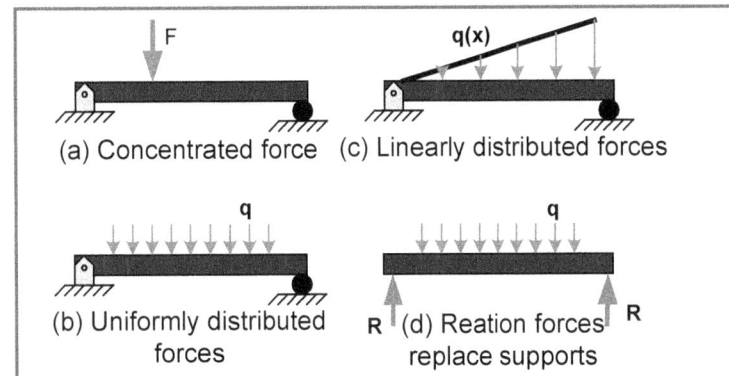

Fig. 1.5 Examples of different types of external forces applied to the structure (beam).

Internal forces develop within a structural member due to the action of the applied external forces. These internal forces are not visible; they are represented by making imaginary cuts through a structural member. Examine the bar subjected to two external forces F applied at each end as shown in Fig. 1.6a. A section cut is made in the central region of the bar, which is perpendicular to its axis. This cut is imaginary, not a real cut. The cut permits visualization of the interior of either segment of the bar. Examine the segment on the left, and find a **normal stress** σ, which is uniformly distributed over the area exposed by the section cut. When this stress is integrated over the area of the bar, an internal force **P** is generated that acts along the axis of the bar. The magnitude of **P** is given by:

$$P = \int \sigma \, dA \qquad (1.6)$$

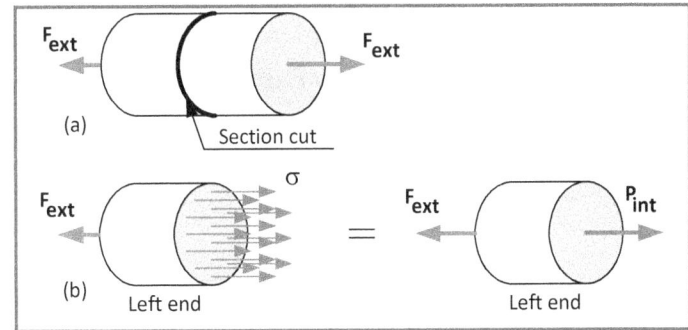

Fig. 1.6 (a) A tension bar with external loads and a section cut dividing it into two parts.
(b) Two different representations of the left side of the bar.

The left end of the bar must be in equilibrium, which implies that:

$$\Sigma F_x = 0 \qquad (1.2b)$$

Summing the forces in the x direction gives:

$$P = F \qquad (1.7)$$

In this elementary example of a rod in tension, the relation between the internal force **P** within the bar and the external force **F** applied to its ends has been determined. The same three-step approach will be used

10 — Chapter 1
Review of Statics

throughout this book in solving for the internal forces in much more complex problems. The tree steps are:

1. Make an appropriate section cut
2. Use the equations of equilibrium
3. Solve for internal forces in structural members

1.9 MOMENTS

When using a wrench or screwdriver to tighten a bolt or screw, a moment is generated. The moment M_o is produced when a force **F**, as shown in Fig. 1.7, is applied in such a manner that it tends to cause a body to rotate about point O. The magnitude of a moment produced by a force is dependent on the location of the point O, and is given by:

$$M_o = (F)(d) \qquad (1.8)$$

where d is the perpendicular distance from the point O to the line of action of the force **F**.

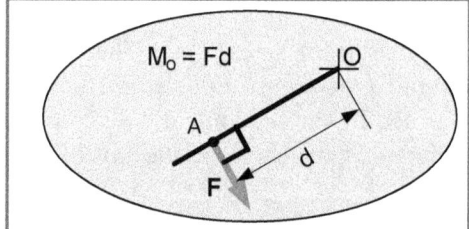

Fig. 1.7 The moment M_0 produced by a force **F** depends on its position relative to point O.

The units of a moment M_o are always (force) times (length), such as N-m or ft-lb. M_o is a vector quantity that must be specified with both magnitude and direction. The magnitude of the vector M_0 is given by Eq. (1.8) and its direction is perpendicular to the plane in which **F** and the line OA lie. As shown in Fig. 1.8, the moment M_o has a sense of direction. In this illustration, the moment M_o tends to rotate the body in a **counterclockwise** direction and is **positive**.

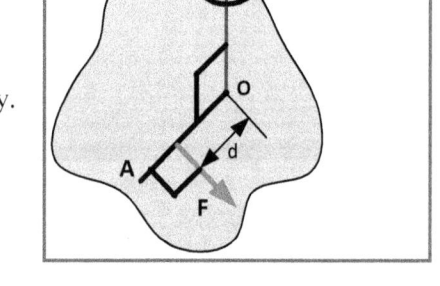

Fig. 1.8 A graphic illustration of the moment M_o as a vector quantity.

To determine the sign of the moment, use the **right hand rule**. In applying this rule, place the palm of your right hand along the axis of rotation and point your fingers in the direction of the force and rotate your hand. If the direction of the force causes you to rotate **counterclockwise**, with your thumb pointing up from the plane in which both **F** and d lie, then the moment is **positive**. However, if you must rotate your hand **clockwise**, with your thumb pointing downward, the moment is **negative**.

1.10 FREE BODY DIAGRAMS AND EQUILIBRIUM

Equilibrium is an extremely important concept, because it provides an excellent approach for determining the magnitude of the unknown forces that act on a body. It is understood that a body is in equilibrium if it is at rest or traveling at a constant velocity. For example the beam, shown in Fig. 1.9, is at rest, because the loads acting downward are resisted by the two reactive forces produced by the supports. The entire assembly is fixed to bedrock.

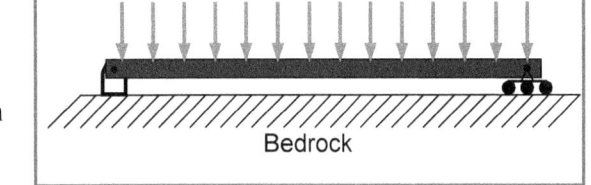

Fig. 1.9 A simply supported beam that is at rest and in equilibrium.

Bodies that move may or may not be in equilibrium. If a body is in equilibrium, it moves with a constant velocity and a constant **momentum**[5]. The effect of unbalanced forces ($\Sigma \mathbf{F} \neq 0$) acting on a body is to change its momentum by altering its velocity. The rate of that change is proportional to the unbalanced force. Therefore, when a body moves at constant velocity (zero acceleration) it possesses constant momentum and is in equilibrium.

The two vector equations of equilibrium $\sum \mathbf{F} = 0$ and $\sum \mathbf{M} = 0$ are equivalent to the six scalar equations of equilibrium.

$$\Sigma F_x = 0; \qquad \Sigma F_y = 0; \qquad \Sigma F_z = 0$$
$$\Sigma M_x = 0; \qquad \Sigma M_y = 0; \qquad \Sigma M_z = 0 \qquad (1.9)$$

Solutions for many problems arising in both statics and mechanics of materials are executed by using Eqs. (1.9). The decision, whether to use the vector or scalar form of the equilibrium equations, is left to your discretion, although in many cases the scalar equations are easier to apply.

Although there are six scalar equations of equilibrium, it is not always necessary to employ all of them to solve an equilibrium problem. By classifying different systems of forces acting on a body, it is possible to identify only those equations that provide relevant information. This classification greatly simplifies the approach used to solve equilibrium problems. Four different force systems are classified as:

- **Non-coplanar and non-concurrent.**
- **Non-coplanar and concurrent.**
- **Coplanar and non-concurrent.**
- **Coplanar and concurrent.**

Non-coplanar, Non-concurrent Force Systems

When a body is subjected to non-coplanar and non-concurrent forces and is in equilibrium, the directions and magnitudes of $\mathbf{F}_1, \mathbf{F}_2 \ldots \mathbf{F}_n$ must satisfy the **six** Cartesian component equations of equilibrium given in Eq. (1.9).

---

[5]Momentum, M a vector quantity, is defined as M = m**v**, where **v** is the velocity of the body.

Non-coplanar, Concurrent Force Systems

When a body in equilibrium is subjected to non-coplanar, but concurrent forces, the directions and magnitudes of $F_1, F_2 \ldots F_n$ must satisfy the **three** Cartesian component force equations, which are:

$$\Sigma F_x = 0 \qquad \Sigma F_y = 0 \qquad \Sigma F_z = 0$$

Due to concurrency, it is evident that moments do not occur about point O and the equations $\Sigma M_x = \Sigma M_y = \Sigma M_y = 0$ are satisfied regardless of the magnitude of the forces.

Coplanar, Non-concurrent Force Systems

When a body in equilibrium is subjected to coplanar but non-concurrent forces, the directions and magnitudes of $F_1, F_2 \ldots F_n$ must satisfy **three** Cartesian component equations of equilibrium, which are:

$$\Sigma F_x = 0 \qquad \Sigma F_y = 0 \qquad \Sigma M_z = 0$$

The remaining three equations of equilibrium are satisfied independently of the solution for the three possible unknowns.

Coplanar, Concurrent Force Systems

This category is the simplest of the four force systems because equilibrium is satisfied by using only the following **two** equations:

$$\Sigma F_x = 0 \qquad \text{and} \qquad \Sigma F_y = 0$$

The other four equilibrium relations are satisfied automatically and they do not provide useful information. Due to concurrency, it is evident that moments do not occur about point O and the equations $\Sigma M_x = \Sigma M_y = \Sigma M_y = 0$ are satisfied regardless of the magnitude of the forces. Also, because the forces are coplanar, they all lie in the x-y plane and $\Sigma F_z = 0$.

1.11 FREE BODY DIAGRAMS (FBDs)

The equations of equilibrium described in the previous section may be applied to:

- A single body or member.
- A structure made of several members.
- A portion of a multi-member structure formed by a section cut.
- A part of a body or structure that has been formed by two or more section cuts.
- An element removed from a body.

Before utilizing the equations of equilibrium, it is essential that a **free body diagram (FBD)** be constructed of the structural element being analyzed. A FBD is a model. Its purpose is to simplify the physical representation of the structure by omitting its fine details that are not necessary in solving an equilibrium problem.

An example, for constructing a FBD, is illustrated in Fig. 1.10. The beam is subjected to a uniformly distributed load of magnitude **q** (N/m), and supported near each end with simple supports. In Fig. 1.10a, a uniformly distributed load is represented with a shaded rectangle placed over the span of the beam. The uniformly distributed load may also be represented with a series of arrows, as illustrated in Fig. 1.10b. Both of these graphical techniques are used to model uniformly distributed loads. The simple supports are modeled with a pin and clevis on the left side and a roller on the right side. The pin and clevis holds the beam in a fixed position at its left end, while the roller permits the beam to freely expand and/or contract by very small amounts due to changes in temperature.

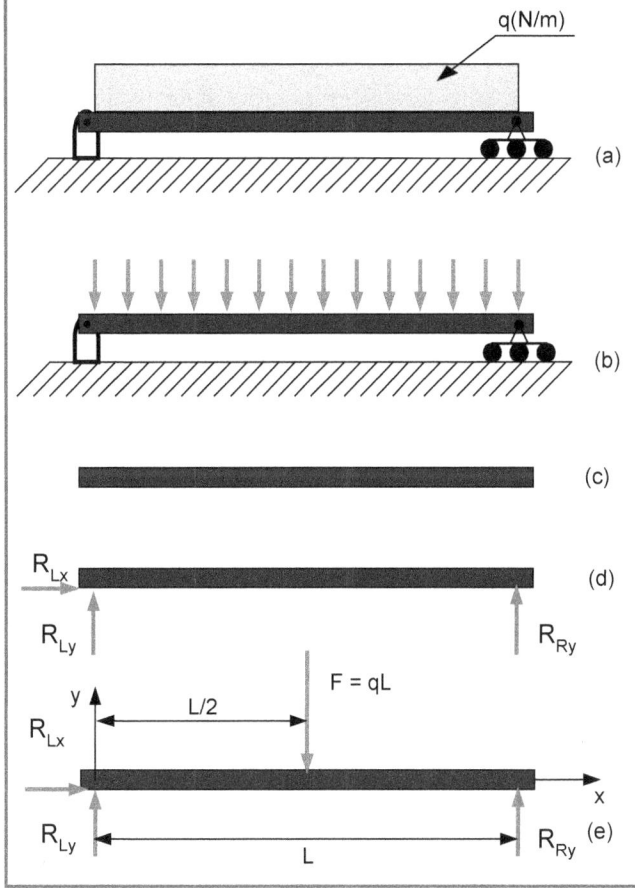

Fig. 1.10 Construction of a FBD for a simply supported uniformly loaded beam.

A four-step procedure is employed in drawing the free body of the beam.

1. Isolate the body (a beam in this case), by removing the supports and the uniformly distributed load. The isolated beam is shown in Fig. 1.10c.
2. The supports are replaced with the **reaction loads** R_{Lx}, R_{Ly} and R_{Ry}. The pin and clevis at the left support produces reactions in both the x and y directions, as shown in Fig. 1.10d. However, the roller, which is free to move horizontally, produces only a single reactive force normal to the surface of the beam (the y direction). At this stage of the analysis, the magnitudes of these forces are not known.
3. The uniformly distributed load **q** applied over the length L of the beam is replaced with a concentrated force **F** located in the center of the beam. The magnitude of the force is F = (qL) and it acts downward, as shown in Fig. 1.10e.
4. Finally, dimensions are added to the FBD, and a coordinate system established to facilitate the equilibrium analysis. In dimensioning the FBD, the concentrated force is placed at the centroid of the shaded rectangular area representing the uniformly distributed load (e.g. at L/2 from the left support).

1.11.1 Modeling Loads

The FBD is a model of a structure or some part it. To prepare the FBD, the loads that act on the structure and its supports are modeled. First consider modeling the loads due to gravity, which is one of the most commonly encountered forces. Examine the block with a mass m as shown in Fig. 1.11a. The gravitational force due to the mass of the block is modeled with a concentrated force F = mg, which is applied at the center of the block (its centroid). Because the force is due to gravity, the direction of the force is downward (in the negative y direction).

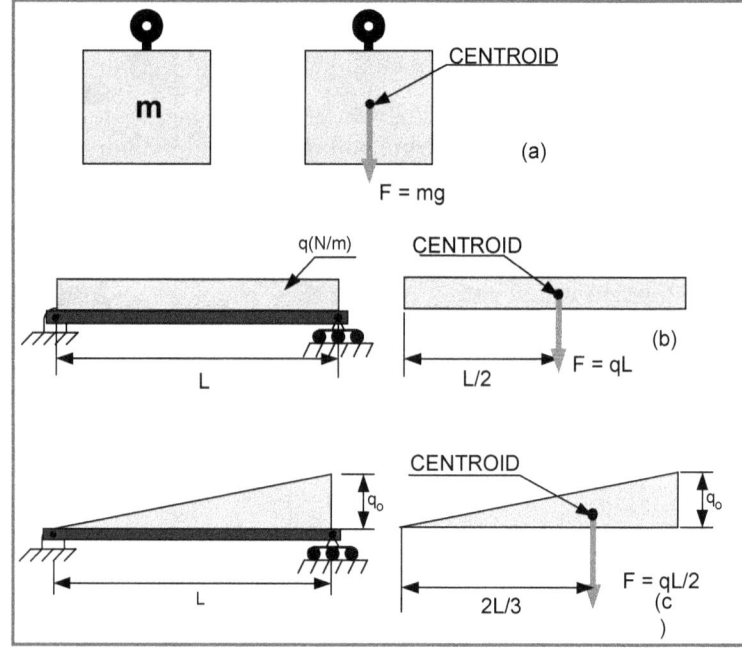

Fig. 1.11 Modeling gravitational and distributed loads.

In Fig.1.11b, a uniform load of magnitude q is distributed over the span of a beam. The magnitude q is expressed in terms of force/unit length (N/m or lb/ft). This uniformly distributed load is represented with a rectangular area of height q and length L. A concentrated force F, the static equivalent of the uniformly distributed load, is given by the area of the rectangle as F = qL. The concentrated force is applied at the centroid of the rectangular area (x = L/2) and acts downward.

In Fig. 1.11c, the load is again distributed over the length of the beam, but it increases as a linear function of x (e.g. $q(x) = q_o x/L$). This load is represented with a triangular area having an altitude q_o and a base L. A concentrated force, the static equivalent of the distributed load, is given by the area of the triangle as $F = q_o L/2$, which is applied at the centroid of the triangle (x = 2L/3) acting downward.

When modeling to solve for reaction forces at structural supports, distributed loads are replaced with statically equivalent concentrated forces. However, later in Chapters **7** and **8** when determining internal moments M and shear forces V, it is not possible to use this simplified modeling technique.

1.11.2 Modeling Supports

The pin/clevis and roller are often used to support beams. The pin/clevis is fixed and does not permit the beam to move in either the x or y directions, although it does permit the beam to rotate as the beam deflects. On the other hand, the roller permits the beam to expand or contract by very small amounts with changes in temperature. When the pin and the roller are removed in the construction of a FBD, they are replaced with reaction forces shown in Fig. 1.12. For a pin/ clevis support, which restrains motion in the x and y directions, reaction forces R_{Ly} (perpendicular to the beam's axis) and R_{Lx} (parallel to the beam's axis) are required to represent this support. For a roller support (on a frictionless surface), a single reaction force R_{Ry} perpendicular to the beam's axis is sufficient.

Fig. 1.12 Modeling the pivot and roller supports.

There are many support conditions and connections to structures. In constructing FBDs, the structure is modeled by removing these supports and replacing them with one or more reactive forces or reactive moments. Several different types of supports or connections and their reactive forces and moments are shown in Table 1.1. The structure is always constrained by supports at one or more locations. These supports are modeled in FBDs with some combination of concentrated forces and moments that effectively provide the constraint the supports provide.

**Table 1.1
Different types of supports or connections and their reactive forces and moments.**

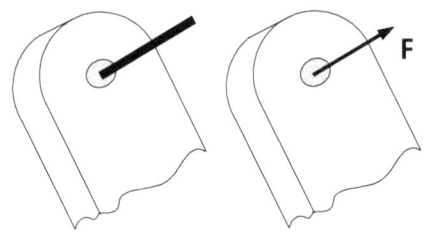

A. A cable connection is modeled with a single force acting in the direction of the cable and away from the structure.

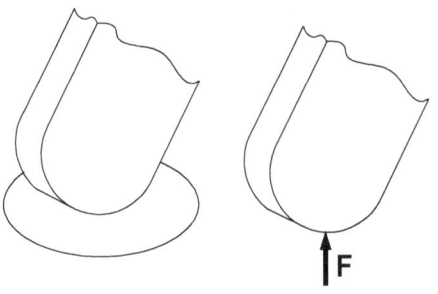

B. A rocker on a frictionless surface is modeled with a concentrated force perpendicular to the surface at the point of contact.

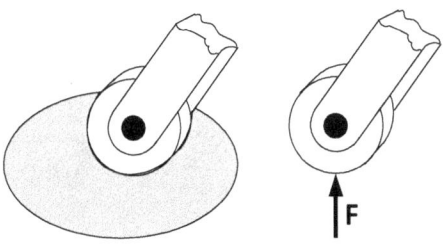

C. A roller on a flat surface is represented with a concentrated force perpendicular to the surface at the point of contact.

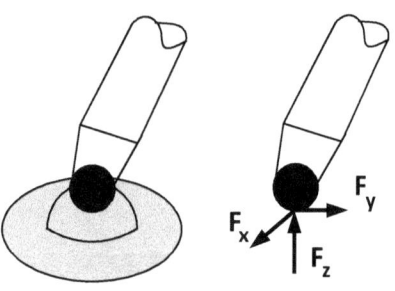

D. A ball and socket joint is modeled with three Cartesian force components.

E. A journal bearing is represented with two forces perpendicular to the shaft and two moments about axes perpendicular to the shaft.

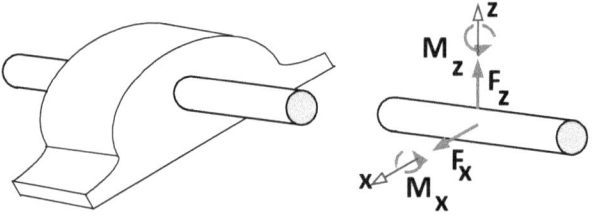

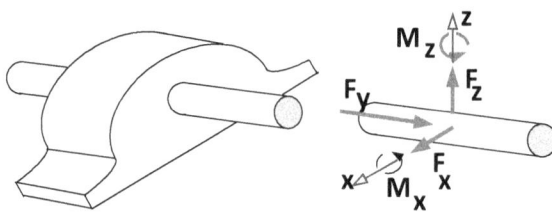

F. A thrust bearing is represented with three Cartesian forces and two moments about axes perpendicular to the shaft.

G. A pin and clevis connection is represented with three Cartesian forces and two moments about axes perpendicular to the pin.

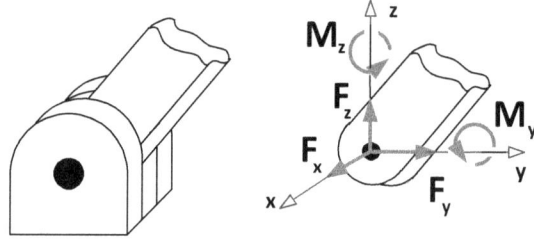

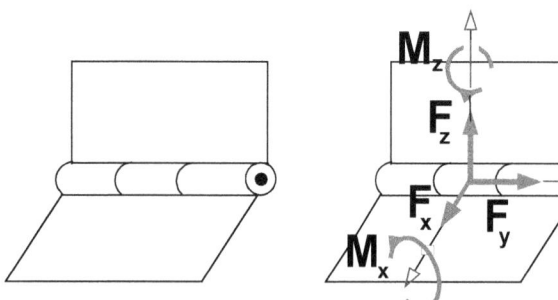

H. A hinge connection is represented with three Cartesian forces and two moments about axes perpendicular to the hinge pin.

I. A fixed support is modeled with six possible reactions (three Cartesian forces and three Cartesian moments).

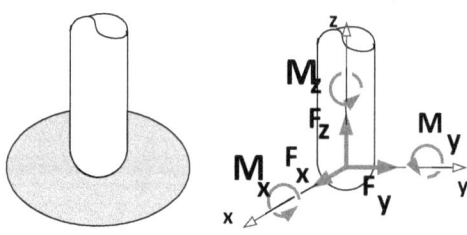

1.11.3 FBDs of Partial Bodies

It is very important to determine internal forces in structural members, because these forces produce stresses and deflections that might cause the structure to fail. The approach followed in determining the magnitude of internal forces is to:

1. **Make an imaginary cut through the structural member being studied.**
2. **Draw a FBD of one part of this member.**
3. **Account for the effect of the portion of the cut away member by applying internal forces and moments to the area exposed by the section cut.**
4. **Solve for the internal forces or moments using the appropriate equations of equilibrium.**

A FBD provides a model to use in writing the equilibrium equations employed in the solution for unknown forces and reactions. The FBD indicates the position and direction of the known and unknown forces. It also provides the dimensions needed to compute relevant moments and a reference coordinate system.

1.12 CENTROID OF AN AREA

In Fig. 1.11, we showed that the forces acting on a body pass through the centroid of the area of the force distribution. For this reason it is important to understand methods for determining the location of centroids. The centroid of an area A is defined by point C located relative to an arbitrary coordinate system Oxy in Fig. 1.13. A centroid is defined as the point that locates the center of gravity of a line, an area or a volume.

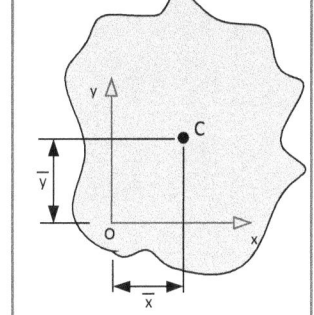

Fig. 1.13 The centroid C of an area A is located with coordinates $\bar{x}$ and $\bar{y}$.

The coordinates $\bar{x}$ and $\bar{y}$ locating the centroid of an area are determined from the first moments as:

$$Q_y = \int_A x\,dA = A\bar{x}$$
$$Q_x = \int_A y\,dA = A\bar{y}$$
(1.10)

where $\bar{x}$ and $\bar{y}$ are dimensions locating the centroid as shown in Fig. 1.13.

Let's illustrate the method for determining the first moment of the area and the location of a centroid by considering a few elementary shapes, in the examples presented below.

EXAMPLE E1.1

Consider the rectangular area illustrated in Fig E.1.1, with the origin of the Oxy coordinate system positioned at its lower left-hand corner. Determine the first moments of the rectangular area and the location of its centroid relative to the Oxy coordinate system.

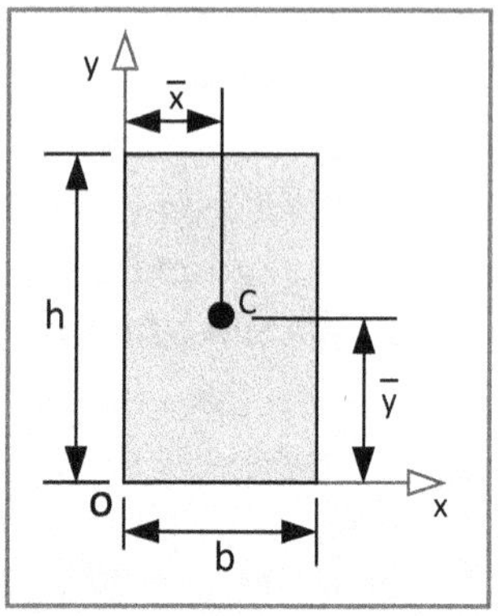

Fig E1.1 A rectangle with the coordinate system located along its edges.

Solution:

For the rectangle area, presented in Fig. E1.1, the first moments of the area about the x and y axes are given by Eq. (1.10) as:

$$Q_x = A\bar{y} = (bh)\frac{h}{2} = \frac{bh^2}{2}$$

$$Q_y = A\bar{x} = (bh)\frac{b}{2} = \frac{b^2 h}{2} \tag{1.11}$$

It was possible to quickly solve for the first moments, Q_x and Q_y, because we recognized the location of the centroid for the rectangular area. When an axis of symmetry exists for a given area, the centroid is located somewhere on this axis of symmetry. With the rectangle, two axes of symmetry exist; hence, the location of the centroid is at the intersection of its two symmetric axes.

For the circular cross section shown in Fig. 1.14, the center of the circle clearly locates the centroid. The center also serves as the origin C for a special set of axes known as the centroidal axes x_c and y_c.

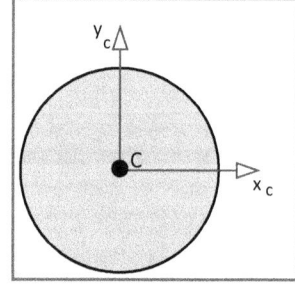

Fig. 1.14 The centroid serves as the origin for the centroidal axes x_c, y_c.

For cross sectional shapes such as ellipses, circles, squares and rectangles, the center may be located by inspection because these geometries have two axes of symmetry. However, for non-symmetric figures, such as triangles, portions of circles, parabolic areas, etc., locating the center of the area is not obvious. We will demonstrate a method for determining the centroid's location for an area that does not exhibit two axes of symmetry.

With respect to a centroidal coordinate system, the first moment of the area must vanish for both axes. Therefore:

$$Q_{\bar{x}} = \int_A y\,dA = 0 \qquad Q_{\bar{y}} = \int_A x\,dA = 0 \qquad (1.12)$$

These relations are employed to locate the centroid of an area of any shape providing its boundary can be defined with some mathematical function.

EXAMPLE E1.2

For a right triangle, determine the first moment of the area about its base and vertical side, and the position of its centroid relative to these two sides. The right triangle with a base b and a height h is illustrated in Fig. E1.2.

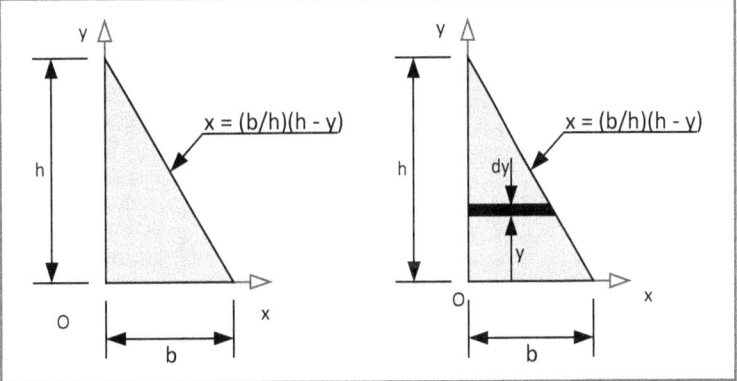

Fig E1.2 A right triangle with a coordinate system coincident with its base and vertical side.

Solution:

To begin, let's determine the first moment of the area of the right triangle relative to the x-axis (its base side). Writing Eq. 1.10 gives:

$$Q_x = \int_A y\,dA = \int_y y(x\,dy) \qquad (a)$$

where $dA = x\,dy$ is located a constant distance y from the x axis.

Note, the equation for the inclined boundary of the triangle is given by:

$$x = (b/h)(h - y) \qquad (b)$$

The limits on the integral go from 0 to h to encompass the area of the triangle. We substitute Eq. (b) as the limits on y for the integral into Eq. (a), and write:

$$Q_x = \frac{b}{h}\int_0^h (h - y)y\,dy \qquad (c)$$

Integrating Eq. (c) gives:

$$Q_x = \frac{b}{h}\left[\frac{hy^2}{2} - \frac{y^3}{3}\right]_0^h \qquad (d)$$

Evaluating Eq. (d) gives:

$$Q_x = bh^2/6 \qquad (1.13)$$

By using Eq. 1.10 and following the same procedure, we find the first moment about the vertical side of the triangle is given by:

$$Q_y = b^2h/6 \qquad (1.14)$$

Equation (E.8) gives the first moment of the area of a right triangle about its base. This is an interesting exercise in calculus, but what does it have to do with determining the location of the centroid of the right triangle? The result presented in Eq. (E.8) is an intermediate step. We continue the solution by combining the results of Eqs. (E.8) and (E.9) with Eqs. (E.5) to obtain:

$$Q_x = bh^2/6 = A\bar{y} = (bh/2)\bar{y} \qquad (e)$$

$$Q_y = b^2h/6 = A\bar{x} = (bh/2)\bar{x} \qquad (f)$$

where $\bar{x}$ and $\bar{y}$ locate the C, x_c, y_c coordinates relative to the Oxy coordinates (see Fig. E1.3).

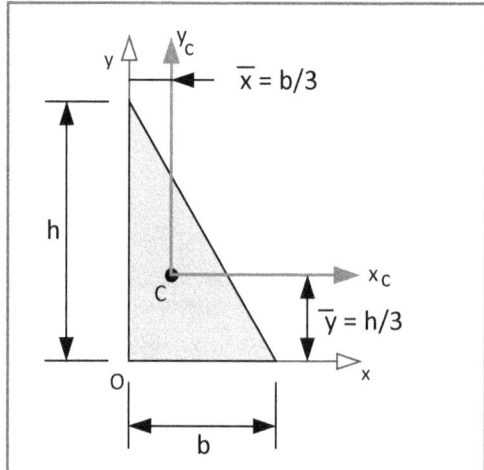

Fig. E1.3 A right triangular area with a base axes Oxy and centroidal axes C, x_c, y_c.

To determine the position of the centroid, let's solve Eqs. (e) and (f) for $\bar{x}$ and $\bar{y}$ to obtain:

$$\bar{y} = h/3 \qquad \bar{x} = b/3 \qquad (1.15)$$

We employ Eqs. (1.10) and (1.11) to determine the first moment of the area Q relative to either the x or y axes. The location of the centroid is then established from Eq. (1.15). The location of the centroid for common shapes has been determined, and the results are presented together with the definition of the shape of the area in Table 1.2.

Table 1.2

The location of the centroid for common shapes

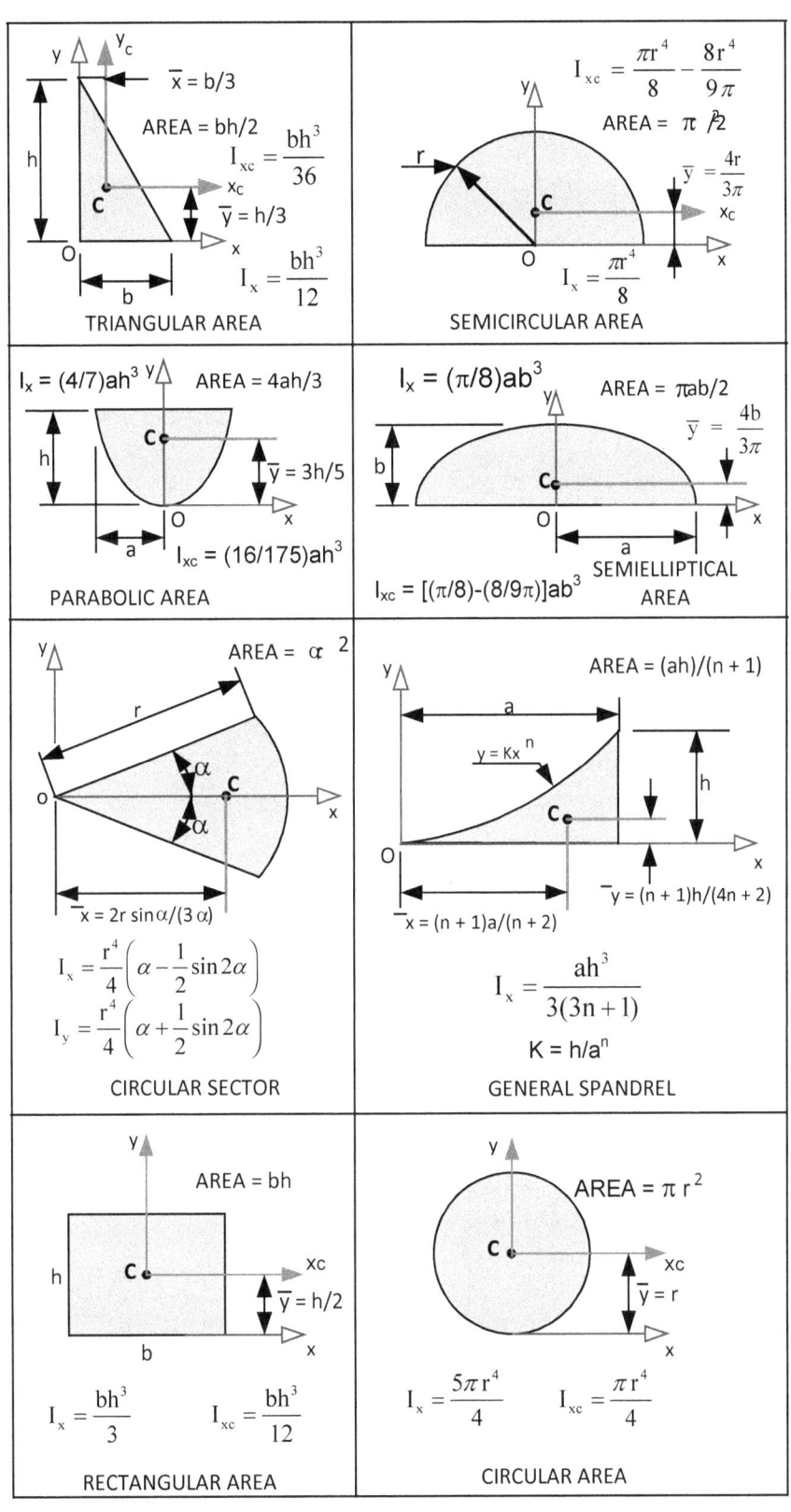

1.13 SUMMARY

This first chapter reviews some of the important topics that were covered in **Statics**. **Mechanics of Materials** is an extension of statics. In the study of Mechanics of Materials we account for the effect of **material deformations** on the **internal stresses** generated in a body. By using the deformation of the body under consideration, we are able to determine the magnitude and the distribution of the stresses due to specified application of forces. The method for determining the deformation and the stresses is outlined in Section 1.2.

Strain, on the other hand, is a **geometric concept**. Strain is determined by the change in geometry that occurs when a body deforms under load. When considered individually, both stress and strain are independent of the material from which the body is fabricated. It is only when we attempt to write a relation connecting stress and strain that the material properties (modulus of elasticity and Poisson's ratio) must be considered.

Newton's laws of motion and his law that governs gravitational forces are described. The difference among scalars, vectors and tensors is discussed. External forces that produce internal forces in a body are defined and section cuts made to expose the internal forces and the resulting internal stresses were explained.

Moments and their direction and magnitude were introduced. The equations of equilibrium were given in both vector and scalar formats. A classification system for force systems that simplifies the equations of equilibrium was described in detail.

Of critical importance is the discussion of **free body diagrams**. This topic is your first of many contacts with modelling, which is of major importance in engineering. When you draw a free body diagram (FBD) you are modeling a structural element prior to beginning you analysis. You will be sketching models of many items and/or processes during the remaining years of your studies.

Finally we introduced centroids, which is one of several properties of areas that is used in the analyses of forces and stresses in deformable bodies.

PROBLEMS

1.1 What are the two fundamental assumptions made in deriving equations in mechanics of materials.
1.2 List the steps followed in solving for deformation and stresses in bodies subjected to external forces.
1.3 What would you weigh if you were on the surface of the moon? Give the answer in SI and U. S. Customary units.
1.4 What are the characteristics of a scalar, vector and tensor?
1.5 Cite three examples of externally applied forces.
1.6 How do you determine the sign of a moment?
1.7 Sketch a non-coplanar and non-concurrent force system.
1.8 Sketch a non-coplanar and concurrent force system.
1.9 Sketch a coplanar and non-concurrent force system.
1.10 Sketch a coplanar and concurrent force system.
1.11 Briefly describe the four-step procedure that is employed in drawing the free body of a beam.
1.12 Verify the location of the centroid of the semi-circular area shown in Table 1.2.
1.13 Verify the location of the centroid of the general spandrel area shown in Table 1.2.

CHAPTER 2

STRESSES AT A POINT

2.1 INTRODUCTION

This chapter deals with stresses produced in a body due to external and body-force loadings. There are two basic types of force, which act on a body to produce stresses. Forces of the first type are called **surface forces** for the simple reason that they act on the surfaces of the body. Surface forces are generally exerted when one body comes in contact with another. Forces of the second type are called **body forces,** because they act on each element within the body. Body forces are commonly produced by centrifugal, gravitational, or other force fields. The most common body forces are gravitational, being present to some degree in almost all cases. For many practical applications, however, they are usually so small compared with the surface forces that they can be neglected without introducing serious error. Body forces are included in the following analysis for the sake of completeness.

2.2 THREE DIMENSIONAL STRESSES AT A POINT

Consider an arbitrary internal or external surface, which may be plane or curvilinear, as shown in Fig. 2.1. Over a small area ΔA of this surface in the neighborhood of an arbitrary point P, a system of forces acts, which has a resultant represented by the vector $\Delta \mathbf{F_n}$, in Fig. 2.1. It should be noted that the line of action of the resultant force vector $\Delta \mathbf{F_n}$ does not necessarily coincide with the outer normal vector **n** associated with the element of ΔA. If the resultant force $\Delta \mathbf{F_n}$ is divided by the increment of area ΔA, the average stress that acts over the area is obtained. In the limit as ΔA approaches zero, a quantity defined as the resultant stress $\mathbf{T_n}$ acting at the point P is obtained. This limiting process is shown below:

$$\mathbf{T_n} = \lim_{\Delta A \to 0} \frac{\Delta F_n}{\Delta A} \qquad (2.1)$$

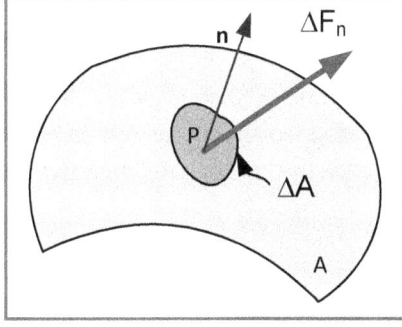

Fig. 2.1 An arbitrary surface (either internal or external) showing the resultant of all forces acting over the element ΔA.

The line of action of the resultant stress $\mathbf{T_n}$ coincides with the line of action of the resultant force $\Delta \mathbf{F_n}$, as illustrated in Fig. 2.2. It is important to note that the resultant stress $\mathbf{T_n}$ is a function of both the position of the point P in the body and the orientation of the plane, which is passed through point P and identified by its outer normal vector **n**. In a body subjected to an arbitrary system of loads, both the magnitude and the direction of the resultant stress T_n at any point P change, as the orientation of the plane under consideration is varied.

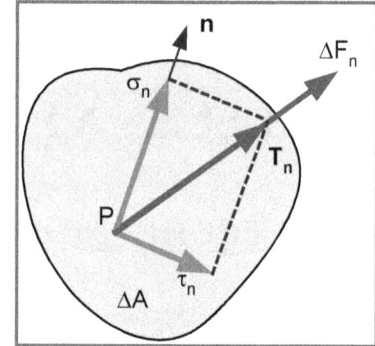

Fig. 2.2 Resolution of the resultant stress T_n into normal and tangential components σ_n and τ_n.

As illustrated in Fig. 2.2, it is possible to resolve T_n into two components: σ_n normal to the surface is known as the resultant normal stress, while the component τ_n is known as the resultant shearing stress.

Cartesian components of stress for any coordinate system can also be obtained from the resultant stress. Consider first a surface whose outer normal is in the positive z direction, as shown in Fig. 2.3. If the resultant stress T_n associated with this particular surface is resolved into components along the x, y, and z-axes, the Cartesian stress components τ_{zx}, τ_{zy}, σ_{zz} are obtained. The components τ_{zx} and τ_{zy} are shearing stresses, because they act tangent to the surface under consideration. The component σ_{zz} is a normal stress, because it acts normal to the surface.

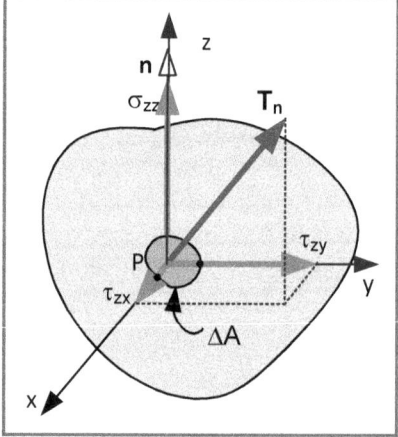

Fig. 2.3 Resolution of the resultant stress T_n into its three Cartesian components τ_{zx}, τ_{zy} and σ_{zz}.

If the same procedure is followed using surfaces with outer normal vectors in the positive x and y directions, two more sets of Cartesian components, τ_{xy}, τ_{xz}, σ_{xx} and τ_{yx}, τ_{yz}, σ_{yy}, respectively, are obtained. The three different sets of three Cartesian components for the three selections of the outer normal are summarized in the array below:

$$\begin{array}{ccc} \sigma_{xx} & \tau_{xy} & \tau_{xz} \\ \tau_{yx} & \sigma_{yy} & \tau_{yz} \\ \tau_{zx} & \tau_{zy} & \sigma_{zz} \end{array} \quad \begin{array}{l} \text{outer normal parallel to x-axis} \\ \text{outer normal parallel to y-axis} \\ \text{outer normal parallel to z-axis} \end{array}$$

From this array, it is clear that nine Cartesian components of stress exist. These components can be arranged on the faces of a small cubic element, as shown in Fig. 2.4. The sign convention employed in placing the Cartesian stress components on the faces of this cube is as follows: If the outer normal defining the cube face is in the direction of increasing x, y, or z, then the associated normal and shear stress components are also in the direction of positive x, y, or z. If the outer normal is in the direction of negative x, y, or z, then the normal and shear stress components are also in the direction of negative x, y, or z. As for subscript convention, the first subscript refers to the outer normal and defines the plane upon which the stress component acts, and the second subscript gives the direction in which the stress acts. Finally, for normal stresses, positive signs indicate tension and negative signs indicate compression.

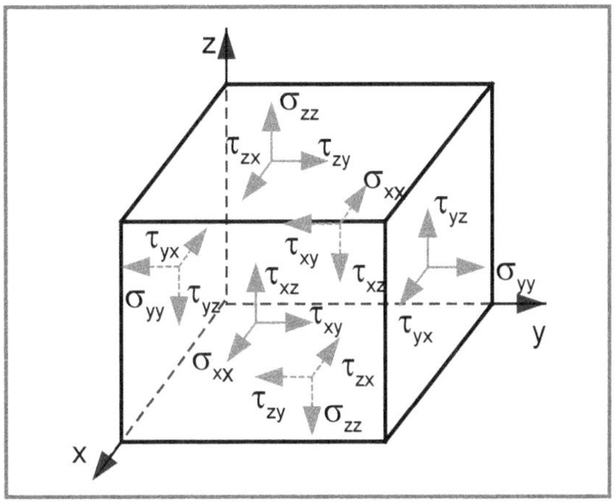

Fig. 2.4 Cartesian components of stress acting on the faces of a small cubic element.

At a given point of interest within a body, the magnitude and direction of the resultant stress $\mathbf{T_n}$ depend upon the orientation of the plane passed through the point. Thus an infinite number of resultant-stress vectors can be used to represent the resultant stress at each point, because an infinite number of planes can be passed through each point. It is easy to show, however, that the magnitude and direction of each of these resultant $\mathbf{T_n}$ stress vectors can be specified in terms of the nine Cartesian components of stress acting at the point. This fact can be established by considering equilibrium of the elemental tetrahedron, shown in Fig. 2.5. In this figure, the stresses acting over the four faces of the tetrahedron are represented by their average values. The average value is denoted by placing a ~ sign over the stress symbol. In order for the tetrahedron to be in equilibrium, the following condition must be satisfied. First consider equilibrium in the x direction and write:

$$\widetilde{T}_{nx}A - \widetilde{\sigma}_{xx}A\cos(\mathbf{n}, x) - \widetilde{\tau}_{yx}A\cos(\mathbf{n}, y) - \widetilde{\tau}_{zx}A\cos(\mathbf{n}, z) + \widetilde{F}_{x}hA/3 = 0 \qquad (2.2)$$

where: h = altitude of tetrahedron; A = area of base of tetrahedron; $\widetilde{F}_x$ = average body-force intensity in x direction; and $\widetilde{T}_{nx}$ = component of resultant stress in the x direction, and A cos($\mathbf{n}$, x), A cos($\mathbf{n}$, y), and A cos($\mathbf{n}$, z) are the projections of the area A on the yz, xz, and xy planes, respectively.

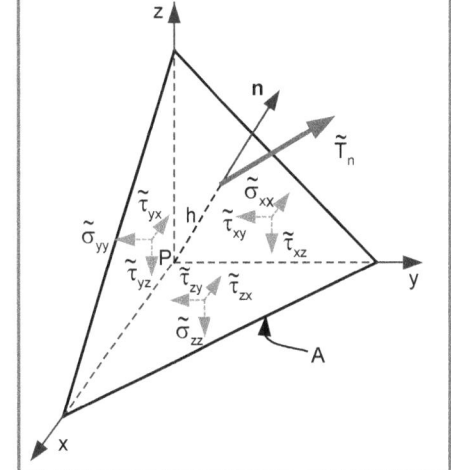

Fig. 2.5 Elemental tetrahedron at a point P showing the average stresses that act over its Cartesian faces

26 — Chapter 2 — Stress at a Point

By letting the altitude h $\Rightarrow$ 0, after eliminating the factor A from each term in eq. (2.2), it is evident that the body-force term vanishes, the average stresses become exact stresses at the point P, and Eq. (2.2) becomes:

$$T_{nx} = \sigma_{xx} \cos(\mathbf{n}, x) + \tau_{yx} \cos(\mathbf{n}, y) + \tau_{zx} \cos(\mathbf{n}, z) \tag{2.3a}$$

Two similar expressions are obtained by considering equilibrium in the y and z directions:

$$T_{ny} = \tau_{xy} \cos(\mathbf{n}, x) + \sigma_{yy} \cos(\mathbf{n}, y) + \tau_{zy} \cos(\mathbf{n}, z) \tag{2.3b}$$

$$T_{nz} = \tau_{xz} \cos(\mathbf{n}, x) + \tau_{yz} \cos(\mathbf{n}, y) + \sigma_{zz} \cos(\mathbf{n}, z) \tag{2.3c}$$

After the three Cartesian components of the resultant stress have been determined from Eqs. (2.3), the resultant stress $\mathbf{T_n}$ can be determined by using the expression:

$$|\mathbf{T_n}| = \sqrt{T_{nx}^2 + T_{ny}^2 + T_{nz}^2} \tag{2.4}$$

The three direction cosines, which define the line of action of the resultant stress $\mathbf{T_n}$ are:

$$\cos(T_n, x) = \frac{T_{nx}}{|\mathbf{T_n}|} \qquad \cos(T_n, y) = \frac{T_{ny}}{|\mathbf{T_n}|} \qquad \cos(T_n, z) = \frac{T_{nz}}{|\mathbf{T_n}|} \tag{2.5}$$

The normal stress σ_n and the shearing stress τ_n which act on the plane under consideration can be obtained from the expressions:

$$\sigma_n = |\mathbf{T_n}| \cos(T_n, \mathbf{n}) \quad \text{and} \quad \tau_n = |\mathbf{T_n}| \sin(T_n, \mathbf{n}) \tag{2.6}$$

The angle between the resultant-stress vector $\mathbf{T_n}$ and the normal to the plane $\mathbf{n}$ can be determined by using the well-known relationship:

$$\cos(T_n, \mathbf{n}) = \cos(T_n, x) \cos(\mathbf{n}, x) + \cos(T_n, y) \cos(\mathbf{n}, y) + \cos(T_n, z) \cos(\mathbf{n}, z) \tag{2.7}$$

It should also be noted that the normal stress σ_n could be determined by considering the projections of T_{nx}, T_{ny}, and T_{nz} onto the normal to the plane under consideration. Thus,

$$\sigma_n = T_{nx} \cos(\mathbf{n}, x) + T_{ny} \cos(\mathbf{n}, y) + T_{nz} \cos(\mathbf{n}, z) \tag{2.8}$$

When σ_n has been determined, τ_n can easily be found because:

$$\tau_n = \sqrt{T_n^2 - \sigma_n^2} \tag{2.9}$$

2.3 STRESS EQUATIONS OF TRANSFORMATION

In developing the stress equations of transformation, we will treat the case of plane stress where $\sigma_{zz} = \tau_{xz} = \tau_{yz} = 0$. Suppose we know the state of stress (σ_{xx}, σ_{yy} and τ_{xy}) in a body relative to a set of coordinates say Oxy. Is this state of stress the most important combination of stresses that will determine the failure loads? Or is some other state of stress relative to a different coordinate system the critical state? To answer these questions it is necessary to explore all of the admissible states of stress. We accomplish this task by using the stress equations of transformation.

Consider an elemental area with stresses σ_{xx}, σ_{yy} and τ_{xy} as shown in Fig. 2.6a.

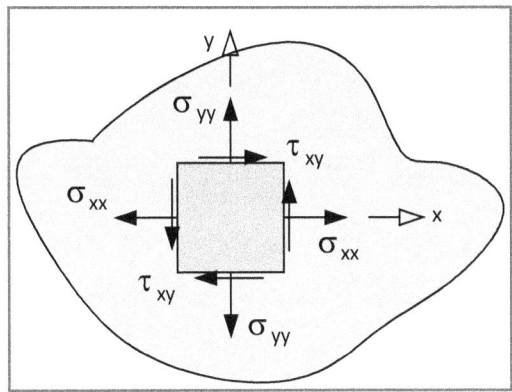
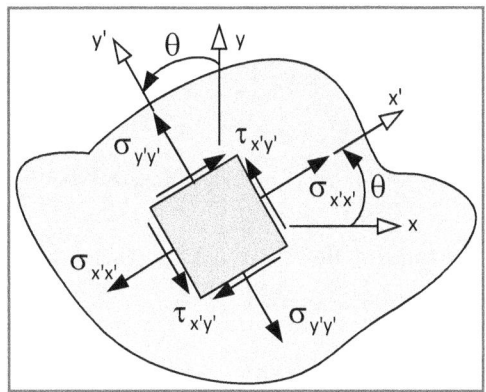

Fig. 2.6a An element with x-y orientation.

Fig. 2.6b An element with x'-y' orientation.

As we rotate the element about its center through some angle θ, the stresses on the element faces change from σ_{xx}, σ_{yy} and τ_{xy} to $\sigma_{x'x'}$, $\sigma_{y'y'}$ and $\tau_{x'y'}$. To explore stress space, it is necessary to establish relations for $\sigma_{x'x'}$, $\sigma_{y'y'}$ and $\tau_{x'y'}$ in terms of σ_{xx}, σ_{yy} and τ_{xy} and the angle θ.

Let's cut the left corner from the element, shown in Fig. 2.6a, at an angle θ to produce a wedge like element with an inclined surface, which is parallel to the y' axis and perpendicular to the x' axis. Consider the element thickness in the z direction to be unity, and define the area of the inclined surface as ΔA as indicated in Fig. 2.7a. The area of the short side of the wedge is given by $\Delta A \sin \theta$ and the area of the longer side of the wedge is $\Delta A \cos \theta$. These areas are important, because they factor into the equilibrium equations that we will employ later in this derivation.

Stresses $\tau_{x'y'}$ and $\sigma_{x'x'}$ act on the inclined surface of the wedge and τ_{xy}, σ_{xx} and σ_{yy} act on the other sides of the wedge. We multiply these stresses with the areas of the three sides of the wedge to obtain the forces shown in Fig. 2.7b. No forces are shown on the face of the wedge because the element was removed from a body in a plane state of stress where $\sigma_{zz} = \tau_{xz} = \tau_{yz} = 0$.

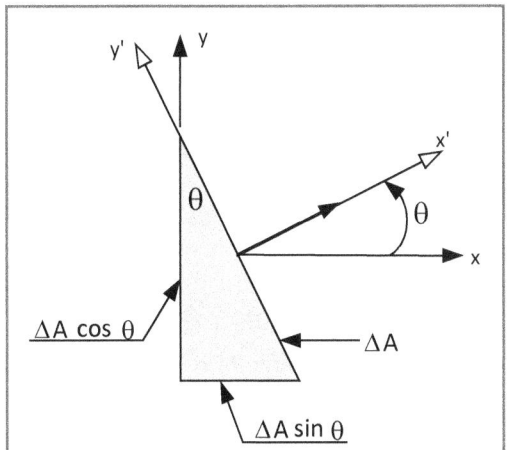
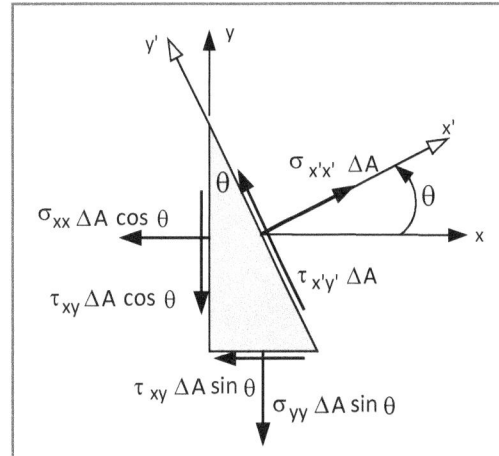

Fig. 2.7a A wedge like element cut from the x-y element at an angle θ.

Fig. 2.7b Forces acting on the three sides of the wedge like element.

Next we write two equilibrium relations summing forces in the x' and y' directions:

28 — Chapter 2 — Stress at a Point

$\Sigma F_{x'} = 0$
$$\sigma_{x'x'}\Delta A - \sigma_{xx}(\Delta A \cos\theta)\cos\theta - \tau_{xy}(\Delta A \cos\theta)\sin\theta$$
$$- \sigma_{yy}(\Delta A \sin\theta)\sin\theta - \tau_{xy}(\Delta A \sin\theta)\cos\theta = 0$$

$\Sigma F_{y'} = 0$
$$\tau_{x'y'}\Delta A + \sigma_{xx}(\Delta A \cos\theta)\sin\theta - \tau_{xy}(\Delta A \cos\theta)\cos\theta$$
$$- \sigma_{yy}(\Delta A \sin\theta)\cos\theta + \tau_{xy}(\Delta A \sin\theta)\sin\theta = 0$$

(2.10)

Simplifying these two equations and solving for $\sigma_{x'x'}$ and $\tau_{x'y'}$ gives:

$$\sigma_{x'x'} = \sigma_{xx}\cos^2\theta + \sigma_{yy}\sin^2\theta + 2\tau_{xy}\sin\theta\cos\theta \qquad (2.11)$$

$$\tau_{x'y'} = -(\sigma_{xx} - \sigma_{yy})\sin\theta\cos\theta + \tau_{xy}(\cos^2\theta - \sin^2\theta) \qquad (2.12)$$

These relations enable you to search solution space to determine the angle θ, which produces maximum and minimum values of both the normal and shear stresses.

By introducing four trigonometric identities, it is possible to convert Eqs. (2.11) and (2.12) into a double angle (2θ) format as shown below.

$$\sigma_{x'x'} = \frac{\sigma_{xx} + \sigma_{yy}}{2} + \frac{\sigma_{xx} - \sigma_{yy}}{2}\cos 2\theta + \tau_{xy}\sin 2\theta$$

$$\tau_{x'y'} = -\frac{\sigma_{xx} - \sigma_{yy}}{2}\sin 2\theta + \tau_{xy}\cos 2\theta$$

(2.13)

A similar relation for $\sigma_{y'y'}$ may be derived by increasing the angle of the cut for the wedge from θ to $90° + \theta$. Alternatively, you may convert Eq. (2.11) or the first of Eq. (2.12) to an expression for $\sigma_{y'y'}$, by letting $\theta = \theta + 90°$ to obtain:

$$\sigma_{y'y'} = \frac{\sigma_{xx} + \sigma_{yy}}{2} - \frac{\sigma_{xx} - \sigma_{yy}}{2}\cos 2\theta - \tau_{xy}\sin 2\theta \qquad (2.14)$$

The form of the relations for $\sigma_{x'x'}$ and $\sigma_{y'y'}$ is almost identical except for the signs of the last two terms. It is evident from inspection that we can add Eq. (2.13) and Eq. (2.14) to obtain:

$$\sigma_{x'x'} + \sigma_{y'y'} = \sigma_{xx} + \sigma_{yy} \qquad (2.15)$$

Equation (2.15) indicates that the sum of the normal stress components is independent of the angle of θ. The sum of the normal stresses is an invariant that does not depend on the orientation of the planes upon which the stresses act. This is one of three stress invariants.

EXAMPLE E2.1

We represent a plane state of stress at a point by a small elemental area ΔA shown in Fig. E2.1. If the stresses are $\sigma_{xx} = 35$ MPa, $\sigma_{yy} = -22$ MPa and $\tau_{xy} = 16$ MPa, determine the stresses on an element that is rotated 30° counterclockwise relative to the element shown in Fig. E2.1.

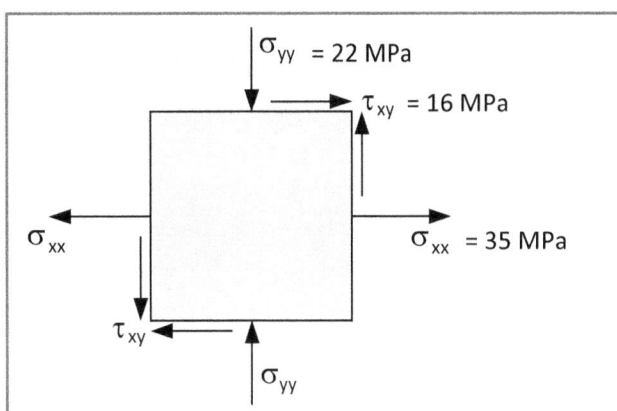

Fig. E2.1

Solution: Using Eqs. (2.13) and (2.14) and setting θ = 30°, we write:

$$\sigma_{x'x'} = \frac{\sigma_{xx} + \sigma_{yy}}{2} + \frac{\sigma_{xx} - \sigma_{yy}}{2} \cos 2\theta + \tau_{xy} \sin 2\theta$$

$$\sigma_{x'x'} = \frac{35 - 22}{2} + \frac{35 + 22}{2} \cos 60 + 16 \sin 60 = 34.61 \text{ MPa}$$

$$\tau_{x'y'} = -\frac{\sigma_{xx} - \sigma_{yy}}{2} \sin 2\theta + \tau_{xy} \cos 2\theta$$

$$\tau_{x'y'} = -\frac{35 + 22}{2} \sin 60 + 16 \cos 60 = -16.68 \text{ MPa}$$

$$\sigma_{y'y'} = \frac{\sigma_{xx} + \sigma_{yy}}{2} - \frac{\sigma_{xx} - \sigma_{yy}}{2} \cos 2\theta - \tau_{xy} \sin 2\theta$$

$$\sigma_{y'y'} = \frac{35 - 22}{2} - \frac{35 + 22}{2} \cos 60 - 16 \sin 60 = -21.61 \text{ MPa}$$

EXAMPLE E2.2

We represent a plane state of stress at a point by a small elemental area ΔA shown in Fig. E2.2. If the stresses are σ_{xx} = 15 ksi, σ_{yy} = 12 ksi and τ_{xy} = − 10 ksi, determine the stresses on an element that is rotated 40° clockwise relative to the element shown in Fig. E2.2.

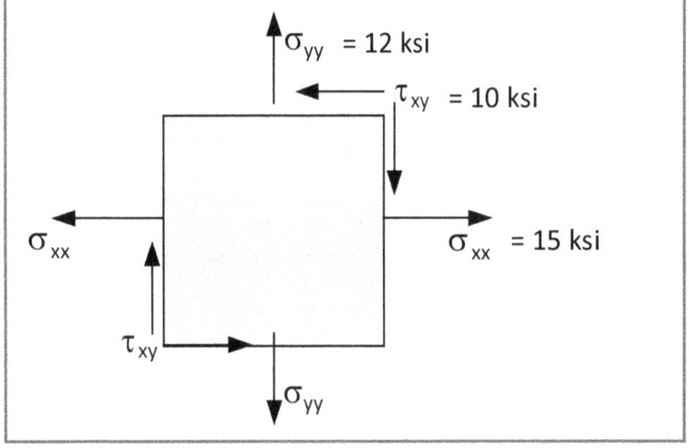

Fig. E2.2

Solution: Using Eqs. (2.13) and (2.14) and setting θ = − 40°, we write:

$$\sigma_{x'x'} = \frac{\sigma_{xx}+\sigma_{yy}}{2} + \frac{\sigma_{xx}-\sigma_{yy}}{2}\cos 2\theta + \tau_{xy}\sin 2\theta$$

$$\sigma_{x'x'} = \frac{15+12}{2} + \frac{15-12}{2}\cos(-80) - 10\sin(-80) = 23.608 \text{ ksi}$$

$$\tau_{x'y'} = -\frac{\sigma_{xx}-\sigma_{yy}}{2}\sin 2\theta + \tau_{xy}\cos 2\theta$$

$$\tau_{x'y'} = -\frac{15-12}{2}\sin(-80) - 10\cos(-80) = -0.2593 \text{ ksi}$$

$$\sigma_{y'y'} = \frac{\sigma_{xx}+\sigma_{yy}}{2} - \frac{\sigma_{xx}-\sigma_{yy}}{2}\cos 2\theta - \tau_{xy}\sin 2\theta$$

$$\sigma_{y'y'} = \frac{15+12}{2} - \frac{15-12}{2}\cos(-80) + 10\sin(-80) = 3.391 \text{ ksi}$$

These two examples show the application of the stress equations of transformation for determining the Cartesian components of stress on an arbitrary element rotated an angle θ with respect to the x-y coordinate system. We will next use the stress equations of transformation to locate the planes, where the normal and shear stresses are a maximum or minimum.

2.4 PRINCIPAL STRESSES

When considering all of the possible planes upon which normal stresses act, we seek those planes, where the normal stresses are either a maximum or a minimum. We will define these planes by determining the angle θ_p, which designates the principal plane. To determine this angle, let's employ the standard calculus approach by differentiating the normal stress $\sigma_{x'x'}$ with respect to θ and set the result equal to zero.

$$\frac{d}{d\theta}\sigma_{x'x'} = -(\sigma_{xx}-\sigma_{yy})\sin 2\theta_p + 2\tau_{xy}\cos 2\theta_p = 0$$

$$\tan 2\theta_p = \frac{2\tau_{xy}}{\sigma_{xx}-\sigma_{yy}}$$

(2.16)

From Eq. (2.16), it is easy to show that:

$$\cos 2\theta_p = \frac{(\sigma_{xx}-\sigma_{yy})/2}{\sqrt{\left(\frac{\sigma_{xx}-\sigma_{yy}}{2}\right)^2 + \tau_{xy}^2}}$$

$$\sin 2\theta_p = \frac{\tau_{xy}}{\sqrt{\left(\frac{\sigma_{xx}-\sigma_{yy}}{2}\right)^2 + \tau_{xy}^2}}$$

(2.17)

Substitute Eq. (2.17) into Eq. (2.13) and simplifying yields:

$$\sigma_{Max, Min} = \frac{\sigma_{xx} + \sigma_{yy}}{2} + \frac{\sigma_{xx} - \sigma_{yy}}{2} \cos 2\theta_p + \tau_{xy} \sin 2\theta_p$$

$$\sigma_{Max, Min} = \frac{\sigma_{xx} + \sigma_{yy}}{2} \pm \sqrt{\left(\frac{\sigma_{xx} - \sigma_{yy}}{2}\right)^2 + \tau_{xy}^2} \qquad (2.18)$$

$$\tau_{pp} = -\frac{\sigma_{xx} - \sigma_{yy}}{2} \sin 2\theta_p + \tau_{xy} \cos 2\theta_p = 0$$

The maximum normal stress $\sigma_{Max} = \sigma_1$ is the largest algebraic value of the two values obtained in the solution of Eq. (2.18). The minimum normal stress $\sigma_{Min} = \sigma_2$ is the smaller of the two values. Both σ_1 and σ_2 are principal stresses, which act on orthogonal planes, defined by Eq. (2.16). The shear stress on the principal planes $\tau_{pp} = 0$. We illustrate a typical principal state of stress in Fig. 2.8. In this illustration, the principal element is inscribed in a traditional Cartesian element, although the Cartesian components of stress are not shown to avoid clutter. On the principal element, we have shown σ_1 and σ_2 as well as the shear stress τ_{pp} vanishing on its edges.

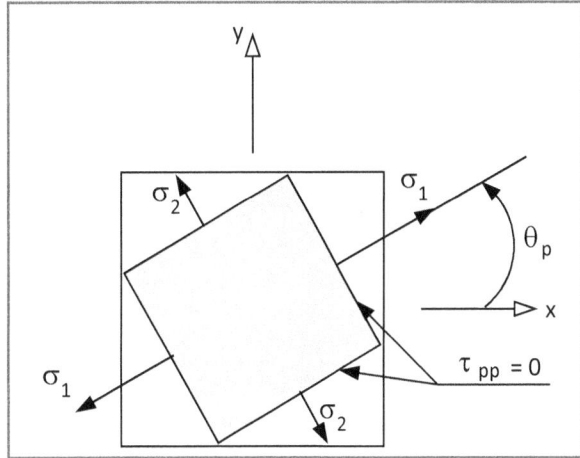

Fig. 2.8 A principal element inscribed in a traditional Cartesian element.

EXAMPLE E2.3

Consider a point, shown here as an elemental area, with Cartesian components of stress defined in Fig. E2.3. Determine the orientation of the principal planes and the stresses that act on the principal planes.

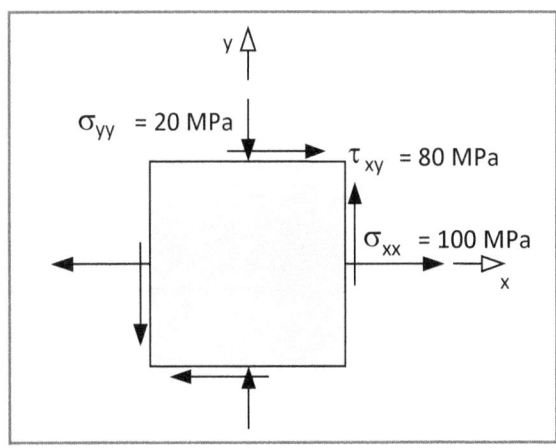

Fig E2.3

Solution: The principal angle θ_p is determined from Eq. (2.16) as:

$$\tan 2\theta_p = \frac{2\tau_{xy}}{\sigma_{xx} - \sigma_{yy}} = \frac{2(80)}{100+20} = 1.333$$

$$2\theta_p = \tan^{-1}(1.333) = 53.13°$$ (a)

$$2\theta_p = 180° + 53.13° = 233.13°$$

$$\theta_p = 26.57° \text{ and } 116.57°$$ (b)

The first principal angle (26.57°), defines the plane upon which the maximum principal stress σ_1 acts and the second principal angle (116.57°), defines the plane upon which the minimum principal stress σ_2 acts.

Let's determine the magnitude of the principal stresses by using Eq. (2.18) as:

$$\sigma_{Max,\,Min} = \frac{\sigma_{xx}+\sigma_{yy}}{2} \pm \sqrt{\left(\frac{\sigma_{xx}-\sigma_{yy}}{2}\right)^2 + \tau_{xy}^2}$$

$$\sigma_{Max,\,Min} = \frac{100-20}{2} \pm \sqrt{\left(\frac{100+20}{2}\right)^2 + (80)^2}$$ (c)

$$\sigma_{Max,\,Min} = 40 \pm 100 \text{ MPa}$$

$$\sigma_1 = 140 \text{ MPa and } \sigma_2 = -60 \text{ MPa}$$ (d)

We show these stresses acting on the principal element in Fig. E2.3a.

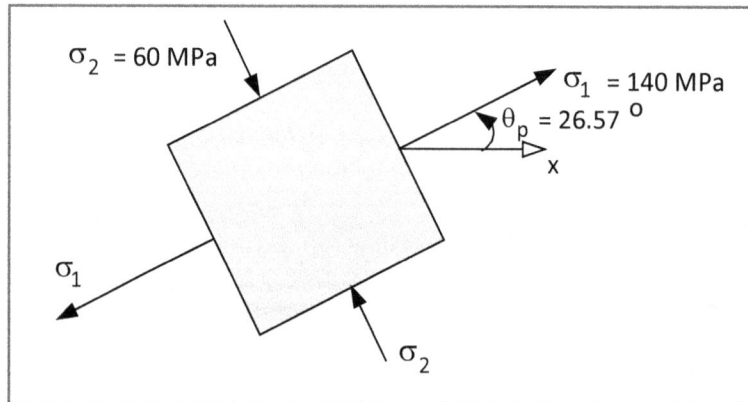

Fig. E2.3a

Of course the shear stresses on the principal element vanish in accordance with Eq. (2.18).

EXAMPLE E2.4

Consider a point, shown here as an elemental area, with Cartesian components of stress defined in Fig. E2.4. Determine the orientation of the principal planes and the stresses that act on the principal planes.

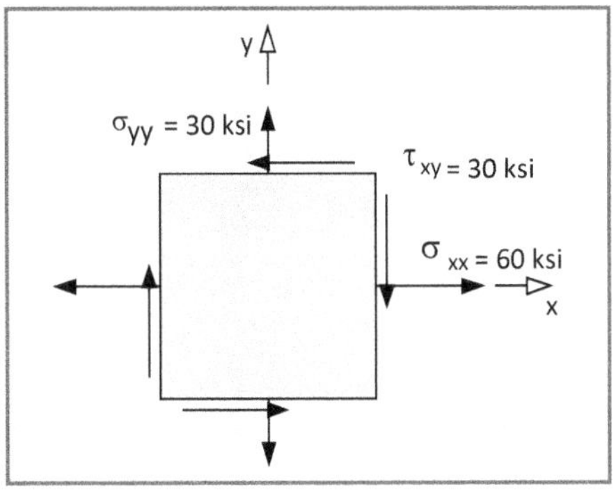

Fig. E2.4

Solution: The principal angles θ_p are determined from Eq. (2.16) as:

$$\tan 2\theta_p = \frac{2\tau_{xy}}{\sigma_{xx} - \sigma_{yy}} = \frac{2(-30)}{60-30} = -2.00$$

$$2\theta_p = \tan^{-1}(-2.00) = -63.43° \quad \text{(a)}$$

$$2\theta_p = 180° - 63.43° = 116.57°$$

$$\theta_p = -31.72° \text{ and } 58.29° \quad \text{(b)}$$

The first principal angle (31.72°), defines the plane upon which the maximum principal stress σ_1 acts and the second principal angle (121.72°), defines the plane upon which the minimum principal stress σ_2 acts.

Let's determine the magnitude of the principal stresses by using Eq. (2.18) as:

$$\sigma_{\text{Max, Min}} = \frac{\sigma_{xx} + \sigma_{yy}}{2} \pm \sqrt{\left(\frac{\sigma_{xx} - \sigma_{yy}}{2}\right)^2 + \tau_{xy}^2}$$

$$\sigma_{\text{Max, Min}} = \frac{60+30}{2} \pm \sqrt{\left(\frac{60-30}{2}\right)^2 + (-30)^2} \quad \text{(c)}$$

$$\sigma_{\text{Max, Min}} = 45 \pm 33.54 \text{ ksi}$$

$$\sigma_1 = 78.54 \text{ ksi and } \sigma_2 = 11.46 \text{ ksi} \quad \text{(d)}$$

We show these stresses acting on the principal element in Fig. E2.4a.

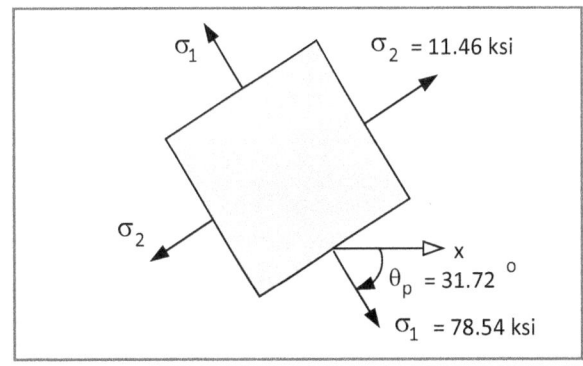

Fig. E2.4a

2.4.1 Maximum Shear Stresses

We used differential calculus to ascertain the principal planes defined by θ_p, which enabled us to determine the maximum and minimum normal stresses. Let's follow the same approach to determine the planes upon which the maximum shear stress occurs. Differentiate the second of Eq. (2.13) with respect to θ, to obtain:

$$\frac{d}{d\theta}\tau_{x'y'} = -\frac{\sigma_{xx}-\sigma_{yy}}{2}(2\cos 2\theta) - 2\tau_{xy}\sin 2\theta = 0 \qquad (a)$$

Solving Eq. (a) for θ_s gives:

$$\tan 2\theta_s = -\frac{\sigma_{xx}-\sigma_{yy}}{2\tau_{xy}} \qquad (2.19)$$

where the angle θ_s defines the planes upon which the maximum shear stresses occur.

Comparing Eq. (2.19) for $2\theta_s$ with Eq. (2.16) for $2\theta_p$, shows that they are negative reciprocals. This fact implies that the angles $2\theta_s$ and $2\theta_p$ are orthogonal. Hence, the angles θ_s and θ_p are separated by 45°. Clearly, the planes of maximum shear and the principal planes are oriented at 45° relative to one another.

Note from Eq. (2.19) that $\cos 2\theta_s$ and $\sin 2\theta_s$ are given by:

$$\sin 2\theta_s = \frac{-(\sigma_{xx}-\sigma_{yy})/2}{\sqrt{\left(\frac{\sigma_{xx}-\sigma_{yy}}{2}\right)^2 + \tau_{xy}^2}}$$

$$\cos 2\theta_s = \frac{\tau_{xy}}{\sqrt{\left(\frac{\sigma_{xx}-\sigma_{yy}}{2}\right)^2 + \tau_{xy}^2}} \qquad (2.20)$$

Substituting Eq. (2.20) into Eq. (2.13) and simplifying yields:

$$\tau_{Max} = \sqrt{\left(\frac{\sigma_{xx}-\sigma_{yy}}{2}\right)^2 + \tau_{xy}^2} \qquad (2.21)$$

EXAMPLE E2.5

Consider a point, shown here as an elemental area, with Cartesian components of stress defined in Fig. E2.5. Determine the orientation of the planes upon which the maximum shear stresses act and also determine the normal stresses that act on these planes.

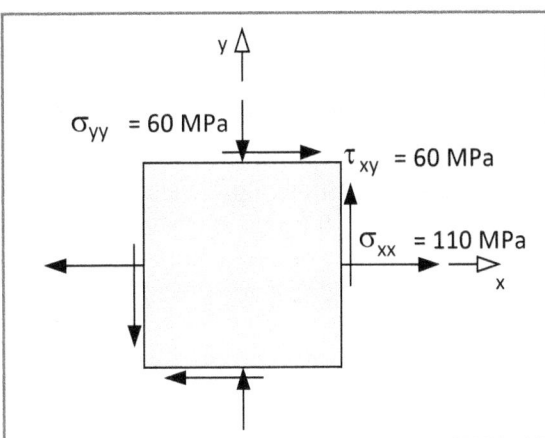

Fig. E2.5

Solution:

The plane upon which the maximum shear stress acts is given by Eq. (2.19) as:

$$\tan 2\theta_s = -\frac{\sigma_{xx} - \sigma_{yy}}{2\tau_{xy}} = -\frac{(110+60)}{2(60)} = -1.417 \quad (a)$$

$$2\theta_s = \tan^{-1}(-1.417) = -54.79°$$

$$\theta_s = -27.39° \quad (b)$$

The maximum shear stress is calculated from Eq. (2.21) as:

$$\tau_{Max} = \sqrt{\left(\frac{\sigma_{xx} - \sigma_{yy}}{2}\right)^2 + \tau_{xy}^2} = \sqrt{\left(\frac{110+60}{2}\right)^2 + (60)^2} = 104.0 \text{ MPa} \quad (c)$$

To determine the normal stress acting on this plane, substitute the value of $2\theta_s = -54.79°$ into Eq. (2.13) to obtain:

$$\sigma_{x'x'} = \frac{\sigma_{xx} + \sigma_{yy}}{2} + \frac{\sigma_{xx} - \sigma_{yy}}{2}\cos 2\theta + \tau_{xy}\sin 2\theta$$

$$\sigma_n = \frac{110-60}{2} + \frac{110+60}{2}\cos(-54.79) + (60)\sin(-54.79) \quad (d)$$

$$\sigma_n = 25 + 85(0.5766) + 60(-0.8170) = 25 \text{ MPa}$$

Using Eq. (2.13) and the angle θ_s is a cumbersome method to determine σ_n. We will show in the next section that on planes of maximum shear stress, σ_n is given by:

$$\sigma_n = (\sigma_{xx} + \sigma_{yy})/2 \quad (2.22)$$

2.5 MOHR'S CIRCLE

A graphical technique, known as Mohr's circle, was employed for solving for $\sigma_{x'x'}$, $\sigma_{y'y'}$ and $\tau_{x'y'}$ prior to the development of efficient electronic computers. Today the equations of stress transformation are easy to solve; however, Mohr's circle remains useful because it enables one to visualize how the Cartesian components of stress vary with the angle θ that defines the plane upon which the stresses act. Also the magnitude and orientation of σ_1, σ_2 and τ_{Max} are apparent from a cursory inspection of Mohr's circle.

To establish the relation between a circle and the stress state, consider the equation of a circle with a radius R centered at x = C, as shown in Fig. 2.9.

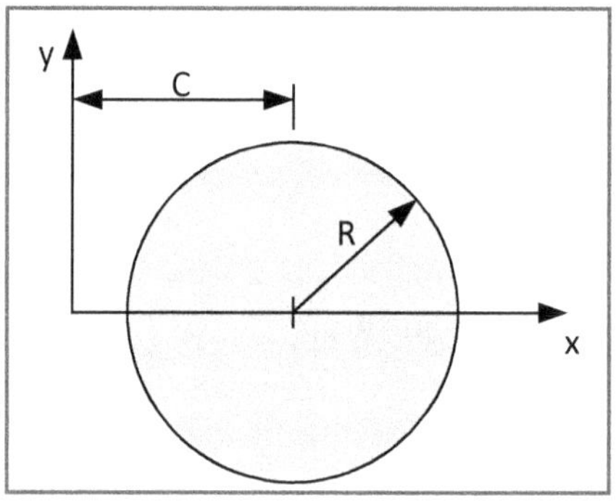

Fig. 2.9 Circle with radius R centered at x = C.

The equation for the circle is:

$$(x - C)^2 + y^2 = R^2 \tag{2.23}$$

To show that the stress equations of transformation map into a similar circle, rewrite Eq. (9.10) to obtain:

$$\sigma_{x'x'} - \frac{\sigma_{xx} + \sigma_{yy}}{2} = \frac{\sigma_{xx} - \sigma_{yy}}{2} \cos 2\theta + \tau_{xy} \sin 2\theta \tag{2.24}$$

$$\tau_{x'y'} = -\frac{\sigma_{xx} - \sigma_{yy}}{2} \sin 2\theta + \tau_{xy} \cos 2\theta$$

Square both sides of these two equations and add the two to obtain:

$$\left(\sigma_{x'x'} - \frac{\sigma_{xx} + \sigma_{yy}}{2}\right)^2 + \tau_{x'y'}^2 = \left(\frac{\sigma_{xx} - \sigma_{yy}}{2}\right)^2 + \tau_{xy}^2 \tag{2.25}$$

Comparing the form of Eq. (2.23) with that of Eq. (2.25), permits us to establish the several facts about Mohr's circle.

1. The correspondence between x and $\sigma_{x'x'}$ indicates σ will be found along the abscissa.
2. Mohr's circle will be centered at C where:

$$C = (\sigma_{xx} + \sigma_{yy})/2 \tag{2.26}$$

3. The correspondence between y and $\tau_{x'y'}$ indicates τ will be found along the ordinate.
4. The radius of Mohr's circle R is given by:

$$R = \sqrt{\left(\frac{\sigma_{xx} - \sigma_{yy}}{2}\right)^2 + \tau_{xy}^2} \qquad (2.27)$$

We present an example of Mohr's circle in Fig 2.10, showing the σ and τ axes, the center point located by C, the radius R, the principal stresses σ_1 and σ_2 and the maximum shear stress τ_{Max}.

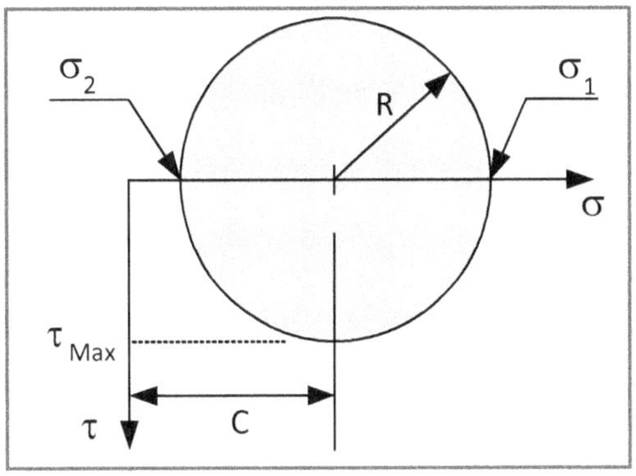

Fig. 2.10 Mohr's circle showing the principal stresses, and maximum shear stress.

We have shown the τ axis as positive downward in this representation. This choice is consistent with the definition of θ as a positive quantity for counterclockwise rotation and for the sign convention for τ_{xy}. Let's consider two examples to demonstrate the usefulness of Mohr's circle in visualizing a compete state of stress at a point.

EXAMPLE E2.6

Consider a point, shown here as an elemental area, with Cartesian components of stress defined in Fig. E2.6. Determine the principal stresses and the orientation of the principal planes. Also determine the maximum shear stresses and their orientation.

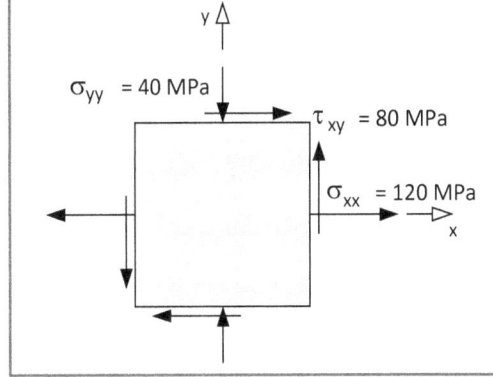

Fig. E2.6

Solution:

Step 1: Use Eq. (2.26) and Eq. (2.27) to determine the position of the center of the circle and its radius R as:

$$C = (\sigma_{xx} + \sigma_{yy})/2 = (120 - 40)/2 = 40 \text{ MPa} \qquad (a)$$

$$R = \sqrt{\left(\frac{\sigma_{xx} - \sigma_{yy}}{2}\right)^2 + \tau_{xy}^2} = \sqrt{\left(\frac{120 + 40}{2}\right)^2 + (80)^2} = 113.1 \text{ MPa}$$

Step 2: Draw the σ and τ axes, locate the center of the circle on the σ axis at C = 40 MPa and construct a circle with the radius R = 113.1 MPa, as shown in Fig. E2.6a. The principal stresses σ_1 and σ_2 are located, where the circle intersects the σ axis and given by:

$$\sigma_1 = C + R \qquad \sigma_2 = C - R \qquad (2.28)$$

The maximum shear stress is given by:

$$\tau_{Max} = R \qquad (2.29a)$$

The normal stress acting on the plane associated with τ_{Max} is given by:

$$\sigma_n = C \qquad (2.29b)$$

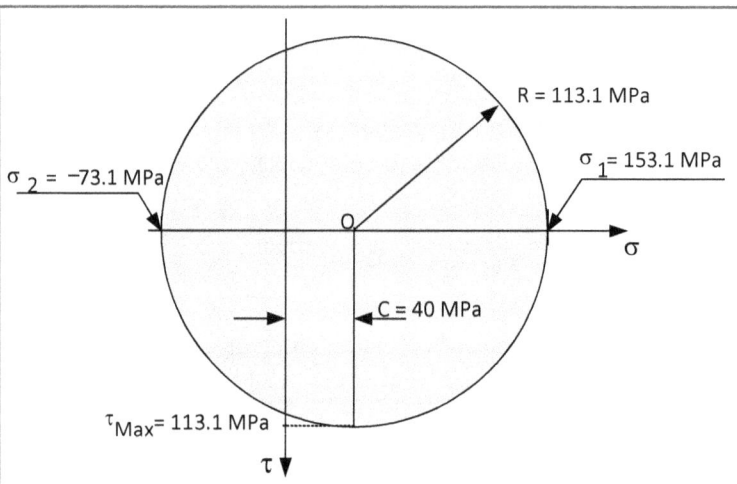

Fig. 2.6a

Step 3: To determine the orientation of the planes, upon which the principal stresses and the maximum shear stress acts, let's establish the location of the x axis and y axes on Mohr's circle. Locate σ_{xx} = 120 MPa on the σ axis and draw a perpendicular line through this point. Measure down the perpendicular a distance equal to the value of τ_{xy} = 80 MPa. We measure downward because τ_{xy} is a positive quantity (counterclockwise) and the positive τ axis is downward. Point A, shown in Fig. E2.6b, represents the point with coordinates σ = 120 MPa and τ = 80 MPa. This point also falls on the circle that was drawn in Step 2. Clearly the line O – A represents the radius of Mohr's circle.

We have extended line O – A beyond the circle and labeled it as the x –axis. We have also extended this line upward and to the left until it intersect the circle at point B. If we drop a perpendicular line from point B to its intersection with the σ axis, the intersection occurs at σ_{yy} = – 40 MPa verifying the accuracy of the construction of the circle. The line O – B represents the y-axis. Note on this axis that τ_{xy} is negative (clockwise) and the negative τ axis is upward. Having established the positions of the x and y axes relative to the circle, the next step is to measure the angles defining the principal planes and the plane, upon which the maximum shear stress acts.

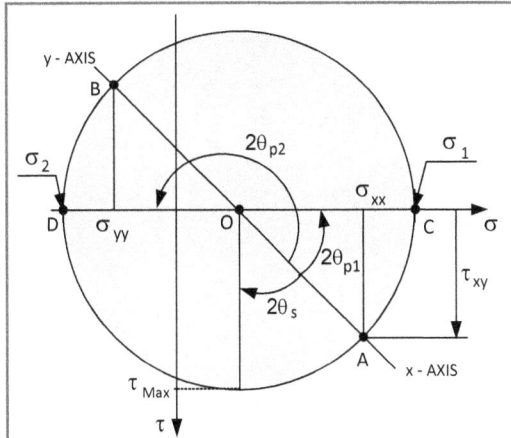

Fig. E2.6b

Step 4: From line O-A, we measure the angle AOC as illustrated in Fig. E2.6b, to obtain $2\theta_{p1}$ = 45°. This observation indicates that the plane, upon which σ_1 acts, is located at θ_{p1} = 22.5°. We also note that the angle $2\theta_{p2} = 2\theta_{p1} + 180°$. Thus, σ_2 acts on the plane located by θ_{p2} = 22.5° + 90° = 112.5°. The angle $2\theta_{p1}$ can be verified by substituting the values for σ_{xx}, σ_{yy} and τ_{xy} into Eq. (2.16) to obtain:

$$\tan 2\theta_{p1} = \frac{2\tau_{xy}}{\sigma_{xx} - \sigma_{yy}} = \frac{2(80)}{120+40} = 1 \tag{d}$$

$$2\theta_{p1} = \tan^{-1}(1) = 45°$$

The angle $2\theta_s$ is measured in the clockwise (negative) direction. The measurement gives $2\theta_s$ = −45° as expected because the expression for $2\theta_s$ is the negative reciprocal of Eq. (2.16).

Step 5: Finally, let's locate the principal planes relative to the x and y axes as shown in Fig. E2.6c.

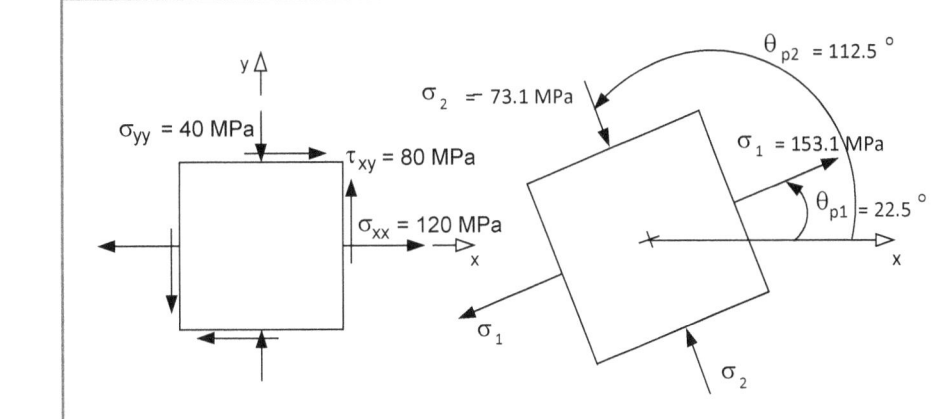

Fig E2.6c

EXAMPLE E2.7

Consider a point, shown here as an elemental area, with Cartesian components of stress defined in Fig. E2.7. Using Mohr's circle, determine the principal stresses and the orientation of the principal planes. Also determine the maximum shear stresses and their orientation.

40 — Chapter 2 — Stress at a Point

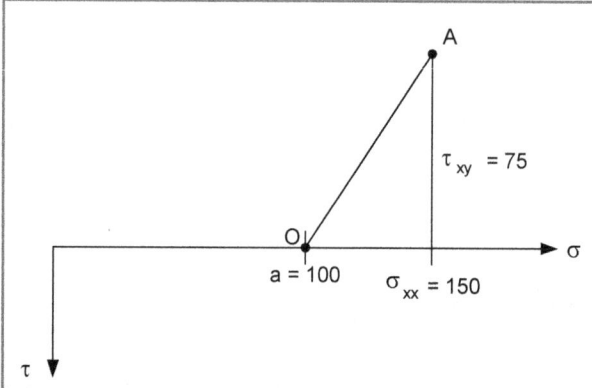

Fig. E2.7

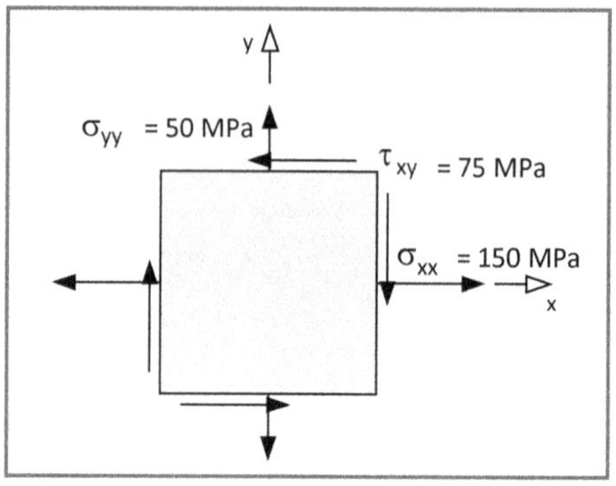

Solution:

Step 1: Use Eq. (2.26) to determine the position of the center of the circle as:

$$C = (\sigma_{xx} + \sigma_{yy})/2 = (150 + 50)/2 = 100 \text{ MPa} \quad (a)$$

Step 2: Layout the axes for Mohr's circle and locate the center point O at $\sigma = C$. Then locate $\sigma_{xx} = 150$ MPa along the σ axis and construct a perpendicular line upward from this intersection. Scale the length of the perpendicular line to be proportional to $\tau_{xy} = 75$ MPa, as shown in Fig E2.7a. Because τ_{xy} is a negative quantity, it is plotted in the negative τ direction. Draw the line O – A, which is the radius of Mohr's circle. This line also locates the x – axis relative to Mohr's circle. The radius can be determined from Eq. (2.27) as:

$$R = \sqrt{\left(\frac{\sigma_{xx} - \sigma_{yy}}{2}\right)^2 + \tau_{xy}^2} = \sqrt{\left(\frac{150 - 50}{2}\right)^2 + (75)^2} = 90.14 \text{ MPa} \quad (b)$$

Fig. E2.7a

Step 3: Using the length of the line O - A as the radius R and the center of the circle at $\sigma = C$, we construct Mohr's circle as shown in Fig. E2.7b. The axes are scaled and the locations of σ_1, σ_2 and τ_{Max} are marked. From the scale imposed on the σ and τ axes, we determine the approximate values of the principal and maximum shear stresses as:

$\sigma_1 = 190$ MPa $\quad\quad \sigma_2 = 10$ MPa $\quad\quad \tau_{Max} = 90$ MPa $\quad\quad \sigma_n = 100$ MPa

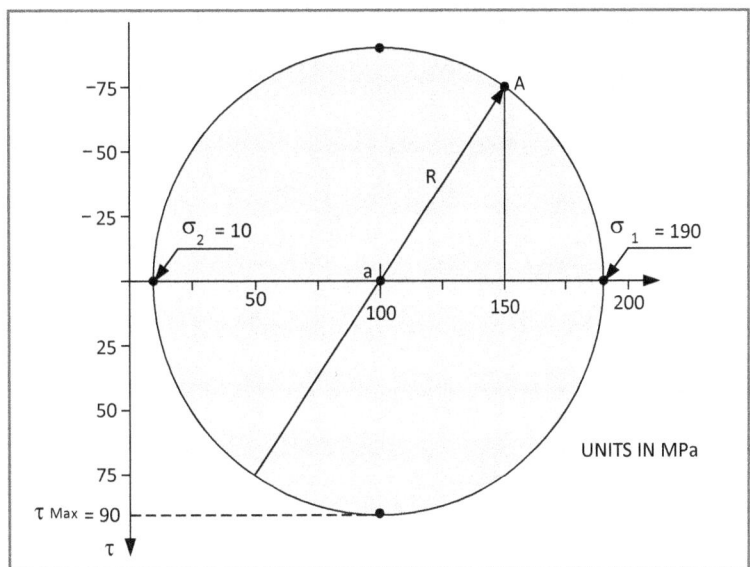

Fig. E2.7b

You can verify the accuracy of these results by employing Eq. (2.18) and Eq. (2.21).

Step 4: The location of the principal planes and the planes of maximum shear stress is determined by measuring the angles $2\theta_p$ and $2\theta_s$ as illustrated in Fig. E2.7c. Measurements of the angles AOC and AOD, both in the clockwise (negative) direction, shows that $2\theta_p = -56°$ and that $\theta_s = 146°$.

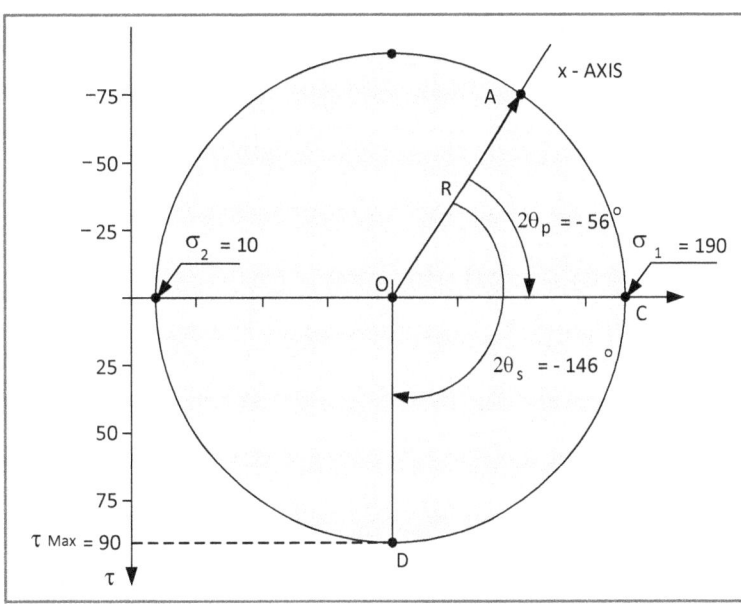

Fig. E2.7c

Step 5: Construct elements showing the planes, upon which σ_1 and σ_2 act. We rotate a line through an angle θ_p measured in the clockwise direction from the x-axis to obtain the element shown in Fig. E 2.7d. Construct a similar element showing the plane upon which τ_{MAX} acts. Again we rotate a line an angle θ_s measured in the clockwise direction from the x-axis to obtain the element shown in Fig. E 2.7e.

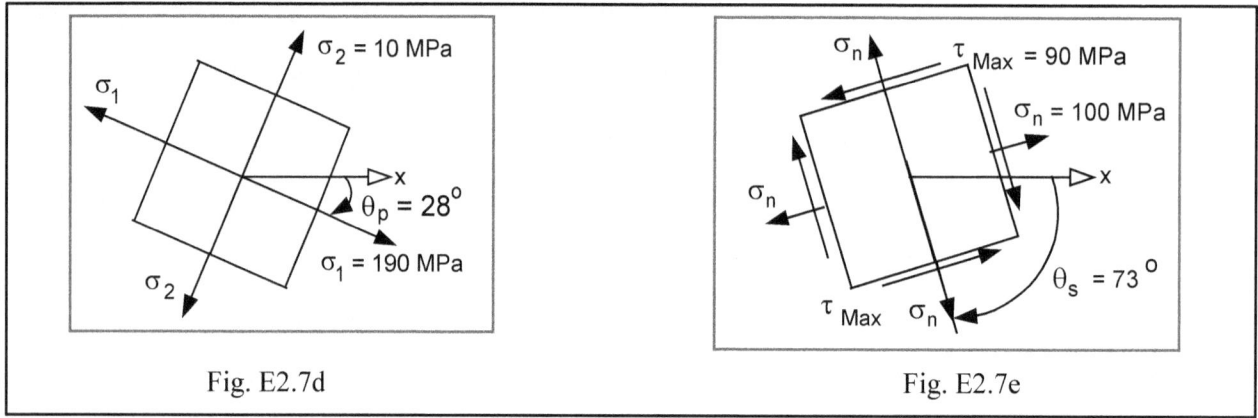

Fig. E2.7d   Fig. E2.7e

2.6 MOHR'S CIRCLE IN THREE-DIMENSIONS

A useful aid for visualizing the complete state of stress at a point is the three-dimensional Mohr's circle shown in Fig. 2.11. This representation, which is similar to the familiar two-dimensional Mohr's circle, shows the three principal stresses, the maximum shearing stresses, and the range of values within which the normal and shear stress components must lie for a given state of stress.

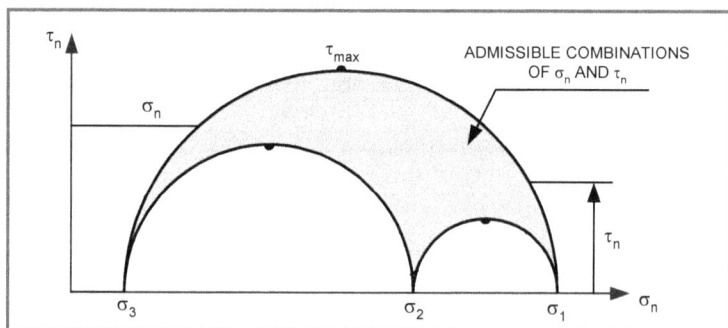

Figure 2.11 Mohr's circle for the three-dimensional state of stress.

2.7 FAILURE THEORIES

When considering the safety of structural members—beams, truss members, tie rods, etc., we often predict that failure occurs when the normal stress σ imposed on the member in question exceeded the strength of the material from which the member was fabricated. The equation for failure for these structural members is given by:

$$\sigma_{xx} = \sigma_1 \geq S_y \qquad \sigma_{xx} = \sigma_1 \geq S_u \qquad (2.30)$$

where the choice of yield or ultimate tensile strength, S_y or S_u, depends on the admissible mode of failure.

For uniaxial members, the stress state is also uniaxial with $\sigma_{yy} = \sigma_{zz} = 0$ and $\sigma_2 = \sigma_3 = 0$; hence, the conditions for structural failure given in Eq. (2.30) are valid. However, for more complex structural components such as plates and shells or machine components, the state of stress is multi-dimensional and more involved theories of failure are required. We will consider a two-dimensional state of stress—plane stress— in this section where the principal stresses are ordered so that:

$$\sigma_1 \geq \sigma_2 \quad\quad \sigma_3 = 0 \quad\quad (2.31)$$

Several different theories of failure are used today, but in the interest of brevity, we will describe only the three most popular ones.

1. The maximum principal stress theory.
2. The maximum shear stress theory.
3. The maximum distortional energy theory.

We will also assume that the failure occurs upon yielding and that the yield strength in tension is the same as the yield strength in compression.

2.7.1 The Maximum Principal Stress Theory

The maximum principal stress theory states that a structural component fails by yielding, when the maximum principal stress σ exceeds the yield strength of the material from which it is fabricated.

$$|\sigma_1| \geq S_y \quad\quad |\sigma_2| \geq S_y \quad\quad (2.32)$$

If we consider all possible combinations of σ_1 and σ_2, it is possible to construct the yielding diagram presented in Fig. 2.12.

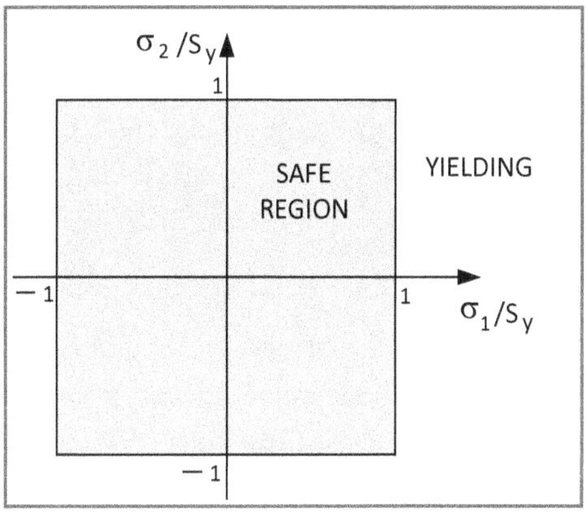

Fig. 2.12 Failure diagram for the maximum principal stress theory for yielding.

If we plot a point with coordinates σ_1/S_y and σ_2/S_y and it falls within the square in Fig. 2.12, the structural component will not yield. However, if the point falls on the boundary or outside the boundary of the box, yielding will occur. This theory is employed in the analysis of uniaxial structural members, where the influence of the second principal stress σ_2 is small. Unfortunately, failure of ductile materials subjected to biaxial states of stress occurs due to shear on planes of maximum shear stress. In these cases, the maximum principal stress theory is not conservative. Yielding by shear occurs at stress states lower than predicted by the maximum principal stress theory.

2.7.2 The Maximum Shear Stress Theory

The maximum shear stress theory for yielding, sometimes called the Tresca theory, is often employed to predict the onset of yielding in machine components fabricated from ductile materials. These ductile materials fail on maximum shear planes that are oriented at 45° to the principal planes. This theory indicates that yielding will occur when τ_{Max} in a machine component equals the maximum shear stress achieved at yield in a simple tensile test of the material. Consider first the simple tension test where:

44 — Chapter 2 — Stress at a Point

$$\sigma_1 = S_y \qquad \sigma_2 = \sigma_3 = 0 \qquad \tau_{Max} = (\sigma_1 - \sigma_2)/2 = S_y/2 \qquad (2.33)$$

Next, consider the maximum shear stress in the structural component. Two possibilities must be considered:

1. When σ_1 and σ_2 are of opposite signs, τ_{Max} is given by:

$$\tau_{Max} = (\sigma_1 - \sigma_2)/2 \qquad (2.34)$$

Substituting Eq. (2.33) into Eq. (2.34) gives the equation governing yielding as:

$$|\sigma_1 - \sigma_2| \geq S_y \qquad (2.35)$$

2. When σ_1 and σ_2 are the same sign, τ_{Max} is given by:

$$\tau_{Max} = (\sigma_1 - \sigma_3)/2 = \sigma_1/2 \qquad (2.36)$$

Substituting Eq. (2.33) into Eq. (2.36) gives the equation governing yielding as:

$$|\sigma_1| \geq S_y \qquad (2.37)$$

If we consider all possible combinations of σ_1 and σ_2, it is possible to construct the yielding diagram presented in Fig. 2.13.

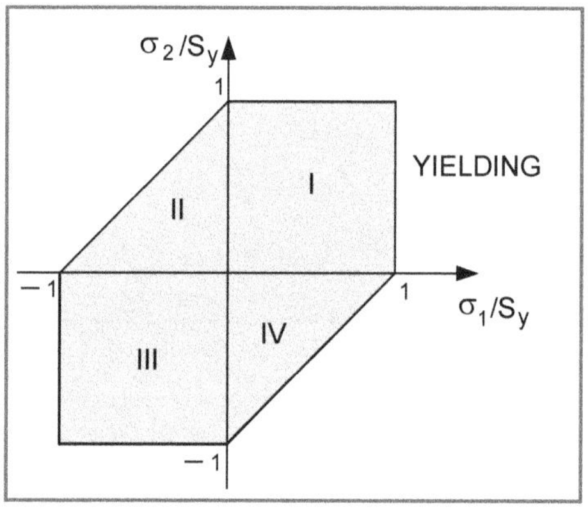

Fig. 2.13 Failure diagram for the Maximum shear stress theory of yielding.

When the principal stresses σ_1 and σ_2 are of like sign, the stress state corresponds to quadrant I or III in Fig. 2.13, and the maximum shear theory for yielding corresponds to the maximum principal stress theory. When the signs of σ_1 and σ_2 are opposite, the stress state corresponds to quadrant II and IV. In this instance, there is a marked difference between the two theories, and the maximum shear stress theory is much more conservative.

2.7.3 Maximum Distortion Energy Theory

The maximum distortion energy theory, sometimes called the von Mises theory, predicts yielding in a machine component when the distortion energy per unit volume equal the distortion energy per unit volume achieved at yielding of the material in a simple tensile test.

For a two-dimensional stress state with $\sigma_3 = 0$, the distortion energy per unit volume U_d is given by:

$$U_d = (1/6G)[\sigma_1^2 - \sigma_1\sigma_2 + \sigma_2^2] \qquad (2.38)$$

In a tensile test with $\sigma_1 = S_y$ and $\sigma_2 = \sigma_3 = 0$, the distortion energy per unit volume is:

$$U_d = (1/6G)S_y^2 \qquad (2.39)$$

By substituting Eq. (2.39) into Eq. (2.38), we obtain the von Mises equation for yielding.

$$(\sigma_1/S_y)^2 - (\sigma_1/S_y)(\sigma_2/S_y) + (\sigma_2/S_y)^2 = 1 \qquad (2.40)$$

If we consider all possible combinations of σ_1 and σ_2, it is possible to construct the yielding diagram presented in Fig. 2.14. The boundary of the failure diagram is an ellipse that is slightly less conservative than the maximum principal stress theory for yielding in quadrants I and III. In quadrants II and IV where the principal stresses are of opposite sign the von Mises theory is less conservative than the maximum shear theory but more conservative that the maximum principal stress theory. The distortional energy theory is widely accepted as the most accurate theory for predicting yielding in ductile materials that are commonly used to fabricate structural elements and machine components.

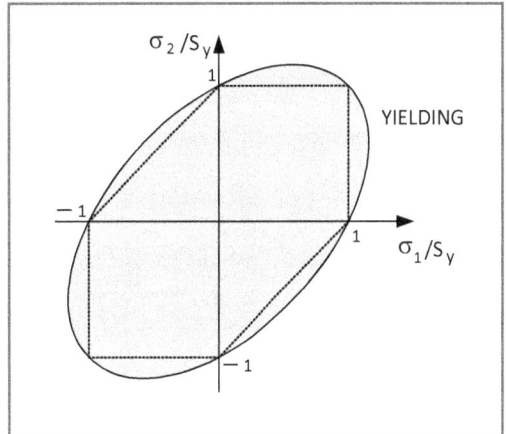

Fig 2.14 Failure diagram for the maximum distortion energy theory for yielding.

EXAMPLE E2.8

Your analysis of the principal stresses in a machine component indicates that $\sigma_1 = 38$ ksi and $\sigma_2 = -20$ ksi. If the component is fabricated from 1020 HR steel, will it yield in service?

Solution: Because $S_y = 42$ ksi for 1020 HR steel, the prediction of yielding will depend on the theory of yielding employed in the analysis. Let's use the three different theories described in the previous subsections. For the maximum principal stress theory for yielding and Eq. (2.32), it is clear that yielding will not occur because:

$$\sigma_1 = 38 \text{ ksi} < S_y = 42 \text{ ksi} \qquad |\sigma_2| = |20| \text{ ksi} < S_y = 42 \text{ ksi} \qquad (a)$$

Because σ_1 and σ_2 are of opposite signs, the maximum shear stress theory for yielding and Eq. (2.35) gives:
$$(\sigma_1 - \sigma_2) = (38 + 20) = 58 \text{ ksi} > S_y = 42 \text{ ksi} \qquad (b)$$

From the results of Eq. (b), we predict that yielding will occur.

Using the maximum distortion energy theory and applying Eq. (2.40), we write:

$$(\sigma_1/S_y)^2 - (\sigma_1/S_y)(\sigma_2/S_y) + (\sigma_2/S_y)^2 \geq 1$$

$$(38/42)^2 - (38/52)(-20/42) + (-20/42)^2 = 1.476 > 1 \qquad (c)$$

The maximum distortional energy theory also predicts yielding. Because the two conservative theories of failure that are widely accepted to predict yielding in ductile materials provide the same result, we indicate that the machine component will fail by yielding. We indicate the same result on the graphic shown in Fig. E2.8, where the operating point $\sigma_1/S_y = 0.905$ and $\sigma_2/S_y = -0.476$ is plotted.

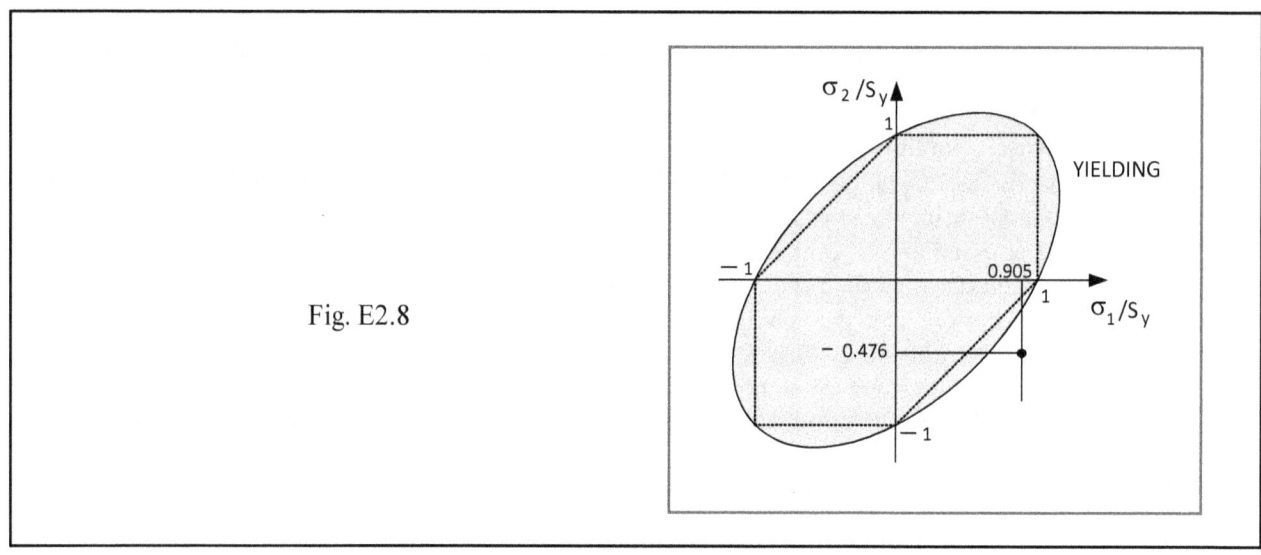

Fig. E2.8

2.8 SUMMARY

To define stresses we introduced a resultant force vector $\Delta \mathbf{F_n}$ that acts over an incremental area with an outer normal vector **n**. We then took the limit of $\Delta \mathbf{F_n}/\Delta A$ as ΔA vanishes to obtain the resultant stress vector $\mathbf{T_n}$. By resolving $\mathbf{T_n}$ onto the normal and tangential coordinates we were able to define normal and shear stresses, respectively. Following a similar procedure with Cartesian coordinates enabled us to define the Cartesian components of stress.

In this chapter multi-dimensional states of stress were introduced that give rise to nine Cartesian components of stress. However, equilibrium conditions yield a shear stress relation that reduced the number of unique Cartesian components of stress from nine to six.

$$\tau_{xy} = \tau_{yx} \qquad \tau_{yz} = \tau_{zy} \qquad \tau_{xz} = \tau_{zx} \qquad (9.1)$$

The equations of stress transformation were derived to provide expressions for $\sigma_{x'x'}$, $\sigma_{y'y'}$ and $\tau_{x'y'}$ in terms of σ_{xx}, σ_{yy} and τ_{xy}.

$$\sigma_{x'x'} = \frac{\sigma_{xx} + \sigma_{yy}}{2} + \frac{\sigma_{xx} - \sigma_{yy}}{2}\cos 2\theta + \tau_{xy}\sin 2\theta$$

$$\sigma_{y'y'} = \frac{\sigma_{xx} + \sigma_{yy}}{2} - \frac{\sigma_{xx} - \sigma_{yy}}{2}\cos 2\theta - \tau_{xy}\sin 2\theta \qquad (2.14)$$

$$\tau_{x'y'} = -\frac{\sigma_{xx} - \sigma_{yy}}{2}\sin 2\theta + \tau_{xy}\cos 2\theta$$

After establishing the stress equations of transformation, we showed that the principal stresses were given by:

$$\sigma_{Max,\,Min} = \frac{\sigma_{xx} + \sigma_{yy}}{2} + \frac{\sigma_{xx} - \sigma_{yy}}{2}\cos 2\theta_p + \tau_{xy}\sin 2\theta_p$$

$$\sigma_{Max,\,Min} = \frac{\sigma_{xx} + \sigma_{yy}}{2} \pm \sqrt{\left(\frac{\sigma_{xx} - \sigma_{yy}}{2}\right)^2 + \tau_{xy}^2} \qquad (2.18)$$

Also, we showed that the principal planes were located at θ_p:

$$\tan 2\theta_p = \frac{2\tau_{xy}}{\sigma_{xx} - \sigma_{yy}} \qquad (2.16)$$

In a similar manner, we derived the relation for the maximum shear stress and the location of the plane θ_s upon which this stress acts as:

$$\tau_{Max} = \sqrt{\left(\frac{\sigma_{xx} - \sigma_{yy}}{2}\right)^2 + \tau_{xy}^2} \qquad (2.21)$$

$$\tan 2\theta_s = -\frac{\sigma_{xx} - \sigma_{yy}}{2\tau_{xy}} \qquad (2.19)$$

A graphical technique, Mohr's circle, was described for visualizing how the Cartesian components of stress vary with the angle of the plane upon which these stresses act. Examples were presented demonstrating the use of Mohr's circle for determining principal stresses, principal planes and maximum shearing stresses.

Three different failure theories were introduced including:

1. The maximum principal stress theory.
2. The maximum shear stress theory.
3. The maximum distortional energy theory.

The maximum principal stress theory indicates that yielding occurs when:

$$|\sigma_1| \geq S_y \qquad |\sigma_2| \geq S_y \qquad (2.32)$$

The maximum shear stress theory indicates yielding occurs when:

$$|\sigma_1 - \sigma_2| \geq S_y \quad \text{Same sign stresses} \qquad (2.35)$$
$$|\sigma_1| \geq S_y \quad \text{Opposite sign stresses} \qquad (2.37)$$

The maximum distortional energy theory indicates yielding occurs when:

$$(\sigma_1/S_y)^2 - (\sigma_1/S_y)(\sigma_2/S_y) + (\sigma_2/S_y)^2 \geq 1 \qquad (2.41)$$

The maximum distortional energy theory is widely accepted as the most accurate method for determining the onset of yielding in ductile materials commonly used to fabricate structural element and machine components.

PROBLEMS

2.1 Prepare a scale drawing representing the stresses associated with a resultant stress $\mathbf{T_n}$, where $|\mathbf{T_n}| = 20$ units and its direction is described by $\cos(\mathbf{T_n}, x) = 0.6$ and $\cos(\mathbf{T_n}, y) = 0.7$. The normal to the surface upon which $\mathbf{T_n}$ acts is given by $\mathbf{n} = (1/3)\mathbf{i} + (2/3)\mathbf{j} + (2/3)\mathbf{k}$. In this drawing show σ_n and τ_n to scale.

2.2 We represent a plane state of stress at a point by a small elemental area ΔA, shown in the figure below. The stresses are σ_{xx}, σ_{yy} and τ_{xy} are given in the table below. Determine the stresses $\sigma_{x'x'}$, $\sigma_{y'y'}$ and $\tau_{x'y'}$ on an element that is rotated θ degrees (counterclockwise is positive) relative to the element shown in the figure below.

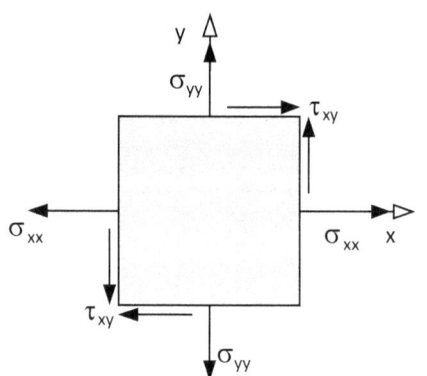

Prob. No.	σ_{xx}	σ_{yy}	τ_{xy}	Angle, θ
2.2a	15 ksi	21 ksi	−18 ksi	25°
2.2b	80 MPa	−45 MPa	35 MPa	−35°
2.2c	18 ksi	−12 ksi	−10 ksi	65°
2.2d	−120 MPa	−88 MPa	−25 MPa	−42
2.2e	−32 ksi	−19 ksi	12 ksi	18°

2.3 Consider a point, shown in the figure below as an elemental area, with Cartesian components of stress defined in the table below. Determine the orientation of the principal planes and the stresses that act on these planes.

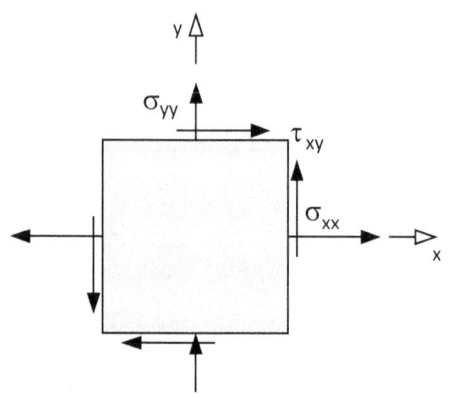

Prob. No.	σ_{xx}	σ_{yy}	τ_{xy}
2.3a	25 ksi	31 ksi	−28 ksi
2.3b	100 MPa	−55 MPa	45 MPa
2.3c	28 ksi	−22 ksi	−20 ksi
2.3d	−112 MPa	−98 MPa	−45 MPa
2.3e	−42 ksi	−29 ksi	32 ksi

2.4 For the Cartesian stress components given in Problem 2.3, determine the maximum shear stresses and the planes upon which they act. Also determine the normal stress that acts on this plane.

2.5 Construct Mohr's circle for the stress state defined in Problem 2.2 and determine the stresses $\sigma_{x'x'}$, $\sigma_{y'y'}$ and $\tau_{x'y'}$ on an element that is rotated θ degrees (counterclockwise is positive) relative to the element shown in Problem 2.2.

2.6 Construct Mohr's circle for the stress state defined in Problem 2.3 and determine the principal stresses and the planes upon which they act. Also determine the maximum shear stress, the corresponding normal stress and the planes upon which they act.

2.7 Using the maximum principal stress theory, predict yielding if the yield strength of a structural component is given in the table below. Note the Cartesian stresses are also given in this table.

Prob. No.	σ_{xx}	σ_{yy}	τ_{xy}	S_y
2.7a	25 ksi	31 ksi	– 28 ksi	36 ksi
2.7b	100 MPa	– 55 MPa	45 MPa	125 MPa
2.7c	28 ksi	– 22 ksi	– 20 ksi	32 ksi
2.7d	– 112 MPa	– 98 MPa	– 45 MPa	162 MPa
2.7e	– 42 ksi	– 29 ksi	32 ksi	52 ksi

2.8 Using the maximum shear stress theory, predict yielding if the yield strength of a structural component is given in the table for Problem 2.7. Note the Cartesian stresses are also given in this table.

2.9 Using the maximum distortional energy theory, predict yielding if the yield strength of a structural component is given in the table for Problem 2.7. Note the Cartesian stresses are also given in this table.

REFERENCES

1. Boresi, A. P., and P. P. Lynn: Elasticity in Engineering Mechanics, Prentice-Hall, Inc., Englewood Cliffs, N.J., 1974.
2. Love, A. E. H.: A Treatise on the Mathematical Theory of Elasticity, Dover Publications, Inc., New York, 1944.
3. Sokolnikoff, I. S.: Mathematical Theory of Elasticity, 2d ed. McGraw-Hill Book Company, New York, 1956.
4. Southwell, R. V.: An Introduction to the Theory of Elasticity, Oxford University Press, Fair Lawn, N.J., 1953.
5. Timoshenko, S. P., and J. N. Goodier: Theory of Elasticity, 2d ed., McGraw-Hill Book Company, New York, 1951.

CHAPTER 3

STRAIN AND THE STRESS-STRAIN RELATIONS

3.1 INTRODUCTION

In Chapter 2, we discussed the state of stress, which develops at an arbitrary point within a body as a result of applied forces. The equations obtained were based only on the conditions of equilibrium. Because no assumptions were made regarding body deformations or physical properties of the material from which the body was composed, the results for stresses are valid for any material and for any amount of body deformation. In this chapter, body deformation and associated strain will be discussed. Because strain is a pure geometric quantity, no restrictions on body material will be required. However, in order to obtain linear equations relating displacement to strain, restrictions must be placed on the magnitude of the allowable deformations. However, when the stress-strain relations are developed, the elastic constants of the body material must be considered.

3.2 DEFINITIONS OF DISPLACEMENT AND STRAIN

If a body is subjected to a system of forces, individual points in or on the body undergo movement. This movement of an arbitrary point is a vector quantity known as a **displacement**. If the various points in the body undergo different movements, each point can be represented by a unique displacement vector. Each vector can be resolved into components parallel to a set of Cartesian coordinate axes such that u, v, and w are the displacement components in the x, y, and z directions, respectively. Motion of the body may be considered as the sum of two parts:

1. The translation and/or rotation of the body as a whole.
2. The movement of points within the body relative to each other.

The translation or rotation of the body as a whole is known as **rigid-body motion**. This type of motion is applicable to either the idealized rigid body or to a real deformable body. The movement of the points of the body relative to each other is known as a **deformation**. Rigid-body motions can be large or small, but deformations are usually small, except when rubber-like materials or specialized structures such as very long, slender beams are involved.

Strain is a geometric quantity, which depends on the relative movements of two or three points in the body. Therefore strain is related only to the deformation displacements. Because rigid-body displacements do not produce strains, they will be neglected in all further discussion. In Chapter 2, two types of stresses were discussed—normal and shear stress. This same classification will be used for strains. A normal strain is defined as the change in length of a line segment between two points, divided by the original length of this line segment. A shearing strain is defined as the angular change between two line segments that were originally perpendicular. The relationships between strains and displacements can be determined by considering the deformation of an arbitrary cube in a body that is subjected to some forces. This deformation is illustrated in Fig. 3.1, where a general point P is moved through a distance u in the x direction, v in the y direction and w in the z direction. The other corners of the cube are also displaced and, in general, they will be displaced by amounts that differ from those at point P. For example, the displacements u*, v* and w* associated with point Q can be expressed in terms of the displacements u, v and w at point P by means of a Taylor-series expansion. Thus:

$$u^* = u + \frac{\partial u}{\partial x}\Delta x + \frac{\partial u}{\partial y}\Delta y + \frac{\partial u}{\partial z}\Delta z + \cdots$$

$$v^* = v + \frac{\partial v}{\partial x}\Delta x + \frac{\partial v}{\partial y}\Delta y + \frac{\partial v}{\partial z}\Delta z + \cdots \qquad (3.1)$$

$$w^* = w + \frac{\partial w}{\partial x}\Delta x + \frac{\partial w}{\partial y}\Delta y + \frac{\partial w}{\partial z}\Delta z + \cdots$$

The terms shown in Eq. (3.1) are the only significant terms, if it is assumed that the cube is sufficiently small for higher-order terms such as $(\Delta x)^2$, $(\Delta y)^2$, $(\Delta z)^2$, ….to be negligible. Under these conditions, planes will remain plane and straight lines will remain straight lines when the cube deforms, as shown in Fig. 3.1.

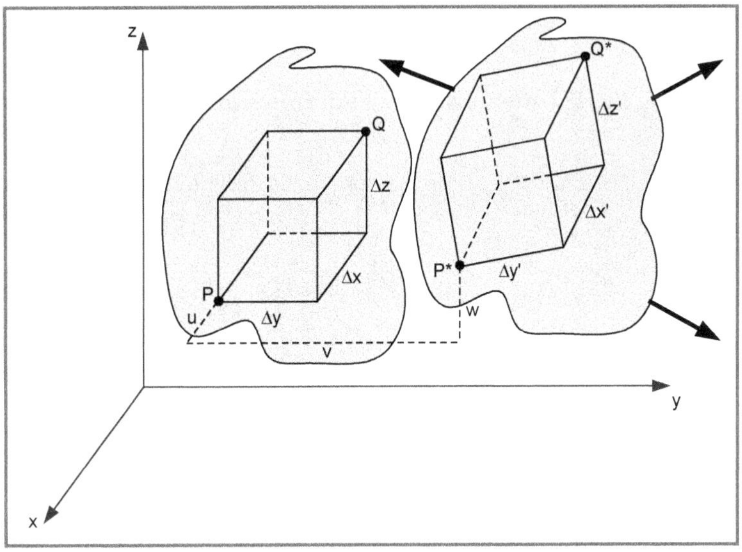

Figure 3.1 The distortion of an arbitrary cube within a body after applying a system of forces.

The average normal strain along an arbitrary line segment was previously defined as the change in length of the line segment divided by its original length. This normal strain can be expressed in terms of the displacements experienced by points at the ends of a line segment. For example, consider the line PQ originally oriented parallel to the x-axis, as shown in Fig.3.2. Because y and z are constant along PQ, Eq. (3.1) gives the following displacements for point Q, if the displacements for point P are given by u, v, and w:

$$u^* = u + \frac{\partial u}{\partial x}\Delta x \qquad v^* = v + \frac{\partial v}{\partial x}\Delta x \qquad w^* = w + \frac{\partial w}{\partial x}\Delta x$$

From the definition of normal strain:

$$\varepsilon_{xx} = \frac{\Delta x' - \Delta x}{\Delta x} \qquad (a)$$

or:

$$\Delta x' = (1 + \varepsilon_{xx})\Delta x \qquad (b)$$

From Fig. 3.2, it is evident that the deformed length Δx' can be expressed in terms of the displacement gradients as:

$$(\Delta x')^2 = \left[\left(1 + \frac{\partial u}{\partial x}\right)\Delta x\right]^2 + \left(\frac{\partial v}{\partial x}\Delta x\right)^2 + \left(\frac{\partial w}{\partial x}\Delta x\right)^2 \qquad (c)$$

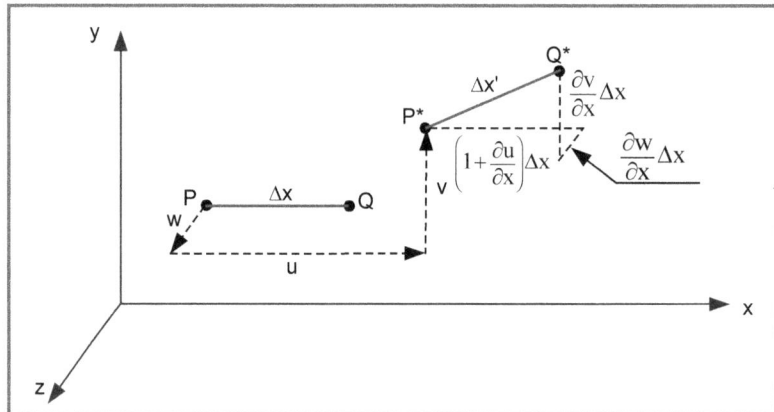

Figure 3.2 Displacement gradients associated with the normal strain ε_{xx}.

Squaring Eq. (b) and substituting Eq. (c) yields:

$$(1 + \varepsilon_{xx})^2 (\Delta x)^2 = \left[1 + 2\frac{\partial u}{\partial x} + \left(\frac{\partial u}{\partial x}\right)^2 + \left(\frac{\partial v}{\partial x}\right)^2 + \left(\frac{\partial w}{\partial x}\right)^2\right](\Delta x)^2$$

or

$$\varepsilon_{xx} = \sqrt{1 + 2\frac{\partial u}{\partial x} + \left(\frac{\partial u}{\partial x}\right)^2 + \left(\frac{\partial v}{\partial x}\right)^2 + \left(\frac{\partial w}{\partial x}\right)^2} - 1 \qquad (3.2a)$$

In a similar manner, considering line segments originally oriented parallel to the y and z-axes leads to:

$$\varepsilon_{yy} = \sqrt{1 + 2\frac{\partial v}{\partial y} + \left(\frac{\partial v}{\partial y}\right)^2 + \left(\frac{\partial w}{\partial y}\right)^2 + \left(\frac{\partial u}{\partial y}\right)^2} - 1 \qquad (3.2b)$$

$$\varepsilon_{zz} = \sqrt{1 + 2\frac{\partial w}{\partial z} + \left(\frac{\partial w}{\partial z}\right)^2 + \left(\frac{\partial u}{\partial z}\right)^2 + \left(\frac{\partial v}{\partial z}\right)^2} - 1 \qquad (3.2c)$$

In most engineering problems, the displacements and strains produced by the applied forces are very small. Assuming small displacements and strains, it is evident that products and squares of displacement gradients will be small with respect to the displacement gradients and can be neglected. Applying the small displacement and strain assumption to Eqs. (3.2), enables the simplification of the normal strain-displacement equations to:

$$\varepsilon_{xx} = \frac{\partial u}{\partial x}$$

$$\varepsilon_{yy} = \frac{\partial v}{\partial y} \qquad (3.3)$$

$$\varepsilon_{zz} = \frac{\partial w}{\partial z}$$

The shear strain components can also be related to the displacements, by considering the changes in right angle experienced by the edges of the cube during deformation. For example, consider lines PQ and PR, as shown in Fig. 3.3. The angle θ* between P*Q* and P*R* in the deformed state can be expressed in terms of the displacement gradients to obtain:

$$\cos\theta^* = \left[\left(1+\frac{\partial u}{\partial x}\right)\frac{\Delta x}{\Delta x'}\right]\left[\left(\frac{\partial u}{\partial y}\right)\frac{\Delta y}{\Delta y'}\right] + \left(\frac{\partial v}{\partial x}\frac{\Delta x}{\Delta x'}\right)\left[\left(1+\frac{\partial v}{\partial y}\right)\frac{\Delta y}{\Delta y'}\right] + \left(\frac{\partial w}{\partial x}\frac{\Delta x}{\Delta x'}\right)\left(\frac{\partial w}{\partial y}\frac{\Delta y}{\Delta y'}\right) \qquad \text{(a)}$$

From the definition of shear strain,

$$\gamma_{xy} = [(\pi/2) - \theta^*] \qquad \text{(b)}$$

therefore:

$$\sin\gamma_{xy} = \sin[(\pi/2) - \theta^*] = \cos\theta^* \qquad \text{(c)}$$

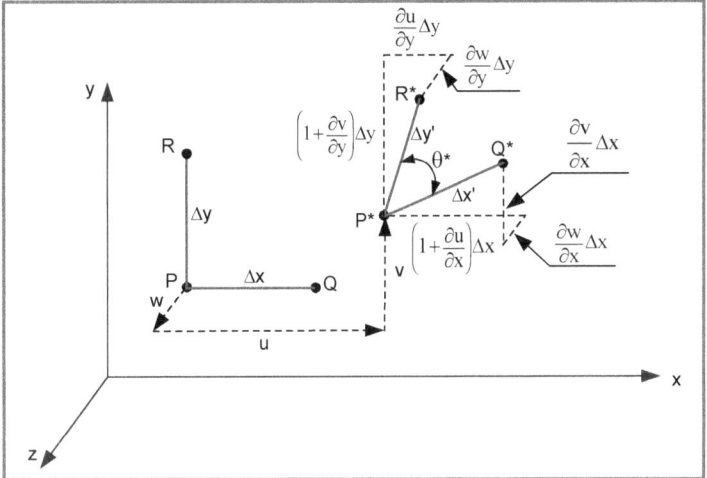

Figure 3.3 Displacement gradients associated with the shear strain γ_{xy}.

Applying the small displacement and strain assumption to Eqs. (a) and (c), enables the simplification of the shear strain-displacement equations to:

$$\gamma_{xy} = \frac{\partial v}{\partial x} + \frac{\partial u}{\partial y}$$

$$\gamma_{yz} = \frac{\partial w}{\partial y} + \frac{\partial v}{\partial z} \qquad (3.4)$$

$$\gamma_{zx} = \frac{\partial u}{\partial z} + \frac{\partial w}{\partial x}$$

Equations (3.3) and (3.4) are used to convert a displacement field into a strain field.

3.3 STRAIN EQUATIONS OF TRANSFORMATION

In Chapter 2, we derived the stress equations of transformation and showed that the stresses $\sigma_{x'x'}$, $\sigma_{y'y'}$ and $\tau_{x'y'}$ varied with the angle of rotation θ of the Ox'y' coordinate system relative to the Oxy coordinate system. A remarkably similar set of transformation equations exist for strain. When deriving the stress equations of transformation, we considered a state of plane stress where $\sigma_{zz} = \tau_{xz} = \tau_{yz} = 0$. This derivation was based solely on the equilibrium equations.

To derive the strain equations of transformation, we will consider a state of plane strain where $\varepsilon_{zz} = \gamma_{xz} = \gamma_{yz} = 0$. This derivation is based only on the geometry of the deformation field. Let's consider the plane body illustrated in Fig 3.4, where the line segment AB is scribed prior to imposing plane strain deformations. After the deformations are imposed, the line segment moves and is designated as A*B*. Point A moves to a new location designated by A*. The displacement in the x direction associated with this movement is u and that in the y direction is v. Point B moves to a new location designated by B*, with displacements u + Δu in the x direction and v + Δv in the y direction.

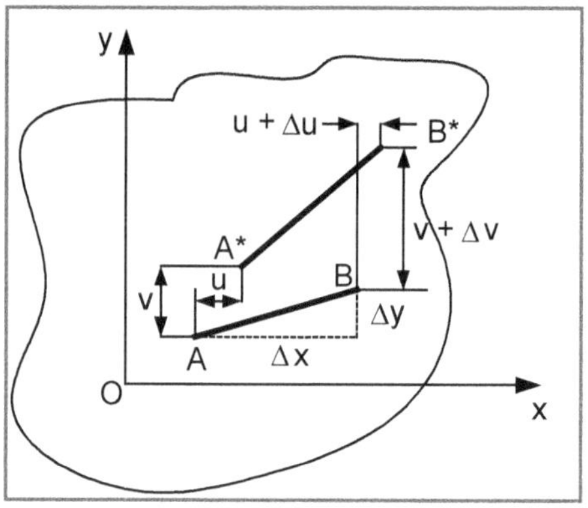

Fig. 3.4 A line segment AB before deformation and A*B* after deformation.

By definition the strain along the line segment AB is given by:

$$\varepsilon_{AB} = (A^*B^* - AB)/AB \quad (a)$$

From the geometry of Fig. 3.4, it clear that the length of the line segments are given by:

$$(AB)^2 = (\Delta x)^2 + (\Delta y)^2 \quad (b)$$

$$(A^*B^*)^2 = (\Delta x^*)^2 + (\Delta y^*)^2 \quad (c)$$

In general, the component Δx^* will have a different length than Δx, because of the deformation of the body in the x direction. We express this difference in both Δx^* and Δy^* by writing:

$$\Delta x^* = \left(1 + \frac{\partial u}{\partial x}\right)\Delta x + \frac{\partial u}{\partial y}\Delta y$$
$$\Delta y^* = \frac{\partial v}{\partial x}\Delta x + \left(1 + \frac{\partial v}{\partial y}\right)\Delta y \quad (d)$$

The length of the line segment A*B* can be determined, by substituting Eq. (d) into Eq. (c). Because the deformations are extremely small, the products and squares of the derivatives can be neglected. From the substitution and simplification, we obtain:

$$(A^*B^*) = \left(1+2\frac{\partial u}{\partial x}\right)(\Delta x)^2 + \left(1+2\frac{\partial v}{\partial y}\right)(\Delta y)^2 + 2\left(\frac{\partial u}{\partial y}+\frac{\partial v}{\partial x}\right)(\Delta x)(\Delta y) \tag{e}$$

Let's rewrite Eq. (a) as:

$$(\varepsilon_{AB}+1)^2 = \left(\frac{A^*B^*}{AB}\right)^2 \tag{f}$$

Substituting Eq. (b) and Eq. (e) into Eq. (f) and simplifying gives:

$$(\varepsilon_{AB}+1)^2 = \cos^2\theta + \cos^2\phi + 2\frac{\partial u}{\partial x}\cos^2\theta + 2\frac{\partial v}{\partial y}\cos^2\phi + 2\left(\frac{\partial u}{\partial y}+\frac{\partial v}{\partial x}\right)\cos\theta\cos\phi \tag{g}$$

where θ and ϕ are defined in Fig. 3.5.

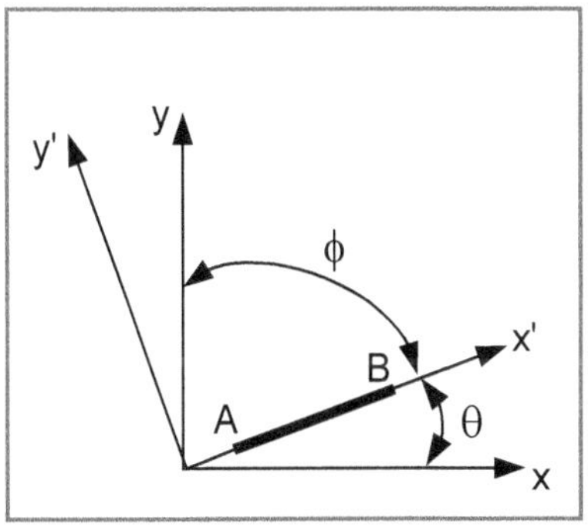

Fig. 3.5 Definition of angles θ and ϕ.

Expand the left side of Eq. (g) and neglect ε_{AB}^2 because it is a second order term and note that

$$\cos^2\theta + \cos^2\phi = 1 \tag{h}$$

We can then express the strain along the line segment AB as:

$$\varepsilon_{AB} = \frac{\partial u}{\partial x}\cos^2\theta + \frac{\partial v}{\partial y}\cos^2\phi + \left(\frac{\partial u}{\partial y}+\frac{\partial v}{\partial x}\right)\cos\theta\cos\phi \tag{i}$$

The relationship between strains and displacements can be expressed as:

$$\partial u/\partial x = \varepsilon_{xx} \qquad \partial v/\partial y = \varepsilon_{yy} \qquad (\partial u/\partial y + \partial v/\partial x) = \gamma_{xy} \tag{3.5}$$

By substituting the strain displacement relations of Eq. (3.5) into Eq. (i), we obtain the relation for ε_{AB} as:

$$\varepsilon_{AB} = \varepsilon_{xx}\cos^2\theta + \varepsilon_{yy}\sin^2\theta + \gamma_{xy}\cos\theta\sin\theta \tag{3.6}$$

If we select the line segment AB so that it is parallel to the x' axis, then it is evident that Eq. (3.6) may be written as:

$$\varepsilon_{x'x'} = \varepsilon_{xx}\cos^2\theta + \varepsilon_{yy}\sin^2\theta + \gamma_{xy}\cos\theta\sin\theta \tag{3.7}$$

Selecting the line segment AB parallel to the y' axis gives:

56 — Chapter 3 — Strain and the Stress-Strain Relations

$$\varepsilon_{y'y'} = \varepsilon_{yy} \cos^2\theta + \varepsilon_{xx} \sin^2\theta - \gamma_{xy}\cos\theta\sin\theta \qquad (3.8)$$

A similar but somewhat more involved derivation can be used to establish the strain equations of transformation for $\gamma_{x'y'}$. If we consider the angular change in an arbitrary right angle formed by two line segments AB_1 and AB_2 and follow the same procedure, the shearing strain $\gamma_{x'y'}$ can be written as:

$$\frac{\gamma_{x'y'}}{2} = -\varepsilon_{xx}\cos\theta\sin\theta + \varepsilon_{yy}\cos\theta\sin\theta + \frac{\gamma_{xy}}{2}\left(\cos^2\theta - \sin^2\theta\right) \qquad (3.9)$$

The strain equations of transformation presented in Eqs. (3.7), (3.8) and (3.9) are the same form as the stress equations of transformation given by Eqs. (2.11) and (2.12). A comparison shows that σ_{xx}, σ_{yy}, $\sigma_{x'x'}$ and $\sigma_{y'y'}$ correspond to ε_{xx}, ε_{yy}, $\varepsilon_{x'x'}$ and $\varepsilon_{y'y'}$. The comparison also shows that τ_{xy} and $\tau_{x'y'}$ correspond to $\gamma_{xy}/2$ and $\gamma_{x'y'}/2$.

It is possible to change Eqs. (3.7), (3.8) and (3.9), which show the angle θ into a double angle representation with suitable trigonometric substitutions. The results of these substitutions are:

$$\varepsilon_{x'x'} = \frac{\varepsilon_{xx} + \varepsilon_{yy}}{2} + \frac{\varepsilon_{xx} - \varepsilon_{yy}}{2}\cos 2\theta + \frac{\gamma_{xy}}{2}\sin 2\theta$$

$$\frac{\gamma_{x'y'}}{2} = -\frac{\varepsilon_{xx} - \varepsilon_{yy}}{2}\sin 2\theta + \frac{\gamma_{xy}}{2}\cos 2\theta \qquad (3.10)$$

$$\varepsilon_{y'y'} = \frac{\varepsilon_{xx} + \varepsilon_{yy}}{2} - \frac{\varepsilon_{xx} - \varepsilon_{yy}}{2}\cos 2\theta - \frac{\gamma_{xy}}{2}\sin 2\theta$$

Because of the correspondence of the stress and strain transformation equations, it is apparent that a Mohr's strain circle can be constructed to graphically represent the state of strain at a point. The procedure for constructing Mohr's circle for strain is exactly the same as that followed previously for the stress circle. The only difference is to change the abscissa from σ to ε and the ordinate from τ to $\gamma/2$. We show a typical Mohr's circle for strain in Fig. 3.6.

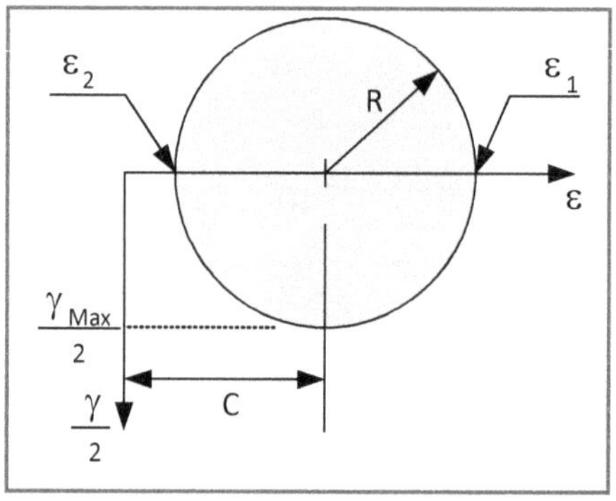

Fig. 3.6 Mohr's circle for strain.

The center of the strain circle is located at position C where:

$$C = (\varepsilon_{xx} + \varepsilon_{yy})/2 \qquad (3.11)$$

The radius R of the circle is given by:

$$R = \sqrt{\left(\frac{\varepsilon_{xx} - \varepsilon_{yy}}{2}\right)^2 + \left(\frac{\gamma_{xy}}{2}\right)^2} \qquad (3.12)$$

It is also evident from Fig. 3.6 that the principal strains are given by:

$$\varepsilon_{1,2} = C \pm R = \frac{\varepsilon_{xx} + \varepsilon_{yy}}{2} \pm \sqrt{\left(\frac{\varepsilon_{xx} - \varepsilon_{yy}}{2}\right)^2 + \left(\frac{\gamma_{xy}}{2}\right)^2} \qquad (3.13)$$

The maximum shearing strain γ_{Max} is given by:

$$\frac{\gamma_{Max}}{2} = R = \sqrt{\left(\frac{\varepsilon_{xx} - \varepsilon_{yy}}{2}\right)^2 + \left(\frac{\gamma_{xy}}{2}\right)^2} \qquad (3.14)$$

Because of the correspondence in the stress and strain equations of transformation, the principal planes are given by:

$$\tan 2\theta_p = \frac{\gamma_{xy}}{\varepsilon_{xx} - \varepsilon_{yy}} \qquad (3.15)$$

The planes upon which the maximum shear strain act are rotated 45° relative to the principal planes. Let's consider an example, which demonstrates the application of these results for determining the state of strain on arbitrary planes.

EXAMPLE E3.1

A state of plane strain exists at point A in a plane body where the Cartesian components of strain are given by:

$$\varepsilon_{xx} = 900 \times 10^{-6} \qquad \varepsilon_{yy} = -400 \times 10^{-6} \qquad \gamma_{xy} = 300 \times 10^{-6}$$

Determine the state of strain on an element rotated counterclockwise 40° relative to the Oxy axes.

Solution: The strains $\varepsilon_{x'x'}$, $\varepsilon_{y'y'}$ and $\gamma_{x'y'}$ are determined from the strain equations of transformation as:

$$\varepsilon_{x'x'} = \frac{\varepsilon_{xx} + \varepsilon_{yy}}{2} + \frac{\varepsilon_{xx} - \varepsilon_{yy}}{2} \cos 2\theta + \frac{\gamma_{xy}}{2} \sin 2\theta$$

$$\varepsilon_{x'x'} = \left(\frac{900 - 400}{2} + \frac{900 + 400}{2} \cos 80° + \frac{300}{2} \sin 80°\right) \times 10^{-6} = 510.6 \times 10^{-6}$$

$$\frac{\gamma_{x'y'}}{2} = -\frac{\varepsilon_{xx} - \varepsilon_{yy}}{2} \sin 2\theta + \frac{\gamma_{xy}}{2} \cos 2\theta$$

$$\gamma_{x'y'} = [-(900 + 400)\sin 80° + 300 \cos 80°] \times 10^{-6} = -1228 \times 10^{-6}$$

$$\varepsilon_{y'y'} = \frac{\varepsilon_{xx} + \varepsilon_{yy}}{2} - \frac{\varepsilon_{xx} - \varepsilon_{yy}}{2} \cos 2\theta - \frac{\gamma_{xy}}{2} \sin 2\theta$$

$$\varepsilon_{y'y'} = \left(\frac{900 - 400}{2} - \frac{900 + 400}{2} \cos 80° - \frac{300}{2} \sin 80°\right) \times 10^{-6} = -10.6 \times 10^{-6}$$

EXAMPLE E3.2

For the state of plane strain described in Example E3.1, determine the principal strains, the maximum shearing strain and the planes upon which they act.

Solution: The magnitudes of the principal strains are given by Eq. (3.13) as:

$$\varepsilon_{1,2} = C \pm R = \frac{\varepsilon_{xx} + \varepsilon_{yy}}{2} \pm \sqrt{\left(\frac{\varepsilon_{xx} - \varepsilon_{yy}}{2}\right)^2 + \left(\frac{\gamma_{xy}}{2}\right)^2}$$

$$\varepsilon_{1,2} = \left(\frac{900 - 400}{2} \pm \sqrt{\left(\frac{900 + 400}{2}\right)^2 + \left(\frac{300}{2}\right)^2}\right) \times 10^{-6}$$

$$\varepsilon_{1,2} = [250 \pm 667.1] \times 10^{-6}$$

$$\varepsilon_1 = 917.1 \times 10^{-6} \quad \text{and} \quad \varepsilon_2 = -417.1 \times 10^{-6}$$

The principal planes are given by Eq. (3.15) as:

$$\tan 2\theta_p = \frac{\gamma_{xy}}{\varepsilon_{xx} - \varepsilon_{yy}} = \frac{300}{900 + 400} = 0.2308$$

$$2\theta_p = 13.00° \quad \theta_p = 6.50°$$

The maximum shear strain is given by Eq. (3.14) as:

$$\frac{\gamma_{Max}}{2} = R = \sqrt{\left(\frac{\varepsilon_{xx} - \varepsilon_{yy}}{2}\right)^2 + \left(\frac{\gamma_{xy}}{2}\right)^2} = \sqrt{\left(\frac{900 + 400}{2}\right)^2 + \left(\frac{300}{2}\right)^2} \times 10^{-6}$$

$$\frac{\gamma_{Max}}{2} = 667.1 \times 10^{-6} \quad \gamma_{Max} = 1334.2 \times 10^{-6}$$

The plane upon which the maximum shearing strain acts is given by:

$$\theta_s = \theta_p \pm 45° = 6.50° \pm 45° = 51.50° \text{ or } -38.50°$$

It is usually worth the time and effort to quickly check the results of a calculation like this one by constructing Mohr's circle. We show the Mohr's circle for this state of strain in Fig. E3.2.

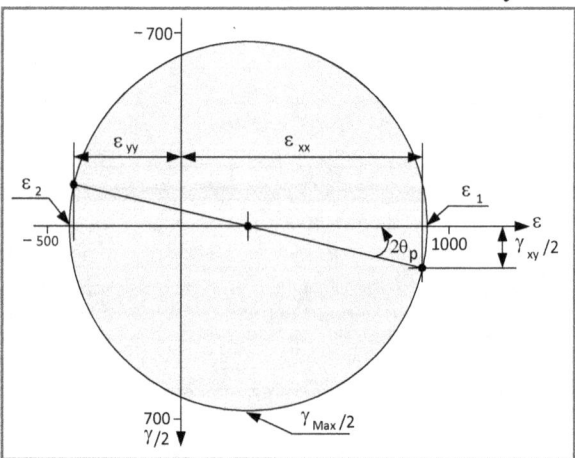

Fig. E3.2 Mohr's strain circle drawn to scale. Multiply strains shown by 10^{-6}.

3.4 STRESS STRAIN RELATIONS

If we consider a general three-dimensional body subjected to forces acting in all three dimensions, the stress system developed contains nine Cartesian components, which include the normal stresses σ_{xx}, σ_{yy} and σ_{zz} and the shear stresses τ_{xy}, τ_{xz}, τ_{yz}, τ_{yx}, τ_{zx}, and τ_{zy} as illustrated in Fig. 3.7.

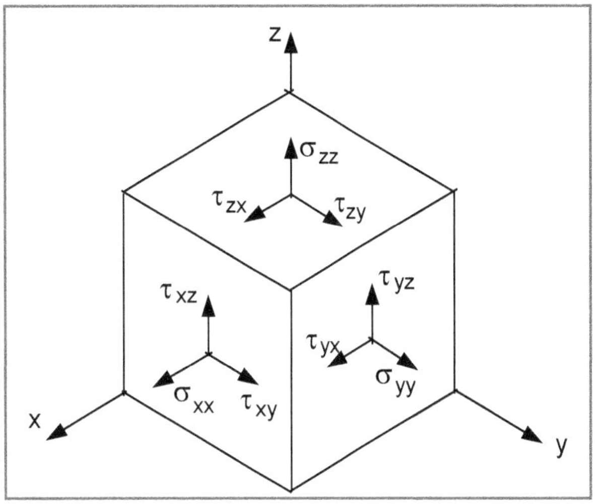

Fig. 3.7 Cartesian components of stress representing a three-dimensional state of stress.

As noted previously, the six Cartesian components of shearing stress may be reduced to three by recognizing that:

$$\tau_{xy} = \tau_{yx} \qquad \tau_{yz} = \tau_{zy} \qquad \tau_{xz} = \tau_{zx} \qquad (a)$$

We have used a double subscript in Fig. 3.7 in designating the normal stresses. The first subscript is to denote the direction of the normal vector of the plane on which the stress acts and the second is to give the direction of the stress. The subscript convention also applies for the shear stress components.

In a general three-dimensional state of stress, it is necessary to specify six Cartesian components of stress (σ_{xx}, σ_{yy}, σ_{zz}, τ_{xy}, τ_{yz}, and τ_{zx}) to completely define the stress state. These general three-dimensional solutions are beyond the scope of this textbook. However, we will develop a special case of plane stress, which is an important state of stress with many applications.

For the general three-dimensional state, the strains are related to the stress by:

$$\varepsilon_{xx} = (1/E)[\sigma_{xx} - \nu(\sigma_{yy} + \sigma_{zz})]$$

$$\varepsilon_{yy} = (1/E)[\sigma_{yy} - \nu(\sigma_{xx} + \sigma_{zz})]$$

$$\varepsilon_{zz} = (1/E)[\sigma_{zz} - \nu(\sigma_{xx} + \sigma_{yy})]$$

$$\gamma_{xy} = \tau_{xy}/G \qquad (3.16)$$

$$\gamma_{yz} = \tau_{yz}/G$$

$$\gamma_{xz} = \tau_{xz}/G$$

Solving Eq. (3.16) for the stresses in terms of the strains gives:

60 — Chapter 3 — Strain and the Stress-Strain Relations

$$\sigma_{xx} = \frac{E}{(1+\nu)(1-2\nu)}\left[(1-\nu)\varepsilon_{xx} + \nu(\varepsilon_{yy}+\varepsilon_{zz})\right]$$

$$\sigma_{yy} = \frac{E}{(1+\nu)(1-2\nu)}\left[(1-\nu)\varepsilon_{yy} + \nu(\varepsilon_{zz}+\varepsilon_{xx})\right] \quad (3.17)$$

$$\sigma_{zz} = \frac{E}{(1+\nu)(1-2\nu)}\left[(1-\nu)\varepsilon_{zz} + \nu(\varepsilon_{xx}+\varepsilon_{yy})\right]$$

$$\tau_{xy} = G\gamma_{xy}; \quad \tau_{yz} = G\gamma_{yz}; \quad \tau_{zx} = G\gamma_{zx}$$

An inspection of Eq. (3.17) shows that Hooke's law ($\sigma_{xx} = E\varepsilon_{xx}$) is not generally valid for three-dimensional states of stress. Only for the case of uniaxial tension (i.e. $\sigma_{yy} = \sigma_{zz} = 0$; $\varepsilon_{yy} = \varepsilon_{zz} = -\nu\varepsilon_{xx}$) does Eq. (3.17) take the simplified form of Hooke's law.

EXAMPLE E3.3

A three-dimensional stress state is defined by six Cartesian components of stress listed below.

$\sigma_{xx} = 10$ ksi, $\sigma_{yy} = 14$ ksi, $\sigma_{zz} = -9.5$ ksi, $\tau_{xy} = 12$ ksi, $\tau_{yz} = 8.4$ ksi, and $\tau_{zx} = -7.5$ ksi

Determine the state of strain, if the body is made from an aluminum alloy.

Solution: From Appendix B-1, we obtain $E = 10.4 \times 10^6$ psi, $G = 3.9 \times 10^6$ psi and $\nu = 0.32$. Substituting the stress components and these values into Eq. (3.16) yields:

$\varepsilon_{xx} = (1/E)[\sigma_{xx} - \nu(\sigma_{yy} + \sigma_{zz})] = [1/(10.4 \times 10^6)][10 - (0.32)(14 - 9.5)] \times 10^3 = 0.8231 \times 10^{-3}$

$\varepsilon_{yy} = (1/E)[\sigma_{yy} - \nu(\sigma_{xx} + \sigma_{zz})] = [1/(10.4 \times 10^6)][14 - (0.32)(10 - 9.5)] \times 10^3 = 1.331 \times 10^{-3}$

$\varepsilon_{zz} = (1/E)[\sigma_{zz} - \nu(\sigma_{xx} + \sigma_{yy})] = [1/(10.4 \times 10^6)][-9.5 - (0.32)(10 + 14)] \times 10^3 = -1.652 \times 10^{-3}$

$\gamma_{xy} = \tau_{xy}/G = (12 \times 10^3)/(3.9 \times 10^6) = 3.077 \times 10^{-3}$

$\gamma_{yz} = \tau_{yz}/G = (8.4 \times 10^3)/(3.9 \times 10^6) = 2.154 \times 10^{-3}$

$\gamma_{xz} = \tau_{xz}/G = (-7.5 \times 10^3)/(3.9 \times 10^6) = -1.923 \times 10^{-3}$

Examination of the results indicates that the strains are small quantities (of the order of 10^{-3}), even when the stress state is of the order of 10^3.

EXAMPLE E3.4

A three-dimensional strain state is defined by six Cartesian components of strain listed below.

$\varepsilon_{xx} = 1.0 \times 10^{-3}$, $\varepsilon_{yy} = 1.4 \times 10^{-3}$, $\varepsilon_{zz} = -0.95 \times 10^{-3}$, $\gamma_{xy} = 1.2 \times 10^{-3}$, $\gamma_{yz} = 0.84 \times 10^{-3}$, and $\gamma_{zx} = -0.75 \times 10^{-3}$

Determine the state of stress, if the body is made from steel.

Solution: From Appendix B-1, we obtain E = 207 GPa, G = 79 GPa and ν = 0.30. Substituting the strain components and these values into Eq. (3.17) yields:

$$\sigma_{xx} = \frac{E}{(1+\nu)(1-2\nu)}\left[(1-\nu)\varepsilon_{xx} + \nu(\varepsilon_{yy} + \varepsilon_{zz})\right]$$

$$\sigma_{xx} = \frac{207 \times 10^9}{(1.3)(0.4)}\left[(0.7)(1.0) + (0.3)(1.4 - 0.95)\right] \times 10^{-3} = 332.4 \text{ MPa}$$

$$\sigma_{yy} = \frac{E}{(1+\nu)(1-2\nu)}\left[(1-\nu)\varepsilon_{yy} + \nu(\varepsilon_{zz} + \varepsilon_{xx})\right]$$

$$\sigma_{yy} = \frac{207 \times 10^9}{(1.3)(0.4)}\left[(0.7)(1.4) + (0.3)(1.0 - 0.95)\right] \times 10^{-3} = 396.1 \text{ MPa}$$

$$\sigma_{zz} = \frac{E}{(1+\nu)(1-2\nu)}\left[(1-\nu)\varepsilon_{zz} + \nu(\varepsilon_{xx} + \varepsilon_{yy})\right]$$

$$\sigma_{zz} = \frac{207 \times 10^9}{(1.3)(0.4)}\left[(0.7)(-0.95) + (0.3)(1.0 + 1.4)\right] \times 10^{-3} = 21.90 \text{ MPa}$$

$$\tau_{xy} = G\gamma_{xy} = (79 \times 10^9)(1.2 \times 10^{-3}) = 94.80 \text{ MPa}$$

$$\tau_{yz} = G\gamma_{yz} = (79 \times 10^9)(0.84 \times 10^{-3}) = 66.36 \text{ MPa}$$

$$\tau_{zx} = G\gamma_{zx} = (79 \times 10^9)(-0.75 \times 10^{-3}) = -59.25 \text{ MPa}$$

3.4.1 Plane Stress

When a body is thin relative to its lateral dimensions, such as that shown in Fig. 3.8, the stresses acting on all of the planes perpendicular to the normal vector z of the thin body are sufficiently small to be disregarded. Also on a free surface of any body, the normal and shear stresses acting on that surface vanish. In these situations, the stress state is called plane stress. For a plane state of stress, we set the stresses acting in the z direction equal to zero to give.

$$\sigma_{zz} = \tau_{xz} = \tau_{yz} = 0 \qquad (3.18)$$

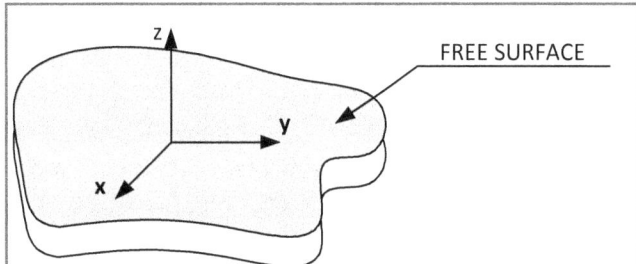

Fig. 3.8 Thin bodies exhibit plane stress conditions.

The remaining stresses σ_{xx}, σ_{yy} and τ_{xy} are specified when defining the stresses associated with a state of plane stress. For a state of plane stress, with $\sigma_{zz} = \tau_{xz} = \tau_{yz} = 0$, Eq. (3.16) reduces to:

$$\varepsilon_{xx} = (1/E)[\sigma_{xx} - \nu(\sigma_{yy})]$$

$$\varepsilon_{yy} = (1/E)[\sigma_{yy} - \nu(\sigma_{xx})] \qquad (3.19)$$

$$\varepsilon_{zz} = (1/E)[-\nu(\sigma_{xx} + \sigma_{yy})]$$

$$\gamma_{xy} = \tau_{xy}/G$$

Similarly, the stress-strain relations given in Eq. (3.16) reduce to:

$$\sigma_{xx} = \frac{E}{(1-\nu^2)}(\varepsilon_{xx} + \nu\varepsilon_{yy})$$

$$\sigma_{yy} = \frac{E}{(1-\nu^2)}(\varepsilon_{yy} + \nu\varepsilon_{xx})$$ (3.20)

$$\tau_{xy} = G\gamma_{xy}$$

While the equations for the shear stresses do not change with the dimension of the state of stress, the expressions for the normal stresses differ markedly. Hooke's law ($\sigma_{xx} = E\varepsilon_{xx}$) is valid only for a uniaxial state of stress where $\sigma_{yy} = \sigma_{zz} = 0$.

EXAMPLE E3.5

The stresses on the external surface of a cylindrical pressure vessel, in the hoop and axial direction, are predicted from a simple analysis to be $\sigma_h = 28$ ksi and $\sigma_a = 14$ ksi when the vessel is pressurized to 125 psi. If a pair of strain gages is placed on the pressure vessel to measure the strains in the hoop and axial directions, predict the strains that will occur, when the pressure of 125 psi is applied. The pressure vessel is fabricated from steel.

Solution: From Appendix B-1, we note $E = 30 \times 10^6$ psi, $G = 11.5 \times 10^6$ psi and Poisson's ratio $\nu = 0.30$. Substituting the hoop and axial stresses in Eq. (3.19) gives:

$$\varepsilon_{xx} = (1/E)[\sigma_{xx} - \nu(\sigma_{yy})] = [1/(30 \times 10^6)][28 - (0.3)(14)] \times 10^{+3} = 0.7933 \times 10^{-3}$$

$$\varepsilon_{yy} = (1/E)[\sigma_{yy} - \nu(\sigma_{xx})] = [1/(30 \times 10^6)][14 - (0.3)(28)] \times 10^{+3} = 0.1867 \times 10^{-3}$$

3.5 DETERMINING PRINCIPAL STRESS WITH STRAIN GAGES

In the laboratory or the field, we usually measure strain with a metal foil resistor called a strain gage. The pattern of the resistor for a typical strain gage is shown in Fig. 3.9. A grid configuration is photoetched from the metallic foil to increase the resistance of the gage and to give the gage a high directional sensitivity. Large tabs at each end of the metal foil conductors are provided to accommodate the lead wire connections. The foil is very thin (about 200 μ in.) and fragile; hence, it requires support to avoid damage in normal handling. This support is provided by a thin plastic carrier (illustrated in Fig. 3.9), which facilitates handling the gage.

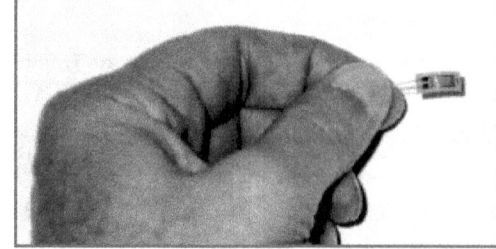

Fig. 3.9 A typical strain gage with a grid pattern and thin film plastic carrier.

To employ a strain gage, we bond it to a mechanical member with an adhesive. The adhesive serves a vital function in the strain measuring system, because it transmits the strain from the mechanical component to the gage-sensing element (the foil) without loss. The singularly unimpressive feat of adhesively bonding a strain gage to a mechanical component is one of the most critical steps in the entire process of measuring strain with a bonded resistance strain gage.

A strain gage acts as a variable resistor, when subjected to a surface strain in the direction of its grid elements. The relation between the change in resistance ΔR with the strain ε is given by:

$$\Delta R/R_0 = (R_f - R_0)/R_0 = (GF)\,\varepsilon_g \qquad (3.21)$$

where R_0 is the initial resistance of the gage, usually 120 or 350 Ω (ohms), R_f is the final resistance of the gage after the application of a strain ε, GF is the gage factor or calibration constant for the gage, usually about 2.0 and ε_g is the strain in the direction of the gage axis.

A strain gage is a sensor that converts strain into a change in the electrical resistance ΔR. Because strains, in structural elements fabricated from common engineering materials, are small quantities (of the order of 10^{-3}), a Wheatstone bridge circuit is employed to transform the change in resistance to a voltage that is proportional to the strain.

3.5.1 Determining Principal Stresses from Strain Measurements

In the analysis of certain structural elements, the directions of the principal stresses are known quantities. Consider axially loaded tension members or beams in bending, where a uniaxial state of stress exists and $\sigma_{yy} = \tau_{xy} = 0$. In this case, the strain gage is oriented with its axis coincident with the x-axis of the structural element. Then the principal stress σ_1 is given by:

$$\sigma_1 = \sigma_{xx} = E\varepsilon_g \qquad \sigma_2 = \sigma_{yy} = 0 \qquad (3.22)$$

where ε_g is the strain measured in the x direction along the axis of the structural member.

Next, consider a spherical pressure vessel, where an isotropic state of stress exists with $\sigma_{xx} = \sigma_{yy} = \sigma_1 = \sigma_2$ and $\tau_{xy} = 0$. In this case the strain gage can be mounted with its axis in any direction, because all directions are principal. The magnitude of the principal stresses is given by:

$$\sigma_{xx} = \sigma_{yy} = \sigma_1 = \sigma_2 = E\varepsilon_g/(1-\nu) \qquad (3.23)$$

where ε_g is the strain measured in any direction in the isotropic stress field.

When less is known beforehand regarding the state of stress in the structural element, it is necessary to employ either two or three strain gages to establish the magnitude and directions of the principal stresses. If the structure has an axis of symmetry, the principal directions are known and the magnitude of the stresses can be determined from the output of two orthogonal strain gages that are mounted with their axes coincident with the axes of symmetry. In this case, the principal stresses are given by:

$$\begin{aligned}\sigma_1 &= \frac{E}{1-\nu^2}(\varepsilon_1 + \nu\varepsilon_2) \\ \sigma_2 &= \frac{E}{1-\nu^2}(\varepsilon_2 + \nu\varepsilon_1)\end{aligned} \qquad (3.24)$$

Special strain gage configurations are available for measurements when either two or three strain determinations are required to establish the strain field. When the principal stress directions are known the two-element rectangular rosette shown in Fig. 3.10 is usually employed in measuring the two principal strains ε_1 and ε_2. When the principal stress directions are not known, there are three unknown quantities σ_1, σ_2 and θ_p. In this instance, a three-element strain gage rosette, illustrated in Fig. 3.10, is employed to completely determine the stress field.

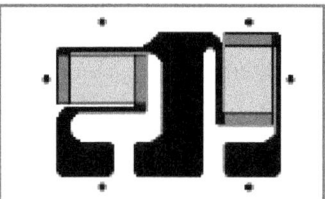
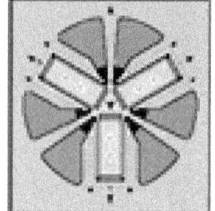

Fig. 3.10 Two and three-element strain gage rosettes. Courtesy of Vishay Inter-technology, Inc.

3.5.2 Strain Gage Rosettes

In the most general situation, the directions of the principal stresses are not known prior to conducting an experimental analysis. Three element rosettes that provide three independent strain measurements are necessary in determining the three unknown quantities σ_1, σ_2 and θ_p. To show that three strain measurements are sufficient, consider three strain gages aligned along axes A, B and C as shown in Fig. 3.11.

Fig. 3.11 Three strain gages placed at arbitrary angles relative to the x and y-axes.

If we recall the strain equation of transformation given by Eq. (3.7), we can write:

$$\begin{aligned}
\varepsilon_A &= \varepsilon_{xx} \cos^2\theta_A + \varepsilon_{yy} \sin^2\theta_A + \gamma_{xy}\cos\theta_A \sin\theta_A \\
\varepsilon_B &= \varepsilon_{xx} \cos^2\theta_B + \varepsilon_{yy} \sin^2\theta_B + \gamma_{xy}\cos\theta_B \sin\theta_B \\
\varepsilon_C &= \varepsilon_{xx} \cos^2\theta_C + \varepsilon_{yy} \sin^2\theta_C + \gamma_{xy}\cos\theta_C \sin\theta_C
\end{aligned} \qquad (3.25)$$

In practice, three-element rosettes with θ_A, θ_B and θ_C fixed at specified values are employed to provide sufficient data to establish the stress field. One of the most popular rosettes is the three-element rectangular rosette that is described in the following subsection.

The Three-Element Rectangular Rosette

The three-element rectangular rosette utilizes gages placed at 0°, 45° and 90° positions, as indicated in Fig. 3.12. With this selection of θ_A, θ_B and θ_C, Eqs. (3.25) reduce to:

$$\varepsilon_A = \varepsilon_{xx} \qquad \varepsilon_B = \tfrac{1}{2}(\varepsilon_{xx} + \varepsilon_{yy} + \gamma_{xy}) \qquad \varepsilon_C = \varepsilon_{yy} \qquad (3.26)$$

From Eq. (3.26), it is clear that:

$$\gamma_{xy} = 2\varepsilon_B - \varepsilon_A - \varepsilon_C \tag{3.27}$$

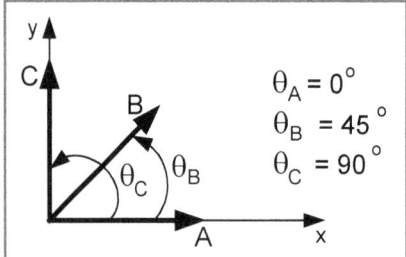

Fig. 3.12 Gage positions in a three-element rectangular rosette.

From Eqs. (3.26) and (3.27), we can determine the Cartesian components of strain from the measurements ε_A, ε_B and ε_C. Next, by employing Eq. (3.13), we can write the expression for the principal strains as:

$$\varepsilon_{1,2} = \frac{\varepsilon_A + \varepsilon_C}{2} \pm \sqrt{\left(\frac{\varepsilon_A - \varepsilon_C}{2}\right)^2 + \left(\frac{2\varepsilon_B - \varepsilon_A - \varepsilon_C}{2}\right)^2} \tag{3.28}$$

The principal angle is established from Eq. (3.15) as:

$$\tan 2\theta_p = \frac{(2\varepsilon_B - \varepsilon_A - \varepsilon_C)}{\varepsilon_A - \varepsilon_C} \tag{3.29}$$

Finally, the expression for the principal stresses is derived by substituting Eq. (3.28) into Eq. (3.20) and simplifying to obtain:

$$\sigma_{1,2} = E\left[\frac{\varepsilon_A + \varepsilon_C}{2(1-\nu)} \pm \frac{1}{2(1+\nu)}\sqrt{(\varepsilon_A - \varepsilon_C)^2 + (2\varepsilon_B - \varepsilon_A - \varepsilon_C)^2}\right] \tag{3.30}$$

By employing Eqs. (3.28) to (3.30) together with the results from the three-element rosette, we can determine the principal strains and stresses and the principal angle.

EXAMPLE E3.6

A two-element rectangular rosette is aligned with the principal stress directions. If the strain measurements are $\varepsilon_1 = 1920 \times 10^{-6}$ and $\varepsilon_2 = -780 \times 10^{-6}$, determine the principal stresses. The structural element upon which the strain gages are mounted is fabricated from 1045 HR steel.

Solution: The principal stresses can be determined from the principal strains by using Eq. (3.24) as:

$$\sigma_1 = \frac{E}{1-\nu^2}(\varepsilon_1 + \nu\varepsilon_2) = \frac{207 \times 10^9}{1-(0.3)^2}[1920 + (0.3)(-780)] \times 10^{-6} = 383.5 \text{ MPa}$$

$$\sigma_2 = \frac{E}{1-\nu^2}(\varepsilon_2 + \nu\varepsilon_1) = \frac{207 \times 10^9}{1-(0.3)^2}[-780 + (0.3)(1920)] \times 10^{-6} = -46.40 \text{ MPa}$$

EXAMPLE E3.7

A three-element rectangular rosette is mounted on a steel structure. The strain readings from each gage element are: $\varepsilon_A = 1800 \times 10^{-6}$, $\varepsilon_B = 600 \times 10^{-6}$ and $\varepsilon_C = -400 \times 10^{-6}$. Determine the principal strains, the principal stresses and the principal angle.

Solution: The principal stresses can be determined by substituting the strain gage readings into Eq. (3.28) to obtain:

$$\varepsilon_{1,2} = \frac{\varepsilon_A + \varepsilon_C}{2} \pm \sqrt{\left(\frac{\varepsilon_A - \varepsilon_C}{2}\right)^2 + \left(\frac{2\varepsilon_B - \varepsilon_A - \varepsilon_C}{2}\right)^2}$$

$$\varepsilon_{1,2} = \left(\frac{1800 - 400}{2} \pm \sqrt{\left(\frac{1800 + 400}{2}\right)^2 + \left(\frac{2(600) - 1800 + 400}{2}\right)^2}\right) \times 10^{-6}$$

$$\varepsilon_{1,2} = [700 \pm 1105] \times 10^{-6} \qquad \varepsilon_1 = 1805 \times 10^{-6} \qquad \varepsilon_2 = -405 \times 10^{-6}$$

To determine the principal stresses, we will substitute the strain gage readings into Eq. (3.30) to obtain:

$$\sigma_{1,2} = E\left[\frac{\varepsilon_A + \varepsilon_C}{2(1-\nu)} \pm \frac{1}{2(1+\nu)}\sqrt{(\varepsilon_A - \varepsilon_C)^2 + (2\varepsilon_B - \varepsilon_A - \varepsilon_C)^2}\right]$$

$$\sigma_{1,2} = 30 \times 10^6 \left[\frac{1800 - 400}{2(1-0.3)} \pm \frac{1}{2(1+0.3)}\sqrt{(1800 + 400)^2 + (2(600) - 1800 + 400)^2}\right] \times 10^{-6}$$

$$\sigma_{1,2} = 30.00 \pm 25.49 \text{ ksi} \qquad \sigma_1 = 55.49 \text{ ksi} \qquad \sigma_2 = 4.51 \text{ ksi}$$

Finally, the principal angle is determined from Eq. (3.29) as:

$$\tan 2\theta_p = \frac{(2\varepsilon_B - \varepsilon_A - \varepsilon_C)}{\varepsilon_A - \varepsilon_C} = \frac{(2(600) - 1800 + 400)}{1800 + 400} = -0.0909$$

$$2\theta_p = -5.19° \qquad \theta_p = -2.60°$$

The minus sign associated with θ_p indicates that the principal element is rotated clockwise relative to the Oxy coordinate system.

3.6 THERMAL STRAINS AND THERMAL STRESSES

Thermal stresses are produced in structural elements by the constraint of the free expansion of a material, subjected to a temperature change. To illustrate this fact, consider a long thin rod of length L_o at a temperature T_o. When the temperature of the rod increases to say T_1, the rod undergoes a free expansion, with an attendant increase in length δ_T given by:

$$\delta_T = L_o \alpha(T_1 - T_o) = L_o \alpha \Delta T \qquad (3.31)$$

where α is the thermal coefficient of expansion.

It is clear from the definition of strain that a thermal strain ε_T accompanies a free expansion.

$$\varepsilon_T = \delta_T / L_o = \alpha \Delta T \tag{3.32}$$

Thermal stresses in a free expansion process are zero, because the temperature change does not generate the internal forces P required for thermal stresses. Thermal strains induce thermal stresses if, and only if, the structural element is constrained in some manner. When the structural element is totally constrained and free expansion is not permitted, the thermal stresses in the uniaxial rod are given by:

$$\sigma_T = E\alpha\Delta T \tag{3.33}$$

To derive Eq. (3.33), consider the free expansion of a long thin rod due to a temperature change ΔT, as shown in Fig. 3.13a.

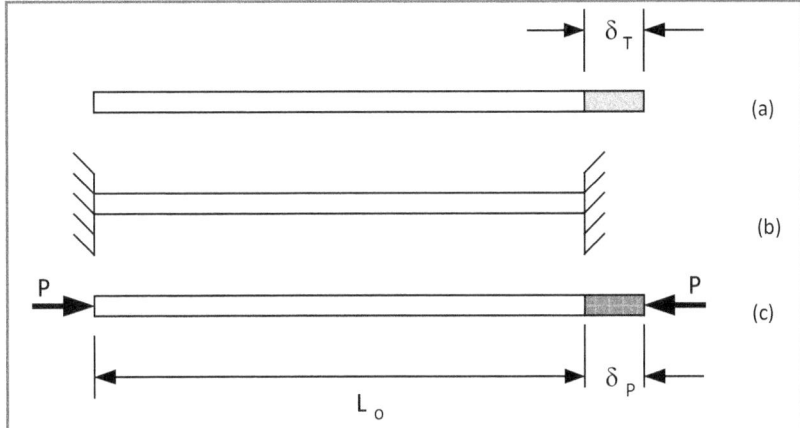

Fig. 3.13 Rod with free expansion, constraint and internal forces P.

If the rod is constrained within rigid walls, as illustrated in Fig. 3.13b, the length remains at L_o after the temperature change ΔT. As the rod attempts to expand against the constraint offered by the rigid walls, an internal force P is produced, as shown in Fig. 3.13c. The internal force P is sufficiently large to produce an axial displacement (contraction) equal to the thermal expansion. Hence, we can write:

$$\delta_T = \delta_P \tag{3.34}$$

Noting that the stress $\sigma = P/A$ and substituting Eq. (3.31) into Eq. (3.34) and simplifying yields:

$$\sigma_T = P/A = E\alpha\Delta T \tag{3.33}$$

The results obtained by applying Eq. (3.33) should be considered as an upper bound on the thermal stresses generated by a temperature change ΔT. The derivation of this relation assumed total constraint of the free expansion of the bar. In practice, total constraint is difficult to achieve and because the displacements are small, even slight relief in the constraint markedly reduces the magnitude of the thermal stresses.

We have included a table showing the coefficient of thermal expansion, modulus of elasticity and thermal stresses for a $\Delta T = 100°$ F or $100°$ C in Table 3.1.

Table 3.1
Coefficient of thermal expansion, modulus of elasticity and thermal stress induced by $\Delta T = 100°$ F or $100°$ C in various engineering materials.

Material	$E \times 10^6$ psi	$\alpha \times 10^{-6}$/°F	σ_T (ksi)	$E \times 10^9$ (Pa)	$\alpha \times 10^{-6}$/°C	σ_T (MPa)
Steel	30	6.3	18.90	207	11.3	233.9
Aluminum	10.4	12.9	13.42	72	23.2	167.0
Brass	16	11.1	17.76	110	20.0	220.0
Stainless Steel	27.5	9.6	26.40	190	17.3	328.7
Titanium	16.5	4.9	8.09	114	8.8	100.3

EXAMPLE E3.8

Two heavy blocks of steel are permanently fastened together by a steel shrink link that is fitted into pockets machined into the two blocks, as illustrated in Fig. E3.8. If a shrink link has a body length of 100.0 mm at a temperature of 275° C, determine the clamping force provided by the shrink link at an ambient temperature of 20° C. Also determine the thermally induced stress in the link after it has cooled to ambient temperature. The body of the link has a square cross section with a side dimension of 50 mm.

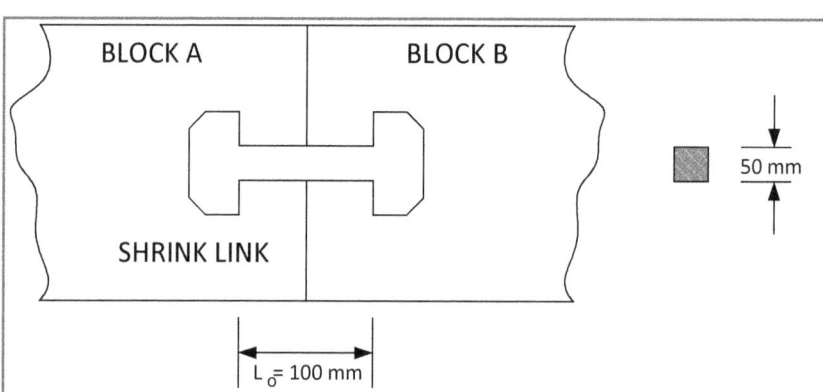

Fig. E3.8

Solution:

Step 1: Upon cooling to ambient temperature the link attempts to shrink an amount δ_T given by Eq. (3.31) as:

$$\delta_T = L_o \alpha \Delta T = 100(11.3 \times 10^{-6})(275 - 20) = 0.2882 \text{ mm} \qquad (a)$$

Step 2: We assume that the free contraction of the link is totally constrained by the two steel blocks. This constraint condition yields:

$$\delta_T = \delta_P \qquad (b)$$

Substituting $\delta = PL/AE$ into Eq. (b) gives the internal force P developed in the shrink link as:

$$P = (\delta_P AE)/L = [(0.2882)(2500)(207 \times 10^3)]/100 = 1491 \text{ kN} \qquad (c)$$

Step 3: The thermal stress induced in the link by the $\Delta T = 255°$ C is given by Eq. (3.33) as:

$$\sigma_T = P/A = E\alpha\Delta T = (1491 \times 10^3)/(0.050)^2 = 596.4 \text{ MPa} \qquad (d)$$

EXAMPLE E3.9

A thin retaining ring, fabricated from stainless steel is fitted over the polished steel journal of a solid shaft 7.000 inches in diameter. The outside diameter of the retaining ring is 7.375 in. and its length is 0.50 in. To fit the ring over the journal it is heated to 350 °F and slid on the shaft to its designated position. The ring becomes snug at a temperature of 295 °F. Determine the stress in the retaining ring and the clamping pressure between the ring and the journal.

Solution:

Step 1: Upon cooling from the snug temperature of 295° F to the ambient temperature of 75° F, the ring attempts to shrink diametrically, by an amount δ_T given by Eq. (3.31) as:

$$\delta_T = d_o\, \alpha \Delta T = (7.000)(9.6 \times 10^{-6})(295 - 75) = 14.78 \times 10^{-3} \text{ in.} \tag{a}$$

Step 2: We assume that the free contraction of the retaining ring is totally constrained by the polished journal of the solid steel shaft. This constraint condition yields:

$$\delta_T = \delta_P \tag{b}$$

Consider the circumference $C = \pi d_o$ of the retaining ring. Then the free contraction ΔC of the circumference of the ring is given by:

$$\Delta C = \pi \Delta d = \pi \delta_T = \pi \delta_P \tag{c}$$

The increase in the circumference of the ring due to the generation of an internal force P is obtained by substituting Eq. (c) into $\delta = PL/AE$ to obtain:

$$\Delta C = (P\pi d)/(AE) = \pi \delta_T \tag{d}$$

Solving Eq. (d) for the internal force P in the ring gives:

$$P = (\delta_T AE)/d_o = (14.78 \times 10^{-3})(0.375/2)(0.5)(27.5 \times 10^6)/7.000 = 5.444 \text{ kip} \tag{e}$$

Step 3: The thermal stress induced in the retaining ring due to $\Delta T = 220°$ F is given by Eq. (3.33) as:

$$\sigma_T = P/A = E\alpha\Delta T = (27.5 \times 10^6)(9.6 \times 10^{-6})(220) = 58.08 \text{ ksi} \tag{f}$$

Step 4: To determine the interfacial pressure between the ring and the journal, consider the FBD of the ring shown in Fig. E3.9.

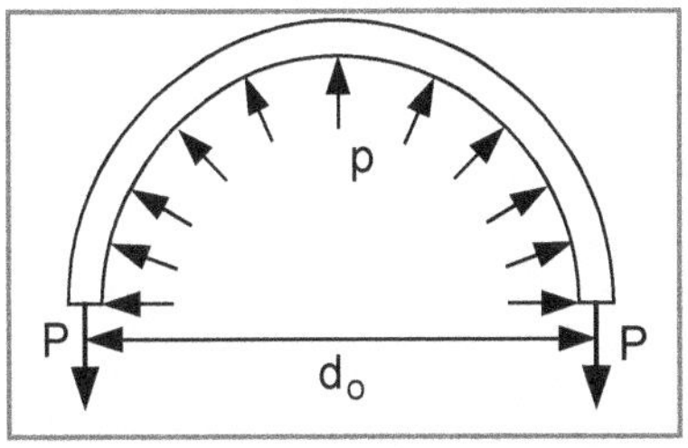

Fig. E3.9

Writing $\Sigma F_y = 0$ gives an expression containing the interfacial pressure p as:

$$\Sigma F = p(db) - 2P = 0 \qquad\qquad p = 2P/(db) \qquad\qquad (g)$$

where b = 0.5 in. is the width of the ring.

The interfacial pressure is then given by:

$$p = 2(5.444 \times 10^3)/(7.0)(0.5) = 3{,}111 \text{ psi} \qquad\qquad (h)$$

PROBLEMS

3.1 Given the displacement field:

$$u = (3x^4 + 2x^2 y^2 + x + y + z^3 + 3)(10^{-3})$$
$$v = (3xy + y^3 + y^2 z + z^2 + 1)(10^{-3})$$
$$w = (x^2 + xy + yz + zx + y^2 + z^2 + 2)(10^{-3})$$

Compute the associated normal and shear strains at point (2, 1, 2)

3.2 Given the displacement field:

$$u = (x^2 + y^4 + 2y^2 z + yz)(10^{-3})$$
$$v = (xy + xz + 3x^2 z)(10^{-3})$$
$$w = (y^4 + 4y^3 + 2z^2)(10^{-3})$$

Compute the associated normal and shear strains at point (1, 2, 3).

3.3 Transform the set of Cartesian strain components:

$$\varepsilon_{xx} = 600 \text{ }\mu\varepsilon \qquad\qquad \varepsilon_{yy} = 400 \text{ }\mu\varepsilon \qquad\qquad \gamma_{xy} = 400 \text{ }\mu\varepsilon$$

into a new set of Cartesian strain components relative to a set of coordinates, where the direction angles associated with the Ox'y'z' axes are:

θ	Case 1	Case 2	Case 3	Case 4
x-x'	π/4	π/2	π/6	π/3

Determine the three principal strains and the maximum shearing strain at the point having the Cartesian strain components given in Problem 3.3.

3.4 A state of plane strain exists at point A in a plane body, where the Cartesian components of strain are given by:

$$\varepsilon_{xx} = 1{,}200 \times 10^{-6} \qquad \varepsilon_{yy} = -600 \times 10^{-6} \qquad \gamma_{xy} = 450 \times 10^{-6}$$

Determine the state of strain on an element rotated counterclockwise 30° relative to the Oxy axes.

3.5 For the state of plane strain described in Problem 3.4, determine the principal strains, the maximum shearing strain and the planes upon which they act.

3.6 A state of plane strain exists at point A in a plane body, where the Cartesian components of strain are given by:

$$\varepsilon_{xx} = 1{,}100 \times 10^{-6} \qquad \varepsilon_{yy} = -300 \times 10^{-6} \qquad \gamma_{xy} = -120 \times 10^{-6}$$

Determine the state of strain on an element rotated counterclockwise 40° relative to the Oxy axes.

3.7 For the state of plane strain described in Problem 3.6, determine the principal strains, the maximum shearing strain and the planes upon which they act.

3.8 A three-dimensional stress state is defined by six Cartesian components of stress listed below.

$$\sigma_{xx} = 12 \text{ ksi}, \sigma_{yy} = 9.2 \text{ ksi}, \sigma_{zz} = -6.2 \text{ ksi}, \tau_{xy} = 10 \text{ ksi}, \tau_{yz} = 4.0 \text{ ksi, and } \tau_{zx} = -4.3 \text{ ksi}$$

Determine the state of strain, if the body is made from an aluminum alloy.

3.9 A three-dimensional stress state is defined by six Cartesian components of stress listed below.

$$\sigma_{xx} = 11.2 \text{ ksi}, \sigma_{yy} = 6.5 \text{ ksi}, \sigma_{zz} = -3.9 \text{ ksi}, \tau_{xy} = 11.3 \text{ ksi}, \tau_{yz} = 8.1 \text{ ksi, and } \tau_{zx} = -3.3 \text{ ksi}$$

Determine the state of strain, if the body is made from a steel alloy.

3.10 A three-dimensional strain state is defined by six Cartesian components of strain listed below.

$$\varepsilon_{xx} = 1.2 \times 10^{-3}, \varepsilon_{yy} = 1.0 \times 10^{-3}, \varepsilon_{zz} = -0.75 \times 10^{-3}, \gamma_{xy} = 0.8 \times 10^{-3}, \gamma_{yz} = 0.73 \times 10^{-3}, \text{ and } \gamma_{zx} = -0.52 \times 10^{-3}$$

Determine the state of stress, if the body is made from steel.

3.11 A three-dimensional strain state is defined by six Cartesian components of strain listed below.

$$\varepsilon_{xx} = 1.4 \times 10^{-3}, \varepsilon_{yy} = 0.7 \times 10^{-3}, \varepsilon_{zz} = -0.62 \times 10^{-3}, \gamma_{xy} = 1.3 \times 10^{-3}, \gamma_{yz} = 0.62 \times 10^{-3}, \text{ and } \gamma_{zx} = -0.32 \times 10^{-3}$$

Determine the state of stress, if the body is made from an aluminum alloy.

3.12 Recall that a state of plane stress is defined by $\sigma_{zz} = \tau_{xz} = \tau_{yz} = 0$. For this state of stress determine the strain ε_{zz}.

3.13 Recall that a state of plane strain is defined by $\varepsilon_{zz} = \gamma_{xz} = \gamma_{yz} = 0$. For this state of strain determine the stress σ_{zz}.

3.14 A cube of steel (E = 207 GPa and ν = 0.30) is loaded with a uniformly distributed pressure of 600 MPa on the four faces having outward normals in the x and y directions. Rigid constraints limit the total deformation of the cube in the z direction to 0.05 mm. Determine the normal stress, if any, which develops in the z direction. The length of a side of the cube is 95 mm.

3.15 Determine the change in volume of a 10-mm cube of aluminum (E = 71 GPa and ν = 0.33) when dropped a distance of 4 km to the ocean floor.

3.16 Determine the stresses at a point in a steel (E = 207 GPa and ν = 0.30) machine component if the Cartesian components of strain at the point are as listed in Problem 3.3.

3.17 The Cartesian components of stress at a point in a steel (E = 207 GPa and ν = 0.30) machine part are:

$$\sigma_{xx} = 440 \text{ MPa} \qquad \sigma_{yy} = 154 \text{ MPa} \qquad \tau_{xy} = 220 \text{ MPa}$$

Determine the principal strains at that point.

3.18 A thick-walled cylindrical pressure vessel will be used to store gas under a pressure of 200 MPa. During initial pressurization of the vessel, axial and hoop components of strain were measured on the inside and outside surfaces. On the inside surface, the axial strain was 1,000 με and the hoop strain was 1,500 με. On the outside surface, the axial strain was 1,000 με, and the hoop strain was 200 με. Determine the axial and hoop components of stress associated with these strains if E = 207 GPa and ν = 0.30.

72 — Chapter 3 — Strain and the Stress-Strain Relations

3.19 The Cartesian components of stress at a point in a steel (E = 207 GPa and ν = 0.30) machine part are as follows:

σ_{xx} = 560 MPa σ_{yy} = –240 MPa τ_{xy} = 560 MPa

Determine the three principal strains.

3.20 A thin rectangular aluminum (E = 71 GPa and ν = 0.33) plate 150 by 200 mm is acted upon by a two-dimensional stress distribution which produces the following uniform distribution of strains in the plate:

ε_{xx} = 2,000 με ε_{yy} = – 500 με γ_{xy} = 2,000 με

(a) Determine the changes in length of the diagonals of the plate.

3.21 For an aluminum (E = 71 GPa and ν = 0.33) body under plane-stress conditions with $\sigma_{zz} = \tau_{yz} = \tau_{zx} = 0$, strains on the surface of the body at a given point are:

ε_{xx} = 800 με ε_{yy} = 600 με

Determine the strain ε_{zz}.

3.22 Determine the stresses σ_{xx}, σ_{yy} and σ_{zz} in a material with ν = ½ if

$\varepsilon_{xx} = \varepsilon_{yy} = \varepsilon_{zz}$ = – 2,000 με

Explain your results.

3.23 A two-element rectangular rosette is aligned with the principal stress directions. If the strain measurements are ε_1 = 1,320 × 10^{-6} and ε_2 = – 640 × 10^{-6}, determine the principal stresses. The structural element upon which the strain gages are mounted is fabricated from 1020 HR steel.

3.24 A three-element rectangular rosette is mounted on a steel structure. The strain readings from each gage element are: ε_A = 1,325 × 10^{-6}, ε_B = 425 × 10^{-6} and ε_C = – 740 × 10^{-6}. Determine the principal strains, the principal stresses and the principal angle.

3.25 Two heavy blocks of steel are permanently fastened together by a steel shrink link that is fitted into pockets machined into the two blocks, as illustrated in Fig. E3.8. If a shrink link has a body length of 150.0 mm at a temperature of 325° C, determine the clamping force provided by the shrink link at an ambient temperature of 22° C. Also determine the thermally induced stress in the link after it has cooled to ambient temperature. The body of the link has a square cross section with a side dimension of 75 mm.

3.26 A thin retaining ring, fabricated from stainless steel is fitted over the polished steel journal of a solid shaft 10.000 inches in diameter. The outside diameter of the retaining ring is 11.000 in. and its length is 1.50 in. To fit the ring over the journal it is heated to 420 °F and slid on the shaft to its designated position. The ring becomes snug at a temperature of 310 °F. Determine the stress in the retaining ring and the clamping pressure between the ring and the journal.

CHAPTER 4

MATERIAL PROPERTIES

4.1 INTRODUCTION

In Chapter 3 we showed that the stress-strain equations for an isotropic material were given by:

$$\varepsilon_{xx} = (1/E)[\sigma_{xx} - \nu(\sigma_{yy} + \sigma_{zz})]$$

$$\varepsilon_{yy} = (1/E)[\sigma_{yy} - \nu(\sigma_{xx} + \sigma_{zz})]$$

$$\varepsilon_{zz} = (1/E)[\sigma_{zz} - \nu(\sigma_{yy} + \sigma_{xx})] \tag{4.1}$$

$$\gamma_{xy} = \frac{2(1+\nu)}{E}\tau_{xy} \qquad \gamma_{yz} = \frac{2(1+\nu)}{E}\tau_{yz} \qquad \gamma_{zx} = \frac{2(1+\nu)}{E}\tau_{zx}$$

The elastic coefficients E and ν shown in Eq. (4.1) are measured in a conventional tension test, where a long, slender bar is subjected to a state of uniaxial stress and the stress state is written as:

$$\sigma_{yy} = \sigma_{zz} = \tau_{xy} = \tau_{yz} = \tau_{zx} = 0 \qquad \sigma_{xx} = \text{applied normal stress}$$

The stress-strain relations for the case of uniaxial stress are written as:

$$\varepsilon_{xx} = \sigma_{xx}/E$$

$$\varepsilon_{yy} = \varepsilon_{zz} = -\nu\sigma_{xx}/E \tag{4.2}$$

where E is the modulus of elasticity and ν is Poisson's ratio, defined as:

$$\nu = -\frac{\varepsilon_{yy}}{\varepsilon_{xx}} \tag{4.3}$$

4.2 THE TENSILE TEST

In Chapter 3 we introduced the material property known as the modulus of elasticity. In this section we will describe the methods employed to determine this property as well as various strengths of the materials used in the analysis of structures. Most of the material properties used in this textbook were determined by conducting standardized tensile tests[1]. For example, the yield and ultimate tensile strength

---

[1] The American Society for Testing Materials (ASTM) publishes standards that define the test specimen and procedures for measuring the material properties described in this chapter. The standard for tension testing of metallic materials is in Section E8 of the *Annual Book of ASTM Standards*.

are measured in this test. Poisson's ratio is also measured by conducting a tensile test. To begin, we prepare **tensile specimens** fabricated from the material under investigation. The standard size tensile specimen is shown in Fig. 4.1.

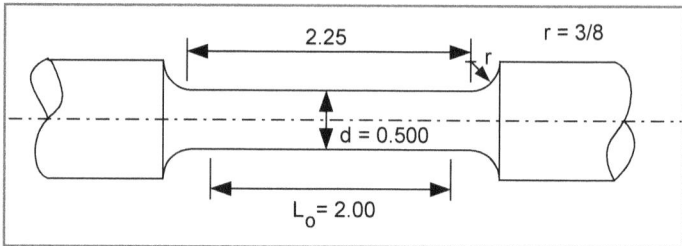

Fig. 4.1 Standard tensile test specimen (dimensions in inches).

The tensile specimen is mounted in a **universal testing machine** similar to the one illustrated in Fig. 4.2. The testing machine may be a mechanical type with one head fixed and the other driven by screws, or it may be a hydraulic machine where the movable head is driven by a hydraulic cylinder. In either instance, the testing machine applies a load F along the axis of the tensile specimen. A monotonic load is applied slowly until the specimen yields and/or fails by fracturing.

The tensile specimen is placed in the load train of the universal testing machine, using wedge grips to hold the specimen, as shown in Fig. 4.3. During the tension test, the applied load P and the elongation δ over the gage length L_o of the specimen are measured. A **load cell** on the universal testing machine measures the applied load and an **extensometer** mounted on the specimen, shown in Fig. 4.4, measures the elongation δ. The electrical signals from the load cell and the extensometer are recorded together on an x-y chart to provide a load-deflection (P-δ) curve that is proportional to the stress-strain (σ-ε) curve.

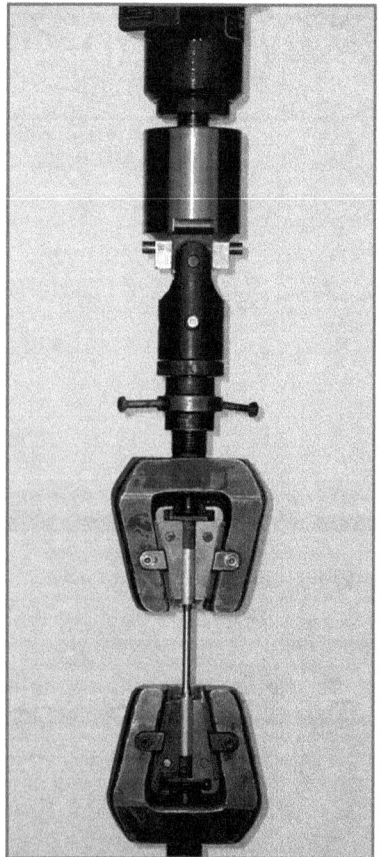

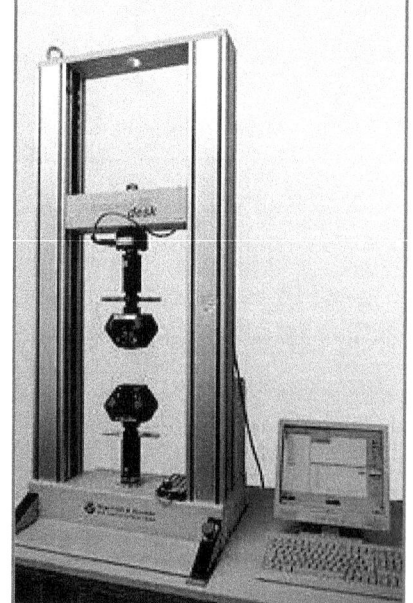

Fig. 4.2 A mechanical type universal testing machine with a screw-driven cross head.

Fig. 4.3 A tensile specimen, wedge-grips, flex-joints and load cell in a testing machine.

The load-deflection curves recorded during the tension tests are converted into stress-strain curves that characterize the tensile behavior of the metallic material by utilizing:

$$\varepsilon = \delta/L_o \qquad (4.4)$$

$$\sigma = P/A_o \qquad (4.5)$$

where L_o and A_o are the initial gage length and cross sectional area of the tensile specimen.

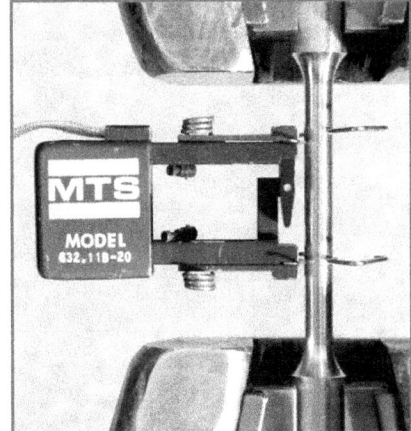

Fig. 4.4 A tension specimen with an extensometer for measuring the elongation δ.

This procedure gives the **engineering stress** and the **engineering strain,** which differ from the **true stress** and the **true strain**. In measuring the true stress and the true strain, the initial values of specimen length and area are replaced with their true values at the instant of the measurement.

For **brittle materials**, which do not exhibit significant plastic deformation before fracture, the stress-strain curve is nearly linear until failure, as indicated in Fig. 4.5. The stress σ_f producing the failure of the brittle specimen is recorded during the tensile test. It is this particular value of the stress that defines the material property known as the ultimate tensile strength S_u for a brittle material.

$$S_u = \sigma_f \text{ for brittle materials} \qquad (4.6)$$

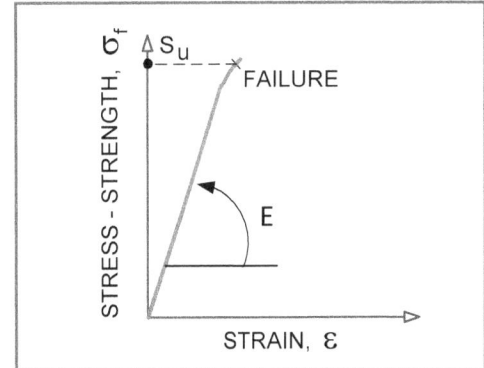

Fig. 4.5 Stress-strain curve for a brittle metallic material.

A photograph of a tension specimen that failed in a brittle manner is depicted in Fig. 4.6. Note the absence of any necking or extensive plastic deformation in the uniform section of the specimen. Brittle failure is dangerous, because it is sudden and catastrophic. The fracture initiates without warning and the cracks propagate across the specimen (structure) in microseconds. Of course, in the selection of materials for our designs, we avoid the use of brittle materials in structures to preclude the possibility of a catastrophic failure. The brittle material gray cast iron is sometimes used in machinery bases, because of its excellent casting properties and for its ability to damp vibrations. However, when using brittle materials, care is exercised to maintain a state of compressive stress in the structure.

Structures are designed with ductile materials, which yield and undergo extensive plastic deformation prior to rupture. A typical stress-strain curve for a ductile material (mild steel) is presented in Fig. 4.7. An inspection of this stress-strain diagram shows that it exhibits four different regions; each region is related to a different material behavior. The first is called the **elastic region,** where the material responds in a linear manner. In this region, Hooke's law ($\sigma = E \varepsilon$) applies. Hooke's law is a mathematical model of material behavior, but it is valid only in the elastic region. The elastic region extends until the low carbon steel (or some other ductile material) begins to yield. When the material yields, the linear response of the material ceases and Eq. (4.2) is no longer valid.

Fig. 4.6 Failure of a brittle material occurs suddenly with little plastic deformation.

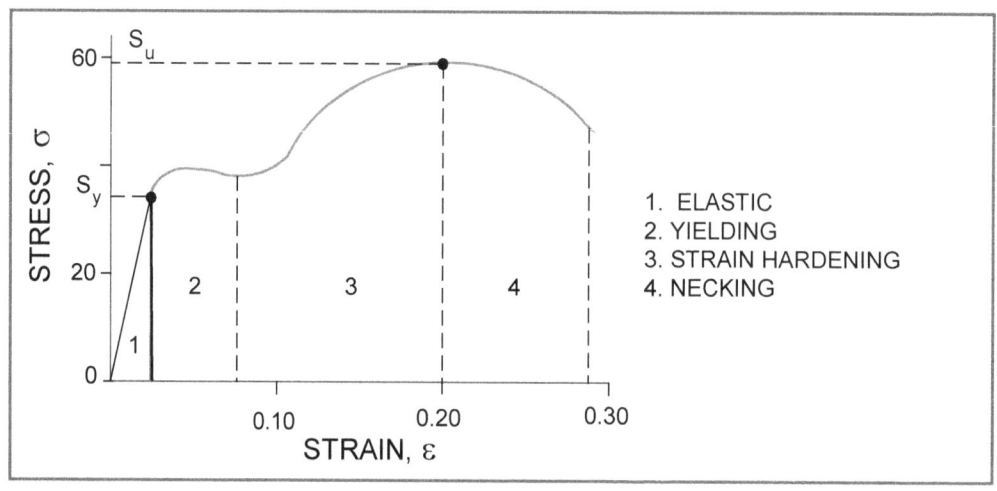

Fig. 4.7 Stress-strain diagram for low carbon steel.

The second region on the stress-strain diagram depicts the initiation of yielding, where the stress in the specimen exceeds the **elastic limit**. Slip planes have developed in the specimen, and deformation by slip along these planes is occurring with small or negligible increases in the stress. The stress-strain relationship is non-linear in this region. In some ductile materials, a small decrease in the applied stress is observed as the tensile specimen yields and continues to elongate by slip.

The third region describes a material behavior known as **strain hardening**. The easy slip that occurred during the initial yielding phase becomes more difficult to induce. As a consequence, higher stresses are required to continue deforming the tensile specimen. The stresses increase with increases in the strain; however, the relationship is not linear and Eq. (4.2) is not valid in this region. In the fourth and final region, the tensile specimen undergoes a dramatic change in appearance; it begins to **neck**. The deformation becomes localized to a small area near the center of the bar. During this phase of the deformation, the region deforming resembles an hourglass. The neck decreases in diameter with increasing deformation until the specimen fails by rupturing. The axial deformation, which occurs as the tensile specimen necks, does not require an increase in the applied load. Indeed, the load may actually decrease significantly during the necking phase of the deformation processes. The appearance of a ruptured tension specimen fabricated from low carbon steel is illustrated in Fig. 4.8. Regions 2, 3 and 4, shown in Fig. 4.7, are often combined and called the **plastic regime** for a ductile material.

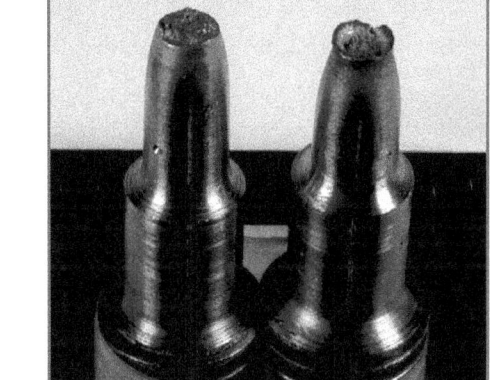

Fig. 4.8 Ductile failure of a tensile specimen.

4.3 <u>MATERIAL PROPERTIES</u>

The tensile test provides several material properties that are important in the analysis of engineering components and structures. These properties include two measures of strength, two measures of ductility, and two elastic constants. Let's discuss strength first.

4.3.1 <u>Measures of Strength</u>

The two measures of strength determined in a tensile test of a ductile material are the yield strength and the ultimate tensile strength. The yield strength, as the name implies, is the stress required to induce yielding:

$$S_y = \sigma_y \tag{4.7}$$

To establish the yield stress σ_y, we examine the stress-strain diagram and attempt to identify the stress when yielding initiated. For some ductile materials with stress-strain diagrams similar to that shown in Fig. 4.7 the precise identification of σ_y is clear. However, the yield behavior of other materials is much less well defined. For example, suppose a material exhibits the stress-strain behavior as indicated in Fig. 4.9a. Where is the yield point?

It is evident in Fig. 4.9a that the stress-strain curve is non-linear, but we might differ in defining the point where slip and yield initiated. To eliminate the ambiguity in the definition of the yield point from a stress-strain diagram, the offset method is employed. We construct a line parallel to the linear portion of the σ-ε curve in the elastic region. This line is offset along the strain axis by 0.2% or ε = 0.002. The intersection of the offset line with the σ-ε curve defines the yield point as illustrated in Fig. 4.9b. With the **0.2% yield stress** σ_y defined, we use Eq. (4.7) to establish the yield strength S_y for the material.

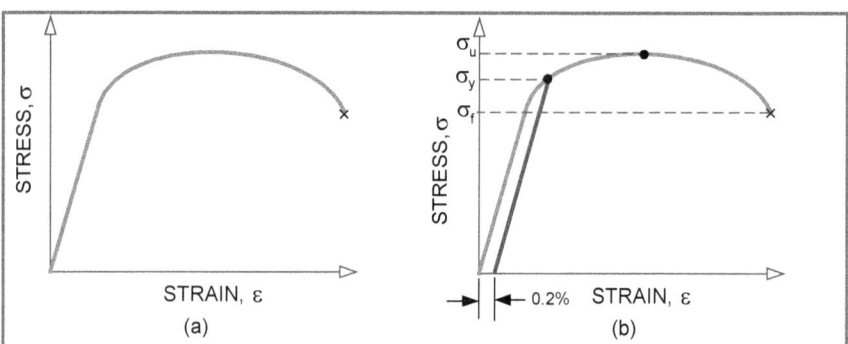

Fig. 4.9 The offset method for determining the stress σ_y for yielding.

The ultimate tensile strength is established from the maximum stress on the specimen that occurs at the onset of necking. We have defined this point on Fig. 4.9b, and have established the ultimate stress σ_u. The ultimate tensile strength S_u for a ductile material is given by:

$$S_u = \sigma_u \qquad \text{for ductile materials}^2 \qquad (4.8)$$

The stress at failure, σ_f for a ductile material, shown in Fig. 4.9b, is only of academic interest. If the structure or machine component is properly designed, the deformation state will probably be limited to the elastic region. In some instances, plastic deformations are tolerated in design, but these exceptions are limited to yielding in small local regions under conditions of constant loading. Global yielding is not tolerated, and yielding under the action of cyclic stresses is a recipe for disaster. For this reason, the stresses imposed on a structure are usually less than σ_y and certainly much less than σ_u.

4.3.2 Measures of Ductility

There are two common measures of ductility for metallic materials. The first is the **percent elongation** given by:

$$\% \, e = \left(\frac{L_f - L_0}{L_0} \right) \times 100 \qquad (4.9)$$

where L_o is the gage length of the tensile specimen — usually 2.0 in. or 50 mm.
L_f is the final deformed length of the specimen between the gage length marks.

For low carbon steels, the percent elongation is usually in the range from 20 to 35%; however, the percent elongation decreases with increasing carbon content and increasing strength in steel. The second measure of ductility is the **percent reduction in area** that is given by:

$$\% \, A = \left(\frac{A_0 - A_f}{A_0} \right) \times 100 = \left[\left(\frac{d_0^2 - d_f^2}{d_0^2} \right) \right] \times 100 \qquad (4.10)$$

where $A_o = \pi d_o^2/4$ is the initial cross sectional area and $A_f = \pi d_f^2/4$ is the final cross sectional area of one of the ruptured ends of the specimen.

---

[2] We use the letter s to denote strength and the Greek letter σ to depict stress. Keep them separate as they represent different concepts

The percent reduction in area for low carbon steel is typically in the range from 60 to 70%. As was the case with percent elongation, the percent reduction in area decreases as the carbon content of steel is increased to enhance its strength.

There is a trade-off between ductility and strength for metallic alloys. To illustrate the loss in ductility with increasing strength, let's examine typical stress-strain curves for three different types of steels that are shown in Fig. 4.10. As the strength increases, the strain to failure decreases with an accompanying decrease in the ductility. For low carbon steels, the strain to failure usually exceeds 60 to 70%. This value decreases to 30 to 40% for higher carbon steels. The very-high-strength steel alloys fail with strains ranging from 10 to 20%.

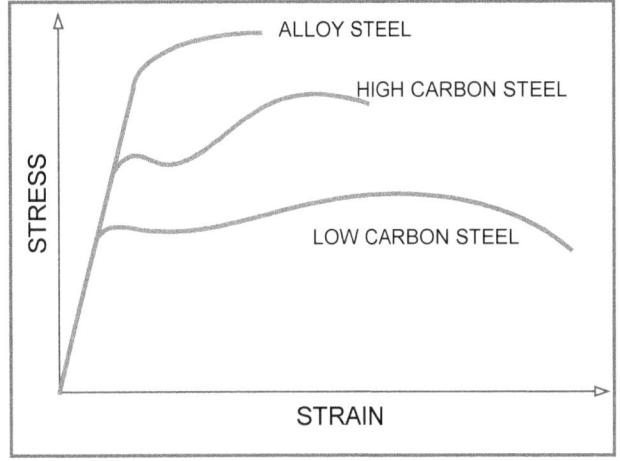

Fig. 4.10 Stress-strain curves for three different types of steels.

EXAMPLE E4.1

A steel supplier provides you with data from a recent series of tensile tests of two different steels that they sell to your corporation. Your manager questions the ductility of both materials, and asks you to determine it. An examination of the supplier's data indicates:

Lower cost steel at $24.00/100lb
L_f = 64 mm and d_f = 9.8 mm

Higher cost steel at $31.00/100lb
L_f = 58 mm and d_f = 10.7 mm

The gage length L_o and diameter d_o for both types of steel were 50 mm and 12.5 mm respectively.

Solution: For the measure of ductility known as the percent elongation, Eq. (4.9) gives:

$$\%e = [(L_f - L_o)/L_o](100) = [(64 - 50)/50](100) = 28\% \Rightarrow \text{lower cost steel.}$$
(a)
$$\%e = [(L_f - L_o)/L_o](100) = [(58 - 50)/50](100) = 16\% \Rightarrow \text{higher cost steel.}$$

For the measure of ductility known as the percent reduction in area, Eq. (4.10) gives:

$$\%A = [(d_o^2 - d_f^2)/d_o^2](100) = \{[(12.5)^2 - (9.8)^2]/(12.5)^2\}(100) = 38.53\% \Rightarrow \text{lower cost steel.}$$
(b)
$$\%A = [(d_o^2 - d_f^2)/d_o^2](100) = \{[(12.5)^2 - (10.7)^2]/(12.5)^2\}(100) = 26.73\% \Rightarrow \text{higher cost steel.}$$

When your manager asks, which steel has the higher ductility, what is your response?

4.3.3 Elastic Constants

Modulus of Elasticity

In this discussion of the elastic constants, we limit their application to characterize deformations in the linear elastic region of the elastic-plastic regime. In this region, the material is elastic and recovers completely, when the load or stress is removed from the specimen. The **slope** of the stress-strain curve is defined as the **modulus of elasticity** or Young's modulus. The slope, illustrated in Fig. 4.5, is determined by rewriting Eq. (4.2) as:

$$\text{Slope} = E = \sigma/\varepsilon \qquad (4.2)$$

EXAMPLE E4.2

After conducting a tensile test with a mild steel specimen, you measure the slope $\Delta P/\Delta \delta$ of the linear region on the load-deflection curve (P-δ) and find:

$$\Delta P/\Delta \delta = (10{,}325 \text{ lb})/(3.465 \times 10^{-3} \text{ in.}) = 2.980 \times 10^6 \text{ lb/in.} \qquad (a)$$

If the specimen diameter is 0.504 in. and the gage length of the extensometer is 2.00 in., determine the modulus of elasticity E.

Solution: From Eqs. (4.2), (4.4) and (4.5), we obtain:

$$E = \frac{\sigma}{\varepsilon} = \frac{PL}{A\delta} = \frac{(\Delta P)L}{(\Delta \delta)A} \qquad (b)$$

$$E = \frac{(10{,}325 \text{ lb})(2 \text{ in})}{(3.465 \times 10^{-3} \text{ in})(\pi)(0.02520)^2 \text{in}^2} = 29.87 \times 10^6 \text{ psi} \qquad (c)$$

Poisson's Ratio

Another elastic constant, **Poisson's ratio**, may be determined in a tensile test; however, strain gages must be attached to the specimen to measure the strain in the axial and the transverse directions. Poisson's ratio is defined as:

$$\nu = -\varepsilon_t/\varepsilon_a \qquad (4.11)$$

where ε_t and ε_a are the strains in the transverse and axial directions, respectively. Note that ε_a and ε_t are both normal strains determined from Eq. (4.4).

When a specimen is subjected to a tensile force, its length extends and its diameter contracts. This contraction is usually ¼ to ½ of the amount of the specimen's axial elongation. The extension of the specimen in the elastic region produces a contraction that is usually too small to be observed during a tensile test. Although too small to be observed, the Poisson effect is a very important phenomenon, because it significantly affects the analysis of structures subjected to multiaxial states of stress, where stresses are imposed in more than one direction.

To describe the Poisson contraction in more detail, consider the rectangular bar of an elastic material with dimensions L, W, and D as shown in Fig. 4.11a. Next, apply an axial strain ε_a and write:

$$\varepsilon_a = \Delta L/L \qquad (a)$$

In the deformed state, the dimensions of the bar change to $L + \Delta L$, $W + \Delta W$, and $D + \Delta D$, as shown in Fig. 4.11b. The strains ε_t in the x and y directions (both x and y are transverse directions) are both due to the Poisson effect. They are equal and given by:

$$\varepsilon_{tx} = \Delta W/W = -\nu\varepsilon_a \qquad \varepsilon_{ty} = \Delta D/D = -\nu\varepsilon_a \qquad (b)$$

If we substitute Eq. (4.11) and Eq. (a) into Eq. (b), we obtain:

$$\varepsilon_t = -\nu\varepsilon_a = -\nu\Delta L/L = \Delta W/W = \Delta D/D \qquad (4.12)$$

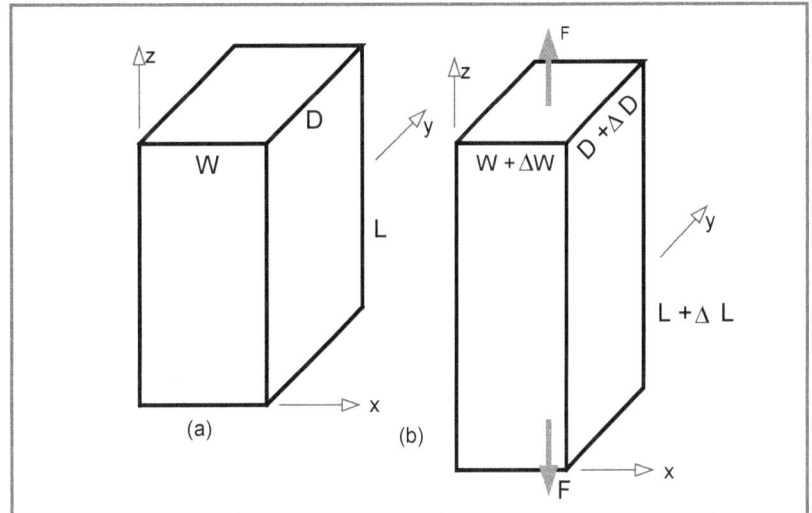

Fig. 4.11 (a) A rectangular bar before imposing an axial force. (b) The deformed bar after application of the force F.

From Eq. (4.12) it is clear that the changes in the transverse dimensions of the bar due to the Poisson effect are given by:

$$\Delta W = -(\nu W \Delta L)/L; \qquad \Delta D = -(\nu W \Delta L)/L \qquad (c)$$

From this elementary analysis of the deformed geometry of a rectangular bar, we note that the material property **Poisson's ratio**, ν enables the transverse deformation to be determined in terms of the axial deformation, when a structural member is subjected to a uniaxial loading. The quantities ΔW and ΔD are negative when ΔL is positive, because the Poisson effect produces a contraction in the transverse (x and y) directions when the body is extended in the axial (z) direction.

EXAMPLE E4.3

Suppose you cut a 90-mm diameter circular hole in a large sheet of dental dam (very thin rubber sheet). If the sheet, originally 900 mm long by 450 mm wide, is extended until it is 1,000 mm long, determine the new width of the sheet and the dimensions of the deformed hole. Assume that the rubber is perfectly elastic with a Poisson's ratio, $\nu = 0.5$.

Solution: Recall Eq. (4.12) and write:

$$\varepsilon_t = -\nu\varepsilon_a = -\nu \Delta L/L_o = -0.5 (1{,}000 - 900)/900 = -0.05556 \quad \text{(a)}$$

$$\varepsilon_a = \Delta L/L_o = (1{,}000 - 900)/900 = 0.1111$$

Because these strains are imposed over the entire sheet, we may determine the deformed width W_{NEW} from:

$$W_{NEW} = W_o + \Delta W = W_o + \varepsilon_t W_o = (1 + \varepsilon_t)W_o = (1 - 0.05556)450 = 425.0 \text{ mm} \quad \text{(b)}$$

The dimensions of the deformed hole are given by:

$$D_a = D_o + \Delta D_a = D_o + \varepsilon_a D_o = (1 + \varepsilon_a)D_o = (1 + 0.1111)90 = 99.9999 \text{ mm} \quad \text{(c)}$$

$$D_t = D_o + \Delta D_t = D_o + \varepsilon_t D_o = (1 + \varepsilon_t)D_o = (1 - 0.05555)90 = 85.00 \text{ mm}$$

where D_a and D_t are the axial and transverse diameters of the hole after deformation.

EXAMPLE E4.4

Determine the change in volume of a rectangular bar subjected to an axial strain of $\varepsilon_a = 2.1 \times 10^{-3}$. The dimensions of the bar before deformation were $L = 4W = 3D = 24$ in. The material from which the bar is fabricated has a Poisson's ratio of $\nu = 0.33$.

Solution: The original volume, V of the bar is given by:

$$V = L \times W \times D = L^3/12 = (24)^3/12 = 1{,}152 \text{ in}^3 \quad \text{(a)}$$

From Eq. (4.12), we determine the new dimensions of the bar as:

$$W_{NEW} = (1 - \nu\varepsilon_a)W; \quad D_{NEW} = (1 - \nu\varepsilon_a)D; \quad L_{NEW} = (1 + \varepsilon_a)L \quad \text{(b)}$$

We rewrite Eq. (b) as:

$$W_{NEW} = (1 - \nu\varepsilon_a)(L/4); \quad D_{NEW} = (1 - \nu\varepsilon_a)(L/3); \quad L_{NEW} = (1 + \varepsilon_a)L \quad \text{(c)}$$

The new volume is given by:

$$V_{NEW} = L_{NEW} \times W_{NEW} \times D_{NEW} = (1 + \varepsilon_a)(1 - \nu\varepsilon_a)(1 - \nu\varepsilon_a)(L^3/12) \quad \text{(d)}$$

From Eqs. (a) and (d), it is evident that:

$$\Delta V = [(1 + \varepsilon_a)(1 - \nu\varepsilon_a)^2 - 1](L^3/12) \quad \text{(e)}$$

Substituting $\varepsilon_a = 2.1 \times 10^{-3}$ and $\nu = 0.33$ into Eq. (e) yields:

$$\Delta V = \{(1 + 2.1 \times 10^{-3})[1 - (0.33)(2.1 \times 10^{-3})]^2 - 1\} V \quad \text{(f)}$$

Performing the calculation gives:

$$\Delta V = 7.116 \times 10^{-4} V = 7.116 \times 10^{-4} (1,152) = 0.8197 \text{ in}^3 \quad \text{(g)}$$

The percentage change in the volume is given by:

$$\Delta V/V = [(1 + \varepsilon_a)(1 - \nu\varepsilon_a)^2 - 1] = 7.116 \times 10^{-4} = 0.07116\% \quad \text{(h)}$$

From these results, it is evident that the change in the volume is extremely small for strains in the elastic region.

Shear Modulus

Another elastic constant that we often use in the analysis of structures subjected to shear stress is the **shear modulus**, G. The shear modulus relates the shear stress to the shear strain by:

$$\tau = G\gamma \quad (4.13)$$

where τ is the shear stress and γ is the shear strain.

The shear strain γ is defined as the change in angle of two perpendicular lines when a body is deformed. For instance, suppose we have a Cartesian coordinate system scribed on a body with the x and y axes serving as the two perpendicular lines. When the body deforms under the action of a shear stress τ, these two lines rotate, and the angle at their intersection changes from 90° to some other angle 90° ± γ as illustrated in Fig. 4.12.

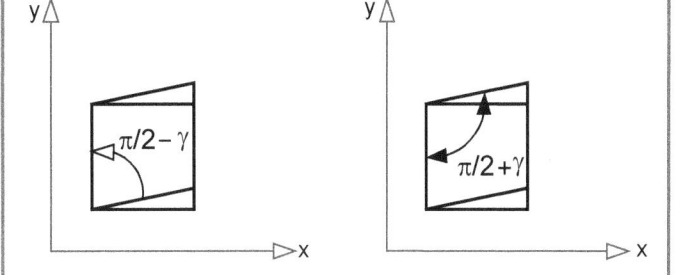

Fig. 4.12 Shear strain is the change in a right angle when a body deforms.

The shear modulus G is related to the modulus of elasticity and Poisson's ratio by:

$$G = \frac{E}{2(1 + \nu)} \quad (4.14)$$

EXAMPLE E4.5

Determine the shear modulus G for steel with the modulus of elasticity $E = 30 \times 10^6$ psi and Poisson's ratio $\nu = 0.30$. Also for an aluminum alloy with $E = 73$ GPa and Poisson's ratio $\nu = 0.33$.

Solution: From Eq. (4.14) we write for steel the following relation:

$$G = (30 \times 10^6) / [2(1 + 0.30)] = 11.54 \times 10^6 \text{ psi for steel.} \quad (a)$$

For the aluminum alloy, the shear modulus is given by:

$$G = (73) / [2(1 + 0.33)] = 27.44 \text{ GPa for an aluminum alloy.} \quad (b)$$

4.4 FATIGUE STRENGTH

Many structures are loaded only once. For example, the structural members in the Sears tower in Chicago were loaded by the dead weight of all of the structural components, as the tower was constructed. The people, furnishings and equipment going into the tower, after it was opened to the public, added some weight, but this additional load was small when compared to the weight of the basic structure. For structures and machine components loaded only once, either the yield strength or the ultimate tensile strength is adequate to predict the safety of the structure.

However, many structures and machine components are subjected to repeated loading. Very heavy tractor-trailer trucks, for example, traverse a bridge. With the passage of each truck, every bridge member is subjected to another cycle of loading. Consider the engine of your automobile. If you are driving along a highway with the tachometer registering 3,000 RPM, you are subjecting the crankshaft in your engine to 3,000 cycles of load for each minute of its operation.

There are two detrimental effects due to the cyclic loading of structures and machine components. First, the design strength of the material from which the structure is fabricated is lowered. We design to a fatigue strength S_f or an endurance limit S_e, which is a function of the magnitude of the cyclic stresses and the number of cycles of load imposed onto the structure.

Second, fatigue failures in **ductile** materials are of a **brittle** nature (i.e., structural failure and collapse occur catastrophically). The failure mechanism in fatigue is markedly different from that observed in yielding or rupture. In fatigue, microscopic cracks are initiated due to accumulated irreversible slip in a very thin layer of material adjacent to the surface of the structural member. A very small and thin layer of material is extruded out of the component, producing a shallow surface crack, as shown in Fig. 4.13. A significant portion of the cycles required to produce a fatigue failure is expended in producing this initial surface crack.

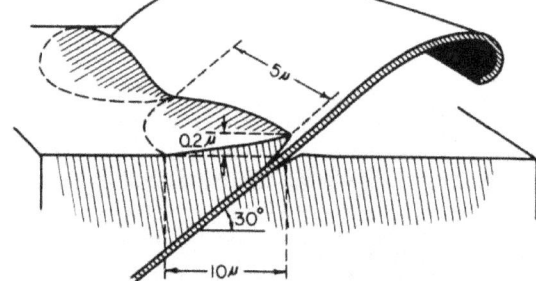

Fig. 4.13 Schematic illustration of generation of a surface crack by extrusion of a thin layer along slip planes. After P. J. E. Forsyth [1].

These surface cracks grow larger with an incremental extension into the component, with each loading cycle. The incremental extensions may be tracked by the striations evident on the fatigue surface. The fatigue crack grows until reaching a critical size. The cracks become unstable and extend at high-speed across the structural member producing sudden and catastrophic collapse. Examination of a surface of a fatigue failure shows the brittle nature of the phenomena, as shown in Fig. 4.14. This figure shows a photograph of the fracture surface of a shaft with a keyway. The fatigue crack initiated near the corner of the keyway and extended with each loading cycle to cover 60 to 70% of the cross section of the shaft. The surface produced by crack extension due to fatigue is relatively smooth, with evidence of many striations. The surface produced by high-speed crack propagation shows roughness typical of more ductile failure. It is clear that a material may be classified as ductile based on the results of a tensile test; however, there is no visible sign of coarse slip or necking associated with failure in fatigue.

Fig. 4.14 A fracture surface produced by fatigue failure of a shaft. Note the smooth surface of the failure surface due to crack extension by fatigue and the rough surface due to the unstable crack propagation across the remainder of the cross section. After P. G. Forest [2]

4.4.1 Fatigue Testing

Fatigue tests of materials are conducted with both rotating beams in bending and with specimens subjected to axial loading. In both testing methods, the specimens are relatively small with a smooth surface finish in the test section. A typical test specimen used in axial loading, with dimensions in mm, is presented in Fig. 4.15 The 30 mm test section is usually polished to provide a flaw free surface.

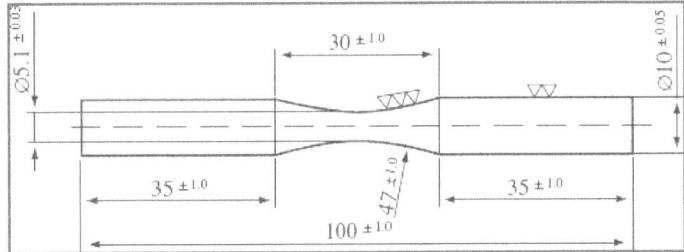

Fig. 4.15 Geometry of a typical fatigue test specimen.

If the fatigue specimen is subjected to alternating tension and compression stresses of equal magnitude a rotating beam testing machine is used. An example of a rotating beam fatigue testing machine is illustrated in Fig. 4.16. The specimen is mounted in chucks that are housed in bearings. Dead weights are placed on a pan located below the specimen. The pan is supported by rods that connect to the supporting bearings. The dead weight produces a bending moment that is uniformly distributed over the length of the specimen. A motor rotating at high speed turns the test specimen, which is subjected to the bending stresses due to the bending moment. The stresses alternate with each rotation of the specimen from maximum tension to maximum compression. The rotating beam testing method is popular, because the tests can be conducted quickly and the test machines are relatively inexpensive.

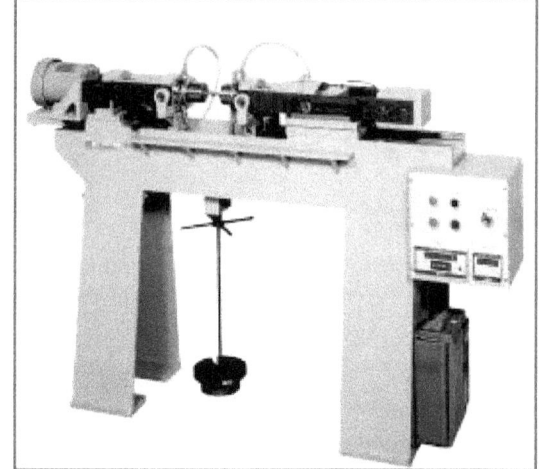

Fig. 4.16 Photograph of a rotating beam fatigue testing machine. Courtesy of Shimadzu Corp.

Another common method of fatigue testing uses axial loading. These tests are usually conducted with hydraulic universal testing machines, as discussed previously. The fatigue specimen is mounted in hydraulic grips and subjected to cyclic axial forces. A typical hydraulic testing machine is presented in Fig. 4.17. With these machines a tensile or compressive preload can be applied to the specimen prior to cycling the alternating load. This testing method enables an engineer to determine the effect of mean stress on fatigue strength.

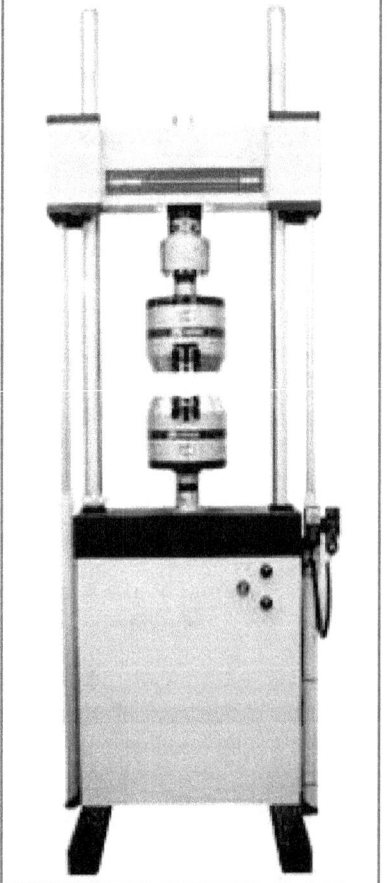

Fig. 4.17 Photograph of a hydraulic fatigue testing machine used to apply static and cyclic loading. Courtesy of the Instron Corp.

The purpose of fatigue testing is to determine the fatigue life measured in the number of cycles to failure at various levels of imposed stress. Several (10 to 20 specimens) are usually tested to failure to establish the data required to draw a **Stress (S) — Life (N)** diagram for a specific material.

4.4.2 The S – N Diagram

The S – N diagrams used in design analysis depend on the material under consideration. To illustrate the marked difference in fatigue behavior with different metals and polymers, we will show typical S – N diagrams for four different materials. Let's first examine the S – N diagram for A-517 steel that is shown in Fig 4.18. The magnitude of the alternating stress is shown on the ordinate and the number of cycles to failure, using a $\log_{10}$ scale, is shown on the abscissa. Inspection of the data shows a curve that slopes downward with increasing $\log_{10}$ (N). We also note that the results are scattered above and below the curve that was drawn though the data points. Even on a $\log_{10}$ scale the scatter is significant and in some case the difference in the number of cycles to failure N_f at 440 MPa is an order of magnitude. Finally we observe that at an alternating stress level of 414 MPa or lower the specimens do not fail after 10^7 cycles. This is an important observation, because it enables us to establish an **endurance limit S_e** for the fatigue strength for the heat of A 517 alloy steel that was tested.

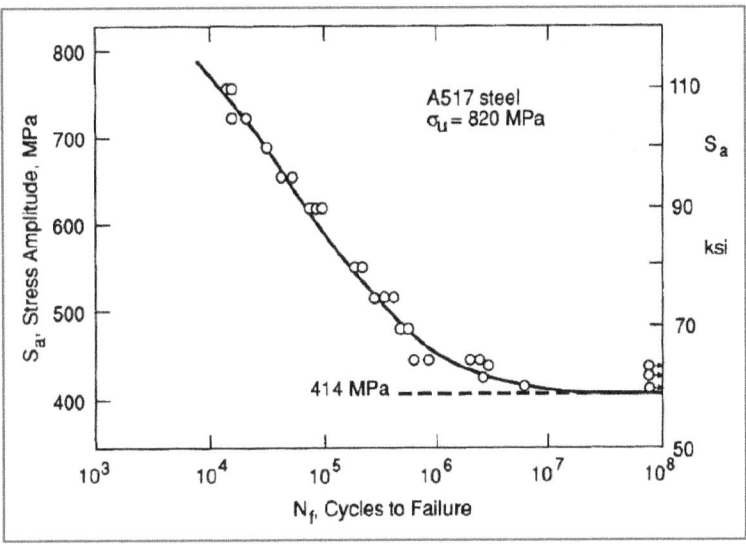

Fig. 4.18 A S–N diagram for A 517 steel tested by rotating bending.
After N. E. Dowling [3]

The behavior of aluminum alloys subjected to fatigue loading is different from that of carbon and alloy steels. Steels exhibit an endurance limit indicating an infinite life at cyclic stresses less than the endurance limit. However, as shown in Fig. 4.19, the aluminum alloy 75S-T does not exhibit an endurance limit. Instead the fully reversed alternating stress required to produce a fatigue failure continues to decrease as the number of cycles of loading increases to 10^9 cycles. This behavior is typical of aluminum and its alloys.

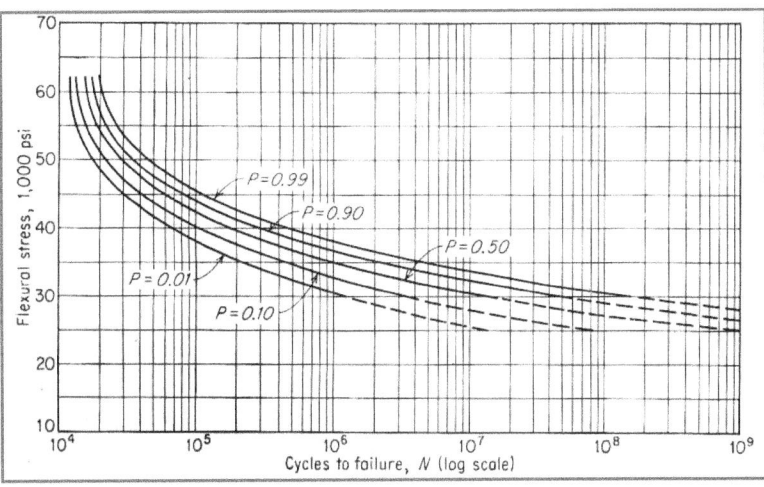

Fig. 4.19 S–N curves for aluminum 75S-T. The aluminum alloy shows a continued decrease in stress required for fatigue failure at 10^9 cycles. The probability of failure P increases with the cyclic stress level.
After T. J. Dolan [4].

The S-N curve for a titanium alloy Ti–6Al–4V is presented in Fig. 4.20. Examination of the data points shown on the graph demonstrates that an endurance limit exists for this titanium alloy. This data is from the U. S. Air Force Research Laboratories. The fatigue tests were conducted using specimens machined from a very pure Ti–6Al–4V alloy. The alternating stress required to produce failure decreases with increasing number of stress cycles until $\sigma_a = 380$ MPa. Several specimens did not fail with $\sigma_a = 400$ MPa at $N_f = 10^8$ or 10^9 cycles.

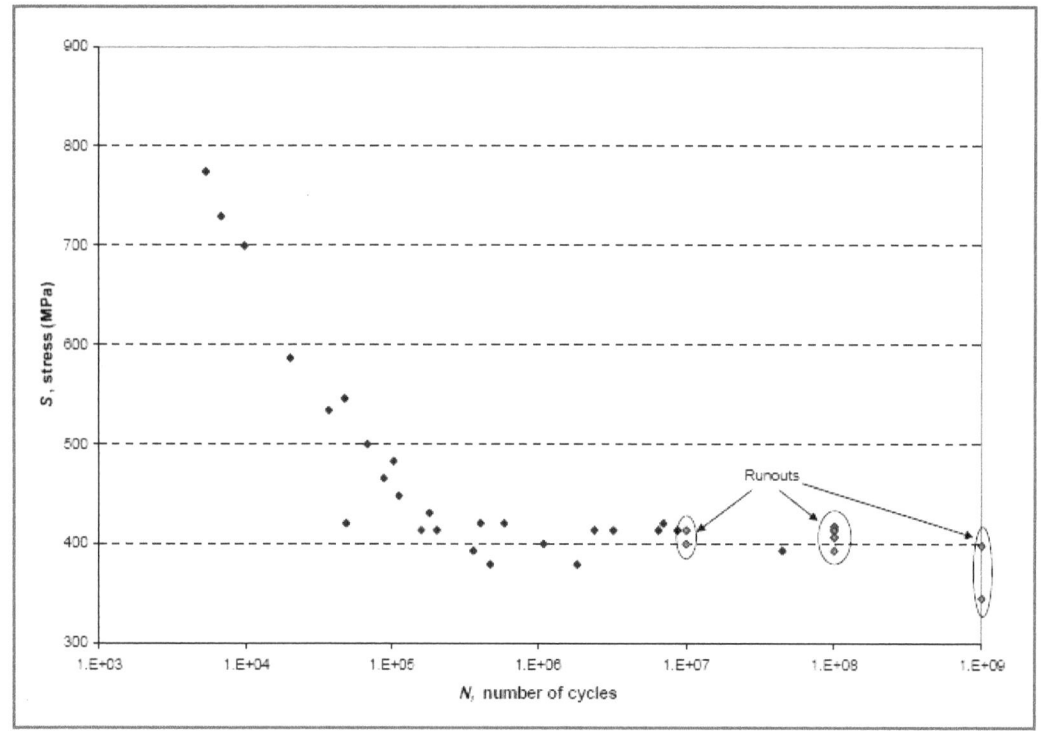

Fig. 4.20 S-N curves for the titanium alloy Ti-6Al-4V. After R. D. Pollak, Air Force Research Laboratory [5].

Today polymers are used in many applications and they perform reliably, if they are not overstresses or exposed to elevated temperatures. In Fig. 4.21 we show the S-N curves for several common polymers used today to manufacture a wide variety of components. The application of polymers has grown immensely over the past 50 years, as new formulations of the polymeric molecular chains have yielded high-strength and tough plastics with long, useful lives.

Fig. 4.21 S-N curves for several different polymers in common usage today. Note that PMMA, PS, PP, PE and PTFE exhibit an endurance limit, but PET and Nylon do not.
PET Polyethylene Terephthalate.
PS Polystyrene.
PP Polypropylene
PE Polyethylene
PTFE Teflon
PMMA Plexiglas

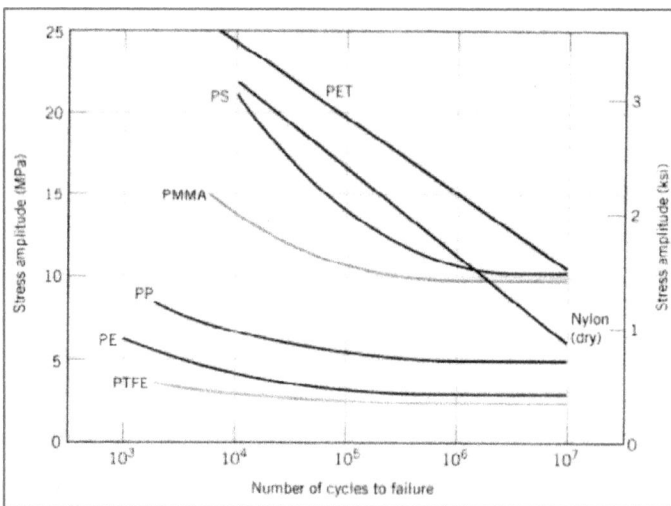

It is evident from Figs. 4.18 to 4.21 that fatigue strength and fatigue behavior varies markedly with the material and with the number of cycles of stress to which the component is subjected. In the following sections, we will outline some of the methods used in fatigue analysis. However, be aware that fatigue failures are a serious concern and there are many factors that affect fatigue life. We will cover two more factors — surface finish and corrosion later in this section.

4.4.3 Using the S – N Diagram in Design Analysis

To accommodate for the degrading effects of cyclic loading, we compare the applied stresses with the **fatigue strength, S_f** or the **endurance limit, S_e** of the material from which the structure is fabricated. A typical example of the fatigue strength for high-strength, low-alloy carbon steel, as a function of the number of cycles of applied load is represented by the S_f-N diagram in Fig. 4.22.

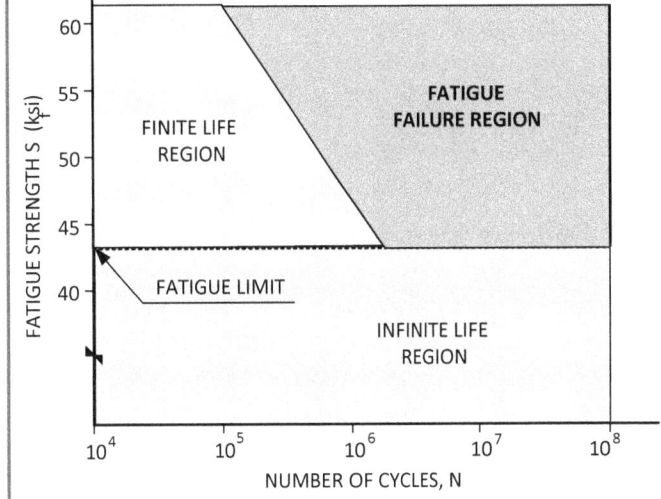

Fig. 4.22 Conceptual S_f – N boundaries for high-strength, low-alloy carbon steel.

The S_f - N diagram is a graphical representation showing the safe and critical states of cyclic stresses. This diagram is divided into three different regions:

1. The failure region (dark) where a sufficient number of cycles of high stresses produce a fatigue crack leading to fracture.
2. Safe regions (light) where component stresses lower than the endurance limit ensure infinite cyclic life.
3. A finite-life region (pale) where a finite (limited) number of cycles can be endured at a specified stress level that is greater than the fatigue limit.

For infinite life, the fatigue strength is often called the endurance limit S_e where:

$$S_e = S_f \qquad \text{for } N > 10^6 \text{ cycles} \qquad (4.15)$$

For finite life, with the number of cycles less than about 10^6, the fatigue strength S_f is larger than S_e. The value of S_f used in a design analysis is determined from the S–N diagram for the material employed to fabricate the structural elements. Example 4.8 demonstrates the technique used to determine the strength S_f associated with finite life. This approach is also valid for those polymers exhibiting an endurance limit such as PMMA, PP, PE and PS.

This approach is not valid for those materials that do not exhibit an endurance limit such as aluminum alloys or polymers such as polyester and nylon. These materials require a different approach, which is illustrated in Fig. 4.23. In this illustration the S-N space is divided into two regions separated by the S-N failure curve. Below the curve, the region is light colored to depict safe design conditions; however, above the failure curve the region is dark colored to depict the combination of stress and loading cycles that will produce component failure.

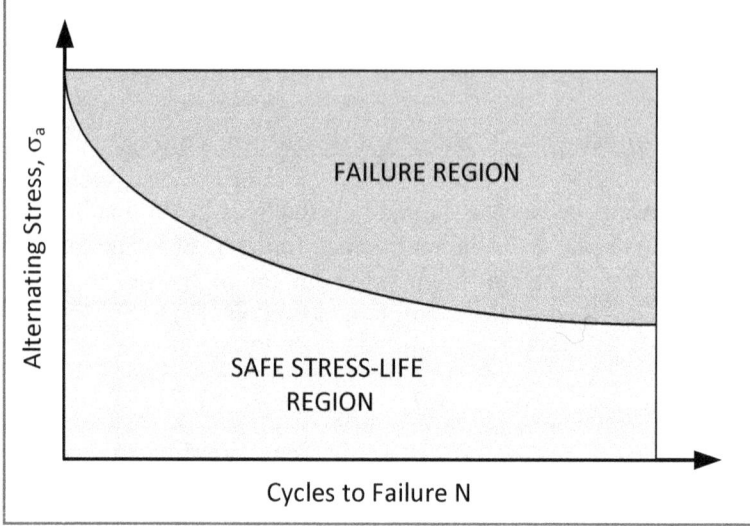

Fig. 4.23 Schematic S_f – N boundary for aluminum alloys and polyester and nylon.

4.4.4 Defining Mean and Alternating Stresses

In comparing the cyclic stresses with the strength, as defined in Figs. 4.22 and 4.23, we use the alternating portion of the applied stresses. The alternating stress, σ_a, and the mean stress, σ_m, for different types of cyclic loading are defined in the stress-time diagrams presented in Fig. 4.24. The mean stress for cyclic loading is computed by identifying the maximum and minimum stress in a given cycle of applied loading. The mean cyclic stress is the average of the maximum and minimum stress, which is given by:

$$\sigma_m = (\sigma_{Max} + \sigma_{Min})/(2) \tag{4.16}$$

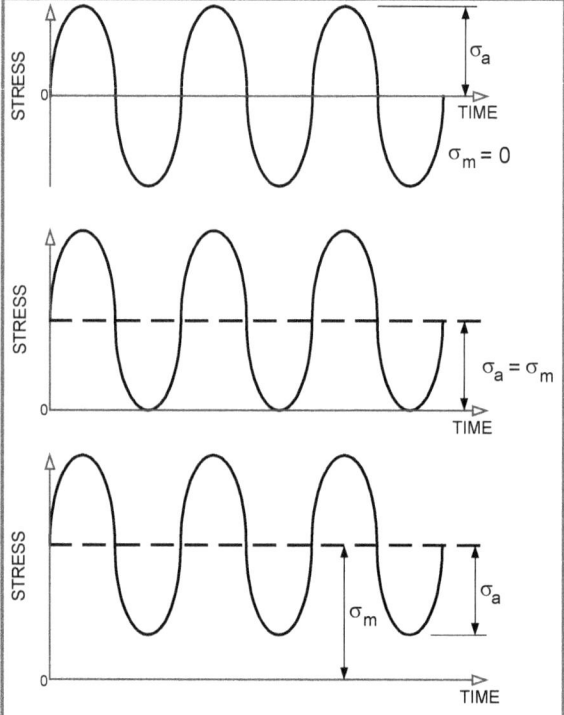

Fig. 4.24 Different types of cyclic stresses imposed on a structure.

The alternating stress is determined from the difference in the maximum and minimum stress during a typical load cycle. The relation used to determine the alternating cyclic stress is given by:

$$\sigma_a = (\sigma_{Max} - \sigma_{Min})/(2) \tag{4.17}$$

The method used to determine the safety of a structural member fabricated from low-alloy steel, when subjected to cyclic loading is described in Examples 4.6 and 4.7.

EXAMPLE E4.6

Consider a cross beam used in constructing a bridge. The dead weight of the bridge structure produces a stress of 28,800 psi in this beam. However, when a fully loaded tractor-trailer crosses over the beam, the maximum stress increases to 74,360 psi. The beam is fabricated from a material with the fatigue properties described in Fig. 4.22. The strengths, yield and ultimate tensile, of this material are 83.6 ksi and 105.6 ksi, respectively. Will the beam fail in fatigue? If failure occurs, predict the cyclic life of the structure. Neglect the effect of the mean stress on fatigue strength in this example.

Solution:

Determine the alternating stresses imposed on the bridge beam from Eq. (4.17).

$$\sigma_a = (\sigma_{Max} - \sigma_{Min})/(2) = (74,360 - 28,800)/(2) = 22.78 \text{ ksi} \tag{a}$$

Next, compare this value to the fatigue limit of 43 ksi that is obtained by reading the S_f - N curve in Fig. 4.22. Clearly $S_f = 43$ ksi $> \sigma_a = 22.78$ ksi. The alternating stresses are less than the endurance limit and the bridge beam is safe. The safety factor is determined from:

$$\mathbf{SF_f} = S_f/\sigma_a \tag{4.18}$$

$$\mathbf{SF_f} = (43)/(22.78) = 1.888$$

The mean stress imposed on the bridge beam during the cyclic loading is determined from Eq. (4.16) as:

$$\sigma_m = (\sigma_{Max} + \sigma_{Min})/(2) = (74,360 + 28,800)/(2) = 51.58 \text{ ksi} \tag{b}$$

This is a relatively large mean stress to impose on the beam. As a consequence, the fatigue strength will be lower than that obtained from the S_f–N curve to account for the effects of the large tensile mean stress. This solution is modified in Example 4.8 to account for the effects of mean stresses in reducing the allowable alternating stresses.

EXAMPLE E4.7

Consider the crankshaft of a high performance race car, which is subjected to maximum and minimum stresses of + 392 MPa and – 392 MPa. If the crankshaft is fabricated from the high-strength, low-alloy steel described in Fig. 4.22, determine its fatigue life.

Solution: Using Eq. (4.17), determine the alternating stress σ_a for the crankshaft as:

$$\sigma_a = (\sigma_{Max} - \sigma_{Min})/(2) = [392 - (-392)]/(2) = 392 \text{ MPa} \qquad (a)$$

Next convert this result from MPa units to ksi units so that we may employ a graphical approach using the fatigue properties of the high-strength, low-alloy steel shown in Fig. 4.22.

$$\sigma_a = (392 \text{ MPa})(1 \text{ ksi})/(6.895 \text{ MPa}) = 56.85 \text{ ksi} \qquad (b)$$

Plot $\sigma_a = 56.85$ ksi at point A on the ordinate of the S_f-N diagram, shown in Fig. E4.7. Extend a line parallel to the abscissa, until it intersects the shaded (fatigue failure) region and plot point B. Drop a vertical line from point B, until it intersects the axis defining the cyclic life to establish point C. Point C gives the anticipated cyclic life, which in this example is about 200,000 cycles. For a crankshaft operating at 5,000 RPM in a high-performance racing car, the anticipated life is less than one hour. Clearly, this life is too short by several orders of magnitude and the design of the crankshaft is inadequate. Either the crankshaft size must be increased to lower the stress level or a different material with a higher endurance limit S_e must be selected.

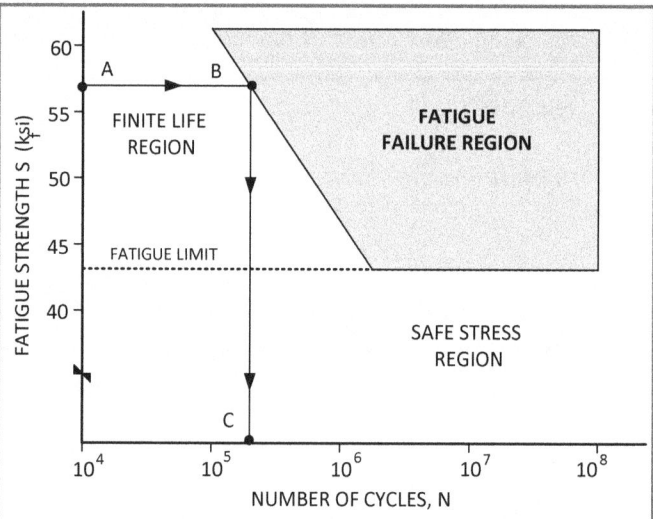

Fig. E4.7

4.4.5 The Effect of Mean Stresses on Fatigue Strength

In some instances, the cyclic loading on a structural member produces a combination of alternating and mean stresses. When mean stresses are superimposed on the alternating stresses, the effect is to decrease the fatigue strength of the material. Goodman, Gerber and others have developed empirical methods to determine a modified fatigue strength that accounts for the detrimental effects of combined alternating and mean stresses. The modified fatigue strength S_a is presented as a function of the cyclic mean stress in Fig. 4.25. Examination of Fig. 4.25 reveals that the modified fatigue strength S_a is equal to S_e when $\sigma_m = 0$ for all the empirical relations. The decrease in the modified fatigue strength is linear with the increase in σ_m, with the Goodman method. The fatigue strength S_a becomes zero when $\sigma_m = S_u$, for the Goodman method and Gerber methods.

Let's consider the Goodman method, because it is the most commonly used technique and it is conservative compared to the Gerber method. We show a load line with a slope R in Fig. 4.25. The slope R is given by the ratio:

$$R = \sigma_a/\sigma_m \qquad (4.19)$$

The load line extends from the origin and intersects the Goodman line. At this point, a horizontal line is drawn to intersect the y axis, defining the allowable alternating strength S_a.

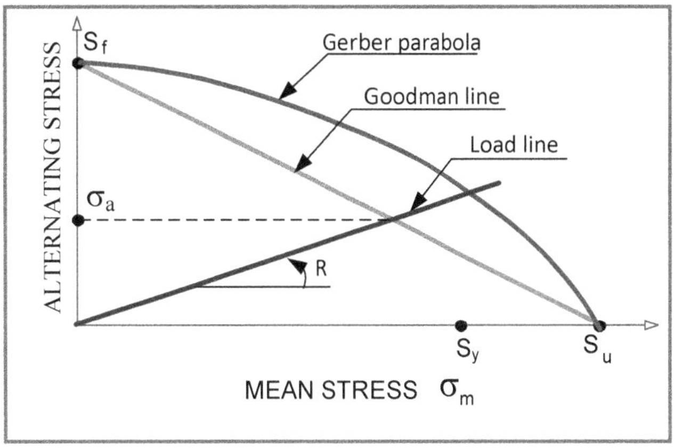

Fig. 4.25 Modified fatigue strength S_a decreases as the cyclic mean stresses increase.

We may also characterize the modified fatigue strength S_a in equation format for the Goodman method as:

$$S_a = S_f [1 - (\sigma_m/S_u)] \quad (4.20a)$$

or

$$(\sigma_a/S_f) + (\sigma_m/S_u) = 1 \quad (4.20b)$$

where S_f is the fatigue strength associated with a specified number of cycle, S_u is the ultimate tensile strength, and σ_m is the mean stress.

A graph of S_a as a function of σ_m for Eqs. (4.20) is shown in Fig. 4.25. The result shows that the Goodman method gives a line. Fatigue test data for both steels and aluminum alloys are usually located in the region between the parabola and the line. The Goodman method is more conservative than the Gerber method.

EXAMPLE 4.8

Let's reconsider the bridge described in Example 4.6. In the previous solution, we neglected the influence of the mean stress on the fatigue strength. In this solution, let's account for the degrading effect of the cyclic mean stress on the fatigue strength by using the Goodman method for adjusting the fatigue strength. Determine the safety factor using the values determined for the allowable alternating and mean fatigue strengths.

Solution: Recall the results for the alternating and mean stresses from Example 4.6.

$$\sigma_a = (\sigma_{Max} - \sigma_{Min})/(2) = (74{,}360 - 28{,}800)/(2) = 22.78 \text{ ksi} \quad (a)$$

$$\sigma_m = (\sigma_{Max} + \sigma_{Min})/(2) = (74{,}360 + 28{,}800)/(2) = 51.58 \text{ ksi} \quad (b)$$

Determine the slope R of the load line from Eq. (4.19) as:

$$R = \sigma_a/\sigma_m = 22.78/51.58 = 0.4416 \quad (c)$$

Next, let's determine the allowable alternating strength. Substituting Eq. (4.19) into Eq. (4.20a) yields:

Solving for S_a gives:
$$S_a = R\, S_e\, S_u/(R\, S_u + S_e) \tag{4.21a}$$

$$S_a = \frac{(0.4416)(43)(105.6)}{(0.4416)(105.6)+43} = 22.37 \text{ ksi} \tag{d}$$

The safety factor is given by setting $S_a = S_f$ in Eq. (4.18) as:

$$SF_f = S_a/\sigma_a = 22.37/22.78 = 0.9820 \tag{e}$$

When accounting for the mean stresses on fatigue strength, the safety factor was reduced to less than 1.0. In the previous solution for Example 4.6, the safety factor **SF_f** = 1.888; whereas, in this analysis the safety factor is reduced to 0.9820. Clearly, the effect of the mean stress is to reduce the allowable fatigue strength S_a. With a safety factor less than 1.0, the beam size must be increased or a substitute material employed with higher fatigue strength.

4.4.6 The Effect of Corrosion and Surface Finish on Fatigue Strength

We have discussed some of the parameters that affect fatigue life including the type of material (steel, aluminum, titanium, and polymers, the mean and alternating stress and stress concentrations. However, there are many other factors that affect fatigue behavior. Two of these factors are corrosion and surface roughness.

Corrosion

Corrosion of many metals including steels and aluminum affect fatigue life. While steels, except for stainless steel, are more susceptible, unclad aluminum alloys are also affected. Structural elements or machine components operating outdoors are often subjected to rain or condensation from humid air. Moisture on the surface initiations corrosion and small corrosion pits are formed. These pits in turn initiate fatigue cracks, which propagate quickly into the component at remarkably low stress levels. The effect of corroding surfaces on the endurance limit for steels is presented in Fig. 4.26. It is clear from this figure that a corroding surface totally eliminates the benefit of employing high-strength steel in fatigue applications.

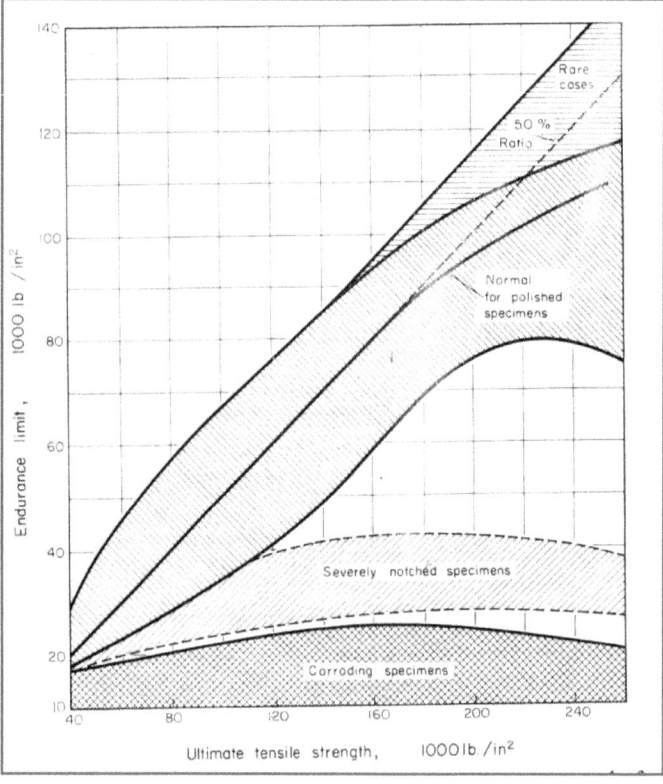

Fig 4.26 Graphical illustration of the effect of corroding surfaces on the endurance limit of steels with varying tensile strength. After Forrest [2].

Aluminum alloys are also affected but to a lesser degree than steel alloys. Fatigue tests showed that atmospheric effects reduced average fatigue life by a factor of 3 times for alloys 7075-T6 and 2024-T3, and about 1.5 times for alclad 7075-T6 but had no effect on alclad 2024-T3. To avoid detrimental effects of corrosion, it is essential to protect the surface of machine components and structural elements from moisture accumulation. Surface treatments include painting with an effective formulation, nitriding surfaces and zinc and chrome plating.

Surface Finish

Most fatigue testing to generate S – N curves involved small specimen with smooth, carefully machined or polished surfaces. However, machine components or structural elements are usually large and often have relatively rough surfaces. The surface finish affects the fatigue strength and must be accommodated in a design analysis.

The effect of surface roughness for tempered (high-strength) and annealed (lower-strength) steels is shown in Fig. 4.27. It is evident from these experimental results that roughness (depth of surface grooves) reduces the fatigue strength. Note that the surface fatigue factor m in Fig. 4.27 is defined as:

m = (Fatigue strength as a specified roughness)/(fatigue strength with smooth surfaces)

In many applications surface roughness cannot be avoided. However, surface treatments including shot peening, surface rolling and plating with zinc are effective in mitigating the effect of surface roughness. Zinc plating is also a cost effect method for eliminating the detrimental effect of corrosion on fatigue strength.

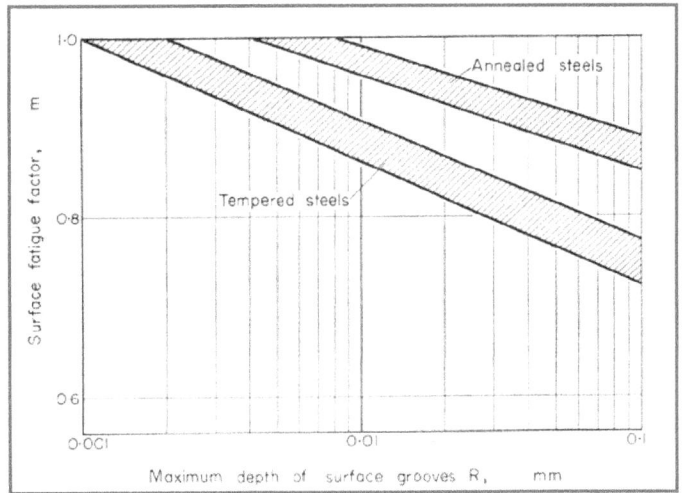

Fig. 4.27 The detrimental effect of surface roughness on fatigue strength of tempered and annealed steels. After Forrest [2].

4.5 SUMMARY

Structures and machine components fail because of fracturing, yielding or excessive elastic deformation. In predicting either fracture or yielding, we compare the stress imposed on the structure with the strength of the material from which it is fabricated. In some analyses, we employ a handbook to provide the strength of the materials, and in more critical designs we conduct standardized tensile tests to establish the strength of the specific materials to be employed in construction.

The standard ASTM tensile test is introduced in Section 4.2. Using a standard tensile specimen and standard test procedures a force-deflection curve is generated. We convert this force-deflection data to a stress-strain curve, and then interpret the curve to give the strength, ductility and the elastic constants. Care is exercised in the interpretation to distinguish between "brittle" and "ductile" materials because the yield strength cannot be measured for brittle materials. For ductile materials, procedures are described for measuring both the yield strength and the ultimate tensile strength. The yield and ultimate tensile strengths are given by:

$$S_y = \sigma_y \tag{4.7}$$

$$S_u = \sigma_u \tag{4.8}$$

Measures of ductility are given by the percent elongation and percent reduction in area:

$$\% e = \frac{L_f - L_0}{L_0} \times 100 \tag{4.9}$$

$$\% A = \frac{A_0 - A_f}{A_0} \times 100 = \left[\left(\frac{d_0^2 - d_f^2}{d_0^2}\right)\right] \times 100 \tag{4.10}$$

Methods for determining both of these quantities have been given. We noted that for most types of steel the ductility decreased with increasing carbon content used for strength enhancement.

Two elastic constants, the modulus of elasticity E and Poisson's ratio ν, are measured in a tensile test. The modulus of elasticity is determined from the slope of the stress-strain curve in the elastic region. Poisson's ratio is determined from the ratio of strain in both the axial and transverse directions measured during a uniaxial tension test.

$$\nu = -\varepsilon_t/\varepsilon_a \tag{4.11}$$

The shear modulus, another elastic constant often used in analysis of shear stresses and torsion loading of circular shafts, is introduced.

$$\tau = G\gamma \tag{4.13}$$

$$G = \frac{E}{2(1+\nu)} \tag{4.14}$$

Clearly, the shear modulus G is not independent, because it is a function of E and ν.

Many structures and machine components are subjected to repeated loading. There are two detrimental effects due to this cyclic loading. First, the design strength of the material from which the structure is fabricated is lowered. Hence, we design to a fatigue strength S_f that is a function of the cyclic stresses and the number of cycles of load imposed onto the structure.

Second, fatigue failures in **ductile** materials are of a **brittle** nature (i.e., structural failure and collapse occur catastrophically). The failure mechanism in fatigue is markedly different from that observed in yielding or rupture. In fatigue, microscopic cracks are initiated due to accumulated irreversible slip in a very thin layer of material adjacent to the component's surface. These cracks grow larger and extend into the material until reaching critical size. At this point, the cracks become unstable and extend at high speed, through the cross section of the structural member, producing sudden and catastrophic collapse.

To accommodate for the degrading effects of cyclic loading, we compare the applied stresses with the **fatigue strength, S_f** of the material, from which the structure is fabricated. A typical example of the fatigue strength for low carbon steels is presented in the S_f-N diagram of Fig. 4.22. This diagram is a graphical representation showing both safe and critical states of cyclic stresses. For infinite life, the fatigue strength is often called the endurance limit S_e where:

$$S_e = S_f \qquad \text{for } N > 10^6 \text{ cycles} \qquad (4.15)$$

In comparing the cyclic stresses with the fatigue strength as defined in Fig. 4.22, we use the alternating portion of the applied stresses. The alternating stress σ_a, and the mean stress σ_m are given by:

$$\sigma_m = (\sigma_{Max} + \sigma_{Min})/2 \qquad (4.16)$$

$$\sigma_a = (\sigma_{Max} - \sigma_{Min})/2 \qquad (4.17)$$

The safety factor in a cyclic loading application, where fatigue is a consideration, is given by:

$$\mathbf{SF_f} = S_f/\sigma_a \qquad (4.18)$$

Tensile mean stresses lower the allowable fatigue strength S_a according to the Goodman and Gerber relations that are given by:

$$S_a = S_e [1 - (\sigma_m/S_u)] \qquad (4.20a)$$

$$S_a = S_e [1 - (\sigma_m/S_u)^2] \qquad (4.20b)$$

Another form of the Goodman relation, which incorporates the load ratio R between the allowable alternating and mean strengths, is given as:

$$S_a = R\, S_e\, S_u/(R\, S_u + S_e) \qquad (9.11a)$$

PROBLEMS

4.1 Using Eq. (4.1) write the equations for the six stresses as a function of the six strains.
4.2 Sketch a stress strain curve for a brittle material.
4.3 Sketch a stress strain curve for a ductile material with strain hardening.
4.4 Determine the percent elongation and the percent reduction in area if a standard tensile specimen exhibited a gage length of 60 mm and a diameter of 10.4 mm after failure.
4.5 Suppose you cut a 25-mm diameter circular hole in a large sheet of dental dam (very thin rubber sheet). If the sheet, originally 400 mm long by 150 mm wide, is extended until it is 475 mm long, determine the new width of the sheet and the dimensions of the deformed hole. Assume that the rubber is perfectly elastic with a Poisson's ratio, $\nu = 0.5$.
4.6 Determine the shear modulus for steel and aluminum.
4.7 Cite the advantages and disadvantages of the rotating beam fatigue testing machines.
4.8 Sketch the principal characteristics of an S-N curve for a steel alloy.
4.9 Sketch the principal characteristics of an S-N curve for an aluminum alloy.
4.10 Consider a cross beam used in constructing a bridge. The dead weight of the bridge structure produces a stress of 32,000 psi in this beam. However, when a fully loaded tractor-trailer crosses over the beam, the maximum stress increases to 41,000 psi. The beam is fabricated from a material with the fatigue properties described in Fig. 4.22. The strengths, yield and ultimate tensile, of this material are 60.0 ksi and 86.0 ksi, respectively. Will the beam fail in fatigue? If failure occurs, predict the cyclic life of the structure. Neglect the effect of the mean stress on fatigue strength in this example.
4.11 Repeat Problem 4.10 taking into account the effect of mean stress on the fatigue behavior.
4.12 Describe the effect of corrosion on fatigue life and provide means for mitigating the effect of fatigue.
4.13 You inspect a machine component and find that it contains surface grooves 0.03 mm deep. If the part is machined from tempered steel determine the fatigue strength reduction factor.

REFERENCES

1. Forsyth, P. J. E., The Mechanism of Fatigue in Aluminum and Aluminum Alloys, <u>Fatigue in Aircraft Structures</u>, Ed. A. M. Freudenthal, Academic Press, New York, 1956.
2. Forest, P. G., <u>Fatigue of Metals</u>, Pergamon Press, New York, 1962.
3. Dowling, N. E., <u>Mechanical Behavior of Materials</u>, Prentice Hall, Upper Saddle River, 1999.
4. Dolan, T. J., Basic Concepts of Fatigue Damage in Metals, <u>Metal Fatigue</u>, McGraw Hill Co., New York, 1959, p. 39-67.
5. Pollack, R. D., Analysis of Methods for determining High Cycle Fatigue Strength of a Material with Investigation of Ti-6Al-4V Giga Cycle Fatigue Behavior, Air Force Institute of Technology, 2005.

CHAPTER 5

AXIALLY LOADED STRUCTURAL MEMBERS

5.1 INTRODUCTION

Large structures like bridges, sports stadiums, skyscrapers, etc. are fabricated from many small structural elements. There are several reasons for constructing big structures from much smaller components. First, manufacturing facilities limit the size (width and thickness) of structural elements that are produced. The rolling mills used in steel and aluminum plants to produce bars, plates and sheets usually are only six to eight feet long. Even more limiting is the requirement for transport from the manufacturing facility to the construction site. If highways are used for transport, state law limits both the width and length of the structural members.

In this chapter we will consider two types of long thin structural members—flexible and stiff. Both types of these structural members will be subjected to axial loading. Wire rope or cables, which are flexible, are capable of supporting only tensile loading that tends to stretch the cable. However, rods or bars, which are stiff[1], are capable of supporting both tensile and compressive loading.

As structural elements, wire rope, cable, rods and bars subjected to axial loading have a significant advantage. They are subjected to a uniform state of stress over their cross sectional area. Also, if they are two-force members loaded at their two ends, the stresses are uniform over their length. Thus, the entire volume of the rod or bar is subjected to the same stress, which is the optimum condition for minimum weight design.

Examples of structural applications of long thin members in bridge construction are numerous. For instance, very high-strength wire rope is used in the construction of suspension bridges and in cable stayed bridges. Also, many bridges are designed with trusses deployed on the two sides of the structure. These trusses are fabricated from many bars or rods connected together at bolted or welded joints. We analyze these rods and bars using the same equations established for wire rope and cables. The difference is that the rods and bars are sufficiently stiff to support compressive stresses.

5.2 CHARACTERISTICS OF CABLES, RODS AND BARS

What characteristics do cables, wire ropes, rods and bars have in common and how do they differ? Cables, rods and bars exhibit lengths that are very long when compared to the dimensions of the cross section. A cable differs from a rod or a bar because it is flexible whereas the rod and bar are stiff. **Cables** have a circular cross section that is uniform along its length. A cable is made from many strands of wire that are twisted together to form a larger diameter very high-strength structural member. Even though a typical cable is larger in diameter than a wire, it is treated as a flexible member capable of supporting only tensile forces. We will use wire, wire rope and cable interchangeably in this text. The cross sectional area of **bars and rods**[2] may be of arbitrary shape.

---

[1] Rods or bars actually stretch or compress under the action of axial loads, but this deformation is small and is neglected in determining the loads and stresses. However, this very small deformation is considered when determining the strain and deflection of the long thin structural members.

[2] In most applications, the cross section of a rod is circular and that of a bar rectangular; however, this distinction is not always maintained and one can expect to find cross sections of arbitrary shape for both rods and bars.

Because cables are flexible, they can be wrapped about pulleys, formed into loops, and sometimes knotted. Indeed, they are so flexible that we cannot push on them because they buckle. It is possible to pull, but not push. This fact means that internal tensile forces develop within a thin flexible member but not compression forces. However, rods and bars are sufficiently stiff to support compressive forces and stresses. They are used as tie rods when loaded in tension, and as columns when loaded in compression. We will assume in this chapter that the cross section of the column is adequate and that failure by buckling is not an issue.

We are considering the simplest structural application—the use of long thin members under axial loading. The geometry of a structural member (long length with small cross sectional dimensions) leads us to constrain the direction of the loading and to make an important assumption about the deformation of the member that significantly simplifies the analysis.

1. The internal and external forces supported by the member act along its length.
2. Plane sections before loading remain plane after loading and subsequent deformation of the member.

Constraining the loads in the uniaxial direction is consistent with the title of the chapter. We are considering only the effects of axial loading on cables, rods and bars. In a later chapter on stresses in beams, we will consider bending induced by transverse loading.

The assumption that plane sections before loading remain plane after loading has also been confirmed by experiment. Suppose we draw a straight scribe line on a long thin wire or bar as shown in Fig. 5.1 a. This line represents the edge of a plane through the cross section of the member. Now apply an axial load F to the wire as indicated in Fig. 5.1b. The wire will stretch a small amount δ, but the line remains straight and the plane through the cross section remains plane (i.e. flat). This is a very important observation because it implies that the normal stresses σ are uniformly distributed across this plane.

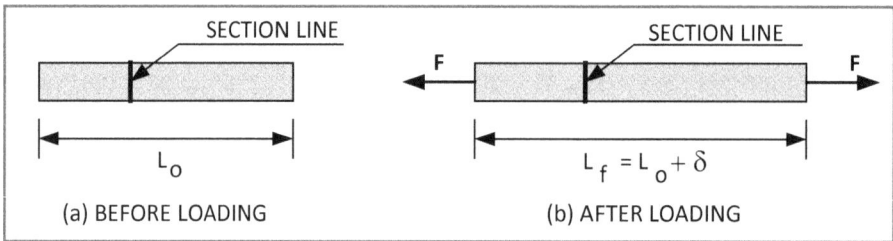

Fig. 5.1 A section line drawn across the member remains straight after loading.

5.3 <u>DESIGN ANALYSIS METHODS</u>

In the design analysis, we compute the axial forces acting on the uniaxial member, the amount it stretches under load, and the stresses developed. Next, we compare the stresses acting on the structural member with its strength to determine a safety factor. The size of the safety factor is evaluated to determine if this member is safe to be employed in a structure used by the public. This chapter describes methods that will enable you to size cables, rods and bars to provide structures with adequate safety margins.

We will divide this presentation into two parts—the first dealing with cables and the second involving rods and bars. The reason for the separation is the difference in the structural applications. The flexible cables are often used with pulleys for lifting and for high-strength tension members. Rods and bars are employed as tie rods or columns in structural applications. While the applications differ, the equations governing the behavior of the flexible and the stiff members are the same.

5.3.1 Design Analysis of Cables and Wire Rope

What happens if we pull on a wire and continue to pull with increasing force? The wire stretches, and stretches still more until it fails by breaking. Let's explore the stretch of the wire, and the consequences of different types of failure of the wire in subsequent sections.

Stretch of Wire under Load

When we pull on a wire it stretches as shown in Fig. 5.1b. We define δ, the **deformation** of the wire, as:

$$\delta = L_f - L_o \qquad (5.1)$$

where L_f is the length of the wire under load and L_o is the original length.

A FBD of a segment of wire is shown in Fig. 5.2a. The equilibrium relation ($\Sigma F_x = 0$) indicates that the internal force P is equal to the external force F for these uniaxial members.

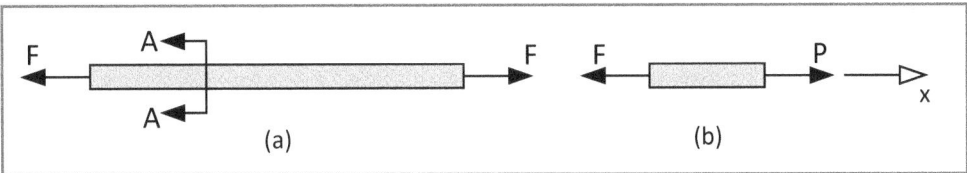

Fig. 5.2 FBD of a long thin member subject to axial load.

We determine the stretch δ from:

$$\delta = (PL)/(AE) \qquad (5.2)$$

where P is the internal force in the wire, in N or lb, L is the length of the wire, in m or in. $A = \pi r^2$ is the cross sectional area of the wire, in m^2 or in^2. r is the radius of the wire, in m or in. E is the **modulus of elasticity**, in GPa or psi.

We will derive Eq. (5.2) later in this section. It is important at this time to understand that the extension of the wire δ is a very small quantity if the wire is fabricated from a metal such as steel or aluminum. To demonstrate the magnitude of the extension, consider the following example.

EXAMPLE E5.1

A No. 30 gage[3] steel music wire (0.080 in. in diameter) is employed to lift a weight of 500 lb. If the wire is 6 ft. in length, determine the amount that the wire stretches under load.

> **Solution:** In Appendix B, we note that the modulus of elasticity for steel is listed as $E = 30 \times 10^6$ psi. Next, let's substitute the values for P, L, $A = \pi r^2$ and E into Eq. (5.2) to obtain:
>
> $$\delta = (PL)/(AE) = (500 \text{ lb})(6 \text{ ft})(12 \text{ in/ft})/[\pi (0.040 \text{ in.})^2 (30 \times 10^6 \text{ lb/in}^2)]$$

---

[3] There are several standards for gage sizes that define the diameter of wire or the thickness of sheet metal. Four of the commonly referenced standards are presented in Appendix A.

> $\delta = 0.2387$ in.
>
> Is 0.2387 in. a small extension of the wire? Small is a relative term and must be compared to another value to be judged. Let's compare the amount of this extension to the original length of the wire by computing the ratio:
>
> $\delta/L_o = 0.2387/(6)(12) = 0.003316$ or 0.3316%
>
> The 500 lb load stretches the wire by less than a one third of one percent. That is a very small extension when compared to the original length of the wire.

Let's conduct a simple experiment and observe the behavior of a wire under the action of a monotonically increasing tensile load. If we measure the load F applied to the wire, the extension δ of the wire, and equate F_{EXT} to P_{INT}, we can construct the graph shown in Fig. 5.3 a.

Note a linear relationship exists between P and δ as indicated by Eq. (5.2). This P-δ curve is a straight line until the wire begins to fail by **yielding**. The linear portion of the P-δ relation is the **elastic region** of the load-deflection response. Equation (5.2) is valid only in this elastic region.

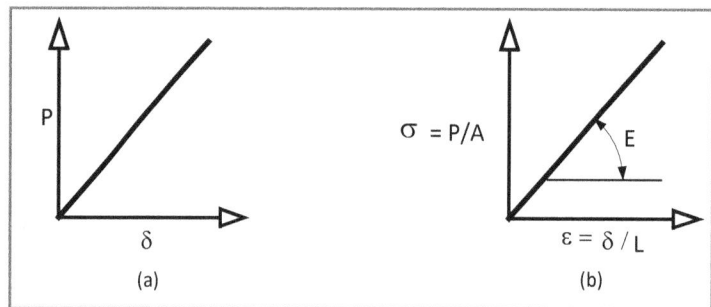

Fig. 5.3 Graph of load versus stretch in (a) and stress versus strain in (b).

Let's modify the axes in Fig. 5.3a by dividing the force P by the area A and the extension δ by the length L as indicated below:

$$\varepsilon = \delta/L \tag{5.3}$$

$$\sigma = P/A \tag{5.4}$$

where ε is the strain and σ is the stress in the wire, respectively.

These relations show that the **normal strain**, ε, is the change in length divided by the original length; strain is a dimensionless quantity. The normal stress σ is the internal force P divided by the area A, over which it acts. The stress is expressed in terms of Pa (Pascal) in SI units or psi (lb/in^2) in U. S. Customary units. We have shown the graph of stress versus strain in Fig. 5.3b. As expected there is a linear relation between stress and strain until the stress is sufficient to cause the wire to yield. The **slope** of the σ - ε line is the **modulus of elasticity** E of the material from which the wire is fabricated. The **modulus of elasticity** is a **material property**, which is independent of the shape of the body.

It is evident from the linear response in the stress-strain diagram that:

$$\sigma = E\varepsilon \tag{5.5}$$

This stress-strain relation is known as **Hooke's law**. It is named after Robert Hooke who is credited with the discovery of elasticity in the 17th century. A word of caution—Hooke's law is valid only for uniaxial states of stress that arise in long thin structural members. We will introduce another more complex form of the stress-strain relation to accommodate multi-axial stress fields in a later chapter.

Let's combine Eqs. 5.3, 5.4 and 5.5 in the manner shown below:

$$\sigma = P/A = E\varepsilon = (E\delta/L)$$

and solve this expression for δ to derive Eq. (5.2):

$$\delta = (PL)/(AE) \qquad (5.2 \text{ bis})$$

EXAMPLE E5.2

Determine the strain that develops in a 7 m long wire, which is stretched by 12 mm.

Solution: Using Eq. (5.3), we write:

$$\varepsilon = \delta/L = 12 \times 10^{-3} \text{ m} / 7 \text{ m} = 0.001714 \text{ dimensionless}$$

Note that strain is dimensionless because we divided a change in length by the original length. Also, observe that we have determined the strain without knowledge of the material used in fabricating the wire. Strain is a geometric concept. The strain imposed on a wire or rod may be determined without knowledge of the load or the modulus of elasticity if the amount of deformation is known.

EXAMPLE E5.3

Determine the strain that develops in a 10 ft long wire that is subjected to a force of 160 lb. The wire size is Gage 10 (0.10189 in. in diameter) and it is fabricated from an aluminum alloy 2024-T4.

Solution: Using Eq. (5.3), we write:

$$\varepsilon = \delta/L$$

Recall Eq. (5.2):

$$\delta = (PL)/(AE)$$

Combining Eqs. (5.2) and (5.3) yields:

$$\varepsilon = P/(AE) \qquad (5.6)$$

From Appendix B, we find the modulus of elasticity E for aluminum alloy 2024-T4 is 10.4×10^6 psi. Next, we compute the cross sectional area of the wire as:

$$A = \pi d^2/4 = \pi(0.1019)^2/4 = 0.008155 \text{ in.}$$

Substituting values for P, A and E into Eq. (5.6) gives:

$$\varepsilon = 160 \text{ lb}/(0.008155 \text{ in.}^2 \times 10.4 \times 10^6 \text{ lb/in.}^2) = 0.001886$$

In this example, it was necessary to use the value of the modulus of elasticity in determining the strain. Why? The modulus of elasticity was needed because the deformation (stretch) of the wire was not specified in the problem statement. It was necessary to determine the deformation of the wire from the load using Eq. (5.2). This step in the analysis required us to introduce the modulus of elasticity.

Internal Forces and Stresses

Internal forces that develop within a structural member when it is subjected to external loads generate stresses. In the study of Statics, we introduced the concept of internal forces and showed that they are determined from the appropriate equations of equilibrium. We also indicated the necessity for making an imaginary section cut to expose the cross sectional area of the member over which the stresses act. In this section, we emphasize the connection between the external forces, internal forces, and the stresses that develop at some point in a structural member.

To illustrate these connections, examine Fig 5.4, which shows an axial member subjected to external axial forces F.

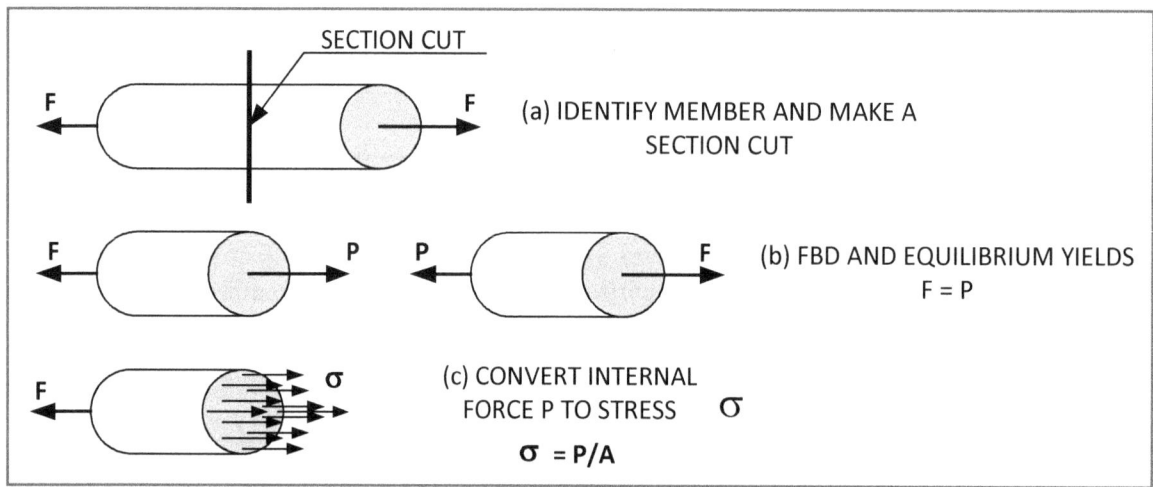

Fig. 5.4 An illustration showing the transition from external forces to internal forces to stresses.

1. We identify the member of interest and the forces acting on the body in Fig. 5.4a.
2. A section cut is made at a location of interest along the length of the member. We then remove one end or the other to construct the FBD depicted in Fig. 5.4b.
3. Next, the equilibrium relation $\Sigma F_x = 0$ is written to prove that $F_{EXT} = P_{INT}$.
4. Finally, from Eq. (5.4) ($\sigma = P/A$), we convert the internal force P into the normal stress σ as indicated in Fig. 5.4c.

In this illustration, the internal force P is a constant over the length of the uniaxial member and the location of the section cut is not important. However, you should be aware that in many applications the internal forces often vary from one location to another. In these situations the location of the section cut is vitally important. Also observe in Fig. 5.4c that the distribution of the stress σ over the cross sectional area exposed by the section cut is uniform. The uniform distribution is a result of two conditions:

- Plane sections remain plane for long thin members subjected to axial loading.
- Internal moments do not develop under the applied external loading.

In the most general case of loading, we relate the internal force and the normal stress over an area A by:

$$P_{INT} = \int \sigma \, dA \qquad (5.7)$$

However, when the stress is uniformly distributed as it is in this illustration Eq. (5.7) reduces to:

$$P_{INT} = \sigma \int dA = \sigma A \qquad (5.8)$$

EXAMPLE E5.4

Determine the stress in a 1.5 mm diameter wire subjected to an axial load of 250 N.

Solution: From Eq. (5.4), we write

$$\sigma = P/A$$

From a FBD identical to the one shown in Fig. 5.4b and the equilibrium relation $\Sigma F_x = 0$, we understand that:

$$F = P = 250 \text{ N}$$

The area $A = \pi d^2/4 = \pi(1.5 \text{ mm})^2/4 = 1.767 \text{ mm}^2$. Substituting these values into Eq. (5.4) gives:

$$\sigma = 250 \text{N}/ 1.767 \text{ mm}^2 = 141.5 \text{ N/mm}^2 = 141.5 \text{ MPa}$$

Note that one N/mm^2 is equivalent to one MPa. Later in this Chapter, we will present an interpretation of this result after we describe in detail the strength of structural materials.

EXAMPLE E5.5

A large diameter wire rope fabricated from steel with a cross sectional area of 6.4 in.^{2} is to support a portion of the roadway on a suspension bridge. The highway engineers have specified that the maximum load imposed on the cable is not to exceed 150 ton. Determine the stress and strain in the wire rope when subjected to the specified load.

Solution: Let's compute the stress from Eq. (5.4) as:

$$\sigma = P/A = (150 \text{ ton})(2000 \text{ lb/ton})/6.4 \text{ in.}^2 = 46{,}880 \text{ psi} = 46.88 \text{ ksi}$$

Next, calculate the strain ε from Eq. (5.6) to obtain:

$$\varepsilon = P/(AE) = 300{,}000 \text{ lb}/[(6.4 \text{ in}^2)(30 \times 10^6 \text{ lb/in}^2)] = 1.563 \times 10^{-3}$$

In determining the strain, we obtained the value of $E = 30 \times 10^6$ psi for steel from Appendix B. We also introduced ksi, a unit in the U. S. Customary system. The conversion factor between psi and ksi is given by $\Rightarrow$ 1 ksi = 1,000 psi.

Failure

Suppose we take a length of wire, grip it in a universal-testing machine and pull until it fails. The behavior of the wire under increasing load depends on the material from which the wire was drawn[4]. Nearly all materials exhibit a **linear elastic** response like that shown in Fig. 5.3 at lower stress levels; however, at higher stresses significantly different behavior is observed for different materials. We have classified these behaviors as:

- **Brittle** with abrupt failure and small, elastic deformations.
- Yielding with **strain hardening** and plastic deformation prior to rupture.
- Yielding with **strain softening** and plastic deformation prior to rupture.

These three behaviors are illustrated graphically in Figs. 5.5 and 5.6.

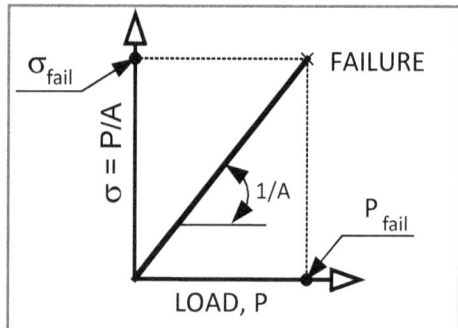

Fig. 5.5 Linear-elastic response until brittle failure.

The graph in Fig. 5.5 indicates the linear elastic response of a brittle material. The stress increases linearly with both axial load and strain. The slope of the stress-load line is the reciprocal of the area (1/A), and the slope of the stress-strain line is the modulus of elasticity E as indicated in Fig. 5.3. When we increase the load to a critical value P_{fail}, the uniaxial member breaks. The failure occurs at a specific value of stress, σ_{fail}. We define this failure stress as the **ultimate tensile strength**, S_u of the material of the wire.

$$S_u = \sigma_{fail} = P_{fail}/A \tag{5.9}$$

The two graphs in Fig. 5.6 illustrate material behavior when yielding occurs. The graph to the left in Fig. 5.6 is typical of a material that yields and then strain hardens. As the axial load on the wire increases, the stress increases linearly until the material begins to yield at σ_{yield}. At that point, the uniaxial member continues to stretch, but the load and stress remain essentially constant. After some degree of post-yield stretch, the material stiffens and the load and stress begin to increase. The stress increases until the wire ruptures at σ_{fail}.

We establish two strengths for the wire from this graph. The **yield strength** S_y given by:

---

[4] Drawing a circular rod through a smaller diameter die produces wire. The drawing process improves the surface finish and enhances the strength of the wire.

$$S_y = \sigma_{yield} = P_{yield}/A \tag{5.10}$$

and the **ultimate tensile strength,** S_u given by Eq. (5.9).

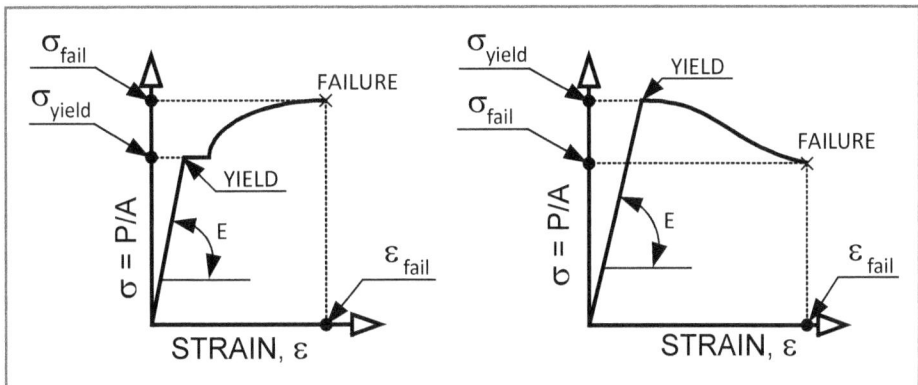

Fig. 5.6 Materials yield with strain hardening (left) and strain softening (right).

The graph to the right side in Fig. 5.6 is typical of a material that yields and then strain softens. As the load increases, the stress increases linearly until the material begins to yield at σ_{yield}. At that point, the uniaxial member continues to stretch, but the load and stress remain essentially constant. After some degree of post-yield stretch, the material softens and the load and stress begin to decrease with a continued increase in the deformation. The strain increases until the wire ruptures at σ_{fail}.

Strength and Safety Factor

We have defined two different strengths (i.e. yield and ultimate tensile) in the previous section. Since a structural member may fail by excessive deformation, we may choose to limit the stress applied to the structure so that it is less than the yield strength S_y. On the other hand in some structures, we can tolerate plastic deformation in one or more members without compromising the structure's function. In these cases, we can tolerate stresses exceeding the yield strength, S_y but they must be less than the ultimate strength, S_u. For example, when designing a bridge, it is important that its shape remain fixed under normal service loads. In this case, the yield strength is specified as the maximum limit for the design stress in the bridge to prevent post-yield deformations. However, in designing a bridge to remain standing during a collision with a freighter or during an earthquake, the ultimate strength of the structural members is the important criterion.

In designing a structural member to carry a specified load, we always size the member so that the **design stress**, σ_{design}, is less than the strength based on either the yield or failure criteria. It would not be prudent to permit the design stress to equal the strength of the member. To size our structural members so they are safe, we usually employ a **factor of safety, SF** in the analysis. We define the safety factor as the ratio of a strength divided by the design stress. This definition leads to the two relations given below:

$$\mathbf{SF_u} = S_u/\sigma_{design} \tag{5.11}$$

$$\mathbf{SF_y} = S_y/\sigma_{design} \tag{5.12}$$

where S_u the ultimate strength of the material, in units of MPa or psi; S_y the yield strength of the material, in units of MPa or psi; σ_{design} the design stress given by P_{design}/A, in units of MPa or psi; P_{design} the load that is specified in designing the structural member, in N or lb.

Again, the value for the factor of safety depends upon the application. Ideally, you would like the factor of safety to be as large as possible. However, this desire for excessive safety factors must be balanced by practical considerations such as economics, aesthetics, functionality, and ease of assembly. In designing, you may be concerned with the cost and weight of its components. The factor of safety you decide to specify should reflect these concerns.

EXAMPLE E5.6

A hoisting cable with a diameter of 7/16 in. is fabricated from many strands of an improved plow steel wire. The manufacturer of the wire certifies its breaking load as 16,500 lb. Determine the strength of the wire used in the manufacture of the cable. Also discuss the assumption pertaining to the wire rope made in the analysis and its implication on the strength of the strands of wire.

Solution: Recall Eq. (5.9), which gives:

$$S_u = \sigma_{fail} = P_{fail}/A$$

The cross sectional area of the cable will be less than $A = \pi d^2/4 = 0.1503$ in^2. Substituting this value for the area into Eq. (5.9) yields:

$$S_u = 16{,}500 \text{ lb}/ 0.1503 \text{ in.}^2 = 109.8 \text{ ksi}$$

We have assumed the cross sectional area of the wire rope to be equivalent to that of a solid wire 7/16 in. in diameter. Wire rope is made of many very small diameter wires that are twisted together to form strands. The strands in turn are formed in a helix about a fiber core. The wire rope in this example has a designation of 6 × 19 (6 strands with 19 small wires in each strand). The cross section of this 6 by 19-wire rope is illustrated below.

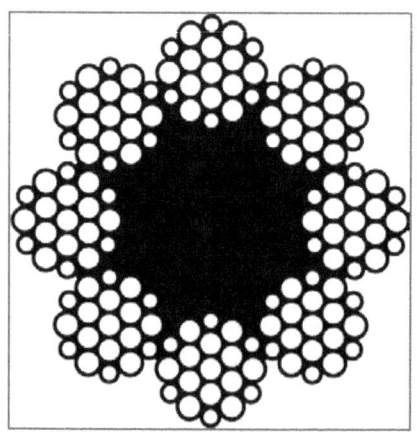

Fig. 5.7 A wire rope is fabricated from many strands of small-diameter high-strength wire.
By Baran Ivo - Own work, Public Domain,
https://commons.wikimedia.org/w/index.php?curid=3132754

The assumption of a solid cross sectional area overestimates the cross sectional area, A by a factor of more than two. The ultimate tensile strength of the small diameter wire used to form the strands of wire for a typical cable is in excess of 200 ksi.

EXAMPLE E5.7

A solid hard drawn copper wire exhibits an ultimate tensile strength of 390 MPa. If the diameter of the wire is 1.6 mm, determine the load required to break the wire.

Solution: Recall Eq. (5.9), which gives:

$$S_u = \sigma_{fail} = P_{fail}/A$$

Rearrange this relation and substitute known quantities to give:

$$P_{fail} = S_u A = (390 \times 10^6 \text{ N/m}^2)(\pi/4)(1.6 \text{ mm})^2 \times 10^{-6} \text{ m}^2/\text{mm}^2 = 784.1 \text{ N}$$

An axial load of approximately 800 N is required to rupture the hard drawn copper wire.

EXAMPLE E5.8

A monofilament, nylon fishing line is rated at 10-lb test. If the line is 0.012 inch in diameter, determine the strength of the nylon in the form of a small diameter line. Comment on the effect that the filament geometry has on the strength of a polymer like nylon-6/6.

Solution: Recall Eq. (5.9), which gives:

$$S_u = \sigma_{fail} = P_{fail}/A$$

$$S_u = (4)(10)/[\pi (0.012)^2] = 88{,}420 \text{ psi}$$

Monofilaments of polymers like nylon-6/6 are drawn from a melt into thin fibers or lines. In this process, the long molecules of the polymer are aligned with the axis of the filament thus enhancing its strength.

EXAMPLE E5.9

A No. 14 gage (0.080 in. diameter) black annealed steel wire is listed in a material handbook with a yield strength of 220 MPa and an ultimate tensile strength of 340 MPa. Determine the axial load that will cause the wire to yield.

Solution: Recall Eq. (5.10), and rearrange it to give:

$$S_y = \sigma_{yield} = P_{yield}/A$$

$$P_{yield} = S_y A \quad\quad\quad (a)$$

Since we have mixed units in the problem statement, let's convert the units for the diameter of the wire to the SI system and determine the cross sectional area A in mm².

$$A = \pi d^2/4 = \pi (0.080 \text{ in.})^2 (25.4 \text{ mm/in.})^2 /4 = 3.243 \text{ mm}^2$$

Next, substitute into Eq. (a) to obtain:

$$P_{yield} = S_y A = (220 \text{ MPa})[(N/(mm^2)/MPa)](3.243 \text{ mm}^2) = 713.4 \text{ N}$$

EXAMPLE E5.10

A 5/8 in. diameter stainless steel (type 304) wire rope with a 6 × 19 configuration is listed in a catalog with a breaking load of 35,000 lb. If a safety factor of 2.4 is to be employed in the design of a structure using this type of cable, specify the design load.

Solution: Recall Eqs. (5.9) and (5.11) and combine them to obtain:

$$\mathbf{SF_u} = S_u / \sigma_{design} = AS_u / P_{design} = P_{fail} / P_{design} \qquad (a)$$

Solving Eq. (a) for the design load P_{design} gives:

$$P_{design} = P_{fail} / \mathbf{SF_u} = 35,000 / 2.4 = 14,580 \text{ lb}$$

EXAMPLE E5.11

A single #20 gage steel wire with a solid cross section is 0.0348 in. in diameter. It is to carry a load of 100 N with a safety factor of 2.8. Specify the steel alloy from which the wire should be manufactured to avoid failure by rupture.

Solution: Recall Eq. (5.11) and solve it to obtain an expression for S_u:

$$\mathbf{SF_u} = AS_u / P_{design}$$

$$S_u = (P_{design})\mathbf{SF_u}/A = 4(100 \text{ N})(2.8)/[\pi (0.0348 \text{ in.})^2 (25.4 \text{ mm/in.})^2] = 456.3 \text{ MPa}$$

It is necessary to refer to a materials handbook to select a suitable material for this application. Reference to Appendix B-2 indicates three steel alloys and four stainless steel alloys with ultimate tensile strengths exceeding the requirement for S_u in this example.

EXAMPLE E5.12

A cold drawn alloy steel wire, with a yield strength of 125,000 psi, is to carry a load of 600 lb while incorporating a safety factor of 2.25. You are to inform the purchasing representative regarding the size of the wire to be ordered.

Solution: To determine the size of the wire in this application, recall Eq. (5.12).

$$\mathbf{SF_y} = S_y / \sigma_{design} = AS_y / P_{design}$$

Solve this relation for the area to obtain:

$$A = \mathbf{SF_y} (P_{design})/S_y = \pi d^2/4$$

$$A = (2.25)(600)/125,000 = 0.0108 \text{ in}^2 = \pi d^2/4$$

$$d = 0.1173 \text{ in.}$$

> Do you call the purchasing representative and tell her to order a cold drawn alloy steel wire with a diameter of 0.1173 in? **No!!!** If you make this mistake, you will soon learn that steel wire is available only with a standard diameter. Economic laws dictate that the most cost effective way to mass-produce materials is by limiting inventories to only those sizes and geometries that satisfy most customers. In Appendix A standard wire sizes and the corresponding gage numbers that are used to order them are listed. For this example, four entries from the listings in Appendix A for standard steel wire are presented below:
>
> | Gage No.[5] | Diameter (in.) | Area (in^2) |
> |---|---|---|
> | 10 | 0.1350 | 0.01431 |
> | 11 | 0.1205 | 0.01140 |
> | 12 | 0.1055 | 0.00874 |
> | 13 | 0.0915 | 0.00658 |
>
> The response to the purchasing representative is to procure Gage No. 11 wire. It is important for you to specify a standard size in ordering structural members. Standard sizes are available with minimum delay at the lowest cost. Analyses rarely give dimensions for structural members that correspond to the available standard sizes. The usual practice is to increase the size determined in the analysis to the next larger dimension available as a standard. This approach enhances safety while minimizing cost and delivery time.

5.3.1 Design Analysis of Rods and Bars

All of the equations [(5.1) to (5.12)], derived in Section 5.3.1 for cables, are also valid for rods and bars. If the external axial forces are tensile (tend to pull the bar apart), the internal forces and stresses are tensile and denoted with a positive sign. On the other hand, if the external axial forces are compressive (tend to squash the bar), the internal forces and stresses are compressive and denoted with a minus sign.

We must qualify the capability of a rod or bar to carry compressive loads. If the rod is very long and slender and the compressive force too high, the rod may buckle. **Buckling** is an unstable condition, and if the critical load is exceeded, the rod fails suddenly and catastrophically. In this chapter, we will assume that the rods or bars loaded in compression are sized to resist buckling; consequently, they fail due to excessive compressive stresses. However, the tendency for these structures to buckle cannot always be ignored. Theories describing the failure of **columns** due to buckling will be introduced in a later chapter.

Centroids

The relations derived for the stresses and deformations of cable assumed that the internal and external forces acted through the **centroid** of its cross sectional area. For cables or wire, with circular cross sections, the location of the centroid is obvious—it is at the center of the circle. However, for cross sections of more complex shapes, the location of the centroid is not obvious.

A centroid is the point coinciding with the center of gravity of a two-dimensional area. For the circular cross section shown in Fig. 5.8, the center of the circle clearly locates the centroid. The center also serves as the origin for the centroidal axes x_c and y_c. For cross sectional shapes such as ellipses, circles, squares and rectangles, the center may be located by inspection because these shapes are all symmetric about both horizontal and vertical axes. The centroid is located at the intersection of the two

---

[5] There are several different standards that refer to gage numbers for wire and steel sheet. In this example we list the American Steel and Wire Co. standard that is commonly used for steel wire.

112 — Chapter 5 — Axially Loaded Structural Members

axes of symmetry. However, for non-symmetric figures, such as a triangle, a portion of a circular area, a parabolic area etc., locating the center of the gravity of the area by inspection is not possible.

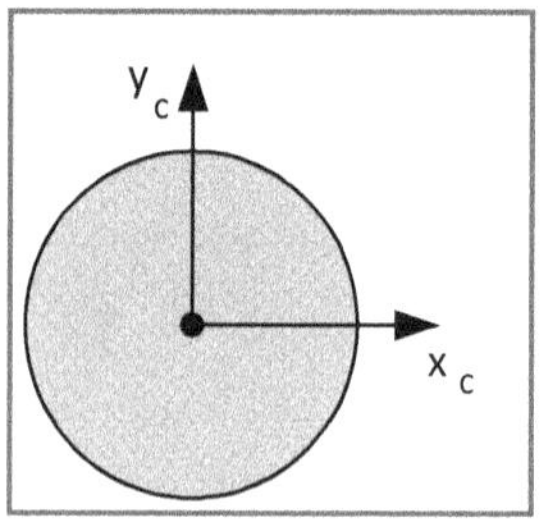

Fig. 5.8 A circular cross section with the center as the centroid and the centroidal axes x_c-y_c.

One approach for determining the location of a centroid is to consider the area as a flat plate positioned in the x-y plane with its outer normal vector in the z direction as illustrated in Fig. 5.9. Let's examine Fig. 5.9a and observe that the weight W of the plate acts downward through the **center of gravity** defined by the point G_c. The center of gravity and the centroid will be located at the same point for a plane body with uniform thickness and density. While the gravitational forces are distributed over the area of the plate, we represent the total weight of the plate as a concentrated force of magnitude W that acts downward though the center of gravity.

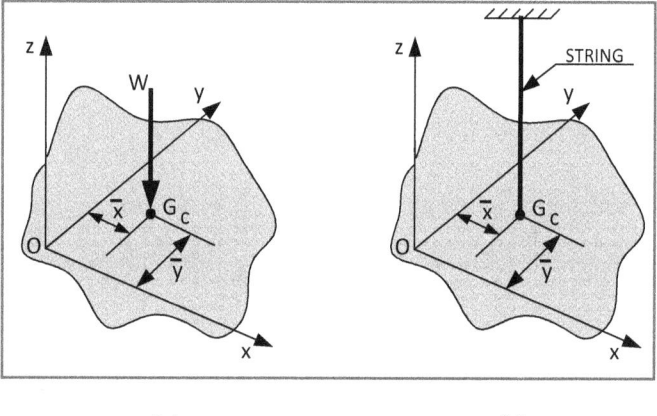

Fig. 5.9 A flat plate of arbitrary area with its weight acting though its center of gravity G_c.

(a)   (b)

The center of gravity may be established experimentally or by analysis. The experimental approach is illustrated in Fig. 5.9b. Suppose we represent the plate with a sheet of cardboard cut to the shape of the arbitrary cross section. We scribe a Cartesian coordinate system on the cardboard to provide a reference for measuring the location of the center of gravity G_c. A string is fixed to the cardboard model at some point P near its center, and the model is suspended in space hanging from this string. In most cases, we cannot precisely estimate the location of the centroid of the model, and the model flops over onto its side. However, when the points P and G_c coincide, the model will hang from the string with its plane horizontal and with its outer normal (the z axis) parallel to the string.

Another experimental method for locating the center of gravity (also the centroid) is depicted in Fig. 5.10. Again we begin with a cardboard model of the area in question. With this approach we punch three small holes at points (A, B and C) located near the boundary of the model as illustrated in Fig. 5.10 a. We hang the model from a pin inserted through the hole at point A, and draw a straight line vertically downward from point A as shown in Fig. 5.10b. A plumb line or a level is used to aid in drawing a perfectly vertical line. We repeat this process with the model suspended from points B and then C as indicated in Figs. 5.10c and 5.10d. The three straight lines intersect; thus, locating the center of gravity or the centroid at a point G_c.

These experimental methods are effective in locating the position of the centroid and give an insight to the meaning of the centroid, center of gravity and the **first moment of an area**. There is one qualification that must be imposed on the fabrication of the model—the sheet of cardboard must be homogenous (of uniform thickness and density).

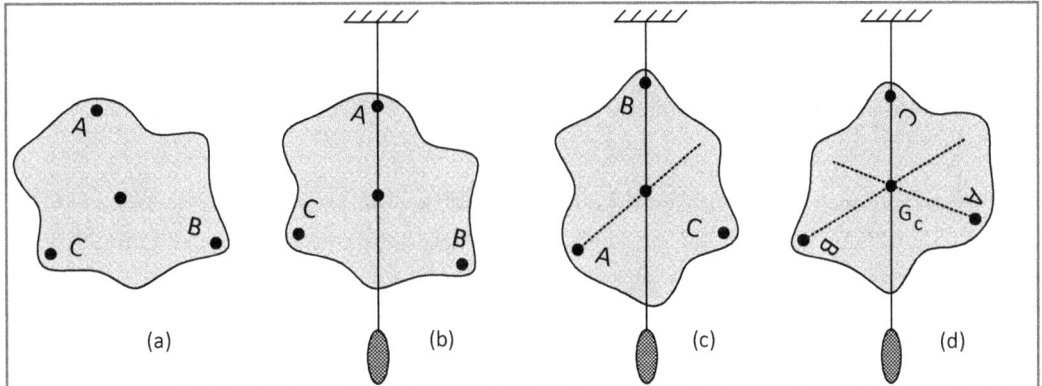

Fig. 5.10 Experimental determination of the centroid of an arbitrary area.

Analytical methods for locating the centroid of an arbitrary shaped area are covered in Appendix C.

Stresses in a Uniform Bar or Rod

The procedure for determining the normal stresses σ is the same as described previously for wires and cables except for the fact that the internal forces may be compressive or tensile. We begin by constructing a FBD to show the point of application and the directions of the internal and external forces. From the equilibrium relations, we determine the internal forces and their sign. Finally, Eq. (5.4) is employed to determine the normal stresses σ. We demonstrate this procedure with the example problems presented below.

EXAMPLE E5.13

Determine the stress in the bar shown in Fig. E5.13 when subjected to a compressive axial force of F that acts through the centroid of the bar. The following numerical parameters define the bar and the applied load: F = 800 kN, L = 0.500 m, w = 100 mm and h = 50 mm.

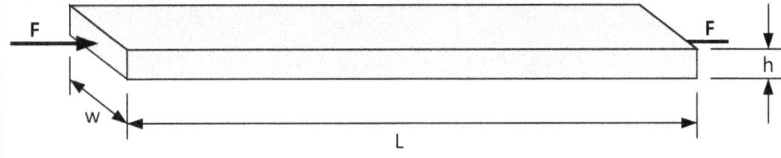

Fig. E5.13

Solution: We make a section cut near the center of the bar, and construct the free body diagram shown below:

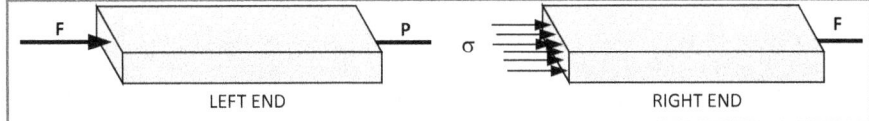

The stresses σ visible on the face exposed by the section cut are due to the internal force P. It is clear from $\Sigma F_x = 0$ that P = F; then from Eq. (5.4), we write:

$$\sigma = P/A = -800 \times 10^3 \text{ N}/(100)(50)\text{mm}^2$$

$$\sigma = -800/5 = -160 \text{ N/mm}^2 = -160 \text{ MPa}$$

The axial stress is a negative number (− 160 MPa) indicating that a compressive stress develops in the bar due to the compressive axial loading. Also observe that the value of the stress is independent of the material from which the bar is fabricated.

EXAMPLE E5.14

Determine the stress in a two-foot long rod that has a diameter of 4 in. The bar is fabricated from mild steel with a yield strength $S_y = 30,000$ psi, and is subjected to an axial compressive load of 100,000 lb. Also, compute the safety factor of the bar against failure by yielding.

Solution: Again, we begin with a drawing of the rod and a free body diagram in Fig. E5.14 showing the forces F and P and the normal stresses σ.

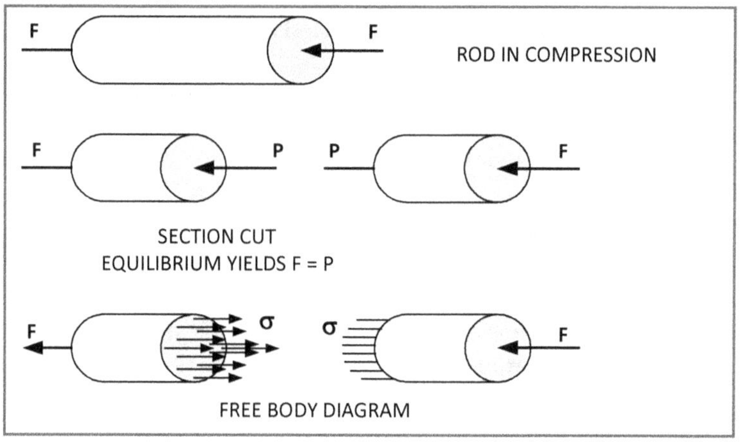

Fig. E5.14

From Eq. (5.4), we write:

$$\sigma = P/A = -100,000 \text{ lb}/[\pi \times (2)^2] = -7,958 \text{ psi}$$

We determine the safety factor against yielding from Eq. (5.12).

$$SF_y = S_y/S_{design} = -30,000/-7,958 = 3.77$$

Note, the safety factor is always a positive quantity. In this problem, we compared a compressive strength of − 30,000 psi (taken as a negative quantity) with the design stress of −7,958 psi.

Deflection of Axially Loaded Bars and Rods

The deformation of rods and bars may be described using Eqs. (5.1) and (5.2), as illustrated in the following examples.

EXAMPLE E5.15

If the bar described in Example 5.13 is fabricated from steel, determine the length of the bar after the application of the compressive force.

Solution: An inspection of the free body diagram shown in Fig E5.13a indicates that the internal force P is constant from one end of the bar to the other and equal to the external force F. Using Eqs. (5.1) and (5.2), we may write:

$$\delta = L_f - L_0 = PL_0/(AE) \tag{a}$$

Solving Eq. (a) for L_f, gives:

$$L_f = L_0 [1 + P/(AE)] \tag{b}$$

Substituting numerical values for the known parameters in Eq. (b) yields:

$$L_f = 0.5 \{1 - [(800 \times 10^3)/(0.05)(0.1)(207 \times 10^9)]\}$$

$$L_f = 0.5[1 - 0.7729 \times 10^{-3}] \text{ m} = (500 - 0.3864) \text{ mm} = 499.6 \text{ mm} \tag{c}$$

Since the new length of the rod is 499.6 mm, it is apparent that it contracted by 0.4 mm under the action of the compressive load. The value of P in Eq. (5.2) is treated as a negative number when the axial load on the bar is compressive. Also, the value of the modulus of elasticity was taken as 207 GPa from Appendix B.

EXAMPLE E5.16

Determine the axial deflection of the rod described in Example 5.14. Repeat the solution for the rod if it is fabricated from an aluminum alloy.

Solution: An inspection of the free body diagram in Fig. E5.14 indicates that the internal force P is constant over the entire length of the rod. From Eq. (5.2), we write:

$$\delta = PL/(AE) \tag{a}$$

Substituting the values for the known quantities for the steel rod in Eq. (a) gives:

$$\delta = -[(100,000)(2)(12)]/[(\pi)(2)^2 (30 \times 10^6)] = -0.006366 \text{ in.}$$

Substituting the values for the known quantities for the aluminum rod in Eq. (a) gives:

$$\delta = -[(100,000)(2)(12)]/[(\pi)(2)^2 (10.4 \times 10^6)] = -0.01836 \text{ in.}$$

Both solutions carry a negative sign indicating that the rod is compressed (shortened) by the action of the compressive force. It is interesting to observe that the rod fabricated from an aluminum alloy exhibited nearly three times the deflection of the steel rod. The reason for this difference is in the lower modulus of elasticity of aluminum relative to steel. The modulus of elasticity for aluminum and steel used in this calculation was 10.4×10^6 and 30×10^6 psi, respectively as cited in Appendix B.

5.4 SHEAR STRESSES

In the preceding discussions, the stresses created by internal forces were normal to the area exposed by the section cut. To emphasize this fact, we called these **normal stresses**. A second type of stress exists—a **shear stress**. As the name implies, the shear stress lies in the plane of the area exposed by the section cut. To show shear stresses in a more graphical manner, consider the stubby beam-like member loaded with the force F in Fig. 5.11.

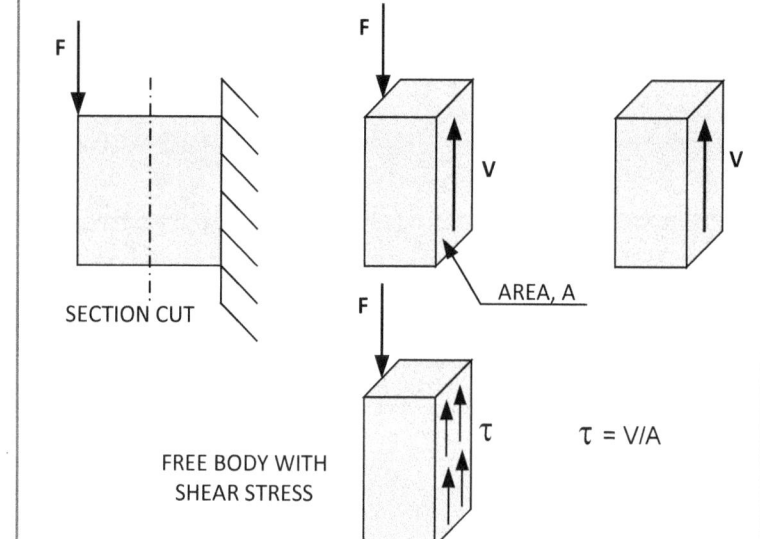

Fig. 5.11 A section cut made in a stubby beam-like member produces a free body.

First, cut the stubby member to create a free body of the left end, and then apply an internal shear force V in the plane of the area exposed by the section cut[6]. From $\sum F_y = 0$, we determine that F = V. The shear force V is produced by a shear stress τ. The relation between shear stress τ and the shear force V is:

$$\tau = V/A \tag{5.13}$$

We have assumed that the shear stress τ is uniformly distributed over the area of the stubby member. Later, in the discussion of beam theory found in Chapter 7, we will note that the shear stress is not uniformly distributed over the cross sectional area of a beam. However, for many block-like members, the assumption of a uniform distribution of shear stresses is a reasonable approximation.

EXAMPLE E5.17

A key is employed to keep a gear from slipping on a shaft when transmitting power. Forces F of 50 kN are created on the key at the locations shown in Fig. E5.17. Determine the shear stresses in the key if it is 8mm wide, 9 mm high and 50 mm long.

> **Solution:** We construct a FBD of the key, presented in Fig. E5.17, showing the equal and opposite forces F and the shear plane between the gear and the shaft. We section the key along the shear plane, and draw another free body diagram of its lower portion. On this second free body diagram, we illustrate the shear stresses τ that occur on the shear plane.
> The shear stresses on the key are given by Eq. (5.13) as:

---

[6]To maintain the focus of the discussion on shear forces, we have not included the internal moment acting at the section cut on the FBD in Fig. 5.11.

$$\tau = V/A = (50 \times 10^3)/(0.008)(0.050) = 125 \text{ MPa}$$

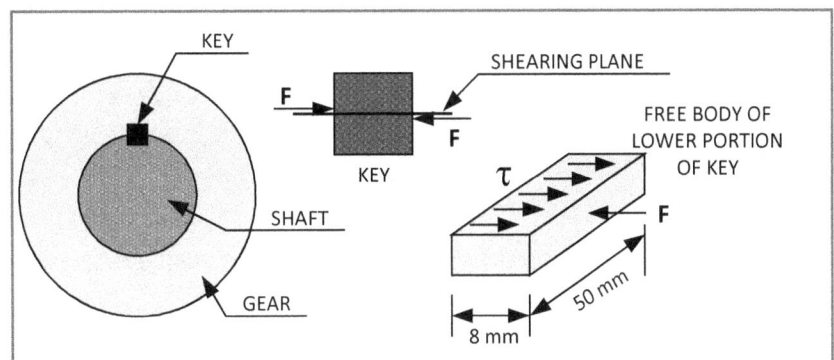

Fig. E5.17 A section cut showing shear forces acting on a key, which prevents the gear from slipping on the shaft

The average shear stress acting on the key is 125 MPa. To analyze the impact of this solution on the design of a structure, it is necessary to compare the imposed stress with the shear strength of the material from which the key is fabricated. The shear strength is usually lower than the yield or ultimate tensile strength of a material. A common practice is to consider either the yield or tensile strength of a material and multiply that value by 0.577 to estimate the shear strength. For example, if the tensile yield strength, S_y of the steel used in manufacturing the key was 320 MPa, the yield strength in shear S_{ys} is estimated as:

$$S_{ys} = 0.577 \, S_y = (0.557)(320) = 184.6 \text{ MPa}$$

The safety factor for the key is determined from:

$$SF = S_{sy}/\tau = 184.6/125 = 1.477$$

In interpreting the results for normal stresses, we recognize the difference between tensile (+) and compressive (−) values for the results. The reason for the careful distinction is the fact that the strength of material subjected to tensile or compressive stresses is often different. However, shear strength is not sensitive to the sign of the shear stress τ. The strength of a material to an imposed shear stress is not dependent on the direction of the imposed shear force. For this reason, we neglect the sign of the shear force in our analysis of stress on machine components.

5.5 STRESSES ON OBLIQUE PLANES

In making section cuts on bars and rods, we have limited our choices to sections perpendicular to the axis of the bar. This restriction was helpful because it simplified the state of stress that was examined. With these perpendicular section cuts, we considered only the normal stresses acting on the area exposed by the cut.

We showed that shear stresses exist in Section 5.4 and computed their magnitude in Example 5.17. In this case, we restricted the section cut to one parallel to the imposed system of equal and opposite forces. With this restriction on the section cut, only shear forces are developed. Clearly, by restricting the section cut we develop special cases where normal stresses exist in the absence of shear stresses and vice versa.

Let's treat the more general case with the bar cut at an arbitrary angle as shown in Fig 5.12. Let's cut the free body of Fig. 5.12 with still another section to produce a triangular shape as shown in Fig. 5.13a. The triangular shape in Fig. 5.13b is defined with the angle θ. The left side of this triangle has an area A; therefore, the hypotenuse of the triangle has an area = A/cos θ. Note, the x-axis is coincident with the axis of the bar; the n axis is coincident with the outer normal to the inclined cut; the t axis is tangential to the inclined cut.

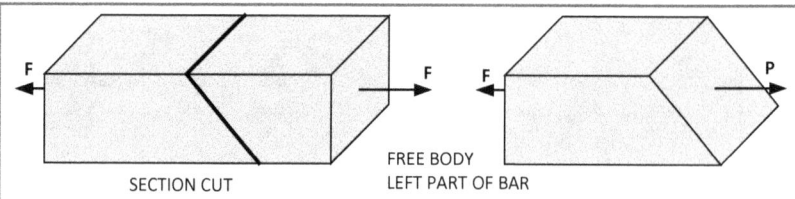

Fig. 5.12 Bar with axial loading with a section cut at an arbitrary angle.

As shown in Fig. 5.13c, the internal force P has been resolved into components along the n and t axes to yield:

$$P_n = P \cos \theta \quad (a)$$

$$V_t = P \sin \theta \quad (b)$$

To ascertain the stresses on the inclined surface, we simply divide either P_n or V_t by the area formed with the inclined section cut in accordance with Eq. (5.4) or Eq. (5.13). The stress σ_θ in the normal direction (n) is determined from Eq. (5.4) as:

$$\sigma_\theta = (P \cos \theta)/(A/\cos \theta) = (P/A) \cos^2 \theta \quad (c)$$

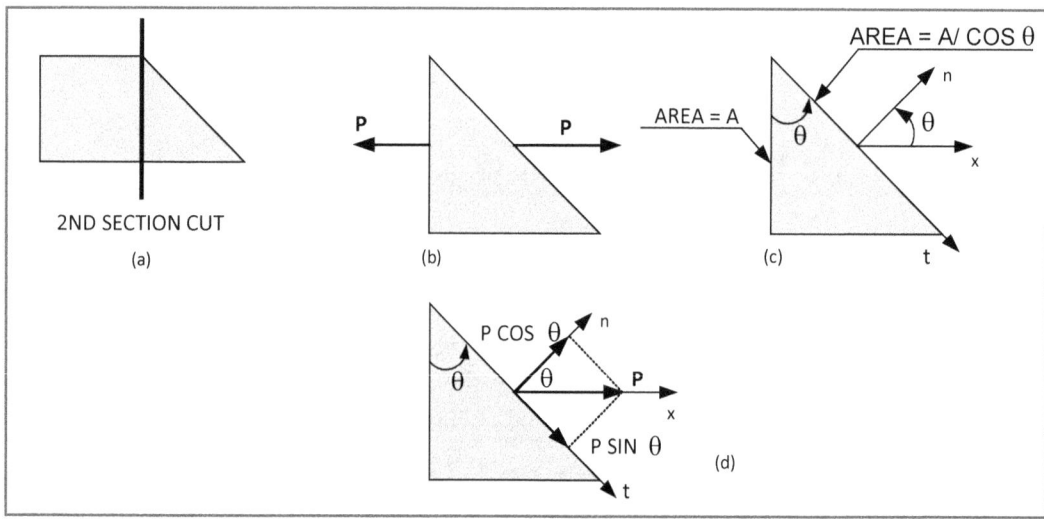

Fig. 5.13 Free body diagrams of the bar with an inclined cut: (a) The section cut is perpendicular to the axis of the bar to produce a triangle; (b) Internal forces acting on both faces of the element; (c) The area of the left side and the hypotenuse is illustrated; (d) The forces on the surface of the inclined cut are resolved into components.

If we define the axial stress $\sigma_x = P/A$, then we may write:

$$\sigma_\theta = (\sigma_x)\cos^2 \theta \quad (5.14)$$

The shear stress τ_θ that acts along the face of the inclined surface is given by Eq. (5.13) as:

$$\tau_\theta = P \sin \theta/(A/\cos \theta) = (P/A) \sin\theta \cos \theta \quad (d)$$

This relation is rewritten as:

$$\tau_\theta = (\sigma_x/2) \sin 2\theta \qquad (5.15)$$

Numerical results are presented in Fig. 5.14 for both σ_θ and τ_θ as the angle of the inclined section varies from zero (a perpendicular cut) to 90° (a parallel cut). In this figure the axial stress $\sigma_x = 100$ units. From Fig. 5.14 it is evident that the normal stress σ is a maximum when $\theta = 0°$ and the section cut is perpendicular to the axis of the bar. For this angle of the cut, $\sigma_\theta = \sigma_x = P/A$. Thus, when employing Eq. (5.4) to compute the normal stress, we determine the maximum possible value of σ_θ. The shear stress is zero on the plane defined by $\theta = 0°$ where the normal stress is a maximum.

Both the normal and shear stress vanish when the section cut is made at 90°. The shear stress is a maximum when the section cut is defined by $\theta = 45°$. When $\theta = 45°$, the normal and shear stresses are equal to each other $\sigma_{45} = \tau_{45} = \sigma_x/2$.

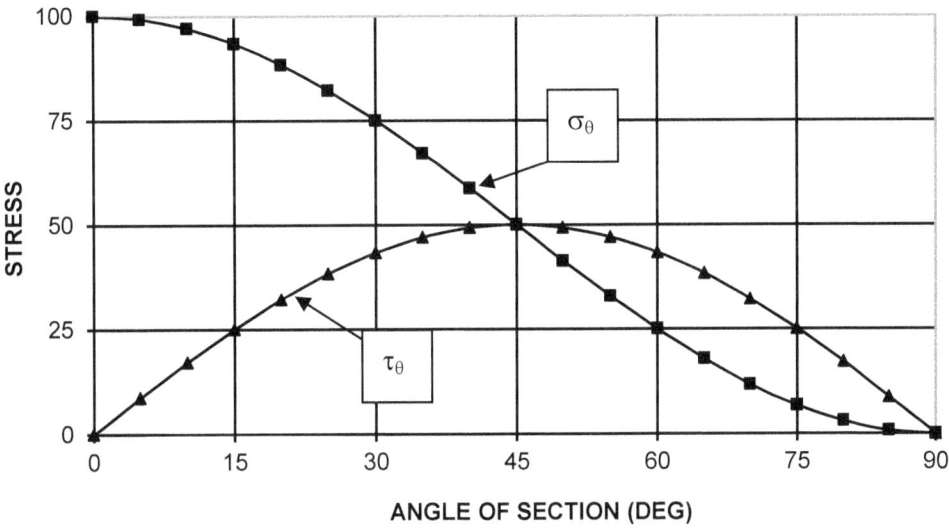

Fig. 5.14 Variation in σ_θ and τ_θ with the angle of the inclined section.

The example of the axially loaded bar, with an inclined section cut, illustrates why stresses must be treated as tensor quantities. As we vary the angle of inclination of the section cut, two parameters are changing relative to either the normal or the shear stresses. First, the magnitude of the force components in the n and t direction changes with the angle θ because forces are vector quantities. Second, the area of the inclined surface exposed by the section cut increases with θ. Both of these parameters affect the magnitude of the normal and shear stresses; therefore stresses must be treated as tensor quantities—not vectors.

EXAMPLE E5.18

A circular rod 30 mm in diameter and 2 m long is subjected to an axial load of 120 kN. Determine:

- The maximum normal stress and the plane upon which it acts.
- The maximum shear stress and the plane upon which it acts.
- Draw a FBD of the section for the maximum normal stress.
- Draw a FBD of the section for the maximum shear stress.

Solution: The maximum normal stress occurs on a plane perpendicular to the axis of the rod where $\theta = 0°$. Equation (5.14) applies.

$$\sigma_{\theta=0} = \sigma_x \cos^2 \theta = (P/A) \cos^2 \theta = (120 \times 10^3)/[\pi (0.015)^2]\cos^2 0° = 169.8 \text{ MPa}$$

The maximum shear stress occurs when $\theta = 45°$ as indicated by the results depicted in Fig. E5.14. From Eq. (5.15), we write:

$$\tau_{\theta = 45} = (\sigma_x/2) \sin 2\theta = (169.8/2) \sin 90° = 84.90 \text{ MPa}$$

The FBDs for the right portion of the bar showing the maximum normal and shear stresses are shown below:

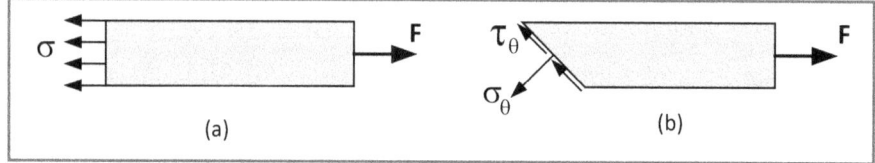

Fig. E5.18 Section cuts for (a) maximum normal stress and (b) maximum shear stress.

The FBD in Fig. E5.18a shows the normal stresses σ acting on a plane area perpendicular to the axis of the rod. The shear stresses are shown in Fig. E5.18b where they act on a surface inclined at a 45° angle to the axis of the rod.

5.6 AXIAL LOADING OF A TAPERED BAR

The discussion of stress and deflection of a rod or bar has been limited to those members with a uniform cross sectional area. Of course, uniform bars and rods are most commonly employed in building structures because they are easy to design with and less expensive to manufacture; however, in some instances it may be more desirable to use members that are not uniform. In these cases, we must accommodate the effect of the changing cross sectional area over the length of the bar on both the stresses and displacement.

5.6.1 Normal Stresses in a Tapered Bar

The normal stresses that occur in a tapered bar or rod are determined using Eq. (5.4). The only consideration made to account for the taper in the bar is to adjust the area A to correspond with the position along the length of the bar. Let's consider the tapered bar presented in Fig. 5.15. The thickness of the bar, b, is constant along its entire length. The height, h_x, of the bar varies with position x according to the relation:

$$h_x = h_1 + (h_2 - h_1)(x/L)$$

The area, A_x, at any position x along the length of the bar is then:

$$A_x = bh_x = b[h_1 + (h_2 - h_1)(x/L)] \qquad (5.16)$$

The normal stresses σ_x due to an axial force are determined by substituting Eq. (5.16) into Eq. (5.4) to obtain:

$$\sigma_x = P/A_x = P/\{b[h_1 + (h_2 - h_1)(x/L)]\} \qquad (5.17)$$

where L is the length of the tapered bar.

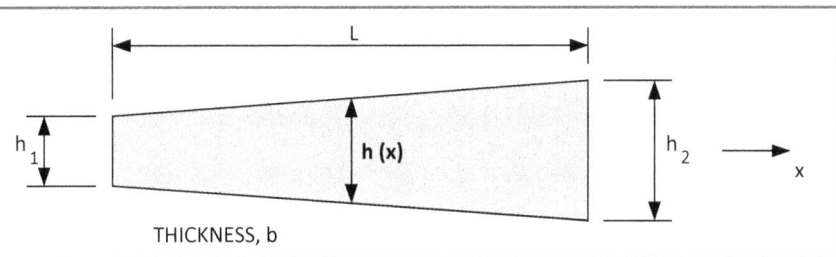

Fig. 5.15 Geometry of a tapered bar with a uniform thickness.

EXAMPLE E5.19

For a tapered bar similar to the one shown in Fig. 5.15, determine the normal stress as a function of position, x along the length of the bar. The geometry of the bar is given by $h_1 = 2$ in., $h_2 = 6$ in., $b = 3$ in. and $L = 36$ in. The axial load imposed on the tapered bar is 5,000 lb.

Solution: We employ Eq. (5.17) and write:

$$\sigma_x = P/\{b[h_1 + (h_2 - h_1)(x/L)]\} = 5,000/\{3[2 + (6 - 2)(x/36)]\} \qquad (a)$$

$$\sigma_x = 15,000/(18 + x)$$

The stress $\sigma_x = 833.3$ psi is a maximum at $x = 0$, and decreases to a minimum value of $\sigma_x = 277.8$ psi when $x = 36$ in.

5.6.2 Deflection of Tapered Bars

The deflection of a tapered rod may be determined from Eq. (5.2) although we must again modify the approach to accommodate for the changing cross sectional area over the length of the bar. To determine the axial deflection, this accommodation is more difficult since the total deformation of the bar is the sum of the incremental deflections at each position x along the entire length of the bar. We begin by considering an incremental length dx at some position x as shown in Fig. 5.16.

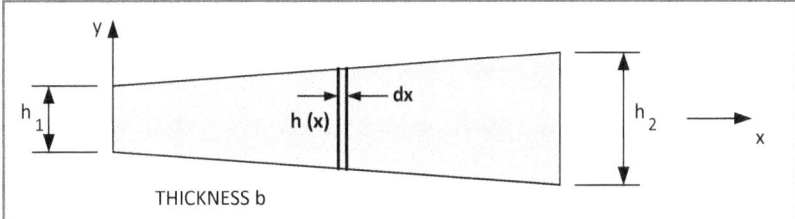

Fig. 5.16 An incremental length (slice) dx at position x along the length of the tapered bar.

To accommodate for the taper in the bar, we determine the incremental deflection dδ of a bar of length dx. Since dx approaches zero, we treat the cross sectional area as a constant over the incremental length. Accordingly, we modify Eq. (5.2) to read:

$$d\delta = P\, dx/(A_x E) \qquad (5.18)$$

where A_x is the cross sectional area of the bar that is a function of x given in Eq. (5.16).

Substitute Eq. (5.16) into Eq. (5.18), simplify and integrate to obtain:

$$\delta = \frac{PL}{Eb}\int_0^L \frac{dx}{h_1 L + (h_2 - h_1)x} \qquad (a)$$

$$\delta = \frac{PL}{Eb}\left(\frac{1}{h_2 - h_1}\right)\ln\left(\frac{h_2}{h_1}\right) \qquad (5.19)$$

EXAMPLE E5.20

Determine the deflection of the tapered bar described in Example 5.19 if the bar is fabricated from an aluminum alloy with a modulus of elasticity $E = 10.4 \times 10^6$ psi.

Solution: Let's solve this example problem by recalling Eq. (5.19).

$$\delta = \frac{PL}{Eb}\left(\frac{1}{h_2 - h_1}\right)\ln\left(\frac{h_2}{h_1}\right)$$

Substituting the parameters describing the geometry and the material constant for the bar into this relation yields:

$$\delta = [(5000)(36)/(10.4 \times 10^6)(3)][1/(6 - 2)]\ln(6/2)] = 1.585 \times 10^{-3} \text{ in.}$$

Examine the magnitude of the deflection of the tapered bar. Is the deflection large or small? What reference do you use to judge? Clearly, the axial extension of the bar is small. For small quantities, we sometimes use the human hair as a reference. The diameter of a single strand of hair is slightly more than 2×10^{-3} in.; hence, the extension is about 75% of a hair diameter.

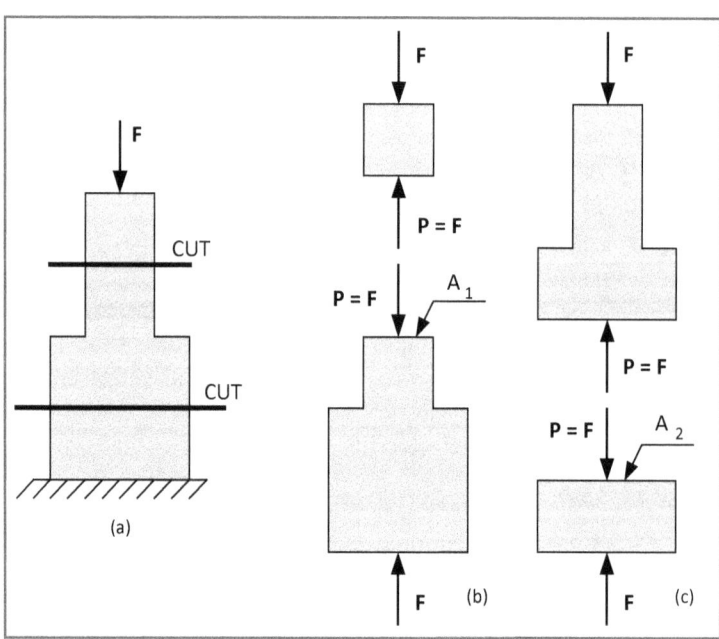

Fig. 5.17 A stepped bar with free body diagrams for each section of the bar.

5.7 AXIAL LOADING OF A STEPPED BAR

In some structures, bars of different cross sectional areas are employed where the area changes abruptly at some position along the length of the bar. An illustration of a stepped bar where the cross section undergoes an abrupt change is presented in Fig. 5.17.

The procedure for determining the normal stresses and deflection remains the same as for a bar with a uniform cross sectional area:

1. Draw a free body diagram of each section of the bar.
2. Use the equilibrium relations to establish the internal axial forces acting on each section.
3. Determine the stresses from Eq. (5.4) using the area of the section of interest.
4. Solve for the deflection of each segment of the bar from Eq. (5.2) and then add them to obtain the total deflection.

5.7.1 Normal Stresses in Stepped Bars

The normal stress varies with the cross sectional area of the bar. For example, in Fig 5.17 the normal stress in the upper section of the bar is given by $\sigma = P/A_1$, and the stress in the lower section of the bar is $\sigma = P/A_2$.

EXAMPLE E5.21

For the stepped bar shown in Fig. E5.21 determine the normal stress in each of the two sections of the bar if F_1 = 40 kN; F_2 = 80 kN; w_1 = 50 mm; b_1 = 60 mm; w_2 = 90 mm; b_2 = 60 mm; L_1 = 200 mm and L_2 = 300 mm

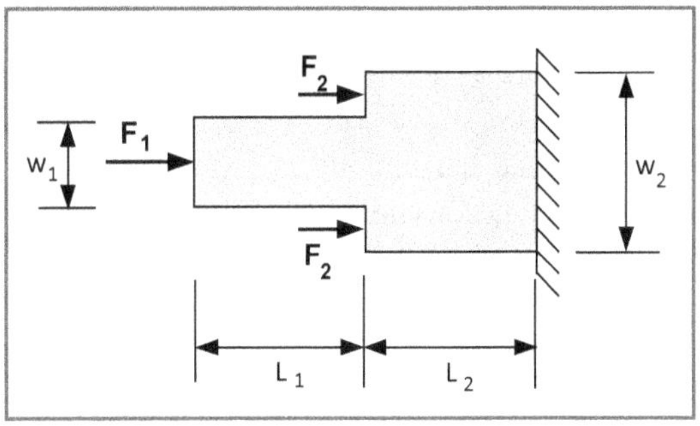

Fig. E5.21 The thickness of the bar is given by b_1 and b_2.

Solution: Let's begin the solution by drawing free bodies of both sections of the bar. We make two section cuts A and B as illustrated in Fig. E5.21a. Then we draw free bodies associated with the portion of the bar to the left side of the section cut as shown in Fig. E5.21a.

From the equilibrium relations, it is clear that the internal force $P_1 = F_1$ in the smaller section of the stepped bar. Also, for the larger section of the bar, the internal force $P_2 = F_1 + 2F_2$.

Finally, from Eq. (5.4) for the smaller section of the bar, we write:

$$\sigma_1 = P_1/A_1 = -(40 \times 10^3)/[(50)(60)] = -13.33 \text{ N/mm}^2 = -13.33 \text{ MPa}$$

For the larger section of the bar, we write:

$$\sigma_2 = P_2/A_2 = -\{[40 + (2)(80)] \times 10^3\}/[(90)(60)] = -37.03 \text{ N/mm}^2 = -37.03 \text{ MPa}$$

Note, the minus sign indicates the stresses in the bar are compressive.

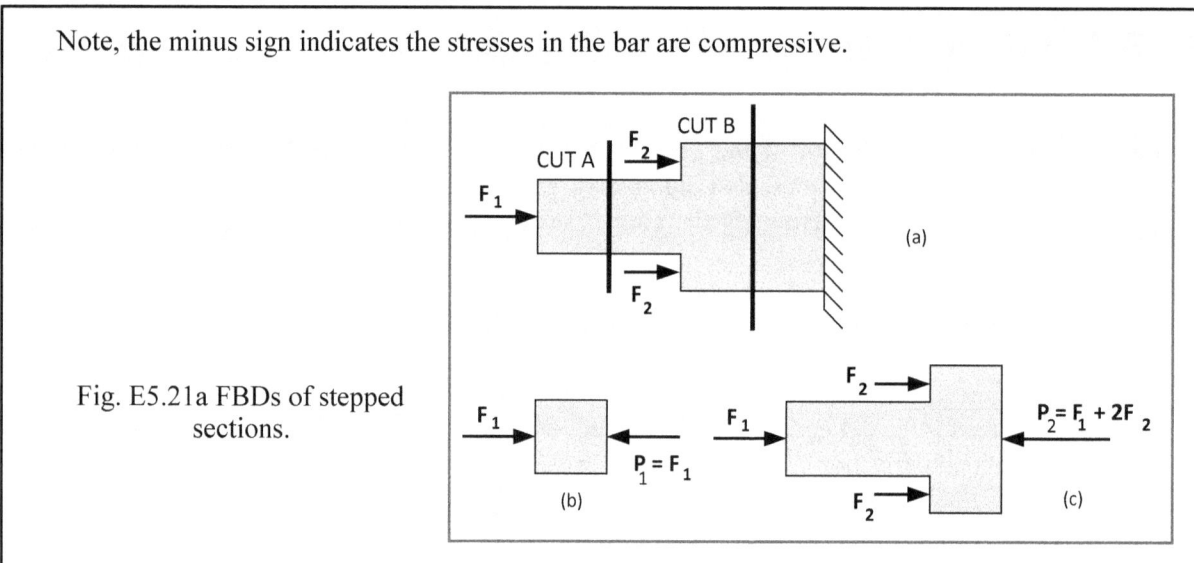

Fig. E5.21a FBDs of stepped sections.

5.7.2 Deflection of Stepped Bars

To compute the axial extension or compression of the stepped bar, we consider each uniform section of the bar separately. Equation (5.2) is valid for each section because each has a uniform cross sectional area. For a stepped bar comprised of n uniform sections, we superimpose the individual deflections δ to obtain:

$$\delta_{total} = \delta_1 + \delta_2 + \delta_3 + \ldots + \delta_n \tag{5.20}$$

EXAMPLE E5.22

If the stepped bar defined in Example 5.21 is fabricated from a titanium alloy with a modulus of elasticity E = 114 GPa, determine the total deflection of the bar.

Solution: We recognize that the deflection of the stepped bar is determined from Eq. (5.20) and Eq. (5.2) as:

$$\delta_{total} = \delta_1 + \delta_2 = (P_1 L_1)/(A_1 E) + (P_2 L_2)/(A_2 E) \tag{a}$$

Since the modulus of elasticity is constant for both sections of the bar, Eq. (a) reduces to:

$$\delta_{total} = \delta_1 + \delta_2 = \frac{1}{E}\left[\frac{P_1 L_1}{A_1} + \frac{P_2 L_2}{A_2}\right] \tag{b}$$

Substituting the numerical parameters for the unknown quantities in this relation yields:

$$\delta_{total} = \delta_1 + \delta_2 = \frac{1}{114 \times 10^9}\left[\frac{(-40 \times 10^3)(0.2)}{(0.05)(0.06)} + \frac{(-200 \times 10^3)(0.3)}{(0.09)(0.06)}\right] = -0.1209 \text{ mm}$$

> Let's interpret this solution. The negative sign indicates the bar was compressed and the deformations reduced its length. The original length of the bar was $L_1 + L_2 = 500$ mm. When the total deformation of the bar is compared to this length, we find the deformation is very small—only 0.024%. This example again emphasizes that deformations of metallic members are usually very small. For this reason, we neglect these deformations and use the original lengths of structural members when substituting into the equilibrium equations.

5.8 STRESS CONCENTRATIONS

In our discussion of stresses in uniform, tapered or stepped bars, we have assumed that the stresses are uniformly distributed over the cross sectional area of the bar. For the uniform and the tapered bars, this is a valid assumption except near the ends of the bar where the external forces are applied. However, for the stepped bar, the stresses are not uniformly distributed across the section in the vicinity of the step. Indeed, whenever we encounter a discontinuity such as a step or a hole in a bar, the stresses tend to concentrate at that discontinuity. As a consequence, the use of Eq. (5.4) to compute the stresses seriously underestimates their actual value. We must account for the effect of the structural discontinuities by determining a suitable stress concentration factor. The topic of stress concentration factors will be introduced in the subsection below. However, since the topic is extremely important, the reader is referred to an excellent book by R. E. Peterson for more details [1].

5.8.1 Stress Concentration Factors Due to Circular Holes in Bars

Let's consider the bar with a centrally located circular hole subjected to an axial tension force as shown in Fig. 5.18.

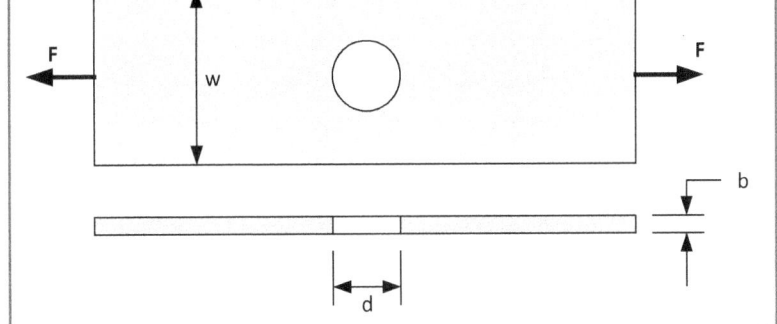

Fig. 5.18 A centrally located circular hole in an axially loaded bar.

The stress distribution in a section removed three or more diameters from the hole is uniform with a magnitude given by Eq. (5.4) as $\sigma_o = P/(bw)$. However, on the section through the center of the hole the stress distribution shows significant variation. The stresses increase sharply adjacent to the discontinuity (the hole) and concentrate at this location. The maximum value of the normal stress occurs adjacent to the hole as indicated in Fig 5.19

We are interested in determining the maximum stress, σ_{MAX} adjacent to the hole. It is convenient to express the maximum stresses in terms of a stress concentration factor by employing:

$$\sigma_{MAX} = K\, \sigma_{NOM} \qquad (5.21)$$

where K is the stress concentration factor and σ_{NOM} is the nominal stress.

The **nominal stress** is the average stress across the net section containing the hole, and is given by:

$$\sigma_{NOM} = P/A_{NOM} = P/[(w - d)b] \qquad (5.22)$$

where b is the thickness of the bar, w is the bar width and d is the hole diameter.

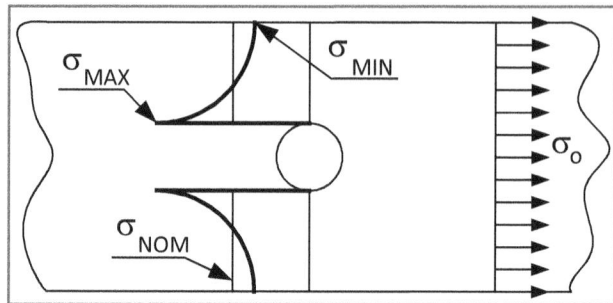

Fig. 5.19 Distribution of stress across a section through the center of the hole shows the concentration of stresses adjacent to the boundary of the hole.

The uniform stress σ_o and the nominal stress σ_{NOM} are related by:

$$P = \sigma_{NOM} A_{NOM} = \sigma_o A_{UNF} \qquad (a)$$

Substituting for the areas in Eq. (a) and simplifying yields:

$$\sigma_{NOM} = [w/(w - d)]\sigma_o \qquad (5.23)$$

The nominal stress is always greater than the uniform stress since the factor $w/(w - d)$ is always greater than one.

The **stress concentration factor** K for a uniform thickness bar with a central circular hole subjected to axial loading is a function of the geometry depending on the ratio of d/w as shown in Fig. 5.20.

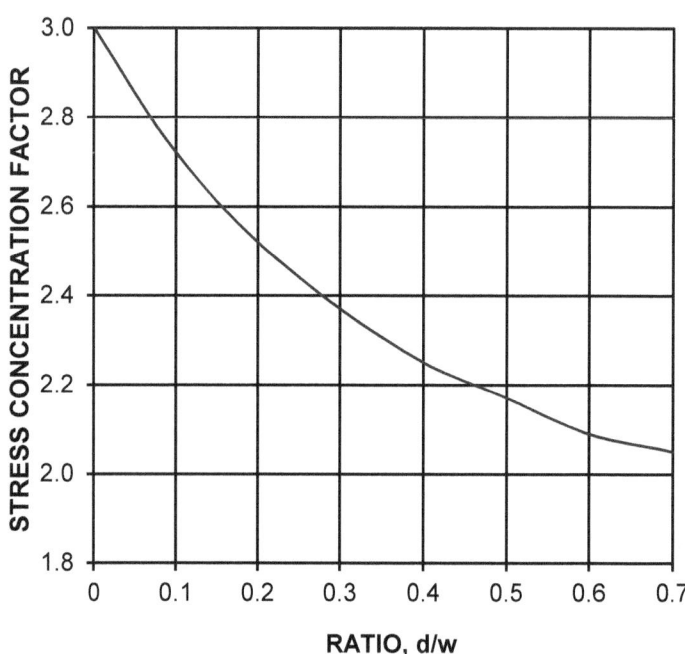

Fig. 5.20 Stress concentration factor for a central circular hole in an axially loaded bar.

EXAMPLE E5.23

A centrally located hole of diameter d = 0.5 in. is drilled in a long thin bar that is subjected to an axial load of 8,000lb. If the bar is defined by w = 1.5 in., and b = 0.50 in., determine the nominal stress, the stress concentration, and the maximum stress.

Solution: The nominal stress is given by Eq. (5.22) as:

$$\sigma_{NOM} = P/[b(w-d)] = 8,000/[(1.5-0.5)(0.5)] = 16,000 \text{ psi}$$

The stress concentration K is determined from Fig. 5.20, by locating the intercept of a vertical line originating at d/w = 0.5/1.5 = 0.3333 with the curve. This intercept gives K = 2.33.

Finally, the maximum stress is given by Eq. (5.21) as:

$$\sigma_{MAX} = K\, \sigma_{NOM} = (2.33)(16,000) = 37.28 \text{ ksi}$$

By drilling a hole in the bar, we increase the maximum stress significantly. The procedure using the stress concentration factor provides a simple yet effective approach for solving a very difficult stress analysis problem. Also, you should be aware that the uniform stress in the bar prior to drilling the hole was σ_o = 10.67 ksi; therefore, the presence of the hole increased the stresses by a factor of 3.50.

5.8.2 Stress Concentrations in Shouldered Bars

The stepped bar is another configuration with an abrupt change in the section of the bar. This geometric discontinuity produces a non-uniform distribution of stresses and a concentration of stress at the fillet used in transitioning from one section of the bar to the other. The geometric parameters involved in characterizing the stress concentration factor are w_2/w_1 and r/w_1 as defined in Fig. 5.21.

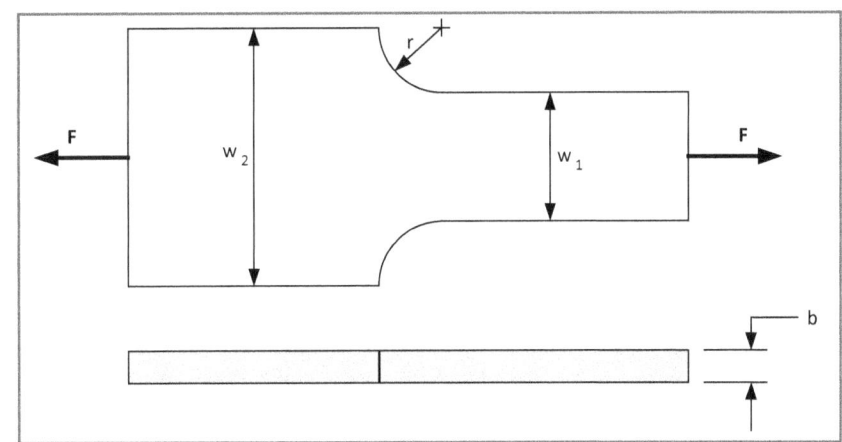

Fig. 5.21 Geometric parameters controlling the stress concentration in a stepped bar.

The maximum stress occurs at the fillet in the transition from the narrow section of the bar (w_1) to the wider section (w_2). The maximum stress is determined from Eq. (5.21). However, for the case of the stepped bar the nominal stress is defined as:

$$\sigma_{NOM} = P/(w_1 b) \tag{5.24}$$

The stress concentration factor K for the stepped bar depends on two different ratios r/w_1 and w_2/w_1. Curves showing the stress concentration factor K as a function of these ratios are presented in Fig. 5.22.

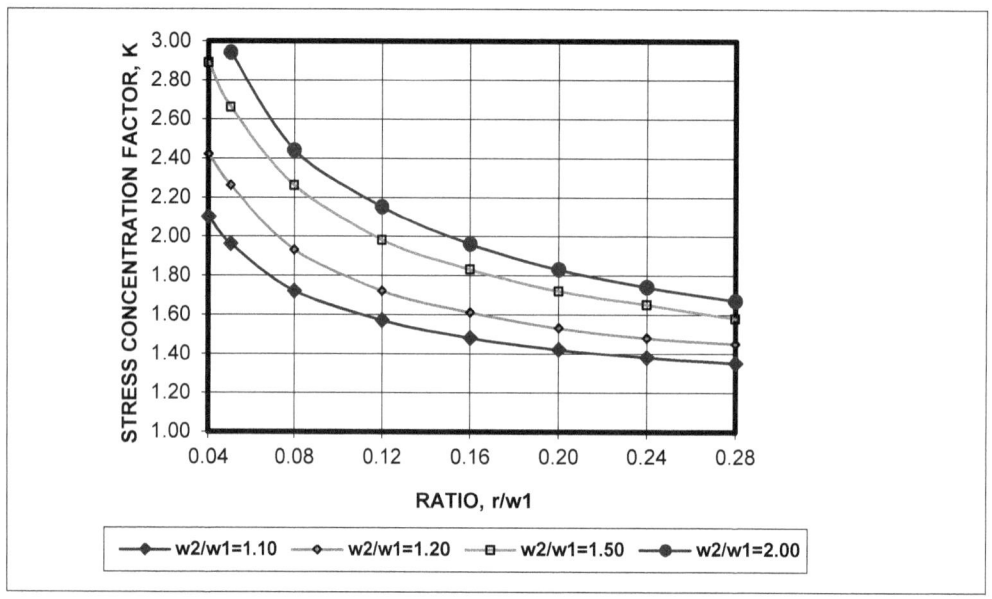

Fig. 5.22 Stress concentration factor for a stepped bar subjected to axial forces.

EXAMPLE E5.24

Consider the axially loaded stepped bar shown in Fig. E5.24, and determine the maximum stress at the fillet in the transition region of the bar.

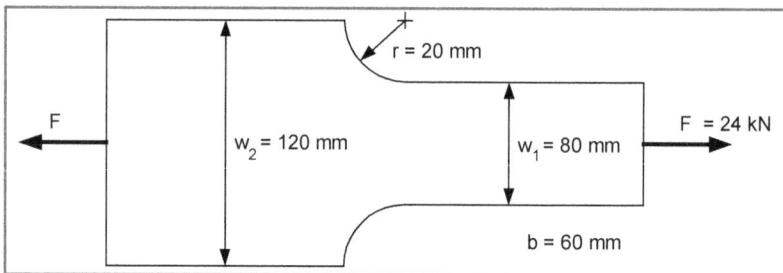

Fig. E5.24

Solution: From Eq. (5.21), we write $\sigma_{MAX} = K\, \sigma_{NOM}$.

From Eq. (5.24), write:

$$\sigma_{NOM} = P/w_1 b = (24 \times 10^3)/[(80)(60)] = 5 \text{ MPa}$$

To determine the stress concentration factor K, we note that $r/w_1 = 20/80 = 0.25$, and $w_2/w_1 = 120/80 = 1.5$. Reading the chart for the stress concentration factor in Fig. 5.22, we find that K is:
$$K = 1.65 - (1/4)(1.65 - 1.58) = 1.63.$$

Finally, Eq. (5.21) yields:

$$\sigma_{MAX} = K\, \sigma_{NOM} = 1.63 \times 5 = 8.15 \text{ MPa}$$

5.9 SCALE MODELS

We have described methods for determining the stresses, strains and deflection of bars and rods subjected to axial tension or compression forces. The analysis required to determine these quantities is relatively simple when the bar can be isolated and the forces acting on it are known. However, if the bar is part of a complex structure, determining the forces acting on the bar may prove to be more difficult. In some cases, scale models of the structure under consideration are constructed and then subjected to various loads to verify structural integrity.

Geometric Scale Factor

Scale models of structures are usually much smaller than the real thing. Let's call the real structure—a bridge, building or stadium—the prototype, and consider a scale model that is 100 times smaller than the prototype. In this case, the geometric scale factor $S = 1/100$. If the same scale factor S, is used for all three dimensions (i.e., length, width and thickness), we may write:

$$L_m = S\, L_p$$

$$w_m = S\, w_p \tag{5.25}$$

$$b_m = S\, b_p$$

where L, w and b represent length, width and thickness and subscripts m and p refer to the model and the prototype respectively.

We note from Eq. (5.25), that the scale model is exactly the same shape as the prototype except for its size. It is geometrically proportional.

Scaling Factor for Stresses

Let's continue with the concept of scaling and consider the stresses occurring in both the model and the prototype. The stress σ in a bar subjected to axial loading is given by Eq. (5.4) as:

$$\sigma = P/A = P/(w\, b) \tag{5.26}$$

where P is the axial load applied to the bar and $A = w\, b$ is the cross sectional area of a bar of width w and depth b.

If the left side of Eq. (5.26) is divided by its right side, a unit dimensionless ratio involving the stress and its controlling parameters is formed:

$$\sigma(w\, b)/P = 1 \tag{5.27}$$

Let's designate a model and a prototype in Eq. (5.27) by writing:

$$\sigma_m\, (w_m\, b_m)/P_m = \sigma_p\, (w_p\, b_p)/P_p \tag{a}$$

Rearranging the terms in Eq. (a) gives:

$$\sigma_m = (P_m/P_p)(w_p/w_m)(b_p/b_m)\sigma_p \qquad (b)$$

Substituting Eq. (5.25) into Eq. (b) yields:

$$\sigma_m = (P_m/P_p)(1/\mathbf{S}^2)\sigma_p \qquad (c)$$

Examining Eq. (c) shows that the stresses in the model and the prototype are related by the geometric scale factor $\mathbf{S}$ and load scale factor $\mathbf{L}$, which is defined by:

$$\mathbf{L} = (P_m/P_p) \qquad (5.28)$$

Finally, we substitute Eq. (5.28) into Eq. (c) and obtain:

$$\sigma_m = (\mathbf{L}/\mathbf{S}^2)\sigma_p \qquad (5.29)$$

Clearly, the stresses produced in the model are related to the scale factors for the load and the geometry. It is also important to recognize that the geometric scale factor usually differs significantly from the load scale factor.

EXAMPLE E5.25

Suppose we construct a scale model of the Golden Gate Bridge using a scale factor S of 1/500. Determine the size of the model if:

- Total bridge length ⇒ 8981 feet
- Length of suspended structure ⇒ 6450
- Length of main span ⇒ 4200 feet
- Length of each side span ⇒ 1125 feet
- Width of bridge ⇒ 90 feet
- Diameter of main cable ⇒ 36-3/8 inch
- Width of roadway between curbs ⇒ 60 feet
- Lanes of vehicular traffic ⇒ 6
- Weight of main span per lineal foot ⇒ 21,300 lb
- Live load capacity per lineal foot ⇒ 4,000 lb

Solution: The model bridge is constructed in three parts including the main span and the two side spans. The length of the main span of the model is given by Eq. (5.25) as:

$$L_m = \mathbf{S}\, L_p = (1/500)(4200) = 8.4 \text{ ft} = 100.8 \text{ in.}$$

The width of the model bridge is determined in the same manner as:

$$w_m = \mathbf{S}\, w_p = (1/500)(90) = 0.18 \text{ ft} = 2.16 \text{ in.}$$

The diameter, D of the main cable on the model is:

$$D_m = \mathbf{S}\, D_p = (1/500)(36.375) = 0.073 \text{ in.}$$

All geometric features of the model of the bridge are determined in this manner.

EXAMPLE E5.26

Let's suppose a very small strain gage is installed on a steel bar of a model used to simulate a critical structural member in a prototype. The gage provides a measurement of the strain equal to 4000×10^{-6} when the model bridge was fully loaded. If the scaling factor for the load $\mathbf{L} = 1/100{,}000$, determine the stress in the main cable of the prototype.

Solution: The stress acting on the bar in the model is given by Hooke's law as:

$$\sigma = E\varepsilon \qquad (5.5 \text{ bis})$$

Substituting numerical values for the modulus of elasticity and the strain into Eq. (5.5) gives the stress in the main cable of the model as:

$$\sigma_m = E_m \varepsilon_m = (30 \times 10^6)(4000 \times 10^{-6}) = 120{,}000 \text{ psi}$$

where the modulus of elasticity $E = 30 \times 10^6$ psi for steel.

Solving Eq. (5.29) for the stress on the prototype gives:

$$\sigma_p = (\mathbf{S}^2/\mathbf{L})\sigma_m \qquad (5.30)$$

Substituting the scale factors for $\mathbf{S}$ and $\mathbf{L}$ into Eq. (5.30) yields:

$$\sigma_p = (\mathbf{S}^2/\mathbf{L})\sigma_m = [(1/500)^2/(1/100{,}000)][120{,}000] = (10/25)(120{,}000) = 48{,}000 \text{ psi}$$

This example illustrates that the scale factors for the load and the geometry of the structure should be selected to limit the stresses induced in the model. In most instances, the scale factor for the load is much smaller than the scale factor for the geometry.

Scaling Factor for Displacements

A model of the structure may also provide displacement measurements that may be used to predict displacements in the prototype. To develop the displacement relation between the model and the prototype, we again seek a unit dimensionless quantity that includes the variables controlling the displacement. Recalling Eq. (5.2) it may be shown that the displacement δ of a rod of length L subjected to an axial load P is given by:

$$\delta = PL/AE = PL/[(w\,b)E] \qquad (5.31)$$

If the left side of Eq. (5.31) is divided by its right side, a unit dimensionless ratio is formed:

$$[\delta(w\,b)E]/(PL) = 1 \qquad (5.32)$$

Let's identify the model and the prototype with appropriate subscripts in Eq. (5.32) and write:

$$\delta_m (w_m\,b_m)\,E_m/(P_m\,L_m) = \delta_p (w_p\,b_p)\,E_p/(P_p\,L_p) \qquad (a)$$

Solving Eq. (a) for δ_p and substituting Eq. (5.25) and (5.28) into the result, yields:

$$\delta_p = [(\mathbf{S}\,\mathbf{E})/\mathbf{L}]\,\delta_m \qquad (5.33)$$

where $\mathbf{E} = E_m/E_p$ is the modulus scale factor between the model and the prototype.

If the same materials are employed in the manufacture of both the model and the prototype the modulus scale factor is one. However, selection of the model materials is not restricted. We may use a wide variety of materials to fabricate the model providing they respond in an elastic and linear manner.

EXAMPLE E5.27

Suppose that a model of a truss type bridge structure is fabricated from members formed from sheet aluminum. The prototype structure is to be fabricated from steel with a span of 600 feet. The model is geometrically scaled so that its span is six feet. The capacity of the live load on the prototype is 15,000 lb/ft, and the model is loaded with 25 lb/ft. If the model deflects a distance of two inches under full load at the center of the span, determine the deflection of the prototype under the design load.

Solution: From Eq. (5.33), we write:

$$\delta_p = (\mathbf{S}\ \mathbf{E}/\mathbf{L})\delta_m$$

Note the scale factors are given by:

$$\mathbf{S} = 6/600 = 1/100$$

$$\mathbf{L} = 25/15{,}000 = 1.667 \times 10^{-3}$$

$$\mathbf{E} = 10 \times 10^6 / 30 \times 10^6 = 1/3$$

Substituting the scale factors into Eq. (5.33) yields:

$$\delta_p = (\mathbf{S}\ \mathbf{E}/\mathbf{L})\delta_m = [(1/100)(1/3)/(1.667 \times 10^{-3})](2) = 4.0 \text{ in.}$$

In this instance, the deflection of the prototype is twice as large as the deflection of the model. The choice of materials for the model and the scaling factors for both the load and geometry influence the differences in the stress, and deflection between the model and the prototype.

5.10 STATICALLY INDETERMINATE AXIAL MEMBERS

Structures are considered to be statically determinate when the unknown forces, either internal forces or external reactions, can be determined using only the equilibrium relations. However, in some structures the number of unknown forces exceed the number of applicable equilibrium equations, and a solution is not possible without the introduction of additional equations. When the number of unknown forces exceeds the number of applicable equilibrium equations, the structure is **statically indeterminate.**

The approach for solving for the unknown forces in statically indeterminate structures is to use the equilibrium relations and then to introduce additional equations based on the deformation of the structure. The solution involves two separate but compatible parts:

1. Prepare the traditional FBD together with all of the applicable equations of equilibrium.
2. Prepare displacement diagrams together with deformation equations that are consistent with the forces shown on the FBD.

Let's consider an example to demonstrate this approach in solving for the unknown forces in a statically indeterminate structure.

EXAMPLE E5.28

A 14 ft. long column supporting the corner of a very large building is fabricated from a steel pipe with an outside diameter of 8 in. and an inside diameter of 7 in. The pipe is filled with medium strength concrete to form a composite column that supports a 100-ton axial load as illustrated in the figures below and to the left. Determine the forces and stresses in the steel pipe and concrete plug.

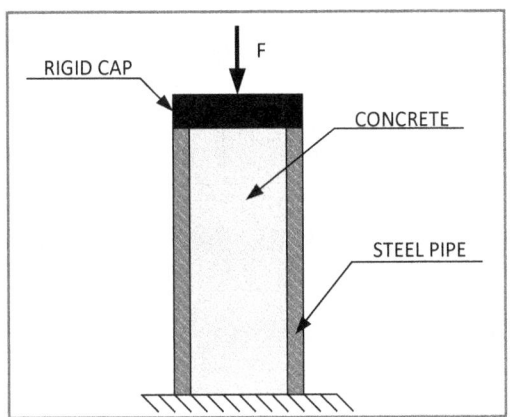
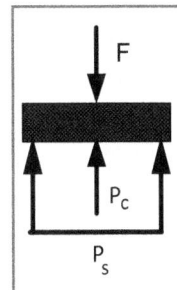

Solution:

Step 1: Prepare FBDs of the rigid cap, steel pipe and concrete plug, as shown in the figure above. E5.28a, and write the only applicable equilibrium equation, which is $\Sigma F_y = 0$.

$$\Sigma F_y = P_c + P_s - F = 0$$

$$P_c + P_s = 100 \text{ ton} \qquad (a)$$

where P_c and P_s are the internal forces in the concrete plug and steel pipe, respectively.

Because we have two unknown forces P_c and P_s and only a single applicable equation of equilibrium, the unknown forces are statically indeterminate.

Step 2: Prepare a drawing shown in Fig. E5.28 that depicts an exaggerated view of the deformations occurring in the concrete plug and the steel pipe when subjected to the internal forces P_c and P_s. Because the two structural elements deform together under the action of the axial loads, we write the deformation equation.

$$\delta_c = \delta_s \qquad (b)$$

Recall Eq. (5.2) and write:

$$\delta_c = P_c L_c / A_c E_c = \delta_s = P_s L_s / A_s E_s \qquad (c)$$

Solve Eq. (c) for P_c noting that $L_c = L_s$, to obtain:

$$P_c = (A_c/A_s)(E_c/E_s)P_s \qquad (d)$$

Equation (d) provides the second relation needed to solve for the unknown internal forces.

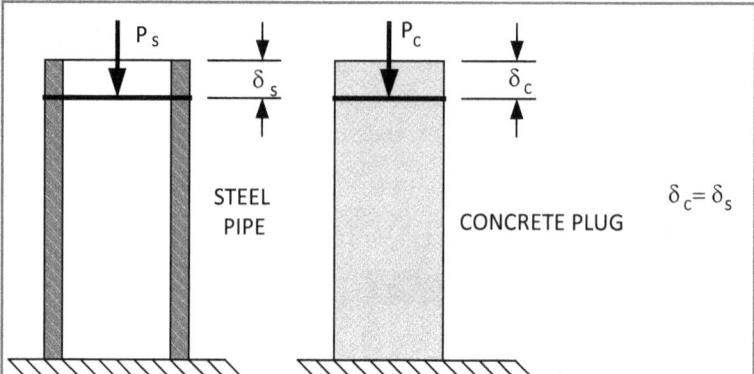

Fig. E5.28 shows an exaggerated view of the deformations

Step 3: Solve for the internal force P_s by substituting Eq. (c) into Eq. (a) to obtain:

$$P_s = 100/[1 + (A_c/A_s)(E_c/E_s)] \quad (e)$$

The area ratio is given by:

$$A_c/A_s = [d_i^2/(d_o^2 - d_i^2)] = (7.0)^2/[(8.0)^2 - (7.0)^2] = 3.267 \quad (f)$$

Using values from Appendix B-1, the modulus ratio is given by:

$$E_c/E_s = 3.6/30 = 0.12 \quad (g)$$

Substituting the results from Eqs. (f) and (g) into Eq. (e) gives:

$$P_s = 71.84 \text{ ton} \quad (h)$$

Then from Eqs. (a) and Eq. (h), it is clear that:

$$P_c = 28.16 \text{ ton} \quad (i)$$

Step 4: Solve for the stresses in the concrete plug and steel pipe as:

$$\sigma_c = P_c/A_c = [4(28.16)(2000)]/[\pi(7.0)^2] = 1{,}463 \text{ psi}$$

$$\sigma_s = P_s/A_s = [4(71.84)(2000)]/\{\pi[(8.0)^2 - (7.0)^2]\} = 12{,}200 \text{ psi}$$

Note that both of these stresses are compressive.

EXAMPLE 5.29

A grade 8 steel bolt ¾ in. in diameter (10 threads/in.) is employed to clamp an aluminum bushing between two rigid platens as illustrated in Fig. E5.29. The aluminum bushing has a 2.0 in. outside diameter and 1.0 in. inside diameter and is 10.00 in. long. After the unit is assembled with a snug fit, the nut is tightened by ½ of a turn. Determine the axial stresses in the bolt and the bushing. Also determine the deflection of the aluminum bushing.

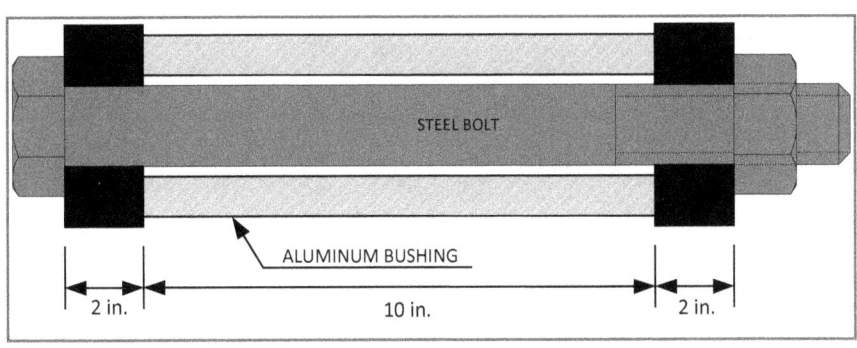
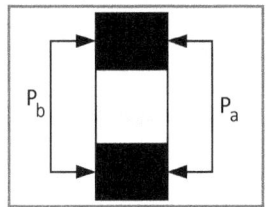

Fig. E5.29   Fig. E5.29a

Solution:

Step 1: Prepare FBDs of one of the rigid platens as shown in Fig. E5.29a, and write the only applicable equilibrium equation, which is $\Sigma F_x = 0$.

$$\Sigma F_x = -P_a + P_b = 0$$

$$P_a = P_b \qquad (a)$$

where P_a and P_b are the internal forces in the aluminum bushing and the steel bolt, respectively. They are equal in magnitude and opposite in sign. Because we have two unknown forces P_a and P_b and only a single applicable equation of equilibrium, the unknown forces are statically indeterminate.

Step 2: Prepare a drawing shown in Fig. E5.29b that depicts an exaggerated view of the deformations occurring in the bolt and the aluminum bushing when subjected to the internal forces P_a and P_b. When the nut is tightened the steel bolt tends to stretch and the aluminum bushing is compressed (shortened). The total of the two deformations must equal the amount of displacement induced by rotating the nut by ½ of a turn, which is $(1/2)(1/10) = 0.050$ in.

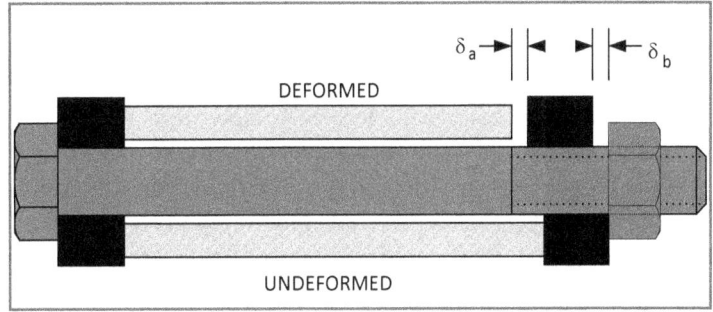

Fig. E5.29b

$$\delta_a + \delta_b = .050 \text{ in.} \qquad (b)$$

Step 3: Recall Eq. (5.2) and Eq. (a) and write:

$$\delta_a = P_a L_a / A_a E_a = P_b L_a / A_a E_a = [(4)(10)/\{\pi[(2)^2 - (1)^2](10.4 \times 10^6)\}] P_b = 0.4081 \times 10^{-6} P_b$$

$$\delta_b = P_b L_b / A_b E_b = \{(4)(14)/[\pi(0.75)^2 (30 \times 10^6)]\} P_b = 1.0563 \times 10^{-6} P_b \qquad (c)$$

> Substitute Eq. (c) into Eq. (b) and solve for P_b.
>
> $$P_b [0.4081 + 1.0563] = 0.050 \times 10^6$$
>
> $$P_a = P_b = 34{,}140 \text{ lb} \qquad (d)$$
>
> **Step 4:** The stresses are given by Eq. (5.4) as:
>
> $$\sigma_b = P_b/A_b = (4)(34{,}140)/[\pi(0.75)^2] = 77.290 \text{ ksi (tension)}$$
>
> $$\sigma_a = P_a/A_a = (4)(34{,}140)/[3\pi] = 14.490 \text{ ksi (compression)}$$
>
> **Step 5:** The amount of the compression of the aluminum bushing is given by Eq. (5.2) as:
>
> $$\delta_a = P_a L_a / A_a E_a = [(34{,}140)(10)(4)]/[(3\pi)(10.4 \times 10^6)] = 0.01393 \text{ in. (shorter)}$$

These two examples demonstrate the approach employed to solve statically indeterminate problems. This approach involves two important steps—writing the applicable equations of equilibrium based on an accurate FBD and writing a deformation equation based on a drawing showing the deformation of the structural elements involved.

5.11 SUMMARY

This chapter treats structural members that are long and thin. Flexible members, such as wire and cable, only support axial tensile loads. They cannot support compressive forces because they buckle under very low loads. In addition, they cannot support transverse loads because their transverse stiffness is also negligible. Under increasing tension forces applied along the axis of these long, thin, and flexible members, they stretch, yield and finally rupture.

Rods and bars are sufficiently stiff to carry compressive forces. In this chapter, we consider only axial loading of the bars and rods. The case of transverse loading is deferred until Chapter 7. The equations derived for stresses and deflections in wire and cable are also applicable to bars and rods. When dealing with compressive forces and stresses, a minus sign is used to indicate the direction of the loading. We also assume compressively loaded bars and rods are sized so they will not fail by buckling. Examples are presented for determining stresses and deflections of uniform axial bars.

Relations have been derived to determine the stretch (axial deformation). Stress and strain have been defined and the stress-strain relation (Hooke's law) has been discussed. Several examples have been provided to guide you in determining, stress, strain, and axial extension.

We have shown that the internal force P is equal to the external force F for these uniaxial members. The internal force is produced by normal stresses that are uniformly distributed over the cross section of the uniaxial member.

We describe failure by yielding and rupture and introduce the stress-strain curve for different types of materials in the process. Ultimate tensile strength and yield strength are defined. Safe design philosophy is discussed together with the use of factors of safety in design. Examples are provided to demonstrate problem-solving techniques dealing with strength and safety factors. The importance of specifying standard sizes of structural members in design is illustrated in one of the examples.

The concept of models that are geometrically similar to actual structures is introduced. The stresses and displacements produced by loading or deforming a model are related to those developed in the prototype (structure). Relations involving scaling factors are employed to determine the stresses and displacements in the structure based on measurements made on a geometrically scaled model.

The equations developed for uniform members subjected to axial forces are summarized below:

$$\delta = L_f - L_o \tag{5.1}$$

$$\delta = (PL)/(AE) \tag{5.2}$$

$$\varepsilon = \delta/L \tag{5.3}$$

$$\sigma = P/A \tag{5.4}$$

$$\sigma = E\varepsilon \tag{5.5}$$

$$\varepsilon = P/(AE) \tag{5.6}$$

$$P_{INT} = \int \sigma \, dA \tag{5.7}$$

$$P_{INT} = \sigma \int dA = \sigma A \tag{5.8}$$

$$S_u = \sigma_{fail} = P_{fail}/A \tag{5.9}$$

$$S_y = \sigma_{yield} = P_{yield}/A \tag{5.10}$$

$$\mathbf{SF_u} = S_u/\sigma_{design} \tag{5.11}$$

$$\mathbf{SF_y} = S_y/\sigma_{design} \tag{5.12}$$

$$\tau = V/A \tag{5.13}$$

Stresses on a section with an inclined cut at an angle θ were determined using the equilibrium relations and the simple definition for normal and shear stresses as:

$$\sigma_\theta = (\sigma_x)\cos^2\theta \tag{5.14}$$

$$\tau_\theta = (\sigma_x/2)\sin 2\theta \tag{5.15}$$

In Fig. 5.14, we note that both the normal stress σ_θ and the shear stress τ_θ vary with the angle θ. The normal stress is a maximum when $\theta = 0°$, and the section cut is perpendicular to the axis of the bar. The shear stress vanishes on this section. The shear stress is a maximum when $\theta = 45°$. On this plane, $\sigma_{45} = \tau_{45} = \sigma_x/2$.

Tapered bars subjected to axial loading were described. For the stresses, Eq. (5.4) was modified to account for the variable cross sectional area as a function of position x along the length of the tapered bar.

138 — Chapter 5 — Axially Loaded Structural Members

$$\sigma_x = P/A_x \tag{5.17}$$

For deflection of the tapered bar, we considered an incremental length dx along the length of the bar and modified Eq. (5.2) to read as:

$$d\delta = P\, dx/(A_x E) \tag{5.18}$$

To determine the deflection δ of the tapered bar, we write an expression for A_x and integrate Eq. (5.18) from zero to L. Examples are provided to demonstrate the computation technique for tapered bars.

We have also described techniques for determining the stresses in stepped bars. The stresses in individual sections of the bar are computed from Eq. (5.4) using the appropriate cross sectional area for the section under consideration. Displacements of each section of the bar are determined from Eq. (5.2) and then superimposed to give the total deflection as shown in Eq. (5.20).

$$\delta_{total} = \delta_1 + \delta_2 + \delta_3 + \ldots + \delta_n \tag{5.20}$$

Stress concentrations that develop at discontinuities in bars and rods have been discussed. These stress concentrations occur for all types of loading: axial, transverse and torsion. The stress concentration increases the maximum stresses by a significant amount. We determine the maximum stresses from:

$$\sigma_{MAX} = K\, \sigma_{NOM} \tag{5.21}$$

The stress concentration factor K is defined for a bar with a central circular hole in Fig. 5.20 and for a stepped bar in Fig. 5.22. The nominal stresses for these two discontinuous bars are given by:

$$\sigma_{NOM} = P/(w-d)b \tag{5.22}$$

$$\sigma_{NOM} = P/w_1 b \tag{5.24}$$

Examples demonstrating the method for determining the stress concentration factor, and the nominal and maximum stresses are provided.

Modeling, an approach to experimentally determine stresses and displacements in large complex structures, was introduced. The relations for the stresses and displacement between the model and the structure (prototype) are given by:

$$\sigma_m = (\mathbf{L}/\mathbf{S^2})\, \sigma_p \tag{5.29}$$

$$\delta_p = [(\mathbf{S}\ \mathbf{E})/\mathbf{L}]\, \delta_m \tag{5.33}$$

Structures are considered to be statically determinate when the unknown forces, either internal forces or external reactions, can be determined using only the equilibrium relations. However, in some structures the number of unknown forces exceed the number of applicable equilibrium equations, and a solution is not possible without the introduction of additional equations. When the number of unknown forces exceeds the number of applicable equilibrium equations, the structure is **statically indeterminate.** The approach for solving for the unknown forces in statically indeterminate structures is to use the equilibrium

relations and then to introduce additional equations based on the deformation of the structure. The solution involves two separate but compatible parts:
1. Prepare the traditional FBD together with all of the applicable equations of equilibrium.
2. Prepare displacement diagrams together with deformation equations that are consistent with the forces shown on the FBD.

PROBLEMS

5.1 A steel music wire is employed to support an axial load P. For the numerical parameters presented in the table below, determine the final length of the wire and the stress and strain in the wire.

Problem No.	Initial Length	Load, P	Gage No.	Diameter, d
5.1a	4.5 m	2.5 kN	00	0.008 in.
5.1b	22 ft	864 lb	10	0.024 in.
5.1c	46 in.	675 lb	30	0.080 in.
5.1d	12.4 m	7.63 kN	6	0.016

5.2 A steel music wire is tightened on a guitar by rotating the wire supporting post through an angle θ. If the post has a diameter d and the length of the wire is L, determine the strain induced in the guitar string. The parameters θ, d and L are defined in the table below:

Problem No.	θ	d	L
5.2a	11°	0.25 in.	32 in.
5.2b	12°	4.0 mm	0.75 m
5.2c	35°	0.22 in.	28 in.
5.2d	24°	5.0 mm	0.9 m
5.2e	43°	0.20 in.	35 in.

5.3 Referring to Problem 5.2, determine the additional stress induced in the guitar string as it is tightened.

5.4 A hemp rope 15 m long and 45 mm in diameter is supported from one of the roof beams in a gymnasium. Three gymnasts climb this rope together. The lead gymnast has a mass of 55 kg, the next 72-kg and the trailing gymnast 52 kg.

(a) Determine the maximum stress in this rope.
(b) Determine the stress in the rope at a location between the lead and intermediate gymnast.
(c) Determine the stress in the rope at a location between the intermediate and trailing gymnast.

5.5 A suspension bridge is to carry a roadway and traffic that may weigh up to 5000 ton. If the safe load that can be imposed on the wire rope to be used in the construction is 175 ton, specify the number of cables to be employed. Justify your answer.

5.6 Describe the constraints on the relation $\sigma = E\varepsilon$.

5.7 Determine the design load that can be specified for various gage numbers of metal wire if the design specification calls for the safety factor listed in the table below:

Problem No.	Gage No.	Failure Criterion	Safety Factor	Strength	Material
5.7a	00	Yield	2.5	530 MPa	Steel
5.7b	30	Ult. Tensile	3.6	600 MPa	Steel
5.7c	10	Yield	2.2	59 ksi	Aluminum
5.7d	24	Ult. Tensile	2.8	75 ksi	Aluminum
5.7e	17	Yield	1.8	1100 MPa	Music
5.7f	7	Ult. Tensile	2.5	1250 MPa	Music

5.8 Discuss the factors to consider in establishing the safety factor in a design specification.

5.9 For the cable-pulley-mass arrangement, as shown below, determine the stresses in the cable. The parameters angle θ, mass m, and effective cable diameter d are defined in the table below:

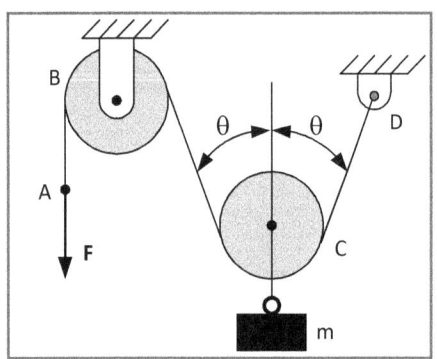

Prob. No.	Angle, θ	Mass, m	Diameter, d
5.9a	10	600 kg	14 mm
5.9b	20	20 slug	0.500 in.
5.9c	30	400 kg	12 mm
5.9d	40	50 slug	1.000 in.
5.9e	50	500 kg	20 mm
5.9f	60	40 slug	1.125 in.

5.10 Cold drawn steel alloy wire exhibits ultimate tensile strength and yield strength shown in the table below. If this alloy is employed, specify the gage of the wire if it is to support a design load of P with the safety factor given in the table. Specify the gage based on both yielding and rupturing as modes of failure.

Problem No.	Tensile Strength	Yield Strength	Load, P	Safety Factor, SF
5.10a	600 MPa	500 MPa	7.5 kN	3.2
5.10b	80 ksi	65 ksi	50 kip	3.0
5.10c	480 MPa	410 MPa	5.5 kN	2.8
5.10d	95 ksi	77 ksi	75 kip	2.3

5.11 Determine the safety factor for the wire C-D in the cable-weight arrangement, as shown below. The steel cables are each L long, exhibit a yield strength of S_y and have an effective cross section area of A. Assume that the forces in the horizontal wires are equal in magnitude and opposite in direction ($F_{CB} = -F_{CA}$). Also assume $F_{CB} = K F_{CD}$. Numerical parameters defining L, A, S_y and K are given in the table below:

Problem No.	Cable Length, L	Cable Area, A	Yield Strength, S_y	Value of K
5.11a	10 m	8 mm^2	500 MPa	0.18
5.11b	22 ft	0.015 in^2	80 ksi	0.25
5.11c	15 m	6 mm^2	600 MPa	0.22
5.11d	18 ft	0.018 in^2	72 ksi	0.15

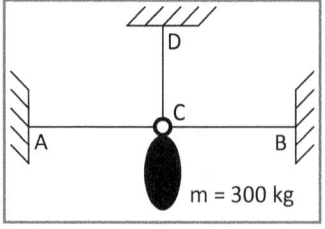

5.12 Determine the mass m in the diagram of Problem 5.11 that is required to yield the cable C-D if the cable has the characteristics described in Problem 5.11a-d. Is the cable-weight arrangement stable after yield? Why?

5.13 The wire-mass system shown below is constructed using metallic wire. Specify a suitable alloy and the Gage Nos. for the wires if the criterion for failure is based on yielding. Use the same diameter for both wires. The numerical parameters for the problem are given in the table below:

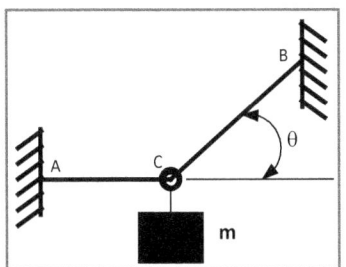

Prob. No.	Safety Factor	Mass, m	Angle, θ	Material
5.13a	2.9	800 kg	20°	steel
5.13b	3.1	10.5 slugs	48°	stainless steel
5.13c	3.5	250 kg	64°	aluminum
5.13d	2.7	16.3 slugs	55°	copper

5.14 The wire-mass system shown below is constructed using metallic wire. Specify a suitable alloy and the Gage Nos. for the wires if the criterion for failure is based on ultimate tensile strength. Use the same diameter for both wires.

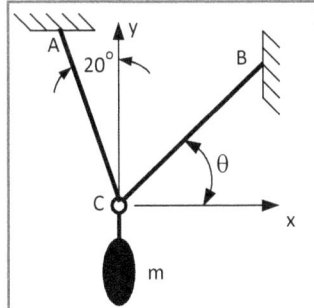

Prob. No.	Safety Factor	Mass, m	Angle, θ	Material
5.14a	2.2	800 kg	45°	steel
5.14b	3.6	15.0 slugs	30°	stainless steel
5.14c	1.8	180 kg	40°	aluminum
5.14d	3.2	12.6 slugs	60°	copper

5.15 The wire-mass system shown below is constructed using steel music wire. A safety factor of 2.9, based on the ultimate tensile strength, is to be employed for both wires. Find the maximum weight that can be lifted (in pounds) and specify the required gage Nos. for each wire. The angle θ and the steel alloy are listed in the table below.

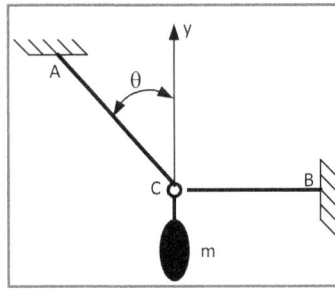

Problem No.	Angle, θ	Alloy
5.15a	15°	1010 A
5.15b	20°	1020 HR
5.15c	30°	1045 HR
5.15d	40°	4340 HR

5.16 Describe in an engineering brief the differences between a wire rope and a rod.

5.17 Describe in an engineering brief the similarities between a wire rope and a rod.

5.18 A long thin bar of length L, width w and thickness b is subjected to a force of F. Determine the tensile stresses and the axial deformation in the bar. Parameters are specified in the table shown below:

Problem No.	F	w	b	L	Material
5.18a	22 kN	50 mm	20 mm	1.8 m	Steel
5.18b	132 lb	0.125 in.	0.1875 in.	70 in.	Aluminum
5.18c	− 42 kN	70 mm	30 mm	0.6 m	Aluminum
5.18d	− 5,520 lb	1.25 in.	0.75 in.	24 in.	Concrete (HS)
5.18e	39.36 kN	90 mm	40 mm	1.8 m	White Pine
5.18f	− 8.64 ton	1.5 in.	1.0 in.	24 in.	Steel

5.19 A bar fabricated from steel with a tensile strength of S_u is subjected to an axial tensile force. The bar is designed with a safety factor of SF. Determine the design stress for the bar and the required cross sectional area. Parameters are specified in the table shown below:

Prob. No.	Strength, S_u	Force, F	Safety Factor
5.19a	54.0 ksi	20 kip	3.2
5.19b	65.2 ksi	32 kip	2.5
5.19c	280 MPa	85 kN	2.4
5.19d	425 MPa	66 kN	1.9

5.20 A bar fabricated from an aluminum alloy with a tensile strength of S_u is subjected to an axial tensile force F. The bar is designed with a safety factor of SF. Determine the design stress for the bar and the required cross sectional area. Parameters are specified in the table shown below:

Prob. No.	Strength, S_u	Force, F	Safety Factor
5.20a	31.0 ksi	15 kip	3.0
5.20b	95.2 ksi	22 kip	2.8
5.20c	400 MPa	75 kN	2.5
5.20d	550 MPa	96 kN	1.5

5.21 A bar fabricated from brass with a tensile strength of 247 MPa is subjected to an axial compressive force of 76 kN. The bar is designed with a safety factor of 2.8. Determine the design stress for the bar and the required cross sectional area. Assume the tensile and compressive stresses are equal.

5.22 A key used to lock a gear onto a shaft is subjected to a shear force of F. If the key is w, wide by h, high and L, long, determine the shear stress acting on the key. Draw a free body diagram showing this shear stress. Parameters are specified in the table shown below:

Prob. No.	Width, w	Height, h	Length, L	Force, F
5.22a	0.25 in.	0.375 in.	1.5 in.	8.0 kip
5.22b	0.375 in.	0.500 in.	2.5 in.	12.0 kip
5.22c	8 mm	10 mm	20 mm	20 kN
5.22d	12 mm	15 mm	25 mm	36 kN

5.23 If the key in Problem 5.22 is machined from a steel alloy with a tensile yield strength of 48,000 psi, determine the safety factor for the key.

5.24 A rectangular bar, shown below, is subjected to an axial tensile force of F. Determine the normal stress and the shear stress on an inclined plane if the angle of the section cut is:
(a) $\theta = 20°$, (b) $\theta = 30°$, (c) $\theta = 45°$, (d) $\theta = 60°$, and (e) $\theta = 75°$

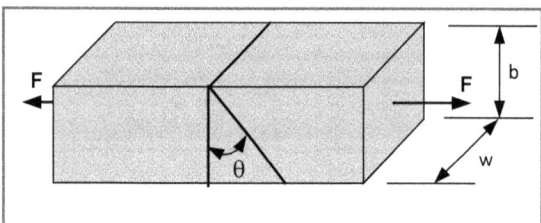

Prob. No.	Width, w	Thickness, b	Force, F
5.24a	0.50 in.	0.375 in.	10.0 kip
5.24b	1.25 in.	0.500 in.	22.0 kip
5.24c	15 mm	10 mm	30 kN
5.24d	20 mm	25 mm	55 kN

5.25 A rectangular bar, as shown in Problem 5.24, is adhesively bonded along a section cut. The normal stress in the bond line is limited to 2.5 ksi and the shear stress to 1.5 ksi. If the bar has a cross sectional area A = 1.75 in.^{2} and a safety factor of 3.0 is specified, determine the largest axial force that can be applied to the bar if the angle of the section cut is:
(a) $\theta = 20°$, (b) $\theta = 30°$, (c) $\theta = 45°$, (d) $\theta = 60°$, and (e) $\theta = 75°$

5.26 A rod with a circular cross section with a diameter d is fabricated from two pieces as illustrated below. If a force F is applied to the bar, derive the expression for the normal and shear stress acting in the plane of the adhesive joint as a function of angle ϕ.

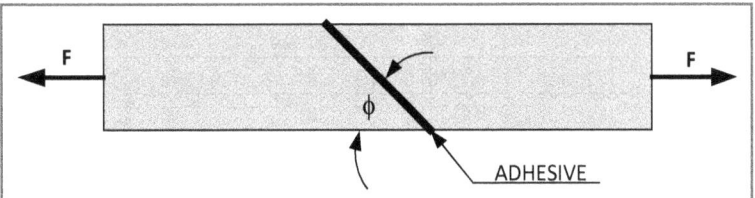

5.27 For the bar illustrated in Problem 5.26, determine the shear stresses in the adhesive joint if the angle ϕ is varied from 0° to 90°. Also compute the normal stresses acting on the adhesive joint. The F and the diameter d are given in the table below: Hint: Use a spreadsheet to determine the shear and tensile stresses in the adhesive joint for the range of ϕ specified.

Problem No.	Force, F	Diameter, d
5.27a	2.0 kN	20 mm
5.27a	2150 lb	1.5 in.
5.27a	5.3 kN	45 mm
5.27a	1950 lb	1.25 in.

5.28 A tensile bar defined below has a cross sectional area of 175 mm^2 and is subjected to a tensile force F. The stresses on the inclined plane A—B are $\sigma_\theta = 81$ MPa and $\tau_\theta = -27$ MPa. Determine the stress σ_x, the angle θ and the force F.

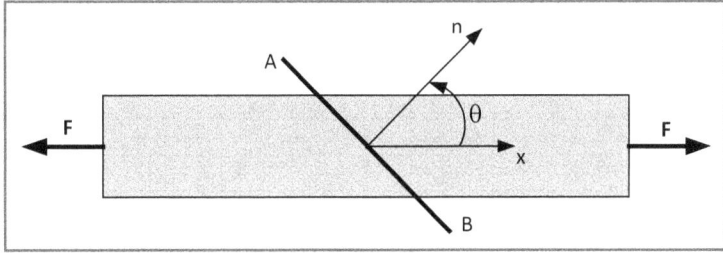

5.29 For the tensile bar described in Problem 5.28, determine the shear and normal stresses acting on an inclined plane with $\theta = 40°$.

5.30 For the tensile bar in Problem 5.28, prepare a graph similar to that shown in Fig. 5.14. Provide two curves that present the stresses σ_θ and τ_θ as a function of the angle of the inclined cut as θ varies from 0 to 90°.

5.31 Consider the tapered bar presented below. Prepare a graph of the axial stress σ as a function of x as it varies from zero to L. The numerical parameters defining the load and the size of the tapered bar are defined in the table below:

144 — Chapter 5 — Axially Loaded Structural Members

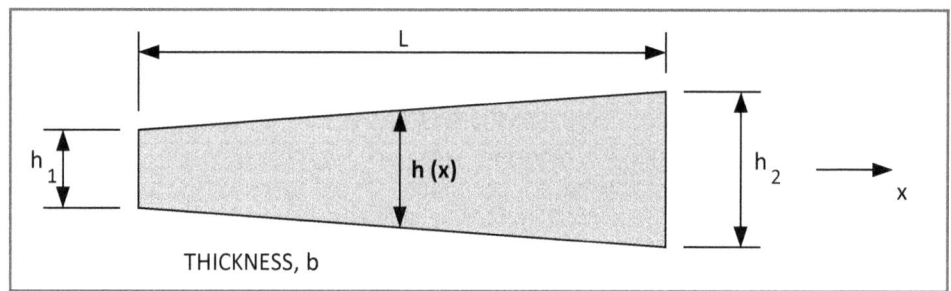

Problem No.	Width, h_1	Width, h_2	Thickness, b	Length, L	Force. F
5.31a	100 mm	500 mm	50 mm	4 m	500 kN
5.32b	6 in.	10 in.	2 in.	72 in.	65,000 lb
5.32c	50 mm	350 mm	40 mm	2.6 m	700 kN
5.32d	4 in.	12 in.	1.25 in.	96 in.	100 kips

5.32 Consider the tapered bar presented above. If the bar is fabricated from a metal alloy, determine the extension of the bar when it is subjected to a tensile force F. The numerical parameters defining the load and the size of the tapered bar are defined in the table below:

Problem No.	Width, h_1	Width, h_2	Thickness, b	Length, L	Force. F	Material
5.32a	100 mm	500 mm	50 mm	4 m	500 kN	Aluminum
5.32b	6 in.	10 in.	2 in.	72 in.	65,000 lb	Steel
5.32c	50 mm	350 mm	40 mm	2.6 m	700 kN	Copper
5.32d	4 in.	12 in.	1.25 in.	96 in.	100 kips	Stainless Steel

5.33 Consider the tapered bar presented in Problem 5.31. If the bar is fabricated from an aluminum alloy, determine the extension of the bar when it is subjected to a tensile force F. In the solution of this Problem, do not employ Eq. (5.19). Instead, use Eq. (5.18) and perform a numerical integration on a spreadsheet. The numerical parameters defining the load and the size of the tapered bar are:

Problem No.	Width, h_1	Width, h_2	Thickness, b	Length, L	Force. F
5.33a	30 mm	90 mm	20 mm	800 mm	120 kN
5.33b	2 in.	10 in.	2 in.	14 in.	85,000 lb
5.33c	25 mm	100 mm	30 mm	1.00 m	85.6 kN
5.33d	3 in.	9 in.	1.5 in.	36 in.	100 kips
5.33e	50 mm	250 mm	35 mm	2.5 m	175 kN

5.34 Describe the procedure employed to solve for the stresses and deflection in a stepped bar subjected to axial loading.

5.35 For the stepped bar illustrated to the right, determine the stresses in both portions of the bar and its total deflection. The numerical parameters defining the loads and the size of the stepped bar are given in the table below:

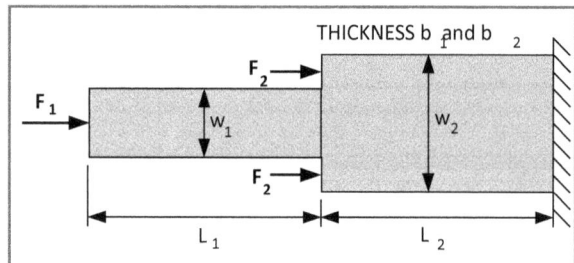

Prob. No.	w_1	w_2	b_1	b_2	L_1	L_2	F_1	F_2	Material
5.35a	2.2 in.	6.5 in.	1.2 in.	2.4 in.	20 in.	35 in.	– 80 kip	52 kip	Steel
5.35b	50 mm	145 mm	60 mm	100 mm	0.7 m	1.0 m	100 kN	– 65 kN	Aluminum
5.35c	2.0 in.	6.0 in.	2.0 in.	4.5 in.	30 in.	48 in.	– 120 kip	75 kip	Cast iron
5.35d	70 mm	140 mm	75 mm	120 mm	1.0 m	1.0 m	150 kN	60 kN	Magnesium
5.35e	2.5 in.	5.5 in.	1.5 in.	3.0 in.	25 in.	40 in.	100 kip	– 30 kip	Titanium

5.36 A tensile bar, shown below, is subjected to an axial force F. Determine: (a) the nominal stress, (b) the stress concentration factor K, and (c) the maximum stress. The numerical parameters defining the load and the size of the bar are given in the table below:

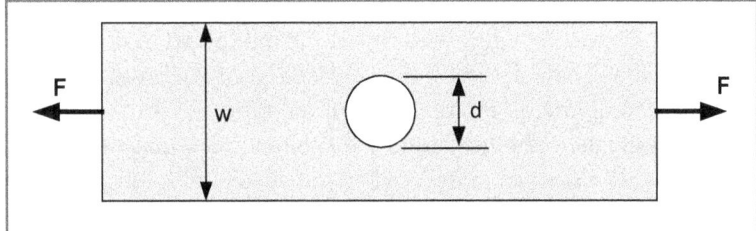

Prob. No.	Width, w	Thickness, b	Length, L	Force, F	Diameter, d
5.36a	3.0 in.	0.25 in.	40 in.	10 kip	1.0 in.
5.36b	100 mm	25 mm	1 m	90 kN	30 mm
5.36c	5.0 in.	0.50 in.	44 in.	15 kip	1.25 in.
5.36d	150 mm	30 mm	0.8 m	120 kN	60 mm
5.36e	5.0 in.	0.375 in.	72 in.	30 kip	2.5 in.

5.37 A stepped tensile bar, illustrated below, with fillets at the transition between the small and the large section is subjected to an axial tensile force F. Determine the maximum stress at the fillets if the numerical parameters defining the geometry of the stepped bar are specified in the table below:

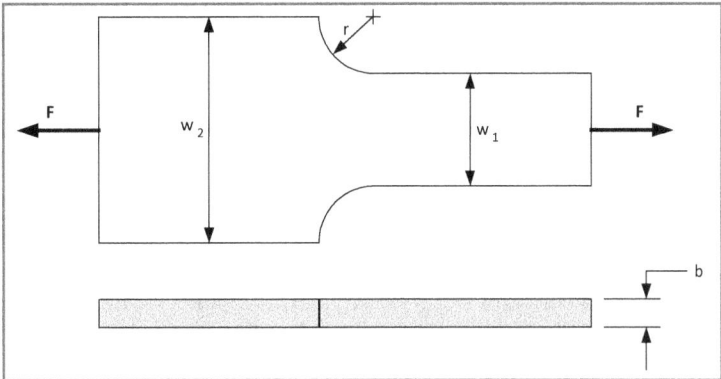

Prob. No.	w_1	w_2	b	F	r
5.37a	60 mm	120 mm	50 mm	75 kN	15 mm
5.37b	3.0 in.	4.5 in.	0.75 in.	60 kip	0.5 in.
5.37c	100 mm	110 mm	40 mm	65 kN	8 mm
5.37d	2.5 in.	3.0 in.	0.5 in.	25 kip	0.35 in.

5.38 A scale model of a large structure has been fabricated from steel and tested. A strain gage on one member of the structure indicated an axial strain of $\varepsilon = 1450$ µm/m. Determine the stress in the corresponding member of the prototype. The numerical parameters defining the scaling factors for the loads and the size of the structural member are given in the table below:

Prob. No.	w_m	w_p	b_m	b_p	L_m	L_p	Scaling Factor, L
5.38a	1.6 mm	80 mm	2.0 mm	100 mm	14 mm	7.0 m	1/10,000
5.38b	0.90 in.	9.0 in.	0.14 in.	1.40 in.	1.0 ft	10.0 ft.	1/5,000.
5.38c	1.0 mm	100 mm	0.40 mm	40 mm	60 mm	6.0 m	1/7,500
5.38d	0.30 in.	6.0 in.	0.10 in.	2.00 in.	3.60 in.	72 in.	1/2,000

5.39 Suppose that a model of a structure is fabricated from members formed from sheet aluminum. The prototype structure is to be fabricated from steel with an open span of 200 feet. The model is geometrically scaled so that its span is four feet. The capacity of the live load on the prototype is 150 lb/ft, and the model is loaded with 2.5 lb/ft. If the model deflects a distance of 0.220 in. under full load at the center of the span, determine the deflection of the prototype under the design load.

5.40 The tensile bar presented below, is fabricated with an adhesive joint inclined at an angle ϕ to the axis of the bar. Determine the optimum angle ϕ for the inclined plane if the stresses in the adhesive are not to exceed either 3.5 ksi in tension or 2.8 ksi in shear. Also determine the maximum force P that can be applied to the member without exceeding the stresses in the adhesive joint.

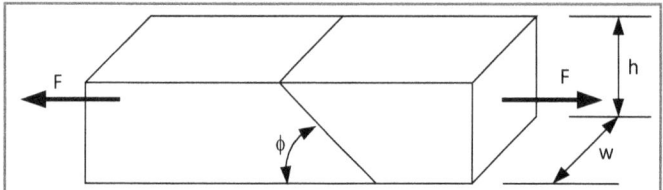

Prob. No.	w	h
5.40a	60 mm	120 mm
5.40b	3.0 in.	4.5 in.
5.40c	100 mm	110 mm
5.40d	2.5 in.	3.0 in.

5.41 A structural member is fabricated from a solid round bar of diameter d 50 mm. If the member is L long, determine the maximum axial force that can be applied if the axial stress is not to exceed 175 MPa and the total elongation is not to exceed 0.14% of L

Prob. No.	L	Material
5.41a	6.0 m	Steel
5.41b	53 in.	Aluminum.
5.41c	900 mm	Titanium
5.41d	2.5 ft.	Brass

5.42 A steel tie rod with a diameter of D and length L_2 is employed to compress a brass bushing with an outside diameter D_o and a length L_1, as shown to the right. The dimensions for the bushing and the tie rod are shown in the table below. Determine the minimum wall thickness of the bushing if the deflection of the tie rod is limited to δ.

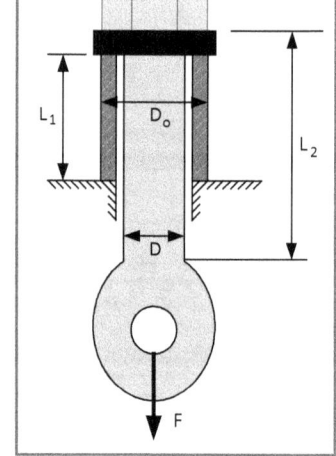

Prob. No.	L_1	L_2	D_o	D	δ	Force, F
5.42a	24.0 in.	34.0 in.	3.0 in.	1.0 in.	0.020 in.	10 kip
5.42b	400 mm	1.0 m	100 mm	15 mm	0.60 mm	25 kN
5.42c	18.0 in.	30.0 in.	3.5 in.	0.75 in.	0.030 in.	65 kip
5.42d	350 mm	800 mm	80 mm	20 mm	0.50 mm	30 kN
5.42e	15.0 in.	27.0 in.	5.0 in.	0.875 in.	0.025 in.	90 kip

5.43 A short column with a height h is fabricated by adhesively bonding aluminum faceplates to a core of plastic foam, as shown below. The aluminum faceplates are t thick and w wide. The foam plastic core has a square cross section with an elastic modulus of 1200 psi. Determine the stresses in the aluminum plates and the plastic foam. Also, determine the displacement of the column under the action of the applied force F. Dimensions for the column are given in the table below.

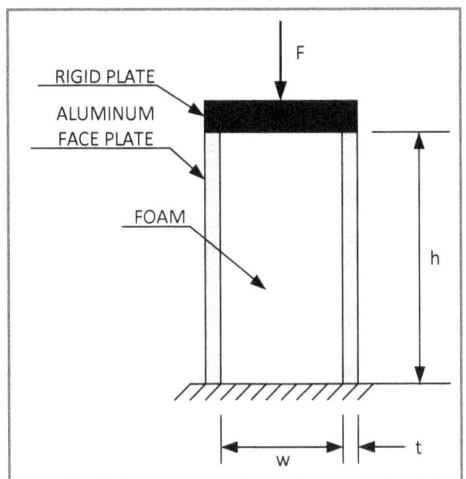

Prob. No.	h	w	t	Force, F
5.43a	5.0 ft	6.0 in.	0.125 in.	15 kip
5.43b	2.0 m	100 mm	1.0 mm	35 kN
5.43c	48.0 in.	3.0 in.	0.060 in.	10 kip
5.43d	2.5 m	80 mm	1.2 mm	20 kN
5.43e	7.5 ft	4.0 in.	0.040 in.	8.0 kip

5.44 A grade 8 steel bolt 1.0 in. in diameter (14 threads/in.) is employed to clamp a brass bushing between two rigid platens, as illustrated below. The brass bushing has a 2.5 in. outside diameter and 1.5 in. inside diameter and is 12.00 in. long. After the unit is assembled with a snug fit, the nut is tightened by 1/3 of a turn. Determine the axial stresses in the bolt and the bushing. Also determine the deflection of the brass bushing.

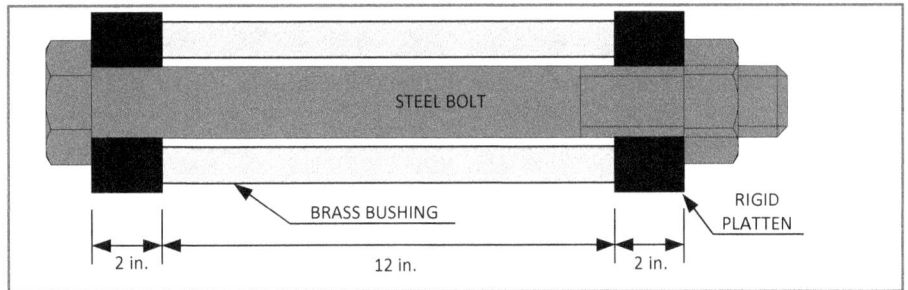

5.45 N steel reinforcing rods with a diameter d are placed within a high strength concrete column with a square cross section that supports an applied force F, as shown below. Determine the stresses in the steel reinforcing bars and the concrete. Also determine the amount of deflection of the column. Dimensions of the steel and concrete in the column are given in the table below.

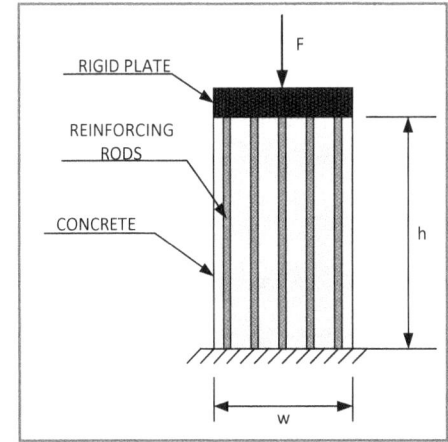

Prob. No.	h	w	d	N	Force, F
5.45a	8.0 ft	6.0 in.	0.25 in.	25	50 kip
5.45b	3.0 m	100 m	8.0 mm	16	100 kN
5.45c	108.0 in.	4.0 in.	0.375 in.	9	70 kip
5.45d	3.5 m	120 mm	10 mm	9	120 kN
5.45e	10.0 ft	5.0 in.	0.50 in.	16	80 kip

5.46 A long aluminum rod is connected to a shorter brass cylinder with a bolted flange as shown below. Prior to assembly a gap of δ occurred between the flange plates. Bolts were inserted and tightened bringing the flange plates together. Determine the stresses induced in both the rod and the cylinder by the assembly operation. Also determine the displacement of the face of each flange.

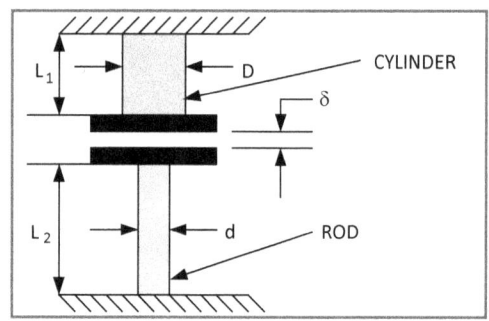

Prob. No.	L_1	L_2	d	D	δ
5.46a	9.0 in.	18.0 in.	1.25 in.	3.0 in	0.010 in.
5.46b	250 mm	750 mm	30 mm	150 mm	0.20 mm
5.46c	7.5 in.	16.0 in.	1.0 in.	4.0 in.	0.012 in.
5.46d	325 mm	900 mm	40 mm	125 mm	0.16 mm
5.46e	6.0 in.	20.0 in.	1.5 in.	2.5 in.	0.008 in.

REFERENCES

1. Peterson, R. E., <u>Stress Concentration Design Factors</u>, 2nd Ed., Wiley & Sons, New York, 1990.

CHAPTER 6

TORSION OF STRUCTURAL ELEMENTS

6.1 TORSION LOADING

Circular shafts are commonly employed to transmit power from a motor or engine to an appliance. For example, when a power drill is used to drive a screw into a sheet of metal, a short circular shaft transmits the power from an electric motor to a chuck holding the screwdriver. When you drive your car, two shafts transmit power from the gearbox to the drive wheels. Recall the relation between the power P transmitted and the torque T applied to the shaft is given by:

$$P = T\omega \qquad (6.1)$$

where ω is the angular velocity of the shaft in radian/s, P is the power in watts W, where $1W = 1$ N-m/s, and T is the torque in N-m.

If Eq. (6.1) is employed, the power and torque are expressed in SI units. In U. S. Customary units, the power is expressed in horsepower HP and the torque in terms of ft-lb. The definition of horsepower[1] is given by:

$$1 \text{ HP} = 33,000 \text{ ft-lb/min} = 550 \text{ ft-lb/s} \qquad (6.2)$$

We are particularly interested in the torsion load (torque) produced, when a shaft transmits power. The torque produces stresses and angular deformations in the shaft. If either of these two quantities becomes excessive, the shaft will fail in service.

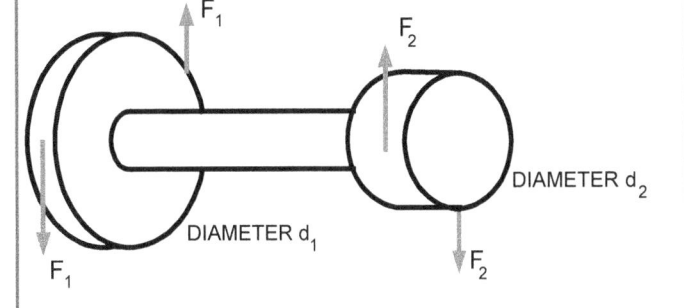

Fig. 6.1 A circular shaft transmitting torque $T_1 = F_1 d_1 = T_2 = F_2 d_2$.

When a motor or engine turns, it acts on a shaft producing a torque T. We illustrate this action in Fig. 6.1, where two circular disks are connected by a circular shaft. A couple with forces F_1 act on the larger diameter disk producing a torque $T_1 = F_1 d_1$. Another couple with forces F_2 act on the smaller diameter disk producing a torque $T_2 = F_2 d_2$. If the shaft is either stationary or rotating at a constant angular

---

[1] James Watt conducted experiments in 1782 to determine that a "brewery horse" was able to produce power equivalent to 32,400 ft-lb per minute. He standardized the conversion factor at 33,000 ft-lb per minute to classify the Boulton and Watt steam engine that his company was manufacturing at the time.

150 — Chapter 6 — Torsion of Structural Elements

velocity, it is in equilibrium and the torques T_1 and T_2 are equal in magnitude and opposite in direction (sign). The shaft transmits the torque from one disk to the other.

In subsequent sections of this chapter, the relations for determining the stresses and strains produced in the shaft by torque are shown. We will also demonstrate the methods for calculating the stresses, strains and deformations when designing a shaft for an infinite service life.

EXAMPLE E6.1

A small gasoline engine powers a string type trimmer used to cut grass and weeds. If the maximum power generated by the engine is 600 W, determine the torque imposed on the shaft from the engine to its string disk, if it is rotating at 520 RPM.

Solution: Using Eq. (6.1), we write:

$$T = P/\omega = 600/[2\pi(520/60)] = 11.02 \text{ N-m} \qquad (a)$$

Note that the angular velocity of the shaft was converted from RPM to radians per second by employing the conversion shown below:

$$\omega \text{ (radian/s)} = 2\pi(n/60) \qquad (b)$$

where n is the angular velocity in RPM.

We will return to this example later in this chapter and determine the stress in the shaft under these operating conditions.

EXAMPLE E6.2

The output shaft from an electric motor drives a compressor on a commercial refrigerator. If the motor operates at 3,560 RPM and is rated at 1.75 HP determine the torque on the shaft.

Solution: Substituting conversion factors into Eq. (6.1) and solving for the torque on the motor-compressor shaft gives:

$$T = 33,000 \text{ HP}/(2\pi n) \qquad (6.3)$$

where HP is the power transmitted by the shaft in horsepower and n is the angular velocity in RPM.

$$T = (33,000)(1.75)/[2\pi(3,560)] = 2.582 \text{ ft-lb} \qquad (a)$$

The torque applied to a shaft is a twisting moment. This moment (torque) acts about the centerline of the shaft and tends to twist it. Plane sections (disks), perpendicular to the axis of the circular shaft, remain plane as the shaft is twisted. This fact is extremely important, because it enables us to determine the stresses and deformation from relatively elementary equations for shafting with circular cross sections. In the next section, we will show how the strains are produced by twisting the shaft.

6.2 DEFORMATION OF A CIRCULAR SHAFT DUE TO TORSION

Let's consider a circular shaft of length L with one end built-in (fixed) and the other end free. A twisting moment (torque) is applied to its free end, as illustrated in Fig. 6.2. A straight line A-B is scribed on the outside surface of the shaft prior to the application of the torque T. Applying the torque, causes the shaft to deform, the scribe line rotates through an angle γ, and point B moves to a new location given by point C.

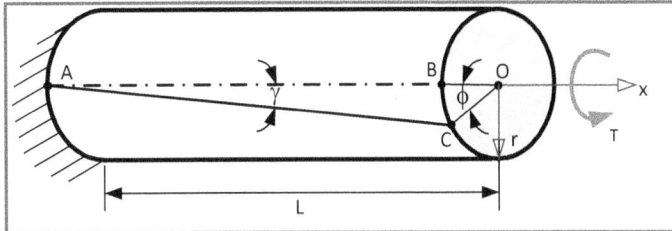

Fig. 6.2 Deformation of a circular shaft due to the application of a torque T at its free end.

The length of the arc B-C is given by:

$$BC = \gamma L \qquad (6.4)$$

The angle ϕ, defined in Fig. 6.2, is known as the angle of twist of the shaft. Let's examine the deformation of the circular shaft more closely, by considering a square scribed on its surface at some arbitrary position along line A-B, as shown on Fig. 6.3.

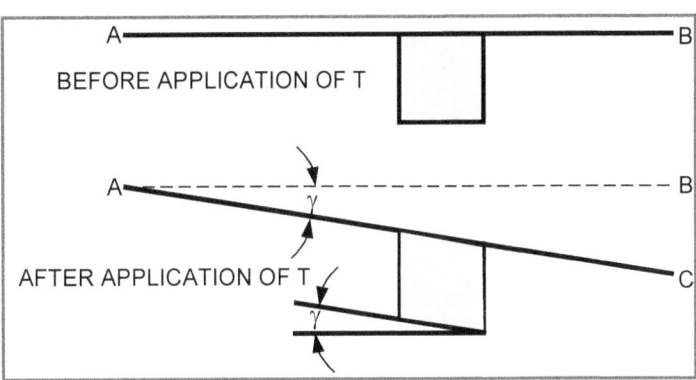

Fig. 6.3 Deformation of a small square scribed on the surface of the shaft before and after application of the torque.

When the shaft deforms under the action of the torque, the line A-B remains straight as it rotates, and is represented by the line A-C. The two sides of the inscribed square, which are perpendicular to the axis of the shaft do not rotate. However, both of the sides of the square that are parallel to the axis of the shaft rotate through the angle γ. Clearly, the element is distorted, as indicated in Fig. 6.3. The included angles of the square have changed 90° ± γ. From these geometric changes, it is clear that a shearing strain γ occurs on the external surface of the circular shaft.

6.2.1 Shear Strain Due to Torque Applied to a Circular Shaft

To determine a relation for the magnitude of the shearing strain γ, consider the free end of a circular shaft subjected to an applied torque T, as indicated in Fig. 6.4. The line OB rotates through an angle ϕ sweeping through an arc BC. The length of the arc B-C is given by:

$$BC = r\phi \qquad (6.5)$$

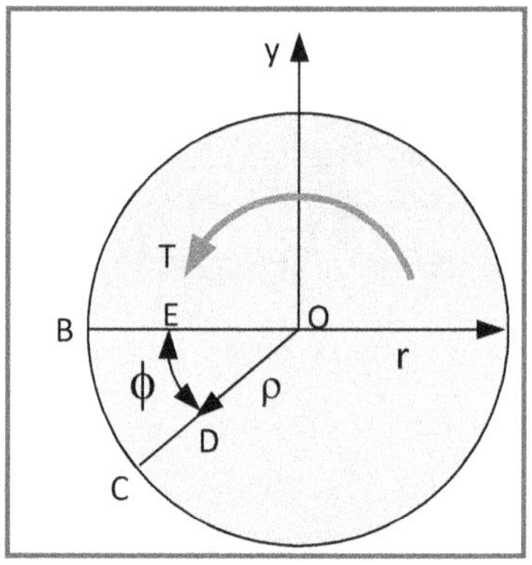

Fig. 6.4 Rotation of line OB on the free end of a circular shaft subjected to a torque T.

Substituting Eq. (6.4) into Eq. (6.5) yields:

$$\gamma_{\rho=r} = r\,\phi/L \tag{6.6}$$

The shear strain on the surface of the shaft, where $\rho = r$, is given by Eq. (6.6). The shear strain at some point D on the interior of the shaft is less, because the arc length D-E is decreased. Following the same procedure, we write the shearing strain γ for an arbitrary point defined by the position radius ρ as:

$$\gamma_\rho = \rho\,\phi/L \tag{6.7}$$

Examination of Eq. (6.7) shows that the shearing strain varies from zero at the centerline of the shaft to a maximum value at the surface that is given by:

$$\gamma_{Max} = r\,\phi/L \tag{6.8}$$

Because the applied torque is constant along the length of the shaft, the strain is also constant along its length.

6.3 STRESSES PRODUCED BY TORSION

To determine the shear stresses acting on the circular shaft, we employ the relation between shear stress and shear strain.

$$\tau = G\,\gamma \tag{6.9}$$

where G is the shear modulus of the shaft material given by:

$$G = \frac{E}{2(1+\nu)} \tag{6.9a}$$

Substituting Eq. (6.7) into Eq. (6.9) yields:

$$\tau_\rho = G\phi\rho/L \tag{6.10}$$

Again, we observe that the shear stress τ increases linearly with position ρ, from zero at the centerline to a maximum value at the surface of the shaft. The linear distribution of shear stress is shown in Fig. 6.5 a.

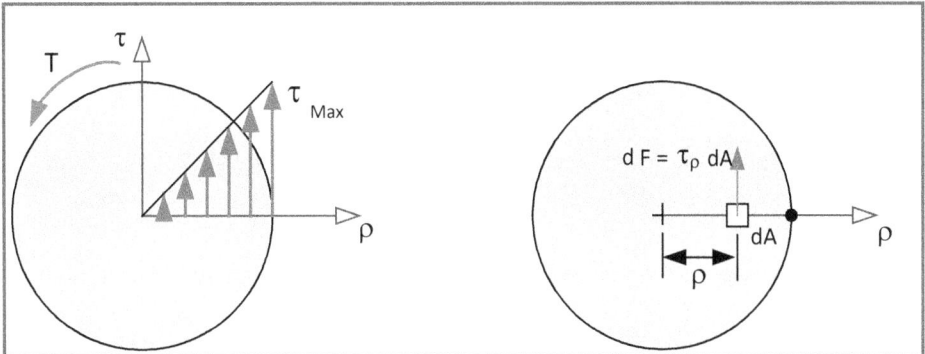

Fig. 6.5a Linear distribution of shear stress.   Fig. 6.5b The incremental force dF = τ dA.

Next, let's consider the equilibrium of the shaft subjected to an applied torque. The shear stress produces an equal and opposite moment to the applied torque at the supported end of the shaft. We write the equilibrium equation $\Sigma M_x = 0$ as:

$$\Sigma M_x = T - \int \rho \tau_\rho \, dA = 0 \tag{6.11}$$

where the x-axis coincides with the axis of the shaft.

The sum of all the incremental forces ($dF = \tau_\rho \, dA$) produce the moment M_x about the centerline of the shaft, as presented in Fig. 6.5b. Substituting Eq. (6.10) into Eq. (6.11) gives:

$$T = (G \phi \int \rho^2 \, dA)/L \tag{6.12}$$

The term $\int \rho^2 \, dA$ is the polar moment of inertia J of a circular area about its center. Hence, we can rewrite Eq. (6.12) as:

$$T = G \phi J/L \tag{6.13}$$

Finally, we note from Eq. (6.10) that $(G\phi/L) = \tau_\rho/\rho$. Accordingly, we can write:

$$\tau_\rho = T \rho/J \tag{6.14}$$

Equation (6.14) is employed to determine the shear stresses in circular shafts when the applied torque is known. Clearly, the maximum shear stress occurs when $\rho = r$. In this instance:

$$\tau_{Max} = T r/J = T d/(2J) \tag{6.15}$$

Because $J = \pi r^4/2 = \pi d^4/32$ for solid circular shafts, we may express Eq. (6.15) as:

$$\tau_{Max} = 2 T/(\pi r^3) = 16 T/(\pi d^3) \tag{6.16}$$

To demonstrate the method for determining stresses in circular shafts, let's consider two examples.

EXAMPLE E6.3

A solid circular shaft with a diameter d is fabricated from 1020 HR steel and subjected to a torque T = 5,400 in-lb. Determine the shear stress τ as a function of the diameter of the shaft.

Solution: To determine the shear stresses in the circular shaft, substitute the numerical parameters describing the shaft and the applied torsion into Eq. (6.16) to obtain:

$$\tau_{Max} = 16\,T/(\pi d^3) = (16)(5,400)/(\pi d^3) = 86.4 \times 10^3 \text{ in-lb.}/(\pi d^3) \tag{a}$$

We now have an equation relating the shear stress on the surface to the shaft diameter. Let's study several cases by permitting the diameter of the shaft to vary from 0.25 to 2.00 in. Using a spreadsheet in performing the calculations, we obtain the results shown below:

d (in.)	d^3 (in.^{3})	Stress, τ (ksi)
0.250	0.0156	1,760
0.375	0.0527	521.5
0.500	0.1250	220.0
0.625	0.2441	112.6
0.750	0.4219	65.19
0.875	0.6699	41.05
1.000	1.0000	27.50
1.250	1.9531	14.08
1.500	3.3750	8.149
1.750	5.3594	5.132
2.000	8.0000	3.438

The numerical values for the shear stress vary from 1,763 ksi for a shaft with a diameter of 0.25 in. to only 3.438 ksi for a 2.0 in. diameter shaft. Obviously, a large range in the results is evident. The very large variations in τ, with small changes in d, are due to the fact that the diameter occurs as a cube in the denominator of Eq. (6.16).

To determine a suitable diameter of the shaft, we must compare the stresses to the strength of the 1020 HR steel. This task raises a question — is it possible to compare shear stress with yield strength or ultimate tensile strength? The answer is **no**. Yield and ultimate tensile strength for materials are measured in a tensile test and cannot be directly compared to shear strength. Let's consider another example to introduce an accepted technique for approximating the shear strengths of most metals.

EXAMPLE E6.4

A shaft diameter of 1.250 in. is selected for the shaft described in Example 6.3. Determine the safety factor for this shaft, if the applied torque is 5,400 in.-lb.

Solution: To account for the effects of shear stress on failure, we determine yield S_{ys} and ultimate S_{us} strengths in terms of shear by multiplying the yield and ultimate tensile strengths, S_y and S_u, by a constant = $\sqrt{3}/3 = 0.5774$:

$$S_{ys} = 0.5774\, S_y \qquad (6.17a)$$

$$S_{us} = 0.5774\, S_u \qquad (6.17b)$$

Reference to Appendix B-2 indicates that $S_y = 42$ ksi and $S_u = 66$ ksi for 1020 HR steel. Substituting numerical values into Eq. (6.17) gives the yield and ultimate shear strength for 1020 HR steel as:

$$S_{ys} = 0.5774\, S_y = (0.5774)(42) = 24.25 \text{ ksi} \qquad (a)$$

$$S_{us} = 0.5774\, S_u = (0.5774)(66) = 38.11 \text{ ksi} \qquad (b)$$

The safety factor is determined by computing the ratio of strength to the stress according to:

$$SF = S_{ys}/\tau_{Max} \qquad (6.18)$$

Substituting the results from Eq. (a) and from the table for Example 6.3 into Eq. (6.18) gives:

$$SF = 24.25/14.08 = 1.722 \qquad (c)$$

We have used the yield strength in shear for the 1020 HR steel in determining the safety factor. Reference to Eq. (c) shows a safety factor of 1.722, which is less than the value often used. It is adequate only if the applied torque is specified accurately and the material properties certified.

EXAMPLE E6.5

A shaft, fabricated from 302 A stainless steel, is to transmit 2,500 W of power while rotating at 1,790 RPM. Specify the diameter of the shaft in millimeters, if the safety factor is to be in the range of 2.25 to 3.00.

Solution: Let's substitute Eq. (6.17 a) into Eq. (6.18) and rearrange terms to obtain:

$$\tau_{Max} = S_{ys}/SF = 0.5774\, S_y / SF \qquad (a)$$

Reference to Appendix B-2 indicates that $S_y = 234$ MPa for 302 A stainless steel.

Consider the lowest safety factor $SF = 2.25$; for this choice, determine τ_{Max} from Eq. (a).

$$\tau_{Max} = S_{ys}/SF = (0.5774)(234)/2.25 = 60.05 \text{ MPa} \qquad (b)$$

Then using Eq. (6.16) we write:

$$d^3 = (16T)/(\pi \tau_{Max}) \qquad (c)$$

> Compute the required diameter d, by determining the applied torque from Eq. (6.1).
>
> $$T = P/\omega = [(2,500)(60)]/[(1,790)(2\pi)] = 13.337 \text{ N-m} = 13,337 \text{ N-mm} \quad (d)$$
>
> Substituting Eq. (d) into Eq. (c) gives:
>
> $$d^3 = (16)(13,337)/[\pi(60.05)] = 1,131 \text{ mm}^3 \quad (e)$$
>
> Specify a shaft with d = 11 mm and note that d^3 = 1,331 mm^3. This choice is a standard size round bar in the SI system that is somewhat larger in diameter than d = 10.42 mm, which is necessary to achieve the required value of d^3 = 1,131 mm^3. Let's check to determine if an 11 mm diameter shaft provides a safety factor of less than 3.0.
>
> From Eq. (6.16), we write:
>
> $$\tau_{Max} = (16T)/(\pi d^3) = (16)(13,337)/[\pi(1,331)] = 51.03 \text{ MPa} \quad (f)$$
>
> Next, substitute the results from Eq. (f) into Eq. (a) to obtain:
>
> $$SF = 0.5774 S_y/\tau_{Max} = (0.5774)(234)/51.03 = 2.648 \quad (g)$$
>
> This result indicates that the diameter of 11 mm is sufficient to provide a safety factor that lies within the range 2.25 to 3.00, specified for this application.

6.3.1 Hollow Circular Shafts

In many engineering applications, weight is an important consideration, because it often detrimentally affects performance and usually increases cost. It is possible to reduce the weight of shafting, by using tubes instead of solid round shafts or bars. The weight penalty associated with solid round bars is due to the reduced stresses in central regions of the bar, which are much smaller than the maximum shear stress that occurs on the outer diameter of the shaft. With a tube, the central region of the shaft is removed and the resulting structure is more uniformly stressed and structurally more efficient.

Equation (6.14), derived for a solid circular shaft, is also valid for a hollow circular shaft. The only difference that arises is in the expression for the polar moment of inertia J. For a solid shaft J = $\pi d^4/32$; however, for a hollow circular shaft, the polar moment of inertia is given by:

$$J = (\pi/32)(d_o^4 - d_i^4) \quad (6.19)$$

where d_i and d_o represent the inside and outside diameters, respectively.

Substituting the results for Eq. (6.19) into Eq. (6.14) yields:

$$\tau_\rho = 32T\rho/[\pi(d_o^4 - d_i^4)] \quad (6.20)$$

The maximum shear stress occurs when $\rho = d_o/2$; hence Eq. (6.20) may be rewritten as:

$$\tau_{Max} = 16\,T\,d_o/[\pi(d_o^4 - d_i^4)] \qquad (6.21)$$

EXAMPLE E6.6

A hollow circular shaft is fabricated from a seamless tube of 1018 HR steel. It is subjected to a torque T = 10,000 in-lb. If the outside diameter of the tube is fixed at 2.75 in., determine the shear stress τ_{Max} as a function of the inside diameter of the shaft.

Solution: Let's employ Eq. (6.21) and substitute numerical values for the known quantities in this relation.

$$\tau_{Max} = (16)(10,000)(2.75)/\{\pi[(2.75)^4 - d_i^4)]\} = 140.1 \times 10^3 /[(57.19) - d_i^4)] \qquad (a)$$

Using a spreadsheet to evaluate Eq. (a) gives:

d_i (in.)	d_i^4 (in.)^4	Stress, τ (ksi)
0.000	0	2.450
0.250	0.0039	2.450
0.500	0.0625	2.452
0.750	0.3164	2.463
1.000	1.0000	2.493
1.250	2.4414	2.559
1.500	5.0625	2.688
1.750	9.3789	2.930
2.000	16.000	3.401
2.250	25.629	4.439
2.500	39.063	7.729
2.750	57.191	∞

Examination of these results shows that the maximum shear stress increases very slowly as the inside diameter of the tube increases. This fact indicates that the material removed from the shaft with increasing diameter of d_i carries very little shear stress. Compared to the shear strength of the 1018 HR steel, which is $S_{ys} = (0.5774)(32\text{ ksi}) = 18.48$ ksi, the maximum shear stresses are lower than this value for every inside diameter considered if $d_1 \leq 2.500$ in. However, the safety factor is only 2.391 when $d_i = 2.50$ in.

6.3.2 Section Modulus in Torsion

When considering beams in bending, we will introduce the concept of section modulus that defines the influence of the shape of the beam's cross-section on the flexural stress. Let's adapt the same concept to circular shafts in torsion by rewriting Eq. (6.21) as:

$$\tau_{Max} = 16\,T\,d_o/[\pi(d_o^4 - d_i^4)] = T/Z_T \qquad (6.22)$$

where Z_T is given by:

$$Z_T = \pi(d_o^4 - d_i^4)/(16 d_o) \qquad (6.23)$$

Clearly, to reduce the maximum shear stress it is necessary to increase the section modulus Z_T. In a hollow shaft, Eq. (6.23) indicates that both the internal and external diameters of the shaft influence the section modulus. To limit the shear stress, we seek to increase Z_T, but without the penalty of a large increase in the weight of the shaft. To establish the most suitable approach to achieve large Z_T values without incurring weight penalties, let's first consider q that is defined as the weight W per unit length L of the shaft.

$$q = W/L = \gamma(\pi/4)(d_o^2 - d_i^2) \qquad (6.24)$$

where γ is the weight per unit volume of the shaft.

Because it is good design practice to increase the section modulus while limiting the increase in q, it is useful to examine the ratio Z_T/q. Dividing Eq. (6.23) by Eq. (6.24) yields:

$$\left(\frac{Z_T}{q}\right)_{Hollow} = \frac{\left(\frac{\pi(d_o^4 - d_i^4)}{16 d_o}\right)}{\left(\frac{\gamma \pi(d_o^2 - d_i^2)}{4}\right)} = \frac{(d_o^2 + d_i^2)}{4 \gamma d_o} \qquad (6.25)$$

For a solid shaft, $d_i = 0$ and Eq. (6.25) reduces to:

$$(Z_T/q)_{Solid} = d_o/4\gamma \qquad (6.26)$$

Comparing the ratio Z_T/q for the hollow and solid shafts provides a method for assessing the improvement in the design by employing hollow shafting. Accordingly, we form the ratio shown below:

$$\frac{\left[\frac{Z_T}{q}\right]_{Hollow}}{\left[\frac{Z_T}{q}\right]_{Solid}} = 1 + \left(\frac{d_i}{d_o}\right)^2 \qquad (6.27)$$

As d_i becomes large and approaches d_o, the ratio in Eq. (6.27) approaches 2, indicating a hollow shaft with a very thin wall has nearly twice the section modulus per unit weight when compared to a solid shaft. Of course, there is a limit selected for the diameter d_i. When the wall thickness of a hollow shaft becomes too small, the shaft will fail by buckling under torsion.

The disadvantage of using hollow shafts in the design of power transmission shafts is cost and availability. It is more expensive to manufacture a tube than it is to manufacture a solid shaft. Sometimes the manufacturing costs are larger than the savings achieved by reducing the weight of the shaft. Also for equal values of Z_T, hollow shafts are larger in outside diameter than solid shafts. The larger diameter tubular shafts require more costly bearings to support their loads.

EXAMPLE E6.7

A 30 in. long shaft, fabricated from cold rolled alloy steel, exhibits a yield strength of 65 ksi and is to support a torque of 510 in.-lb. If a safety factor of at least 3.5 is employed, determine the weight of a solid shaft for this application. Also, determine the weight of a hollow shaft for this same application if $d_i = 0.70\, d_o$.

Solution: The design stress τ_{Design} is given by Eq. (6.17a) and Eq. (6.18) as:

$$\tau_{Design} = (0.5774)S_y/SF \qquad (a)$$

Substituting numerical parameters into Eq. (a) gives:

$$\tau_{Design} = (0.5774)(65)/(3.5) = 10.72 \text{ ksi} \qquad (b)$$

The diameter is determined from Eq. (6.16) as:

$$d^3 = (16T)/(\pi\tau_{Design}) = (16)(510)/[\pi(10{,}720)] = 0.2423 \text{ in.}^3 \qquad (c)$$

Solving Eq. (c) gives d = 0.6234 in. Selecting a commercially available standard size round bar gives:

$$d = 5/8 = 0.625 \text{ in.} \qquad (d)$$

The weight W of the shaft is given by Eq. (6.24) as:

$$W = \gamma(\pi/4)d^2 L = 0.283(\pi/4)(0.625)^2(30) = 2.605 \text{ lb.} \qquad (e)$$

where the weight density of steel is $\gamma = 0.283$ lb./in.^{3}

For the hollow shaft, the diameter is determined from Eq. (6.21) as:

$$[(d_o^4 - d_i^4)]/d_o = (16T)/\pi\tau_{Design} \qquad (f)$$

Substitute $d_i = 0.70\, d_o$ into Eq. (f) and simplify to obtain:

$$d_o^3 = (16)(510)/[(0.7599)(\pi)(10{,}720)] = 0.3189 \text{ in.}^3 \qquad (g)$$

Solving Eq. (g) for the outside diameter gives $d_o = 0.6832$ in. Selecting the next larger standard size tube, which is $d_o = 0.75$ in. gives:

$$d_o = 3/4 = 0.750 \text{ in. and } d_i = (0.70)(0.75) = 0.5250 \text{ in.} \qquad (h)$$

The weight is determined as:

$$W = \gamma(\pi/4)(d_o^2 - d_i^2) L = (0.283)(\pi/4)[(0.750)^2 - (0.5250)^2](30) = 1.913 \text{ lb.} \qquad (i)$$

The results show that the use of a hollow shaft in this application reduces the weight of the shaft from 2.605 lb. to 1.913 lb. This savings in shaft weight of 0.692 lb. (26.56%) reduces the overall weight of the product; however, the costs may increase, because of the need to procure tubing instead of solid rod.

6.3.3 Structural Efficiency of Circular Shafts

The term structural efficiency **SE** is defined as the section modulus of the shaft divided by its cross sectional area. Adapting this definition to torsion of both solid and hollow circular shafts gives:

$$\text{SE} = Z_T/A \tag{6.28}$$

Substituting Eq. (6.23) into Eq. (6.28) and simplifying yields:

$$\text{SE} = \frac{d_o^2 + d_i^2}{4d_o} \tag{6.29}$$

As d_i increases from zero towards the value of d_o, the structural efficiency increases from 0.250 d_o to 0.500 d_o. Of course, d_i must always be less than d_o. In fact, the tubular shaft will buckle in torsion if the wall thickness becomes too small.

6.4 SHEAR STRESSES ON DIFFERENT PLANES

6.4.1 Shear Stresses on Orthogonal Planes

Let's consider an elemental area ΔA scribed on the surface of a circular shaft as shown in Fig. 6.6. The x-axis is parallel to the shaft's longitudinal axis and the y-axis is perpendicular to it. With the x and y axes defined in this manner, the shear stress $\tau_{xy} = \tau_{Max} = 16T/\pi d^3$.

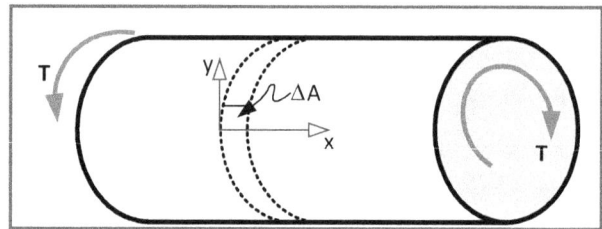

Fig. 6.6 An elemental area ΔA on the surface of a circular shaft subjected to torque T.

Let's remove this elemental area and draw a FBD of the element as shown in Fig. 6.7.

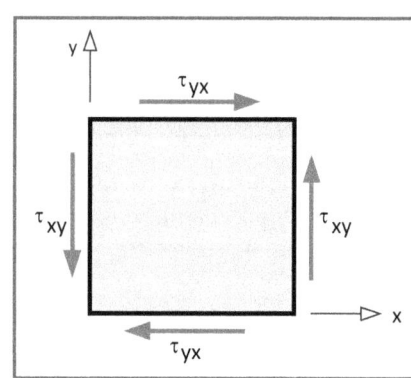

Fig. 6.7 A FBD of an elemental area ΔA from the surface of the shaft. All of the shear stresses shown on the element are considered positive.

In drawing the FBD of the elemental area, we note that both of the edges of the element that are perpendicular to the x-axis are subjected to the shear stress $\tau_{xy} = \tau_{Max}$. The convention for the double subscript placed on τ is:

- The first subscript identifies the outer normal to the surface upon which the stress acts.
- The second subscript shows the direction in which the stress acts.
- The stress is considered positive if the outer normal is positive and its direction is positive.
- The stress is considered positive if the outer normal is negative and its direction is negative.

Let's convert the shear stresses acting on the elemental area to shear forces and then write the equilibrium equations. The element has dimensions Δx, Δy and Δz; hence, the shear forces acting on the two sides of the element perpendicular to the x-axis are given by:

$$V_y = \tau_{xy} \Delta y \Delta z \qquad (6.30\ a)$$

Similarly, the shear forces acting on the sides perpendicular to the y-axis are:

$$V_x = \tau_{yx} \Delta x \Delta z \qquad (6.30\ b)$$

Next, consider the equilibrium equation $\Sigma F_y = 0$. This relation shows that the shear force V_y acting on the right side of the element in the positive y direction must be balanced by an equal and opposite force acting on the left side of the element. Similarly the equation $\Sigma F_x = 0$ indicates that the shear force V_x acting on the top side of the element must be balanced by an equal and opposite force acting on the bottom side of the element. We have shown these shear stresses in Fig. 6.7 with the correct directions, and have satisfied these two of the three applicable equilibrium relations.

Finally, apply the equilibrium equation $\Sigma M_z = 0$ about the origin and write:

$$(\tau_{yx} \Delta x \Delta z)\Delta y/2 = (\tau_{xy} \Delta y \Delta z)\Delta x/2 \qquad (a)$$

This relation reduces to:

$$\tau_{xy} = \tau_{yx} \qquad (6.31)$$

To satisfy the equilibrium equations, the shear stresses acting on orthogonal planes must be equal in magnitude.

6.4.2 Shear Stresses on Oblique Planes

To examine the shear stress acting on planes that are oblique relative to the shaft's axis, let's slice a corner from the elemental area, as shown in Fig. 6.8. If the inclined side of the triangular element has an area ΔA, the vertical and horizontal sides of the triangular element have areas of $\Delta A \cos \theta$ and $\Delta A \sin \theta$ respectively. Two stress components act on the inclined plane — one perpendicular to the surface (σ_n) and the other along the surface (τ_{nt}). Applying the equations of equilibrium to the triangular element gives:

$$\Sigma F_t = 0 \qquad (a)$$

$$\tau_{nt} \Delta A - [\tau_{xy} (\Delta A \cos \theta)]\cos \theta + [\tau_{yx} (\Delta A \sin \theta)] \sin \theta = 0 \qquad (b)$$

162 — Chapter 6 — Torsion of Structural Elements

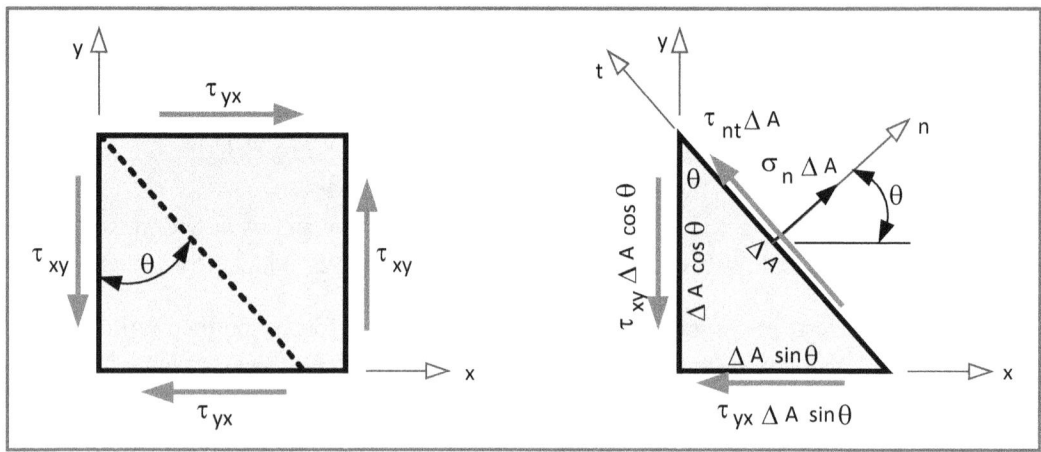

Fig. 6.8 Forces acting on a triangular element cut from the square element.

Solving Eq. (b) and recalling the results of Eq. (6.31) yields:

$$\tau_{nt} = \tau_{xy} (\cos^2 \theta - \sin^2 \theta) = \tau_{xy} \cos (2\theta) \qquad (6.32)$$

$$\Sigma F_n = 0 \qquad (c)$$

$$\sigma_n \Delta A - [\tau_{xy} (\Delta A \cos \theta)] \sin \theta - [\tau_{yx} (\Delta A \sin \theta)] \cos \theta = 0 \qquad (d)$$

Solving Eq. (d) and using Eq. (6.31) yields:

$$\sigma_n = 2 \tau_{xy} \cos \theta \sin \theta = \tau_{xy} \sin (2\theta) \qquad (6.33)$$

Equation (6.33) shows that a normal stress σ_n occurs on all planes, except those perpendicular or parallel to the axis of the shaft. Let's take $\tau_{xy} = 1$ unit and evaluate Eq. (6.33) by permitting θ to vary from 0 to 360°. The results obtained for the normal stress σ_n, which acts on the oblique plane, are presented in Fig. 3.9.

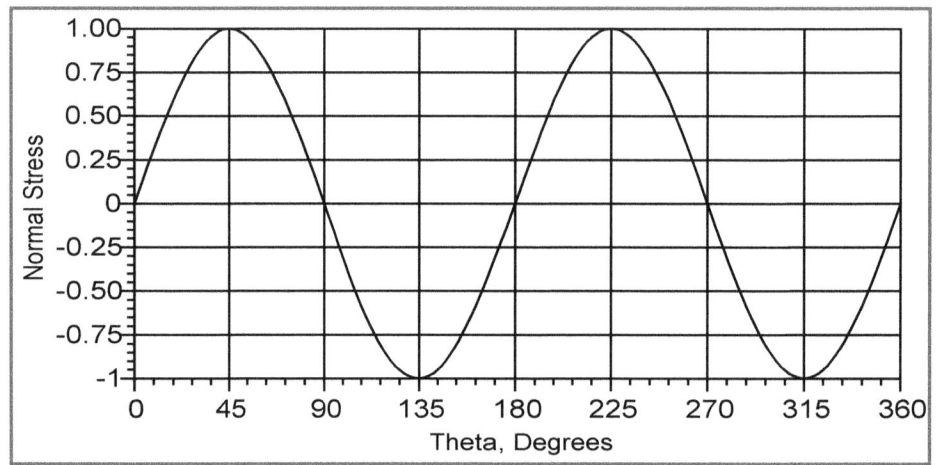

Fig. 6.9 The distribution of σ_n as a function of the orientation of the plane defined by θ.

The interpretation of the result shown in this graph will be deferred until Section 6.5.

EXAMPLE E6.8

If the shear stress $\tau_{xy} = 1$ unit, determine the shear stress τ_{nt} as a function of the orientation of the n-t plane defined by θ.

Solution: Let's employ $\tau_{nt} = \tau_{xy} \cos 2\theta$ and solve for τ_{nt} as a function of θ. We will make use of a spreadsheet and vary θ from 0 to 360° in steps of 5° in this evaluation. The results are shown in Fig. E6.8. It is clear that $\tau_{Max} = \tau_{xy} = 1$ unit when $\theta = 0°$ and 180° and $\tau_{Min} = -\tau_{xy} = -1$ unit when $\theta = 90°$ and 270°. The shear stress vanishes on planes defined by $\theta = 45°, 135°, 225°, 315°$, etc.

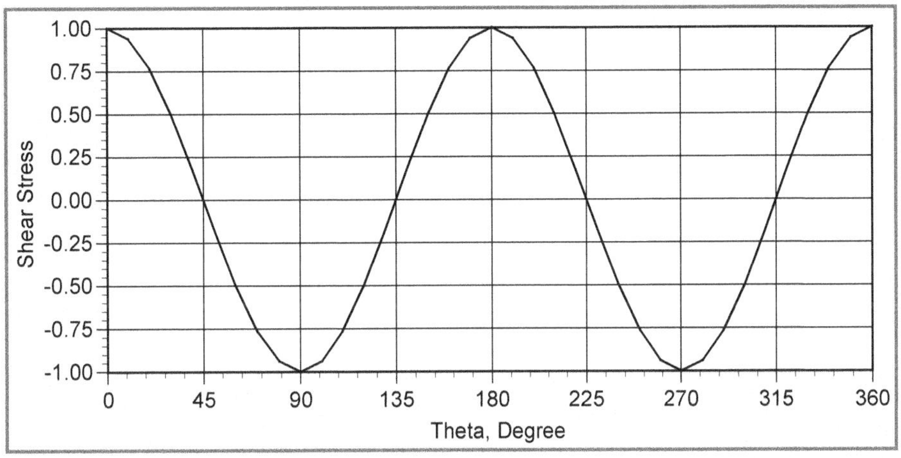

Fig. E6.8 Shear stress τ_{nt} as a function of the inclination angle θ.

6.5 PRINCIPAL STRESSES IN SHAFTS

It is evident from Fig. 6.9 that the normal stress is a maximum ($\sigma_n = \sigma_{Max} = \tau_{xy} = \tau_{Max}$) when $\theta = 45°$ and 225° and a minimum ($\sigma_n = \sigma_{Min} = -\tau_{xy} = -\tau_{Max}$) when $\theta = 135°$ and 315°. Also the shear stress on these four planes is given by Eq. (6.32) as:

$$\tau_{nt} = \tau_{xy} \cos 2\theta = 0 \quad \text{when } \theta = (1 + 2n)\pi/4 \text{ and } n = 0, 1, 2, \text{ etc.} \quad (6.32a)$$

When the normal stress is a maximum or a minimum acting on a plane with zero shear stress, a principal state of stress exists. The planes upon which the normal stress acts are called principal planes. We usually attempt to determine the direction of the principal planes and the magnitude of the principal stresses, because failure often occurs along these planes when the stresses become excessive.

For a circular bar subjected to a torque T, the principal plane makes an angle of 45° with the bar's axis as shown in Fig. 6.10. The principal plane spirals like a helix along the length of the bar. If this bar is made of a brittle material such as gray cast iron or chalk, its failure surface forms a helix as illustrated in Fig. 6.11.

Fig. 6.10 The principal planes for a circular shaft subjected to torque T form a helix about its surface.

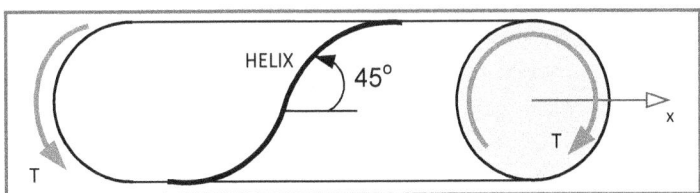

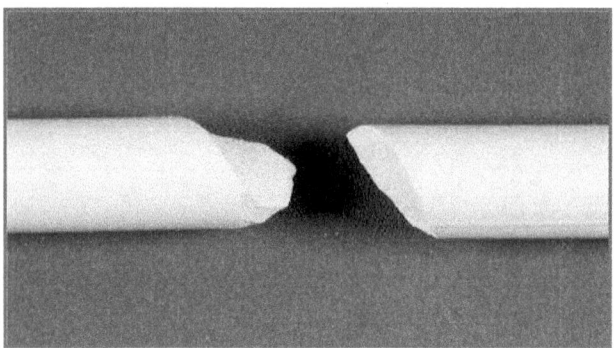

Fig. 6.11 Twisting a piece of chalk produces a helical failure making a 45° angle with its axis.

EXAMPLE E6.9

A circular shaft 15 mm in diameter is transmitting 10.0 kW from an electric motor to an air compressor. If the motor is operating at 1,800 RPM, determine the maximum normal stress acting on the shaft. Also determine the plane upon which this maximum normal stress acts.

Solution: We solve for the torque acting on the shaft by using Eq. (6.1) to obtain:

$$T = P/\omega = (10,000)(60)/(1,800)(2\pi) = 53.05 \text{ N-m} \quad (a)$$

The maximum shear stress may be determined from Eq. (6.16) as:

$$\tau_{Max} = \tau_{xy} = 16T/(\pi d^3) = (16)(53.05 \times 10^3)/[\pi(15)^3] = 80.06 \text{ N/mm}^2 = 80.06 \text{ MPa} \quad (b)$$

The relation between the shear stress and the normal stress is given by Eq. (6.33) as:

$$\sigma_n = \tau_{xy} \sin(2\theta) \quad (c)$$

The normal stress is a maximum when $2\theta = \pi/2$ or $\theta = \pi/4$. This result for the principal angle is consistent with the helix presented in Figs. 6.10 and 6.11.

In this example, the maximum normal stress $(\sigma_n)_{Max}$ is the principal stress σ_1 given by:

$$(\sigma_n)_{Max} = \sigma_1 = \tau_{xy} \quad (6.34)$$

$$(\sigma_n)_{Max} = \sigma_1 = \tau_{xy} = 80.06 \text{ MPa} \quad (d)$$

6.6 ANGLE OF TWIST

When torque is applied to a shaft, it twists through some angle ϕ about its longitudinal axis. This angle of twist of a circular shaft, illustrated in Fig. 6.2, is analogous to the elongation in a tension rod under the action of an axial force. To derive the expression for the angle of twist, rewrite Eq. (6.8) as:

$$\phi = \gamma_{Max} L/r \quad (a)$$

Substituting Eq. (6.9) into Eq. (a) yields:

$$\phi = (\tau_{Max} L)/(G r) \qquad (b)$$

Finally, substituting Eq. (6.15) into Eq. (b) gives:

$$\phi = (T L)/(G J) \qquad (6.35a)$$

Because the polar moment of inertia $J = \pi d^4/32$, we rewrite Eq. (6.35 a) as:

$$\phi = (32 T L)/(\pi d^4 G) \qquad (6.35b)$$

where is the shear modulus defined in Eq. (6.9a).

If T, G or J are not constant along the length of the shaft, then Eq. (6.35a) must be modified to accommodate the different shaft sizes by writing:

$$\phi_{Total} = \sum_{i=1}^{n} \frac{T_i L_i}{G_i J_i} \qquad (6.35c)$$

In this instance, the shaft must be divided into several sections and Eq. (6.35a) used to determine ϕ for each section. The total angle of twist, ϕ_{Total} is calculated by summing the angle ϕ from each section.

Let's consider two examples to demonstrate the application of Eq. (6.35).

EXAMPLE E6.10

A solid steel shaft is 325 mm long and is subjected to a torque T of 445 N-m. If the shaft is 38 mm in diameter, determine the shear stress τ_{Max} and the angle of twist ϕ.

Solution: Let's employ Eq. (6.16) to solve for the maximum shear stress as:

$$\tau_{Max} = (16T)/(\pi d^3) = (16)(445)(1,000)/[\pi(38)^3] = 41.30 \text{ MPa} \qquad (a)$$

The angle of twist is given by Eq. (6.35b) as:

$$\phi = (32TL)/(\pi d^4 G) = (32)(445)(0.325)/[\pi(0.038)^4 (79 \times 10^9)]$$

$$\phi = 0.008943 \text{ radian} = 0.5124° \qquad (b)$$

Inspection of Eq. (b) shows that the angle of twist is given in radians. The result is converted to degrees by multiplying by the conversion factor $180°/\pi$.

EXAMPLE E6.11

A line shaft, shown in Fig. E6.11, fabricated from high-strength steel incorporates five spur gears. Gear A provides the input torque T_A and gears B, C, D and E each remove torque from the shaft. Determine the angle of twist for the entire shaft AE and between gears at A and C. The shaft diameters and lengths are d_{A-B} = 2.00 in., d_{B-C} = 1.75 in., d_{C-D} = 1.50 in. and d_{D-E} = 1.25 in., L_{A-B} = 30 in., L_{B-C} = 24 in., L_{C-D} = 36 in. and L_{D-E} = 48 in. The applied torques are T_A = 900 ft-lb, T_B = 225 ft-lb, T_C = 300 ft-lb, T_D = 262.5 ft-lb and T_E = 112.5 ft-lb. Neglect the contribution of gear deformations.

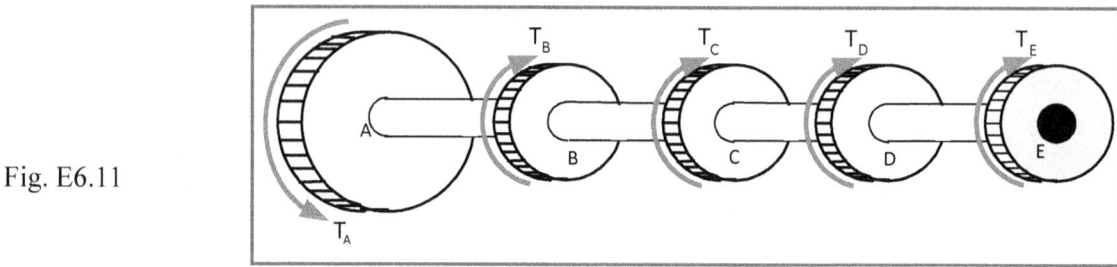

Fig. E6.11

Solution: First, prepare a series of FBDs at each section of the shaft and determine the torque acting on the shaft segments between the gears. Then construct a torque-position diagram that shows the distribution of torque along the line shaft as shown in Fig. E6.11a.

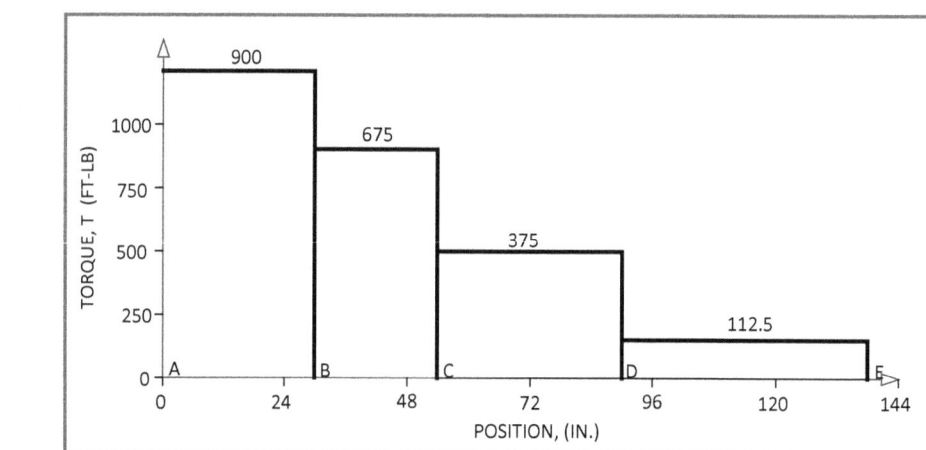

Fig. E6.11a

Next, write Eq. (6.35b) to determine the angle of twist for each segment of the line-shaft.

$$\phi_{A-B} = \frac{32TL}{\pi d^4 G} = \frac{32(-900)(12)(30)}{\pi (2.00)^4 (11.5 \times 10^6)} = -0.01794 \text{ radians} = -1.028°$$

$$\phi_{B-C} = \frac{32TL}{\pi d^4 G} = \frac{32(-675)(12)(24)}{\pi (1.75)^4 (11.5 \times 10^6)} = -0.01836 \text{ radians} = -1.052°$$

$$\phi_{C-D} = \frac{32TL}{\pi d^4 G} = \frac{32(-375)(12)(36)}{\pi (1.50)^4 (11.5 \times 10^6)} = 0.02834 \text{ radians} = -1.624°$$

$$\phi_{D-E} = \frac{32TL}{\pi d^4 G} = \frac{32(-112.5)(12)(48)}{\pi (1.25)^4 (11.5\times 10^6)} = 0.02351 \text{ radians} = -1.347°$$

The angle of twist for the entire shaft AE is given by Eq. (6.35c) as:

$$\phi_{A-E} = \phi_{A-B} + \phi_{B-C} + \phi_{C-D} + \phi_{D-E} = -1.028° - 1.052° - 1.624° - 1.347° = -5.051°$$

The angle of twist between gears A and C can be determined from Eq. (6.35c) as:

$$\phi_{A-C} = \phi_{A-B} + \phi_{B-C} = -1.028 - 1.051° = -2.079°$$

The negative signs describe the sense of rotation (clockwise).

6.6.1 Torsion Bar Springs

When torque is applied to a shaft, it twists through an angle ϕ and strain energy is stored in the shaft. The strain energy E is equal to the work W input to the shaft as it twists. We depict this work as the shaded triangular area in Fig. 6.12.

The strain energy $\mathcal{E}$ is given by:

$$\mathcal{E} = (1/2) T \phi \tag{6.36}$$

Substituting Eq. (6.35a) into Eq. (6.36) gives:

$$\mathcal{E} = (T^2 L)/(2GJ) \tag{6.37}$$

$$\mathcal{E} = (\phi^2 G J)/(2L) \tag{6.38}$$

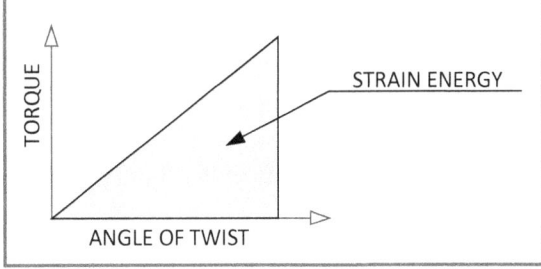

Fig. 6.12 The area under the torque-angle of twist curve is equal to the strain energy stored in a shaft.

The shaft acts as a torsion spring with the applied torque linearly proportional to the angle of twist.

$$T = k_T \phi \tag{6.39}$$

k_T the spring rate is given by:

$$k_T = GJ/L \tag{6.40}$$

EXAMPLE E6.12

A steel torsion bar is incorporated into the chassis of an automobile to serve as a suspension spring. If the bar is fabricated from a steel rod 15 mm in diameter and 1.0 m long, determine its spring rate. Also determine its angle of twist if it stores strain energy $\mathcal{E} = 4.0$ N-m as the automobile traverses a bump.

Solution: We apply Eq. (6.40) to obtain:

$$k_T = GJ/L = [(79)(10^9)\pi(0.015)^4]/[(1.0)(32)] = 392.6 \text{ N-m} \quad (a)$$

From Eq. (6.38) and Eq. (a), it is evident that:

$$\phi = [2\mathcal{E}/k_T]^{1/2} = [(2)(4.0)/(392.6)]^{1/2} = 0.1427 \text{ radians} = 8.179° \quad (b)$$

The result shows that the torsion spring undergoes a relatively large angle of twist. Will it survive over the service life of an automobile?

6.7 DESIGN OF POWER TRANSMISSION SHAFTING

Power transmission shafts connect a power source such as an electric motor or an internal combustion engine to industrial equipment such as a compressor or a machine tool. Power transmission shafts are circular in cross section, because they rotate (often at very high speed), and are supported by bearings. Shafting is usually fabricated from an alloy steel because of its superior strength. Because the shafting rotates and the stresses imposed are usually cyclical, the design is based on the fatigue strength of the steels. Also, stress concentrations are taken into account in a typical design analysis. Methods for designing to accommodate stress concentrations and fatigue loading are described in Section 6.11.

The relations for determining the shearing stress in shafting were introduced in a previous section of this chapter. The purpose of this section is to demonstrate the application of these relations in a design analysis of two challenging problems.

EXAMPLE E6.13

The electric motor shown in Fig. E6.13 is rated at 250 kW and operates at 1,180 RPM. It powers a gear drive with a variable load, which at times requires the motor to operate at its maximum rating. The shaft is supported by four bearings — one on each side of the gear drive, shown in Fig. E6.13, and two bearings within the motor to support the armature. The shaft has a diameter D. Select a material for the shaft and determine its diameter, if the maximum shear stress in the material is not to exceed 0.3 times the yield strength of the material in shear. Another constraint on the design is the angle of twist ϕ, which is not to exceed 6.0° measured from the motor to the gear drive.

Fig. E6.13

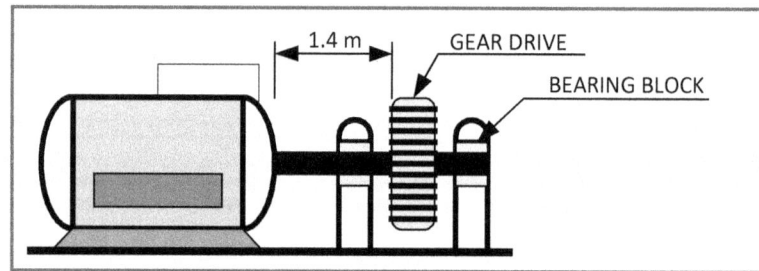

Solution: A step-by step approach will be used in the solution of this example.

Step 1: We begin by selecting the material for the shaft. Alloy steel, 4340 HR, is specified because it exhibits a very high strength. The high strength is important, because it will reduce the diameter of the shaft, and hence the diameter of the bearings and the diameter of the armature in the electric motor. The cost savings in the bearings and armature will offset the added cost of the alloy steel.

From Appendix B-1 and B-2, we determine the material's strength and shear modulus as:

$$G = 79 \text{ GPa} \qquad S_u = 1,041 \text{ MPa} \qquad S_y = 910 \text{ MPa} \qquad (a)$$

Next we use Eq. (6.17a) to determine the yield strength of the 4340 HR alloy steel in shear.

$$S_{ys} = 0.5774 \, S_y = (0.5774)(910) = 525.4 \text{ MPa} \qquad (b)$$

Step 2: Determine the maximum shear stress that can be imposed on the shaft from the problem statement and Eq. (b) as:

$$\tau_{Max} = (0.3)S_{ys} = (0.3)(525.4) = 157.6 \text{ MPa} \qquad (c)$$

Step 3: Determine the torque applied to the shaft when the electric motor is operating at its rated capacity by using Eq. (6.1):

$$T = P/\omega = [(250 \times 10^3)(60)]/[(1,180)(2\pi)] = 2,023 \text{ N-m} \qquad (d)$$

Step 4: If we employ a solid steel shaft, which is the most cost effective choice, Eq. (6.16) can be used together with Eq. (c) and Eq. (d) to determine the diameter D of the shaft as:

$$D^3 = (16T)/(\pi\tau_{Max}) = (16)(2,023)/[\pi(157.6 \times 10^6)] = 65.37 \times 10^{-6} \text{ m}^3$$

$$D = 0.04028 \text{ m} = 40.28 \text{ mm} \qquad (e)$$

Select a standard size available in metric diameters, which is 45 mm.

Step 5: Check to determine if the angle of twist ϕ is within the specified limit of 6.0° by employing Eq. (6.35b).

$$\phi = (32T\,L)/(\pi D^4 G) = [(32)(2,023)(1.4)]/[\pi(0.045)^4(79 \times 10^9)]$$

$$\phi = 0.08905 \text{ radians} = 5.102° \qquad (f)$$

This angle of twist is within the specifications outlined in the problem statement. It is clear that the controlling parameter in the design of this shaft is the maximum shear stress and not the angle of twist. However, the shaft diameter of 45 mm provides a design that is well balanced with the angle of twist within 0.90 degrees of the maximum permitted.

EXAMPLE E6.14

The electric motor shown in Fig. E6.14 is rated at 400 HP and operates at 1,780 RPM. It powers a drive shaft with three output gears. When the motor is operating at rated capacity, gear A removes 50% the power developed by the motor, while gear B and gear C remove 30% and 20%, respectively. The shaft is supported by the bearings shown in Fig. E6.14. The shaft from the motor has a diameter D. If an alloy steel with a yield strength in shear of 70 ksi is used to fabricate the shaft, determine the diameter of the shaft, if the maximum shear stress in the material is not to exceed 40% the yield strength of the material in shear. Another constraint on the design is the angle of twist ϕ, which is not to exceed 5.5° measured from the motor to the gear C.

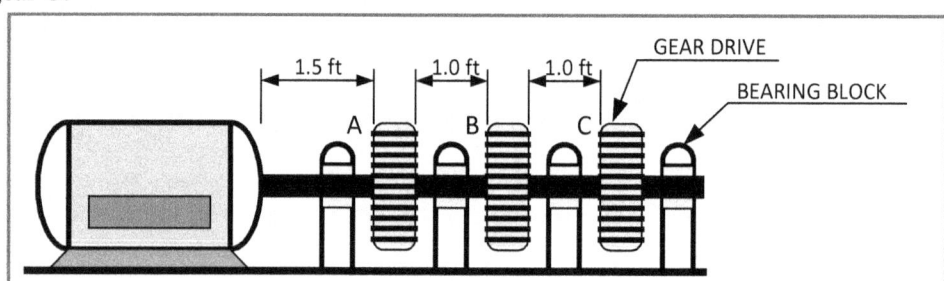

Fig. E6.14

Solution: The shaft is of uniform diameter from the motor to the free end of the shaft. A step-by-step solution is given below:

Step 1: Determine the torque exerted on the shaft when the motor is operated at its rated capacity from Eq. (6.3).

$$T = 33,000 \text{ HP}/2\pi n = (33,000)(400)/(2\pi)(1,780) = 1,180 \text{ ft-lb} \qquad (a)$$

Step 2: Compute the torque removed at each gear by using the fractions listed in the problem statement and construct a torque position diagram as shown in Fig. E6.14a:

$$T_A = 1,180 (0.50) = 590 \text{ ft-lb}, \quad T_B = 1,180 (0.3) = 354.0 \text{ ft-lb}, \quad T_C = 1,180 (0.20) = 236.0 \text{ ft-lb}$$

Fig. E6.14a

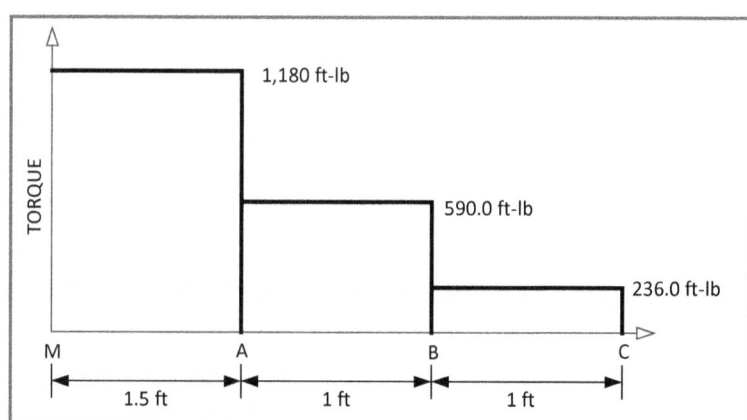

Step 3: Determine the maximum shear stress that is allowable for the shaft from:

$$\tau_{Max} = (0.40)(70) = 28.0 \text{ ksi} \qquad (b)$$

Step 4: If a solid steel shaft is employed, Eq. (6.16) can be used together with Eq. (a) and Eq. (b) to determine the diameter D of the shaft as:

$$D^3 = (16\,T)/(\pi\,\tau_{Max}) = (16)(1{,}180)(12)/[\pi(28.0 \times 10^3)] = 2.576 \text{ in.}^3$$

$$D = 1.371 \text{ in.} \qquad (c)$$

Select a standard size diameter as D = 1.375 in.

Step 5: Determine the total angle of twist by summing the twist of the shaft between each of the gears. We will assume that the twist due to the deflection of the gear teeth is sufficiently small to neglect. From Eqs. (6.35b) and (6.35c), we write:

$$\phi_{M-C} = \phi_{M-A} + \phi_{A-B} + \phi_{B-C} = [(32)/(\pi D^4 G)][T_{M-A} L_{M-A} + T_{A-B} L_{A-B} + T_{B-C} L_{B-C}]$$

$$\phi_{M-C} = \{32/[\pi(1.375)^4(11.5 \times 10^6)]\}\{(1{,}180)(12)(18) + (590.0)(12)(12) + (236.0)(12)(12)\}$$

$$\phi_{M-C} = 0.09263 \text{ radians} = 5.307° \qquad (d)$$

Clearly this angle of twist is less that the allowable twist of 5.5° and it is not necessary to modify the diameter or the material selected for this design:

6.8 TORSION OF NON-CIRCULAR SHAFTS

The discussion of torsion thus far has focused on rotating shafts with circular cross sections, because most applications involve transmitting power from a motor or an engine to an appliance. However, structural engineers sometimes load members such as angle and channel sections with transverse forces that induce torque. The derivations of the equations for the maximum shear stress and the angle of twist in shafts with non-circular cross sections are beyond the scope of this text. For this reason, this section is limited to the definition of the cross section geometry and the presentation of the equations for τ_{Max} and ϕ.

Elliptical Cross Section

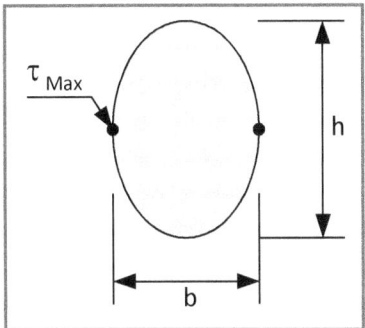

$$\tau_{Max} = 16T/\pi b^2 h$$
$$\phi = (4\pi^2\,T\,L\,J)/(A^4\,G) \qquad (6.41)$$
$$J = (\pi/64)(b\,h^3 + b^3\,h)$$
$$A = \pi\,b\,h/4$$

Equilateral Triangle

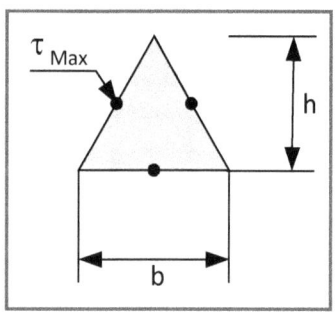

$$\tau_{Max} = 20T/b^3$$
$$\phi = (T\,L)/(0.6\,G\,J) = 46.2T\,L/b^4\,G \qquad (6.42)$$

Hexagon

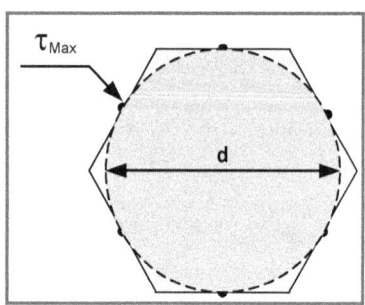

$$\tau_{Max} = T/[0.217\,Ad]$$
$$\phi = (T\,L)/(0.133\,A\,d^2\,G) \qquad (6.43)$$
$$A\ 0.866d^2$$

Rectangle

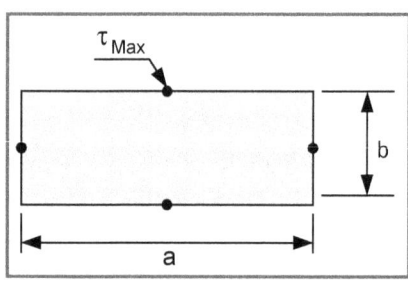

$$\tau_{Max} = T/(\alpha\,a\,b^2) \qquad \text{for } a \geq b$$
$$\phi = (T\,L)/(\beta\,a\,b^3\,G) \qquad (6.44)$$

$$\alpha = 0.3334 - 0.1904(b/a) - 0.2574(b/a)^2 + 1.0255(b/a)^3$$
$$\quad - 1.0946(b/a)^4 + 0.3916(b/a)^5$$
$$\beta = 0.3333 - 0.1997(b/a) - 0.0790(b/a)^2$$
$$\quad + 0.1892(b/a)^3 - 0.1766(b/a)^4 + 0.0739(b/a)^5$$

Thin Rectangle

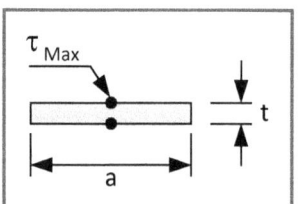

$$\tau_{Max} = 3.0\,T/at^2$$
$$\phi = (3T\,L)/(a\,t^3\,G) \qquad (6.45)$$

Angle Section

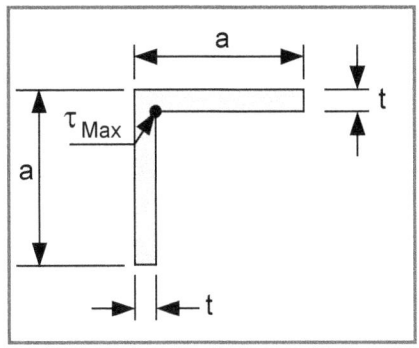

$\tau_{Nom} = 3.0\, T/2a\, t^2$
$\phi = (3T\, L)/(2a\, t^3\, G)$ (6.46)
for $t \ll a$

The maximum shear stress occurs at the reentrant corner of the angle, channel and I sections. The stress concentration associated with the reentrant corner is usually neglected, if the section is fabricated from a ductile material and is not subjected to cyclic loading.

Channel Section

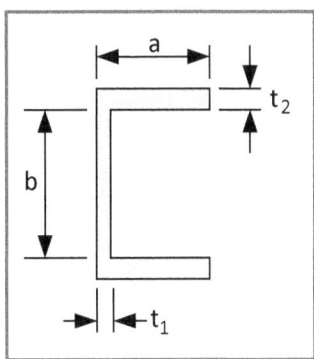

$\tau_{Nom} = 3.0\, T\, t_2 /(b\, t_1^3 + 2a\, t_2^3)$
$\phi = (3TL)/[(b\, t_1^3 + 2a\, t_2^3)G]$ (6.47)

I Section

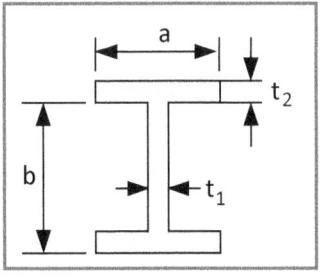

$\tau_{Nom} = 3.0\, T\, t_2 /(b\, t_1^3 + 2a\, t_2^3)$
$\phi = (3T\, L)/[(\, b\, t_1^3 + 2a\, t_2^3)\, G]$ (6.48)

EXAMPLE E6.15

A torque T = 325 N-m is applied to a structural member with a square cross section. If the square section is 25 mm on a side and is 2.8 m long, determine the maximum stress and the angle of twist. Will the structural member yield if it is fabricated from 1010 A carbon steel?

Solution: First calculate the numerical factors α and β from Eq. (6.44):

$$\text{For } (b/a) = 1.0 \quad \Rightarrow\Rightarrow \alpha = 0.2081; \quad \beta = 0.1411$$

Reference Eq. (6.44) and write:

$$\tau_{MAX} = T/(\alpha ab^2) = (325)/[(0.2081)(0.025)^3] = 99.95 \text{ MPa} \tag{a}$$

Next use Eq. (6.17a) to determine the yield strength of the 1010A carbon steel in shear.

$$S_{ys} = 0.5774 S_y = (0.5774)(200) = 115.5 \text{ MPa} \tag{b}$$

It is clear that the structural member will not yield because S_{ys} exceeds τ_{Max} by a factor of 1.156; however, the margin is small for structural applications.

The angle of twist for the square cross section is given by:

$$\phi = (TL)/(\beta a b^3 G) = (325)(2.8)/[(0.1411)(.025)^4 (79 \times 10^9)] = 0.2090 \text{ radians}$$

$$\phi = 11.97° \tag{c}$$

6.9 SHEAR CENTER

The topic of shear center combines the concept of shear stresses induced by transverse forces applied to beams and the angle of twist in torsion. Twisting of beams in bending does not arise for beams with symmetric cross sections. If the cross section of the beam is symmetric about its vertical axis, the shear stresses, which exist, do not produce a twisting moment. However, if the beam's cross-section is not symmetric, transverse loads will cause the beam to twist. To illustrate this point, consider a cantilever beam, fabricated from a channel section loaded at its free end with a concentrated force F. The concentrated force is applied through the center of gravity of the channel section, as shown in Fig. 6.13. When the load is applied, the beam will twist about its longitudinal axis through an angle φ.

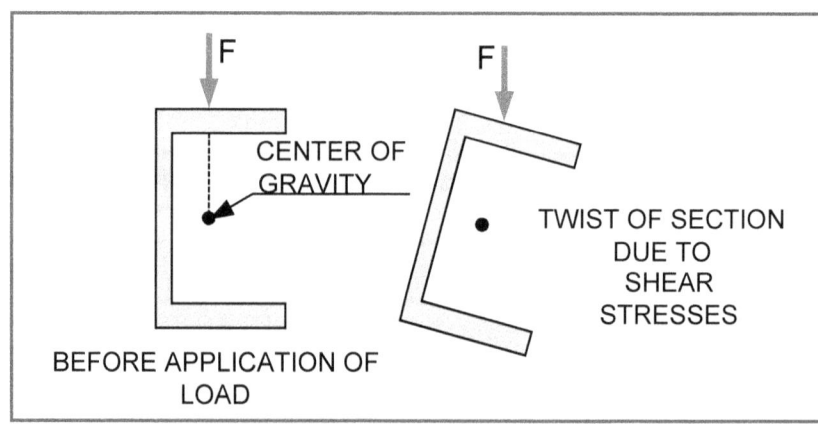

Fig. 6.13 Twisting of unsymmetrical beam section induced by shear stresses.

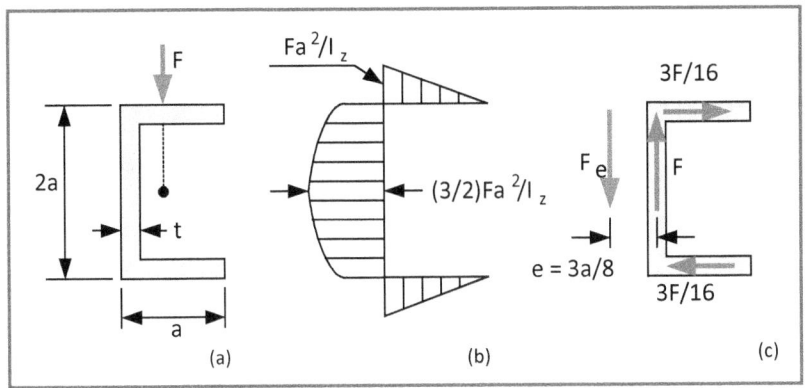

Fig. 6.14 Shear stress distribution in a channel section subjected to shearing forces V = F.

Let's consider a cantilever beam fabricated from a channel section, with the force F applied at its free end through its center of gravity, as illustrated in Fig. 6.14a. A constant shear force V = F occurs over the length L of the beam. This shear force produces shear stresses in both the web and the flanges of the channel section that are given by Eq. (6.49) as[2]:

$$\tau = \tau_V = \tau_H = V\, Q_1/(b\, I_z) \tag{6.49}$$

where
$$Q_1 = \int_{y_1}^{c} y\, dA \tag{6.50}$$

where Q_1 is the first moment of the area from the section cut at position y to the location of the top fiber of the beam at y = c.
V is the internal shear force
I_z is the moment of inertia about the neutral axis of the beam.
b is the width of the beam

If the shear stresses in the channel section are determined from Eq. (6.49), the distribution of τ_V and τ_H are shown in Fig. 6.14b. In this solution of Eq. (6.49) higher powers of the dimension ratio (t/a) have been disregarded, because they are small in relation to the other terms. Integrating the shear stress distribution over the three rectangular areas comprising the channel section and substituting:

$$I_z = 8a^3 t/3 \tag{a}$$

Into Eq. (6.49) yields the three shear forces shown in Fig. 6.14c.

Examination of Fig. 6.14c shows that the shear force V in the vertical portion of the channel is equal to the applied force F as expected. The two horizontal forces 3F/16 are equal and opposite and cancel satisfying $\Sigma F_z = 0$. However, these two forces produce a couple, a torque T = (3/8)Fa, which tends to twist the channel section (in a clockwise direction) producing the twisted shape presented in Fig. 6.13.

The shear center is established from this twisting moment. If a transverse force $F_e = F$ is applied at an eccentric position, e = (3/8)a, from the centerline of the vertical segment of the channel, as shown in Fig. 6.14c, it produces a couple with T = (3/8)Fa in the counter clockwise direction. Thus, the couples cancel and the channel section will not twist, when the transverse force F is applied at the eccentric

---

[2]Equation (6.49) will be derived later in Chapter 7.

176 — Chapter 6 — Torsion of Structural Elements

position. When loading unsymmetrical sections, the transverse force must be applied through the shear center instead of the center of gravity to avoid twisting deformations.

Channel sections and other unsymmetrical sections are often employed in structural applications. In these cases, attachments are welded to the beams to allow for the eccentric placement of the applied forces, as indicated in Fig. 6.15.

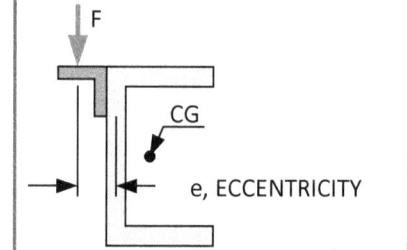

Fig. 6.15 Eccentric placement of the applied force to suppress twisting due to shear stresses.

EXAMPLE E6.16

The channel section presented in Fig. E6.16 is to be employed in fabricating a cantilever beam. Determine the eccentricity e of the transverse force F if the beam is to deflect without twisting.

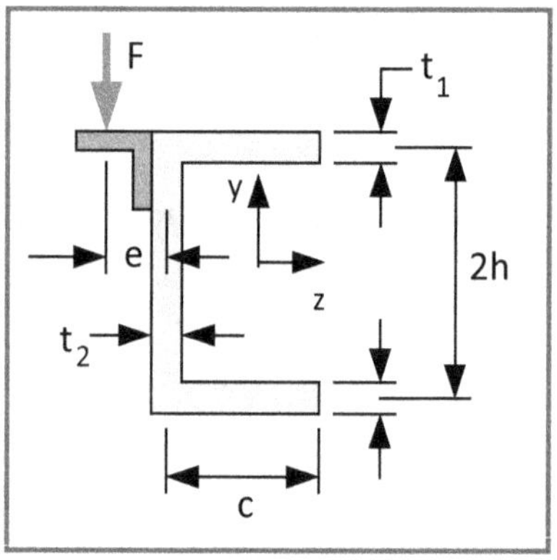

Fig. E6.16

Solution:

Step 1: Prepare FBDs showing the shear stresses introduced by the shear force V = F. Section the top flange of the channel section a distance d from its edge as shown in Fig. E6.16a.

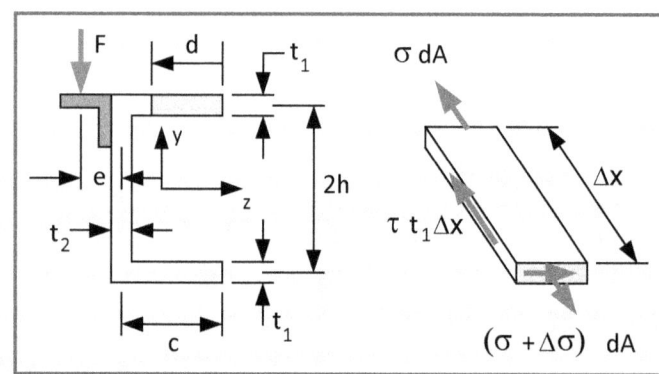

Fig. E6.16a

Step 2: The shear stress τ in the top flange is determined from Eq. (6.49) as:

$$\tau = V Q_1 / I_z b = (F h d / I_z) \tag{a}$$

where $V = F$, $b = t_1$ and $Q_1 = h\,d\,t_1$

The shear stress in the flange is a linear function of d increasing from zero at the edge of the flange to a maximum value where the flange intersects the web of the channel and $d = c$. Clearly τ_{Max} is given by:

$$(\tau_{Max})_{Flange} = F\,h\,c/I_z \tag{b}$$

Step 3: The shear stress at the top of the web is also determined from Eq. (6.49) as:

$$\tau = V\,Q_1/I_z\,b = [(F\,h\,c\,t_1)/(t_2\,I_z)] \tag{c}$$

where $V = F$, $b = t_2$ and $Q_1 = h\,c\,t_1$

Step 4: The shear stress is a maximum at the center of the web where $y = 0$. From Eq. (6.49), we write:

$$\tau = V\,Q_1/I_z\,b = Fh[c(t_1/t_2) + (h/2)]/I_z \tag{d}$$

where $V = F$, $b = t_2$ and $Q_1 = h\,[c\,t_1 + t_2\,(h/2)]$

The shear stress exhibits a parabolic distribution over the height of the web. If this shear stress is integrated over the area of the web, we find:

$$\int \tau\,dA_{Web} \approx F = V \tag{e}$$

Step 5: The shear stress in the flanges varies linearly from zero to a maximum at the junction of the web and the flange. Integrating the shear stresses, given in Eq. (a), over the area of the top flange yields the shear force H_1 as:

$$H_1 = \int \tau\,dA_{Flange} = (1/2)(ct_1)(F\,h\,c/I_z) = (F\,h\,c^2\,t_1)/(2I_z) = -H_2 \tag{f}$$

A second shear force H_2, equal in magnitude and opposite in direction, is induced in the lower flange as shown in Fig E6.16b.

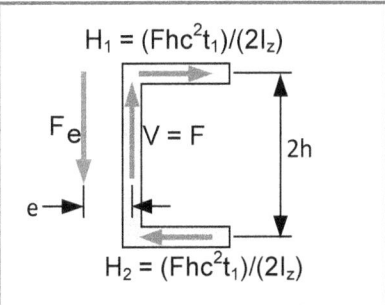

Fig. E6.16b

Step 6: The two equal and opposite shear forces, H_1 and H_2, produce a clockwise couple T equal to:

$$T = (2h)H_1 = (F\,h^2\,c^2\,t_1)/(I_z) \tag{g}$$

Step 7: An equal and opposite moment (torque) is produced by applying the force F with an eccentricity e as shown in Fig. E6.16b. The eccentricity is given by e = T/F as:

$$e = T/F = (h^2 c^2 t_1)/(I_z) \qquad (h)$$

The moment of inertia I_z of the channel section is given by:

$$I_z = (2/3)t_2 h^3 + 2t_1 c h^2 \qquad (i)$$

Step 8: Substituting Eq. (i) into Eq. (h) gives the eccentricity necessary in locating the point of application of the transverse force F to avoid twisting the channel section beam:

$$e = \left(\frac{3}{2}\right)\left[\frac{c^2 t_1}{t_2 h + 3 t_1 c}\right] \qquad (j)$$

6.10 THIN WALLED TUBE WITH ARBITRARY SHAPE

In Section 6.3.1, we determined the shear stress and the angle of twist in circular, thin-walled tubes subjected to torsion loading. We noted that the shear stresses were linearly distributed with only small differences across the thickness of the wall, as shown in Fig. 6.16. It is evident from this illustration that the average shear stress τ_{ave} is a close approximation of τ_{Max} for very thin walled tubes. We will make use of this fact in developing the expression for shear stresses in thin-walled tubes with arbitrary cross sections.

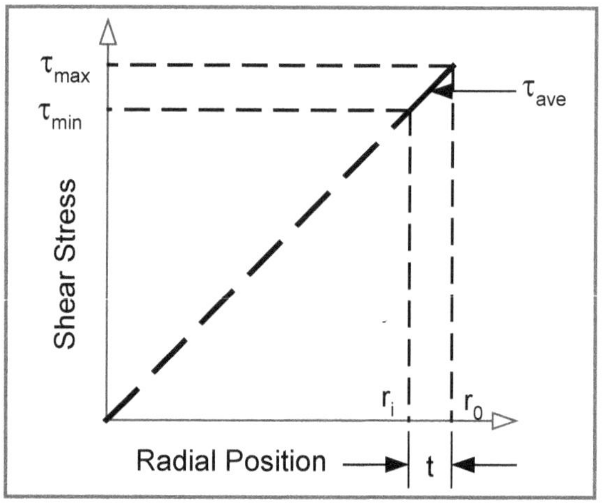

Fig. 6.16 Shear stress distribution across the wall of a thin-walled circular tube.

Consider the thin walled tube, shown in Fig. 6.17, subjected to a torque T. The shape of the cross section is not circular, nor is the wall thickness uniform; however, the thickness of the wall is small relative to the other dimensions of the tube.

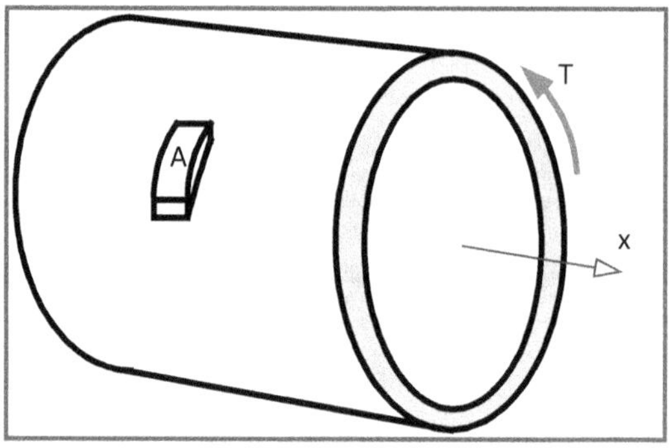

Fig. 6.17 A thin-walled tube with an arbitrary shape.

Next, let's remove element A from the wall of the tube and note the forces F_1 and F_2 acting on the elements longitudinal faces, as shown in Fig. 6.18. These forces are due to the shear stresses τ_{yx} distributed over the longitudinal faces of the element A. The shear stresses in the vertical direction τ_{xz} and τ_{yz} on all four faces vanish due to the proximity of the free boundaries formed by the inside and outside surfaces of the tube.

Considering equilibrium of the element A in the x direction enables us to write:

$$\Sigma F = F_1 - F_2 = 0 \tag{6.51}$$

The forces F_1 and F_2, due to the average shear stresses τ_{1Ave} and τ_{2Ave} acting on the element faces, can be expressed as:

$$F_1 = \tau_{1Ave}\, t_1\, dx \quad \text{and} \quad F_2 = \tau_{2Ave}\, t_2\, dx \tag{6.52}$$

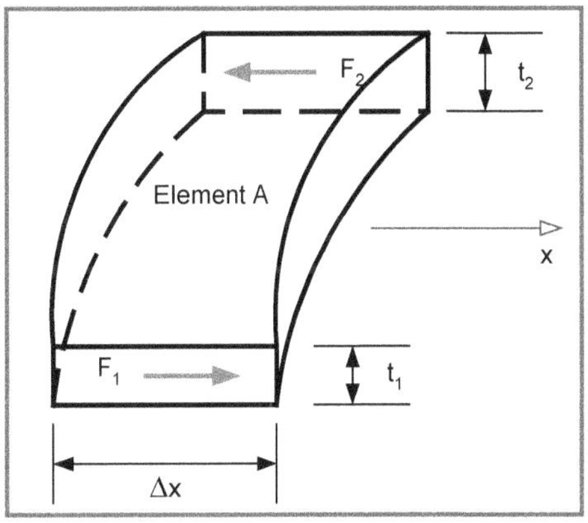

Fig. 6.18 Forces acting on the longitudinal faces of element A.

Substituting Eq. (6.52) into Eq. (6.51) yields:

$$\tau_{1Ave}\, t_1\, dx - \tau_{2Ave}\, t_2\, dx = 0 \tag{a}$$

or

$$\tau_{1Ave}\, t_1 = \tau_{1Ave}\, t_2 \tag{b}$$

If we define the product in Eq. (b) as the shear flow rate q, we may write:

$$q = \tau_{ave}\, t = C \tag{6.53}$$

where C is a constant.

Because the element A can be selected anywhere on the tube, it is evident that the shear flow rate q is constant and independent of the tube's thickness and the position of element A. Next consider a sub-element cut from element A as indicated in Fig. 6.19.

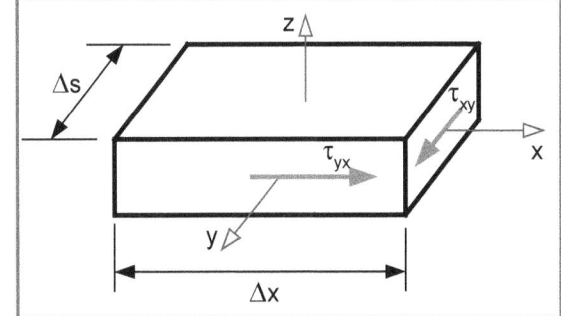

Fig. 6.19 Sub-element showing shear stresses on the longitudinal and transverse faces.

Note that:
$$\tau_{xy} = \tau_{yx} = \tau_{1Ave} \qquad (6.54)$$

Also it is evident that $\tau_{xz} = \tau_{yz} = 0$, because of the thin wall and the proximity of the two free boundaries. It is useful at this stage of the derivation to introduce an analogy involving steady state fluid flow in a closed-rectangular channel with a unit depth but a variable width w. The flow rate Q in this channel is given by:
$$Q = v_{Ave}\, w = \text{Constant} \qquad (6.55)$$

where v_{Ave} is the average flow velocity.

It is clear that the average velocity of the flow will increase or decrease as the width of the channel decreases or increases. This analogy has lead the engineering community to define $q = \tau_{Ave}\, t$ as the shear flow rate, where Q is analogous to q, v_{Ave} is analogous to τ_{Ave}, and w is analogous to t.

Next let's determine the relation between the shear flow rate q and the applied torque by considering an element on the cross section of the tube, as shown in Fig. 6.20.

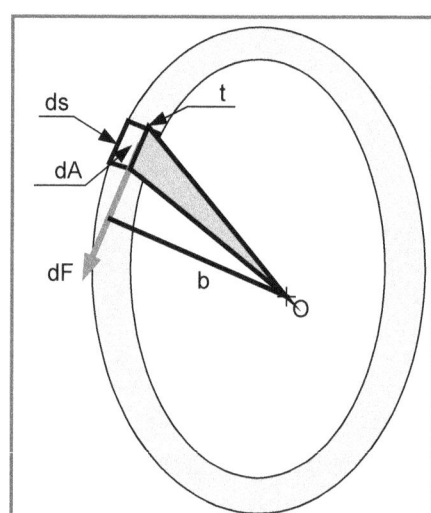

Fig. 6.20 A element on the cross section of the tube showing the incremental force dF.

Using Eq. (6.53) we can write the following relation for the force dF:
$$dF = \tau_{Ave}\, dA = \tau_{Ave}\, t\, ds = q\, ds \qquad (6.56)$$

From Fig. 6.20 and Eq. (6.56), the incremental moment dM_0 produced by dF can be written as:
$$dM_0 = b\, dF = q(b\, ds) \qquad (6.57)$$

To determine the quantity (b ds), let's consider the two triangles shown in Fig. 6.20. For clarity these triangles have been redrawn, as indicated in Fig. 6.21.

The area of the triangle AOC is $A_2 = \frac{1}{2} be$ and the area of the triangle AOB is $A_1 = \frac{1}{2} b(e - ds)$. The area da of the cross hatched triangle is given by:

$$da = A_2 - A_1 = \tfrac{1}{2} b\, ds \qquad (6.58)$$

Substituting Eq. (6.58) into Eq. (6.57) yields:

$$dM_0 = 2q\, da \qquad (6.59)$$

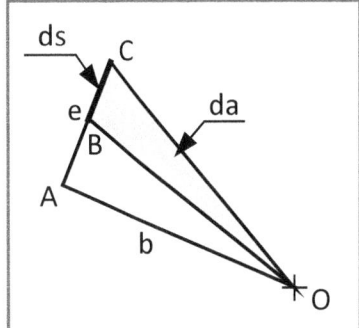

Fig. 6.21 Redrawn triangles from Fig. 6.20 showing the cross hatched area da.

The torque T is obtained by integrating da as the element in Fig. 6.20 is swept around the perimeter of the tube. The integration involves a line integral as indicated below:

$$T = \oint dM_0 = 2q \oint da \qquad (6.60)$$

Integrating Eq. (6.60) gives:

$$T = 2q\, A_i \qquad (6.61)$$

where A_i is the area of the interior of the tube measured to the midpoint of the thickness of the wall.

Substituting Eq. (6.53) into Eq. (6.61) yields:

$$T = 2\, \tau_{ave}\, t\, A_i \qquad (6.62)$$

or

$$\tau_{Ave} = T/(2t\, A_i) \qquad (6.63)$$

The angle of twist ϕ of the tube is obtained using energy methods that are beyond the scope of this book. For a tube of length L, the elastic deformation ϕ due to the torque T is given by:

$$\phi = \frac{TL}{4A_i^2 G} \oint \frac{ds}{t} \qquad (6.64)$$

The shear stress flow approach provides a method for determining the average shear stress about the perimeter of a tube with an arbitrary shape. The shear stress is an average across the thickness of the tube's wall. If the wall is thin, a requirement to apply this method, the average stress is a close approximation of the maximum shear stress. When the wall thickness varies around the perimeter of the tube, the maximum average shear stress occurs where the wall is the thinnest, as indicated in Eq. (6.63).

EXAMPLE E6.17

A square hollow tube with the dimensions shown in Fig. E6.17 is subjected to a torque of 24,000 in.-lb. If the tube is fabricated from steel with a uniform wall thickness and is 8.0 ft long, determine the average shear stress and the angle of twist.

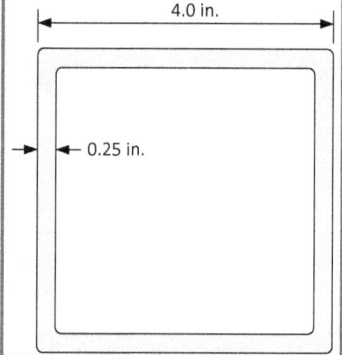

Fig. E6.17

Solution: The average shear stress in the wall of the tube is given by Eq. (6.63) as:

$$\tau_{Ave} = T/(2t\, A_i) \qquad \text{where } A_i = (3.75)^2 = 14.0625 \text{ in.}^2 \qquad (a)$$

Then:

$$\tau_{Ave} = 24{,}000/[(2)(0.25)(14.0625)] = 3{,}413 \text{ psi} \qquad (b)$$

Next calculate the angle of twist ϕ using Eq. (6.64):

$$\phi = \frac{TL}{4A_i^2 G} \oint \frac{ds}{t} \qquad (c)$$

Note that:

$$\oint \frac{ds}{t} = \frac{(4)(3.75)}{0.25} = 60 \qquad (d)$$

Then from Eqs. (c) and (d), we obtain:

$$\phi = \frac{24 \times 10^3 (8)(12)}{4(14.0625)^2 (11.5 \times 10^6)}(60) = 0.01520 \text{ radians} = 0.8707 \text{ degrees} \qquad (e)$$

EXAMPLE E6.18

A rectangular hollow tube, with the dimensions shown in Fig. E6.18, is subjected to a torque of 1,200 N-m. If the tube is fabricated from aluminum and is 2.4 m long, determine the maximum value of the average shear stress and the angle of twist. Note that the wall thickness is not uniform.

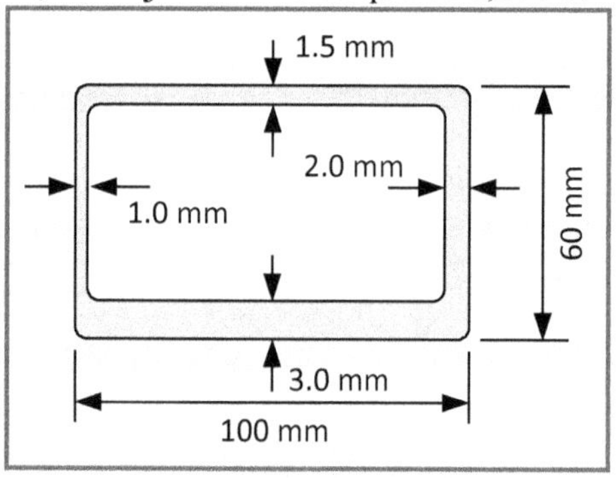

Fig. E6.18

Solution: The maximum value of the average shear stress in the wall of the tube is given by Eq. (6.63) as:

$$(\tau_{Ave})_{Max} = T/(2t_{Min} A_i) \tag{a}$$

where $A_i = (100 - 1.5)(60 - 2.25) = 5,688 \text{ mm}^2$ \hfill (b)

Then:
$$(\tau_{Ave})_{Max} = 1.2 \times 10^6 /[(2)(1.0)(5,688)] = 105.5 \text{ N/mm}^2 = 105.5 \text{ MPa} \tag{c}$$

Next calculate the angle of twist ϕ using Eq. (6.64):

$$\phi = \frac{TL}{4A_i^2 G} \oint \frac{ds}{t} \tag{d}$$

Note that:
$$\oint \frac{ds}{t} = \frac{98.5}{1.5} + \frac{57.75}{2} + \frac{98.5}{3} + \frac{57.75}{1} = 185.1 \tag{e}$$

Then from Eqs. (d) and (e), we obtain:

$$\phi = \frac{1.2 \times 10^6 (2.4 \times 10^3)}{4(5,688)^2 (27 \times 10^3)} (185.1) = 0.1526 \text{ radians} = 8.741 \text{ degrees} \tag{f}$$

The angle of twist is significant but the member is not over stressed.

EXAMPLE E6.19

An irregular shaped hollow tube, with the dimensions shown in Fig. E6.19, is subjected to a torque of 11,000 in-lb. If the tube is fabricated from aluminum and is 1.8 ft long, determine the maximum value of the average shear stress and the angle of twist. Note that the wall thickness of the vertical sides is 1/4 in. and the thickness of the horizontal sides is 1/8 in.

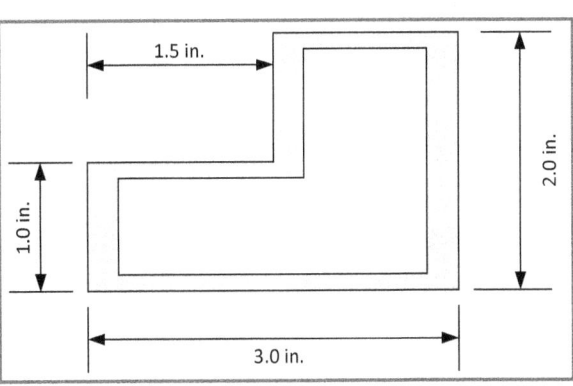

Fig. E6.19

Solution: The maximum value of the average shear stress in the wall of the tube is given by Eq. (6.63) as:

$$(\tau_{Ave})_{Max} = T/(2t_{Min} A_i) \tag{a}$$

where $A_i = (2.75)(0.875) + (1.25)(1.0) = 3.656 \text{ in}^2$ (b)

Then from Eq. (6.63) we write:

$$(\tau_{Ave})_{Max} = (11 \times 10^3)/[(2)(0.125)(3.656)] = 12{,}035 \text{ psi} \tag{c}$$

Next calculate the angle of twist ϕ using Eq. (6.64):

$$\phi = \frac{TL}{4A_i^2 G} \oint \frac{ds}{t} \tag{d}$$

Note that:

$$\oint \frac{ds}{t} = (2.75)(8) + (1.875)(4) + (1.5 + 1.25)(8) + (0.875 + 1.0)(4) = 59.0 \tag{e}$$

Then from Eq. (d) and (e), we obtain:

$$\phi = \frac{(11 \times 10^3)(1.8)(12)}{4(3.656)^2(3.9 \times 10^6)}(59.0) = 0.06723 \text{ radians} = 3.852 \text{ degrees} \tag{f}$$

The angle of twist is reasonable and the stresses are relatively low.

6.11 STRESS CONCENTRATIONS IN CIRCULAR SHAFTS SUBJECT TO TORSION

In many applications, shafts are stepped with different diameters to accommodate gears and bearings. Stress concentrations occur at the fillets adjacent to a shaft's shoulder where one diameter transitions into the other. The geometry leading to a stress concentration of this type is illustrated in Fig. 6.22. In applications where the loads are cyclic, the shafting is subjected to fatigue failure and the stress concentrations must be accounted for in the design analysis to avoid failures.

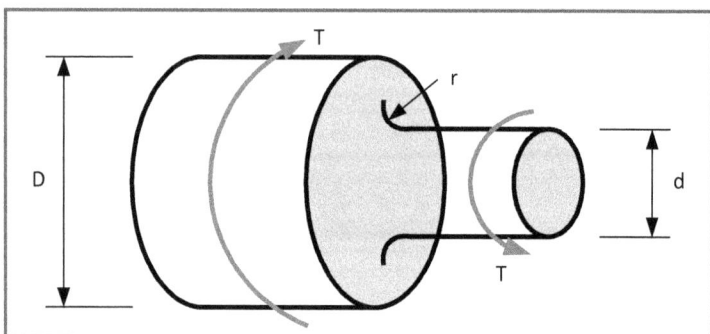

Fig. 6.22 Stress concentrations occur at the fillet located at the transition of the shaft diameters.

The effect of the stress concentration on elevating the shear stresses due to the applied torque is accommodated by introducing a torsion stress concentration factor K_{TS}. In this case, the maximum shear stress at the fillet is given by:

$$\tau_{Max} = K_{TS}\,\tau_{Nom} = K_{TS}\,[16T/(\pi d^3)] \tag{6.65}$$

where the nominal shear stress $\tau_{Nom} = 16T/(\pi d^3)$ occurs in the smaller diameter shaft.

The stress concentration factor depends on the radius of the fillet and the mismatch in the diameters, as shown in Fig. 6.23.

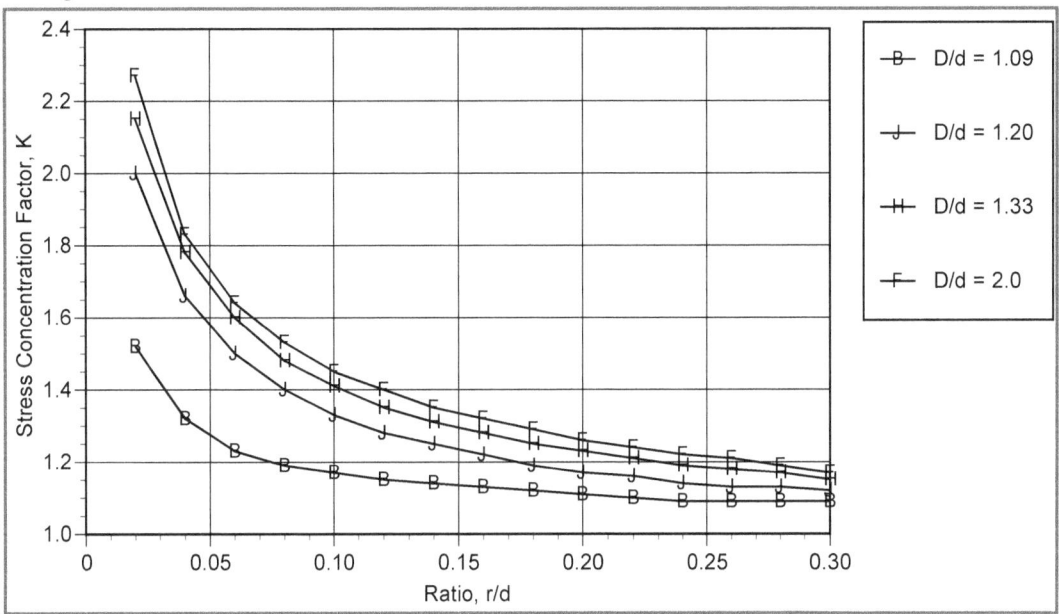

Fig. 6.23 Stress concentration factor for a shouldered shaft subject to torsion.

EXAMPLE E6.20

A shouldered shaft is subjected to a torque T = 700 N-m. The diameters are D = 70 mm and d = 35 mm. The radius of the transition fillet is 7.0 mm. Determine the maximum shear stress at the fillet.

Solution: The nominal shear stress is given by:

$$\tau_{Nom} = 16T/(\pi d^3) = (16)(700)/[\pi\,(0.035)^3] = 83.15 \text{ MPa} \tag{a}$$

To determine the stress concentration factor compute the ratios D/d and r/d as:

$$D/d = (70)/(35) = 2.00 \qquad \text{and} \qquad r/d = (7.0)/(35) = 0.20$$

Reference to the graph for the stress concentration factor K_{TS} shown in Fig. 6.23 gives:

$$K_{TS} = 1.27 \tag{b}$$

Then from Eq. (6.65), we write:

$$\tau_{Max} = K_{TS}\,\tau_{Nom} = (1.27)(83.15) = 105.6 \text{ MPa} \tag{c}$$

Examination of these results indicates that the presence of the shoulder on the shaft increased the shear stresses in the fillet from 83.15 MPa to 105.6MPa. This increase is significant and it is clear that stress concentrations must be taken into account, whenever we are dealing with cyclic loading or components fabricated from brittle materials.

EXAMPLE E6.21

A shouldered shaft, made of steel, is subjected to a cyclic torque that varies from zero to 4,400 N-m. The diameters are D = 210 mm and d = 105 mm. The radius of the transition fillet is 12.0 mm. Determine the maximum shear stress at the fillet. Also determine the allowable alternating shear stress S_{as} in fatigue, if the steel exhibits an ultimate tensile strength of 450 MPa.

Solution: The nominal shear stress is given by:

$$\tau_{Nom} = 16T/(\pi d^3) = (16)(4,400)/[\pi (0.105)^3] = 19.36 \text{ MPa} \qquad (a)$$

To determine the stress concentration factor compute the ratios D/d and r/d as:

$$D/d = (210)/(105) = 2.00 \qquad \text{and} \qquad r/d = (12.0)/(105) = 0.114$$

Reference to the graph for the stress concentration factor, shown in Fig. 6.23, gives $K_{TS} = 1.42$.

Then from Eq. (6.65), we write:

$$\tau_{Max} = K_{TS}\, \tau_{Nom} = (1.42)(19.36) = 27.49 \text{ MPa} \qquad (b)$$

Because the minimum shear stress is zero, the mean and alternating shear stresses are both equal to:

$$\tau_a = \tau_m = \tau_{Max}/2 = (27.49)/(2) = 13.75 \text{ MPa} \qquad (c)$$

and

$$R = \tau_a/\tau_m = 1 \qquad (d)$$

The ultimate strength in shear S_{us} is given by:

$$S_{us} = 0.5774\, S_u \qquad (6.66)$$

$$S_{us} = 0.5774(450) = 259.8 \text{ MPa} \qquad (e)$$

The endurance strength in shear S_{es} is approximated by:

$$S_{es} = S_{us}/2 \qquad (6.67)$$

$$S_{es} = (259.8)/2 = 129.9 \text{ MPa} \qquad (f)$$

Applying the Goodman relation to cyclic shear stresses gives the allowable alternating shear stress S_{as} as:

$$S_{as} = R\, S_{es}\, S_{us}/(R\, S_{us} + S_{es}) \qquad (6.68)$$

From Eqs. (9.14), (9.15) and (9.16), we determine S_{as} as:

$$S_{as} = (1)(129.9)(259.8)/[(1)(259.8) + 129.9] = 86.6 \text{ MPa} \qquad (g)$$

6.12 SUMMARY

The emphasis in this chapter was on the design of shafting fabricated from circular rods. Shafts are commonly employed to transmit power from an electric motor or an internal combustion engine to some appliance. In transmitting power, the shaft is subjected to a torque and it may fail due to shear stresses that produce yielding or due to an excessive angle of twist ϕ. The relation between power transmitted and the torque developed in the shaft was given as:

$$P = T\omega \qquad (6.1)$$

When a shaft is subjected to a torque T, it twisted through an angle ϕ. This deformation in turn produces a shearing strain γ given by:

$$\gamma_\rho = \rho\, \phi/L \qquad (6.7)$$

The shearing stress τ in the shaft is determined from the stress-strain relation for shearing deformation, which is given by:

$$\tau = G\,\gamma \qquad (6.9)$$

where the shear modulus is related to the elastic modulus by:

$$G = \frac{E}{2(1+\nu)} \qquad (6.9a)$$

The shear stress in terms of the torque and radial position ρ is:

$$\tau_\rho = T\,\rho/J \qquad (6.14)$$

The shear stress τ_ρ is a maximum on the outside surface of the shaft where $\rho = r = d/2$.

$$\tau_{Max} = 2\,T/(\pi\,r^3) = 16\,T/(\pi\,d^3) \qquad (6.16)$$

The yield strength in shear is much lower than the yield strength in either tension or compression. We approximate the yield and ultimate strength in shear by:

$$S_{ys} = 0.5774\,S_y \qquad (6.17a)$$

$$S_{us} = 0.5774\,S_u \qquad (6.17b)$$

We introduced the concept of section modulus in torsion Z_T for a hollow circular shaft by defining τ_{Max} as:

$$\tau_{Max} = 16\,T\,d_o/[\pi(d_o^4 - d_i^4)] = T/Z_T \qquad (6.22)$$

$$\text{where } Z_T = \pi(d_o^4 - d_i^4)/(16 d_o) \quad (6.23)$$

Using the section modulus in torsion, we introduced the concept of structural efficiency **SE** of a shaft in torsion following the same approach that we will employ later in the discussion of beams in bending.

$$\mathbf{SE} = Z_T/A \quad (6.28)$$

$$\mathbf{SE} = \frac{d_o^2 + d_i^2}{4d_o} \quad (6.29)$$

We considered shear stress acting on orthogonal planes and established that they are equal in magnitude and opposite in direction.

$$\tau_{xy} = \tau_{yx} \quad (6.31)$$

We examined the shear stresses on oblique planes and found that the shear and normal stresses on the oblique planes oscillate as sinusoids:

$$\tau_{nt} = \tau_{xy}(\cos^2\theta - \sin^2\theta) = \tau_{xy}\cos(2\theta) \quad (6.32)$$

$$\sigma_n = 2\tau_{xy}\cos\theta\sin\theta = \tau_{xy}\sin(2\theta) \quad (6.33)$$

We found that the normal stresses are a maximum on planes where the shear stress vanishes. This fact enabled us to define principal planes and leads to the relation for the principal stress σ_1:

$$(\sigma_n)_{Max} = \sigma_1 = \tau_{xy} \quad (6.34)$$

We considered the deformation of the shaft due to torsion and established the expression for the angle of twist as:

$$\phi = (TL)/(GJ) \quad (6.35\text{ a})$$

$$\phi = (32\,TL)/(\pi d^4 G) \quad (6.35\text{ b})$$

Torsion rods are often employed as springs in suspension systems or to store strain energy. We derived the relation between the torsion and the angle of twist in terms of the spring rate k_T as:

$$T = k_T \phi \quad (6.39)$$

$$k_T = GJ/L \quad (6.40)$$

Torsion of non-circular sections was introduced. When the section is non-circular, the plane sections do not remain plane, when the bar is subjected to torsion. Instead the plane sections warp. This distortion complicates the derivation of the relations for the maximum shear stress and the angle of twist. We have provided these equations without derivation for the following cross sections: elliptical, equilateral triangle, hexagon, rectangle, thin rectangle and angle, channel and I sections.

We introduced the concept of shear stresses produced as a beam bends under the action of transverse forces. If the cross section of the beam is not symmetric about its vertical axis, the beam will twist due to the effect of these shear stresses. The twisting couple produced by bending a cantilever beam fabricated from a channel section was discussed. The twisting of the beam can be avoided if the transverse force is applied through the shear center of the section. Example 6.16 demonstrated the analysis used to determine the location of the shear center.

Finally we introduced a method to approximate the shearing stresses and the angle of twist in tubes with an arbitrary shape and a varying wall thickness. We established that the shear flow rate q was a constant independent of the wall thickness of the tube.

$$q = \tau_{ave} \, t = C \tag{6.53}$$

Integration of the incremental moments produced by the shear stresses enabled us to write the relation for the average shear stresses in the tube produced by the torque T as:

$$\tau_{Ave} = T/(2t \, A_i) \tag{6.63}$$

The angle of twist is determined by using a relation involving the line integral as shown below:

$$\phi = \frac{TL}{4A_i^2 G} \oint \frac{ds}{t} \tag{6.64}$$

The effect of the stress concentration on elevating the shear stresses due to the applied torque is accommodated by introducing a torsion stress concentration factor K_{TS}. In this case, the maximum shear stress at the fillet is given by:

$$\tau_{Max} = K_{TS} \, \tau_{Nom} = K_{TS} \, [16T/(\pi d^3)] \tag{6.65}$$

PROBLEMS

6.1 A shaft is employed to transmit power P from an electric motor to an appliance. Determine the torque imposed on the shaft if the motor is operated at the angular velocity indicated in the table below.

Problem No.	Power, P	Angular Velocity, ω
6.1a	700 W	3600 RPM
6.1b	0.75 HP	1780 RPM
6.1c	1.9 kW	1180 RPM
6.1d	6.0 HP	1780 RPM

6.2 A shaft with a circular cross section is twisted through an angle ϕ. If the shaft has a diameter d and a length L, determine the maximum shearing strain γ.

190 — Chapter 6 — Torsion of Structural Elements

Problem No.	Diameter, d	Length, L	Angle, ϕ
6.2a	0.750 in.	15 in.	2.5°
6.2b	15 mm	1.3 m	3.1°
6.2c	1.25 in.	22 in.	1.5°
6.2d	30 mm	850 mm	2.0°

6.3 Determine the maximum shearing stresses for the conditions described in Problem 6.2 if the shaft is fabricated from one of the materials listed in the table below.

Problem No.	Material
6.3a	Steel
6.3b	Aluminum
6.3c	Brass
6.3d	Stainless Steel

6.4 A shaft with a circular cross section is subjected to a torque T. If the shaft has a diameter d and a length L, determine the maximum shearing stress τ_{MAX}.

Problem No.	Diameter, d	Length, L	Torque, T
6.4a	0.750 in.	15 in.	120 ft-lb
6.4b	25 mm	1.3 m	95 N-m
6.4c	1.25 in.	22 in.	250 ft-lb
6.4d	30 mm	850 mm	180 N-m

6.5 A shaft fabricated from 302 A stainless steel transmits power P while rotating at an angular velocity ω. Specify the diameter of the shaft, in a standard size, if the safety factor with respect to yielding is specified as SF.

Problem No.	Power, P	Angular Velocity, ω	SF
6.5a	4.0 HP	600 RPM	2.0
6.5b	1.5 kW	1180 RPM	2.9
6.5c	10.0 HP	1780 RPM	3.5
6.5d	3.0 kW	3600 RPM	3.2

6.6 A hollow tube is employed in the design of a shaft to transmit power P with a safety factor with respect to yielding of SF. Determine the outside diameter of the tube if its wall thickness is 0.05 d_o. The tube is fabricated from 1020 HR steel.

Problem No.	Power, P	Angular Velocity, ω	SF
6.6a	4.0 HP	600 RPM	2.2
6.6b	1.5 kW	1180 RPM	2.7
6.6c	10.0 HP	1780 RPM	3.4
6.6d	3.0 kW	3600 RPM	3.0

6.7 Clearly a hollow tube is structurally more efficient than a solid rod as a shaft for transmitting power. Cite several reasons for the limited usage of tubes in actual practice.

6.8 Determine the section modulus of the shaft with a circular cross section as specified in the table below.

Problem No.	Diameter, d_o	Diameter, d_i
6.8a	3.0 in.	2.75 in.
6.8b	120 mm	90 mm
6.8c	4.5 in.	4.0 in.
6.8d	150 mm	130 mm

6.9 Determine the structural efficiency **SE** of the shafts described in Problem 6.8.

6.10 For the inclined surface defined in the figure below, determine the stresses τ_{nt} and σ_n on the inclined plane for the conditions described in the table below.

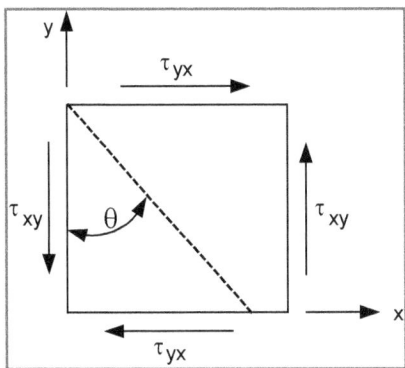

Problem No.	Stress, τ_{xy}	Angle, θ
6.10a	6.5 ksi	22.0°
6.10b	75 MPa	45.0°
6.10c	9.2 ksi	67.0°
6.10d	46 MPa	125°

6.11 For the stress τ_{xy} listed in Problem 6.10, prepare a graph of τ_{nt} and σ_n as θ varies from zero to 360°. Define the principal planes and the planes on which the shear stress is a maximum.

6.12 A solid circular shaft with a diameter d is transmitting a power P from an electric motor to an air compressor. If the motor is operating at 1800 RPM, determine the maximum normal stress acting on the shaft. Also determine the plane upon which this maximum normal stress acts.

Problem No.	Diameter, d_o	Power, P
6.12a	1.0 in.	7.0 HP
6.12b	30 mm	8.0 kW
6.12c	1.5 in.	15 HP
6.12d	50 mm	12 kW

6.13 A steel shaft with a circular cross section is subjected to a torque T. If the shaft has a diameter d and a length L, determine the angle of twist ϕ.

Problem No.	Diameter, d	Length, L	Torque, T
6.13a	0.750 in.	15 in.	120 ft-lb
6.13b	25 mm	1.3 m	95 N-m
6.13c	1.25 in.	22 in.	250 ft-lb
6.13d	30 mm	850 mm	180 N-m

6.14 A line shaft, shown in in the figure below, fabricated from high-strength steel incorporates five spur gears. The gear A provides the input torque T_A and gears B, C, D and E each remove torque from the shaft. Determine the angle of twist for the entire shaft AE and between gears at A and C. The shaft diameters and lengths are given in the table below. The torque at each gear is also listed in this table. Neglect the effect of gear deformation on the angle of twist. Also determine the maximum shear stress in the shaft. Where does this shear stress occur?

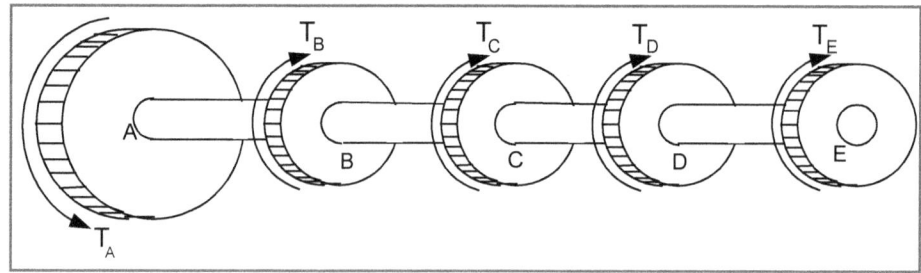

Prob. No.	d_{A-B}	d_{B-C}	d_{C-D}	d_{D-E}	L_{A-B}	L_{B-C}	L_{C-D}	L_{D-E}
6.14a	2.0 in.	1.5 in.	1.2 in.	1.0 in.	12 in.	15 in.	18 in.	21 in.
6.14b	40 mm	35 mm	30 mm	25 mm	0.4 m	0.4 m	0.5 m	0.7 m
6.14c	1.5 in.	1.25 in.	1.0 in.	0.75 in.	15 in.	24 in.	30 in.	36 in.
6.14d	50 mm	40 mm	30 mm	20 mm	0.3 m	0.6 m	0.8 m	1.0 m
	T_A	T_B	T_C	T_D	T_E			
6.14a	2200 ft-lb	400 ft-lb	600 ft-lb	500 ft-lb	700 ft-lb			
6.14b	3.0 kN-m	1.0 kN-m	1.0 kN-m	0.4 kN-m	0.6 kN-m			
6.14c	1800 ft-lb	700 ft-lb	400 ft-lb	500 ft-lb	200 ft-lb			
6.14d	4.5 kN-m	1.5 kN-m	1.2 kN-m	0.8 kN-m	1.0 kN-m			

6.15 A line shaft, shown in Fig. P6.15, fabricated from high-strength steel incorporates five spur gears. The gears A and E provide input torques T_A and T_E. Gears B, C, and D each remove torque from the shaft. Determine the angle of twist for the entire shaft AE and between gears at A and C. The shaft diameters and lengths are given in the table below. The torque at each gear is also listed in this table. Neglect the effect of gear deformation on the angle of twist. Also determine the maximum shear stress in the shaft. Where does this shear stress occur?

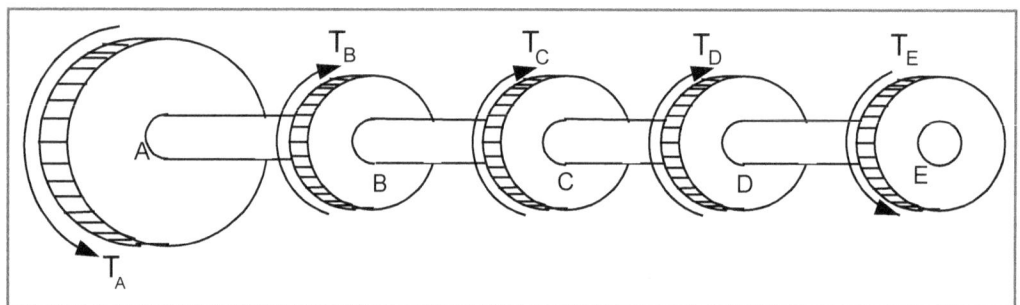

Fig. P6.15

Prob. No.	d_{A-B}	d_{B-C}	d_{C-D}	d_{D-E}	L_{A-B}	L_{B-C}	L_{C-D}	L_{D-E}
6.15a	2.0 in.	1.5 in.	1.2 in.	1.0 in.	12 in.	15 in.	18 in.	21 in.
6.15b	40 mm	35 mm	30 mm	25 mm	0.4 m	0.4 m	0.5 m	0.7 m
6.15c	1.5 in.	1.25 in.	1.0 in.	0.75 in.	15 in.	24 in.	30 in.	36 in.
6.15d	50 mm	40 mm	30 mm	20 mm	0.3 m	0.6 m	0.8 m	1.0 m
	T_A	T_B	T_C	T_D	T_E			
6.15a	1500 ft-lb	1400 ft-lb	1200 ft-lb	600 ft-lb	1700 ft-lb			
6.15b	2.5 kN-m	2.0 kN-m	1.0 kN-m	2.1 kN-m	2.6 kN-m			
6.15c	2100 ft-lb	1000 ft-lb	1000 ft-lb	1000 ft-lb	900 ft-lb			
6.15d	6.5 kN-m	3.5 kN-m	2.2 kN-m	2.6 kN-m	1.8 kN-m			

6.16 A steel torsion bar is incorporated into the chassis of an automobile to serve as a suspension spring. If the bar is fabricated from steel rod with a diameter d and a length L, determine its spring rate. Also determine its angle of twist if it stores strain energy of E as the automobile traverses a bump.

Problem No.	Diameter, d	Length, L	Strain Energy, E
6.16a	0.750 in.	25 in.	5.0 ft-lb
6.16b	15 mm	1.1 m	6.0 N-m
6.16c	1.0 in.	32 in.	7.5 ft-lb
6.16d	20 mm	950 mm	4.5 N-m

6.17 Determine the nominal and maximum shear stresses factor for a shouldered shaft subjected to a torque T with the geometry listed in the table below:

Problem No.	Diameter, D	Diameter, d	Fillet Radius, r	Torque, T
6.17a	1.5 in.	1.25 in.	0.25 in.	500 ft-lb
6.17b	20 mm	10 mm	3.0 mm	700 N-m
6.17c	4.0 in.	3.0 in.	0.375 in.	650 ft-lb
6.17d	55 mm	50 mm	5.0 mm	900 N-m

6.18 A shouldered shaft is designed such that the maximum shear stress τ_{Max} cannot exceed a critical value listed in the table below. Determine the largest torque that the shaft can support. The shaft dimensions are also listed in the table.

Problem No.	Diameter, D	Diameter, d	Fillet Radius, r	τ_{Max}
6.18a	2.0 in.	1.0 in.	0.125 in.	8.0 ksi
6.18b	40 mm	30 mm	6.0 mm	50 MPa
6.18c	3.0 in.	2.5 in.	0.625 in.	10 ksi
6.18d	11 mm	10 mm	0.5 mm	75 MPa

6.19 The electric motor, shown below, has a power rating of P and operates at 1180 RPM. It powers a gear drive with a variable load, which at times requires the motor to operate at its maximum rating. The shaft, fabricated from 4340HR steel, is supported by two bearings—one on each side of the gear drive. The shaft from the motor has a diameter d and the shaft at the gear has a diameter D. The radius at the transition is specified as r/d = 0.15 and the diameter mismatch ratio is specified as D/d = 1.33. Determine the diameters d and D if the maximum shear stress in the material is not to exceed 1/4 of the yield strength of the material in shear. Another constraint on the design is the angle of twist ϕ, which is not to exceed the specified value measured from the motor to the gear drive.

Prob. No.	L	Power, P	Angle, ϕ
6.19a	1.0 m	5.0 kW	2.0°
6.19b	1.1 m	10 kW	3.0°
6.19c	1.2 m	8.0 kW	3.5°
6.19d	1.5 m	25 kW	4.0°

6.20 The electric motor, shown below, has a power rating of P and operates at 1780 RPM. It powers a drive shaft with three gears. When the motor is operating at rated capacity gear A removes 1/2 of the power developed by the motor, while gear B and gear C remove 1/3 and 1/6 of the power, respectively. The shaft is supported by four bearings as shown in Fig. P6.20. The shaft from the motor has a diameter d but it is shouldered with a diameter D at each of the three gear locations. The radius at the transition is specified as r/d = 0.20 and the diameter mismatch ratio is specified as D/d = 2.00. If an alloy steel with a yield strength in shear of 150 ksi is used to fabricate the shaft, determine the diameters d and D if the maximum shear stress in the material is not to exceed 1/3 of the yield strength of the material in shear. Another constraint on the design is the angle of twist ϕ, which is not to exceed a specified value measured from the motor to gear C.

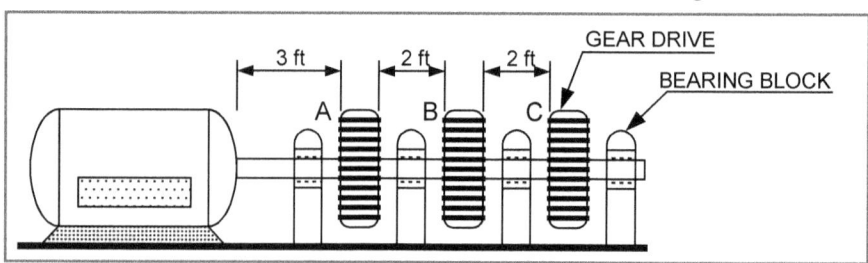

Problem No.	Power, P	Angle, ϕ
6.20a	10.0 HP	2.5°
6.20b	12.0 HP	3.2°
6.20c	15.0 HP	3.6°
6.20d	20.0 HP	4.4°

6.21 A torque T is applied to a structural member with an elliptical cross section that is fabricated from steel. Determine the maximum shear stress and the angle of twist for the structural members defined in the table below:

Problem No.	Torque, T	Height, h	Width, b	Length, L
6.21a	20 ft-lb	1.0 in.	0.35 in.	4.2 ft
6.21b	15 N-m	20 mm	10 mm	1.2 m
6.21c	38 ft-lb	1.5 in.	0.75 in.	10 in.
6.21d	50 N-m	40 mm	10 mm	0.8 m

6.22 A torque T is applied to a structural member with an equilateral cross section that is fabricated from steel. Determine the maximum shear stress and the angle of twist for the structural members defined in the table below:

Problem No.	Torque, T	Width, b	Length, L
6.22a	20 ft-lb	0.50 in.	2.2 ft
6.22b	25 N-m	30 mm	1.0 m
6.22c	48 ft-lb	1.25 in.	30 in.
6.22d	60 N-m	40 mm	1.5 m

6.23 A torque T is applied to a structural member with a hexagonal cross section that is fabricated from steel. Determine the maximum shear stress and the angle of twist for the structural members defined in the table below:

Problem No.	Torque, T	Inscribed Dia., d	Length, L
6.23a	30 ft-lb	0.75 in.	4.2 ft
6.23b	35 N-m	20 mm	1.3 m
6.23c	55 ft-lb	1.35 in.	36 in.
6.23d	60 N-m	60 mm	1.1 m

6.24 A torque T is applied to a structural member with a rectangular cross section that is fabricated from steel. Determine the maximum shear stress and the angle of twist for the structural members defined in the table below:

Problem No.	Torque, T	Width, a	Height, b	Length, L
6.24a	25 ft-lb	0.50 in.	0.50 in.	2.9 ft
6.24b	45 N-m	30 mm	15 mm	1.4 m
6.24c	65 ft-lb	1.00 in.	0.25 in.	45 in.
6.24d	80 N-m	40 mm	40 mm	2.0 m

6.25 Prove that the shear stress τ is zero at all four corners of a square cross section.

6.26 Determine the eccentricity locating the shear center for a beam fabricated from the channel section shown below.

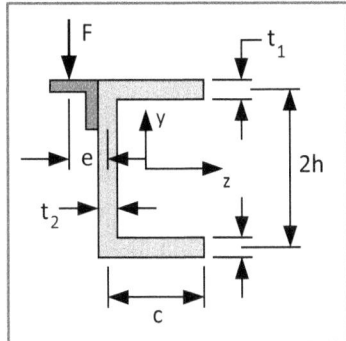

Problem No.	2h	t_1	t_2	c
6.26a	4.0 in.	0.350 in.	0.250 in.	1.75 in.
6.26b	160 mm	20 mm	15 mm	60 mm
6.26c	8.0 in.	0.420 in.	0.350 in.	2.5 in.
6.26d	240 mm	25 mm	15 mm	70 mm

6.27 For the channel section, shown above, show that integrating the parabolic distribution of shear stresses over the area of the web yields a shear force V equal to:

$$V = F\left[\frac{2ct_1 + (3/5)ht_2}{2ct_1 + (2/3)ht_2}\right]$$

Explain why V is slightly less than F. Hint: the shear stress at the top of the web is given by:

$$\tau = F\left[\frac{ct_1 h}{t_2 I_z}\right]$$

CHAPTER 7

STRESSES IN BEAMS

7.1 PURE BENDING OF SYMMETRIC BEAMS

A beam is similar to a rod because both have the same geometry—long and slender. However, the beam is loaded differently. A rod is subjected to tensile or compressive forces applied in the axial direction. A beam is subjected to transverse forces applied in a perpendicular direction to the longitudinal axis of the long, thin member. The difference in the loading is depicted in Fig. 7.1.

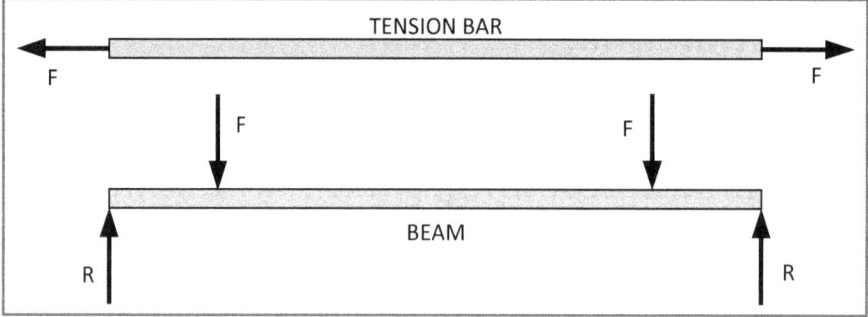

Fig. 7.1 Both the rod and beam are fabricated from long, thin members, but the direction of the applied external loads is different.

Now that we understand the geometry and the loading of a beam, let's consider what the expression "pure" bending implies. A beam is in "pure" bending if it is subjected to a constant moment over its length. When a straight beam is subject to a constant moment, as illustrated in Fig. 7.2, it deforms into a circular arc with a radius of curvature ρ.

Fig. 7.2 A beam subjected to external moments M deforms into a circular arc with a radius ρ.

We limit our analysis to beams with symmetric cross sections. Several examples of symmetric cross sections are shown in Fig. 7.3. Clearly, the symmetry is only about the vertical, or y-axis. The theory for beams in bending accommodates the lack of symmetry with respect to the horizontal or z-axis.

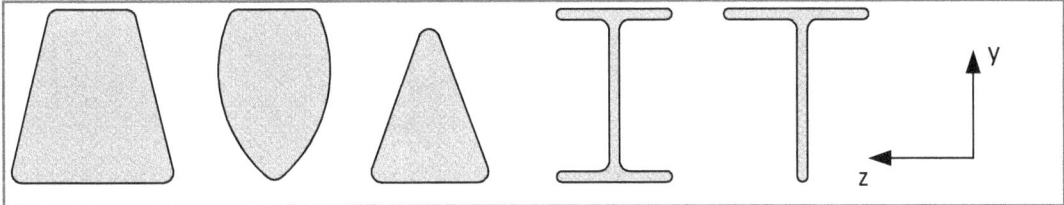

Fig. 7.3 Cross sectional areas of beams exhibiting symmetry about the vertical y-axis.

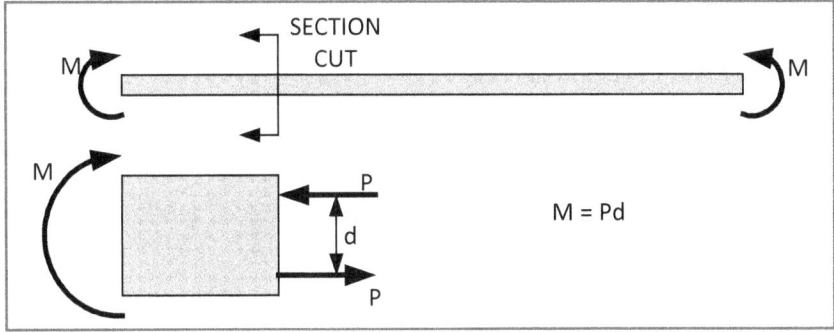

Fig. 7.4 The left-hand portion of the beam showing a couple (P)(d) acting on the area exposed by a section cut.

7.1.1 External Moments Produced by Internal Forces

Suppose we consider a beam with moments applied at its ends as shown in Fig. 7.2. Clearly, the entire beam is in equilibrium, because the equilibrium equation $\Sigma M = 0$ is satisfied. The clockwise moment acting on the left end of the beam is equilibrated by the counterclockwise moment on the right end. Next, let's make a section cut at some position x along the length of the beam, as indicated in Fig. 7.4, and consider the left portion of the beam. In constructing a FBD, we must apply internal forces that will maintain equilibrium of this portion of the beam. Equilibrium is achieved by placing two equal and opposite internal forces P that act on the surface of the cross sectional area exposed by the section cut. Because the internal forces are equal and opposite, the equilibrium equation $\Sigma F_x = 0$ is satisfied. The equation of equilibrium for the moments is satisfied if:

$$\Sigma M = M - P\,d = 0$$

$$P\,d = M \qquad\qquad (a)$$

EXAMPLE E7.1

The concept of a pure moment acting on a beam may be difficult to comprehend. How can a pure moment be generated in a beam?

Solution: We produce moments by applying forces to a member or component in such a way that the forces create a moment. For instance, we pull with a force F on a wrench with a length d to create a moment (torque) on a bolt to screw it into a joint by overcoming resistance to the bolt's rotational motion. A method for producing a pure moment on a beam is to apply two forces to a simply supported beam in what is referred to as a **four-point bend configuration**, as illustrated in Fig. E7.1.

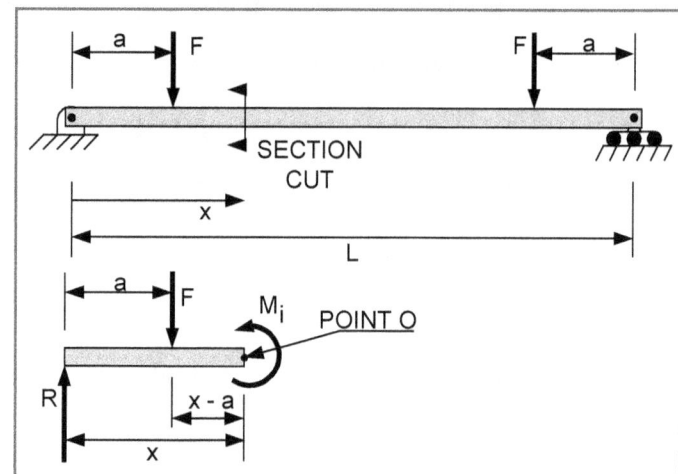

Fig. E7.1.

To show that this arrangement of externally applied forces produce a pure moment, we make a section cut at position x, remove the left end of the beam and construct a FBD. Let's apply the equilibrium equations to the FBD and write:

$$\Sigma F_y = R - F = 0, \quad \Rightarrow \quad R = F \quad (a)$$

Next, consider moments about point O, and write:

$$\Sigma M_O = -Rx + F(x-a) + M_i = 0 \quad \text{for } a < x < (L-a)$$

$$M_i = Fa \quad (b)$$

where the subscript i refers to an internal moment produced by internal forces.

It is evident that only the moment M_i acts on the central segment of the beam. For the segment of the beam defined by $a < x < (L-a)$, we have a state of pure bending. No shear force V exists on the section and the net force in the x direction is zero.

7.1.2 Stress Distributions

We have used the concept of equilibrium to establish that equal and opposite internal forces P act on the surface exposed by the section cut. The forces produce an internal couple, $M_i = (P)(d)$, which counteracts the externally applied moment M. However, at this stage in the analysis, we cannot determine the individual values of either P or d, although we know the direction of the forces. We need the concept of **stress distributions** to continue the analysis.

The internal forces P are due to normal stresses σ_x, which act over the surface of the exposed cross sectional area. Clearly, the upper most force in Fig. 7.4 is compressive because it presses against the surface exposed by the section cut. Similarly, the lower force is tensile. We are concerned with the distribution of the normal stresses that produce these two forces. Several possible stress distributions are suggested in Fig. 7.5.

The distributions of normal stresses presented in Fig. 7.5 produce equal and opposite internal forces P separated by some distance d. Accordingly, each distribution satisfies the equilibrium equations. However, which, if any, of the three possibilities are correct? We have exhausted the equilibrium principle, and to continue it is necessary to consider the deformation of the beam. To begin this second phase of the analysis, let's make an assumption about the geometry of the beam in its deformed state.

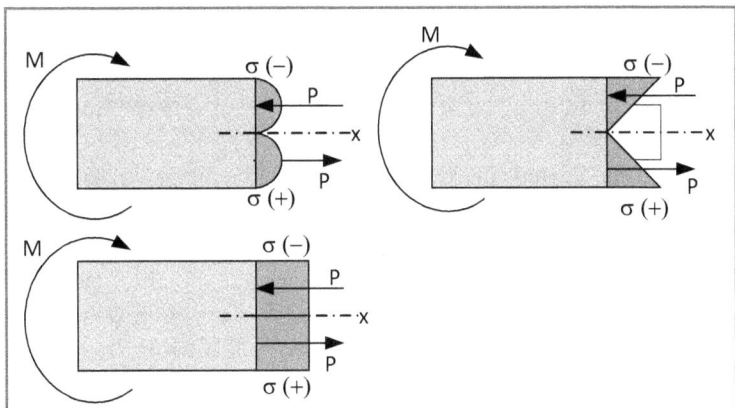

Fig. 7.5 Possible distributions of stress σ_x acting on an internal surface producing the couple $M = P\,d$.

7.2 **DEFORMATION OF A BEAM IN BENDING**

When a beam is subjected to a moment, as shown in Fig. 7.2, it bends into a circular arc. Parallel vertical lines scribed on the sides of the beam remain straight after the beam undergoes deformation from a straight member to one with curvature. We show a segment of the beam before and after deformation into a circular arc in Fig. 7.6.

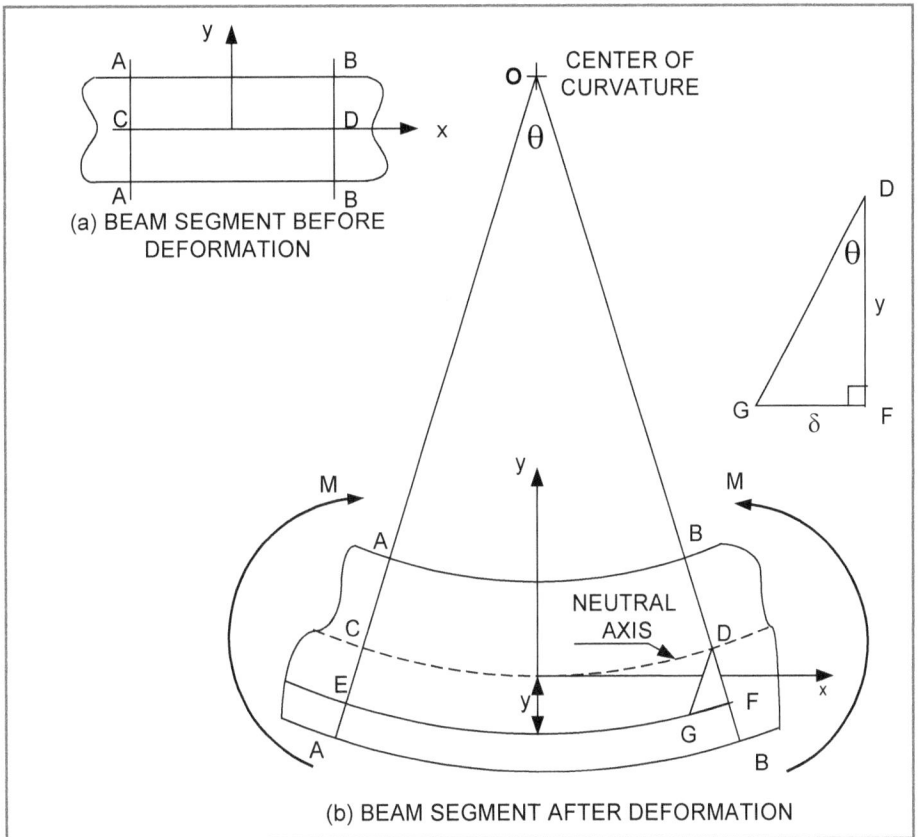

Fig. 7.6 Plane sections A – A and B – B of the beam remain plane after deformation.

The geometry of the deformed segment of the beam is extremely important. The circular arc is centered at point O. The lines A – A and B – B are straight and radial. They have each rotated about axes perpendicular to the plane of bending, and their extensions intersect at the center of curvature to form an angle θ. In addition to the lines A – A and B – B remaining straight, the transverse sections of the beam, originally plane, remain plane after bending.

The dotted line C – D is exactly the same length as it was before deformation. This line defines what is known as the **neutral axis**. The radius of curvature ρ of the circular arc is the distance from the center of curvature O to the neutral axis (the arc C – D). Lines (arcs) above the neutral axis, become shorter after deformation and lines (arcs) below the neutral axis, such as E – F, become longer. These changes in the length of lines (arcs) above and below the neutral axis result in longitudinal strain. The strains are compressive (negative) above the neutral axis when the beam has a positive curvature[1], as illustrated in Fig. 7.6. Similarly, the strains are tensile (positive) below the neutral axis when the beam has a positive curvature.

7.2.1 Strains in Beams with a Constant Radius of Curvature

Let's determine the magnitude of the strains by examining the changes in the length of line E - F. Clearly, the line E - F is lengthened in the deformation process. The distance GF in Fig. 7.6b shows the increase in the length, δ of line E - F. Note that we constructed line D - G parallel to line A – A to produce the shaded triangle in Fig. 7.6b. Recall Eq. 4.4 and write:

$$\varepsilon = \delta/L = GF/EG \qquad (a)$$

The similarity of triangle COD and the triangle GDF, shown in the inset of Fig. 7.6b, enables us to write:

$$CD = EG = \rho\theta; \; GF = -y\theta \qquad (b)$$

$$GF/EG = -y\theta/\rho\theta = -y/\rho \qquad (c)$$

Combining Eqs. (a) and (c) yields:

$$\varepsilon = -y/\rho \qquad (7.1)$$

where y is the distance from the neutral axis (CD) to the fiber in question (EF).

Note, the strain ε is a linear function of the position y. The strain is a minimum (zero) at the neutral axis and a maximum when y locates either the top or bottom fiber of the beam. If we consider a beam with a rectangular cross section and define its height as h, then $y_{max} = -h/2$ and from Eq. (7.1), it is evident that:

$$\varepsilon_{max} = h/(2\rho) \qquad (7.2)$$

The sign of the maximum strain is positive (tensile) for fibers along the bottom surface of the beam if the curvature is positive.

---

[1] We define positive curvature of the beam in two ways. The curvature is positive if the beam bends so it will hold water (concave upward). The curvature is positive if the internal moment acting on the section under consideration is positive.

EXAMPLE E7.2

A thin strip of metal 1.00 mm thick and 10.0 mm wide is wrapped about a mandrel with a radius of 200 mm. Determine the maximum bending strain produced in the strip. Assume that the neutral axis is in the center of the cross sectional area, as shown in Fig. E7.2:

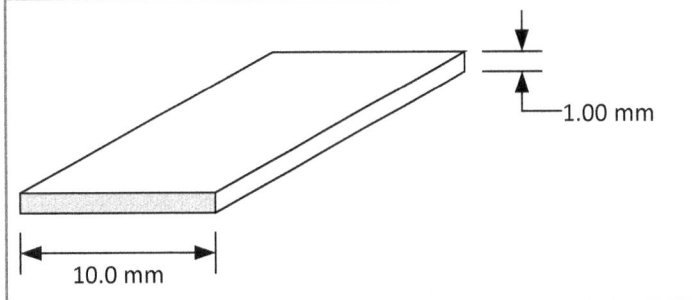

Fig. E7.2

Solution: From Eq. (7.2), we write:

$$\varepsilon_{max} = y_{max}/\rho = h/2\rho = 1.0/[(2)(200.5)] = 0.002494$$

While this is a very simple example, you should note that we have determined a strain for a beam in bending without employing the elastic constants for the material. Indeed, we did not even specify the metal employed in fabricating the strip in the problem statement. Strain is a geometric quantity (a change in length over original length). If we know the geometry before and after deformation, **strain can be determined without knowledge of either the loads or the material properties.**

7.2.2 Stresses in Beams

It is easy to convert the strains in the beam to stresses if we recall Hooke's Law—Eq. (4.5). The longitudinal fibers in the beam are subjected to a uniaxial state of stress with fibers on the convex side of the neutral axis in tension and those on the concave side in compression. By combining Eqs. (7.1) and (4.2), we obtain:

$$\sigma = E\varepsilon = -yE/\rho \qquad (7.3)$$

Examination of Eq. (7.3) indicates that the normal stress σ is linearly distributed with respect to y; a minimum of zero at the neutral axis where y = 0, and a maximum at the outer fibers in the beam where y is a maximum.

The fact that the stresses are not uniformly distributed has significant implications regarding the design of the cross section of the beam. We will discuss this important design consideration in much more detail in Section 7.3.

7.2.3 Locating the Neutral Axis

The results in Eq. (7.3) may be used to calculate the stress on a beam if we know the material from which it is fabricated, the location of the neutral axis, and the radius of curvature. However, in most cases we have knowledge of the applied external forces—not the radius of curvature. Also, we have not yet established the location of the neutral axis relative to the cross section of the beam.

To establish the position of the neutral axis of the beam, let's consider the portion of the beam shown in Fig. 7.7 as a free body. Writing the equilibrium relation $\Sigma F_x = 0$ and using Eq. (7.3) gives;

$$\Sigma F_x = \int \sigma \, dA = -\int (Ey/\rho) \, dA = -(E/\rho) \int y \, dA = 0$$

where A is the cross sectional area over which the stress σ is distributed.

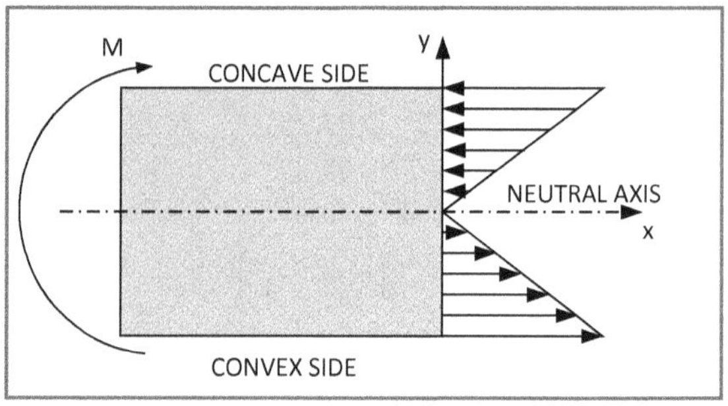

Fig. 7.7 The normal stresses σ are linear in y for a beam in bending.

Because (E/ρ) cannot be zero when the beam is deformed, it is evident that:

$$\int y \, dA = Q_x = 0 \tag{7.4}$$

where Q_x is the first moment of the area A of the cross section of the beam.

For equilibrium of the beam in the x direction to be satisfied, it is necessary that the position of the neutral axis be adjusted so that $Q_x = 0$. If the neutral axis passes through the centroid of the area of the transverse section of the beam, Eq. (7.4) is satisfied. For common cross section shapes such as circular, rectangular, I, H, wide-flange, etc., the location of the centroid is at the intersection of the two axes of symmetry. Procedures for locating the centroid of areas without two axes of symmetry are presented in Appendix C.

7.2.4 The Relationship between Moment and Curvature

Now that a procedure has been presented for locating the neutral axis, it is necessary to develop an approach for determining the radius of curvature. The stresses σ are distributed so as to produce a moment M, as shown in Fig. 7.7. We will use this fact in the following derivation. The moment of the force acting on an element dA, shown in Fig. 7.8, with respect to the neutral axis is:

$$dM = -\sigma \, dA \, (y) = -(-Ey/\rho)(dA)(y) = +(Ey/\rho)(dA)(y) \tag{a}$$

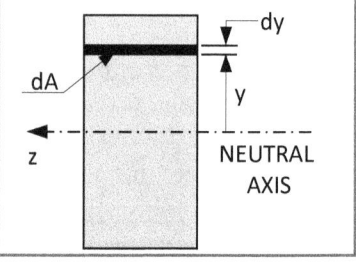

Fig. 7.8 Cross sectional area of a rectangular beam with elemental area dA.

Integrate Eq. (a) to obtain the internal moment M_i:

$$M_i = \int (Ey^2/\rho) \, dA = (E/\rho)\int y^2 \, dA \tag{b}$$

Recall that the second moment[2] of the area A relative to the neutral axis I_z is given by:

$$I_z = \int y^2 \, dA \tag{c}$$

---

[2] The second moment of the area is often called the moment of inertia. Methods for determining this quantity for different shaped cross sections are presented in Appendix C.

Combining Eqs. (b) and (c), and using the equilibrium relation $\Sigma M = 0$ gives:

$$M_i = (EI_z / \rho) = M \tag{7.5}$$

$$\kappa = 1/\rho = M/(EI_z) \tag{7.6}$$

where $\kappa = 1/\rho$ is the curvature of the beam and M is the moment.

From Eq. (7.6), it is evident that the curvature κ is directly proportional to the moment M and inversely proportional with the flexural rigidity (EI_z) of the beam. We have dropped the subscripts on the moments, because it is understood that the internal moment produces the stress σ.
    Finally, by substituting Eq. (7.6) into Eq. (7.3) the relationship between the applied moment and the internal stress distribution can be obtained:

$$\sigma = - My/I_z \tag{7.7}$$

In Eq. (7.7), M is positive when the beam is deflected so that the convex side is on its bottom and y is positive in the upward direction. This relation is considered to be one of the most important of the many equations employed by engineers when designing structures that contain beams, and as such it will become essential to you in the design of structures.

7.2.5 Stresses in Beams with Rectangular Cross Sections

Consider a beam with a rectangular cross section as defined in Fig. 7.9. For a rectangular cross section, the neutral axis coincides with the horizontal axis of symmetry:

$$y_{top} = h/2; \qquad y_{bottom} = - h/2 \tag{a}$$

Substituting Eq. (a) into Eq. (7.7) yields:

$$\sigma_{max} = Mh/(2I_z) \quad \text{and} \quad \sigma_{min} = - Mh/(2I_z) \tag{7.8}$$

The tensile stress is a maximum when y is equal to $- h/2$, and the compressive stress is a maximum when y is equal to $h/2$. The tensile and compressive stresses are equal because the neutral axis for the rectangular section divides the section into two equal parts.

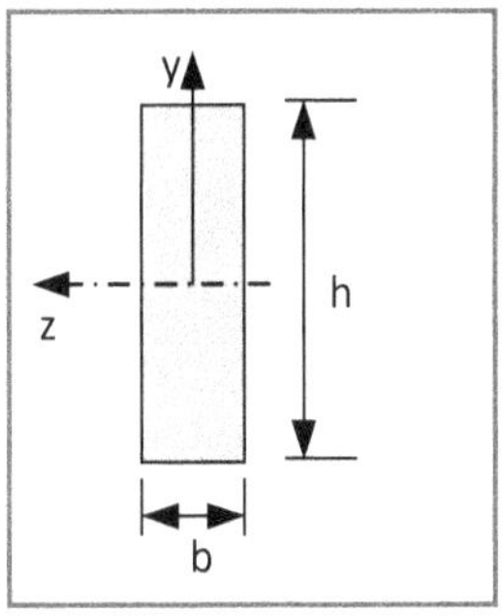

Fig. 7.9 Rectangular cross section of a beam with height h and width b.

Let's define the **section modulus**, Z, for a symmetric cross section of arbitrary geometry as:

$$Z = I_z / c \tag{7.9a}$$

where c is the distance from the centroid to an outer fiber of the beam. For a rectangular cross section, $c = h/2$ and Eq. (7.9a) becomes:

$$Z = 2I_z/h \tag{7.9b}$$

204 —Chapter 7 — Stresses in Beams

Then, we may express the maximum and minimum normal stress σ as:

$$\sigma_{max} = M/Z \quad \text{and} \quad \sigma_{min} = -M/Z \tag{7.10}$$

For a rectangular cross section, the second moment of the area I_z and the section modulus are given by:

$$I_z = bh^3/12 \quad \text{and} \quad Z = bh^2/6 \tag{7.11}$$

We now have the relations necessary to determine the stresses in a beam subjected to internal moments that are constant along its length. Let's consider two examples to illustrate the procedure for determining the stresses in beams with rectangular cross sections.

EXAMPLE E7.3

Consider a beam with a rectangular section with h = 4 in. and b = 0.75 in. Determine the margin of safety for the beam if it is loaded with a moment of 6,000 ft-lb and fabricated from steel with a yield strength of 42,000 psi.

Solution: Let's substitute Eq. (7.11) into Eq. (7.8) to determine the maximum stress.

$$\sigma_{max} = Mh/(2I_z) = 12\,Mh/(2bh^3) = 6M/(bh^2) \tag{7.12}$$

$$\sigma_{max} = (6)(6000)(12)/[(0.75)(4)^2] = 36{,}000 \text{ psi}$$

Finally, the safety factor and the margin of safety are given by:

$$SF = S_y/\sigma = 42{,}000/36{,}000 = 1.167; \quad MOS = SF - 1 = 16.7\%$$

The margin of safety for this beam is extremely small and careful consideration should be given to reducing either the applied moment or to increasing the size of the section.

EXAMPLE E7.4

Consider a beam with a rectangular section with h = 300 mm and b = 100 mm. Determine the largest moment that can be applied to the beam if it is fabricated from steel with a yield strength of 300 MPa. The beam is to have a safety factor of 3.0.

Solution: Determine the allowable design stress by considering the safety factor and the yield strength:

$$\sigma = S_y/SF = 300/3.0 = 100 \text{ MPa} \tag{a}$$

Rearrange Eq. (7.12) to solve for the moment as:

$$M = \sigma(bh^2)/6 = (100)(100)(300)^2/6 = 150 \times 10^6 \text{ N-mm} = 150 \text{ kN-m} \tag{b}$$

In this solution, we used the fact that 1 MPa = 1 N/mm². Alternatively, we could have obtained the same result by expressing the stress in Pa and all lengths in m.

7.3 PROPERTIES OF CROSS SECTIONS

If we examine the stress distribution, shown in Fig. 7.7, it is clear that the beam is not an efficient structural component. Only the top and the bottom fibers of the beam are stressed to the maximum. The interior region of the beam is under utilized. Indeed, near the neutral axis the stresses approach zero. A beam with a rectangular cross section is extremely inefficient, because the average magnitude of stress over the section is only $\sigma_{max}/2$. The closer the average magnitude of stress is to the maximum stress, the more efficient is the cross sectional area of the beam.

To improve the effectiveness of beams, we remove most of the under-stressed, central-region of the cross section to produce I beams and WF (wide-flange) shapes. A typical example of a more efficient section is shown in Fig. 7.10.

For a beam section with a specified height h, we judge its structural effectiveness **SE** by the ratio of its section modulus to its cross sectional area.

$$\mathbf{SE} = Z/A \tag{7.13}$$

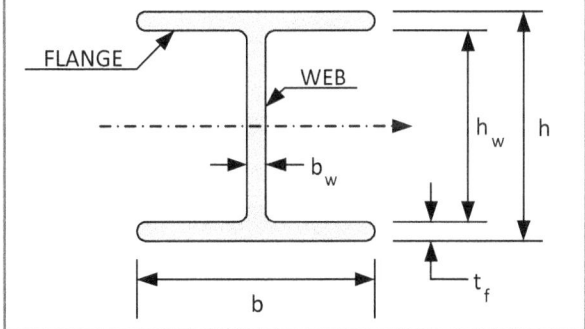

Fig. 7.10 Removing a large part of the under stressed material in a section of a beam increases its structural effectiveness.

The effectiveness of a beam with a rectangular cross section is given by Eq. (7.11) as:

$$\mathbf{SE} = bh^2/(6bh) = h/6 = 0.1667\,h \tag{7.14}$$

For the wide flange section illustrated in Fig. 7.10, the relation for the effectiveness depends on several variables including b, h, b_w and h_w. To derive the relation for **SE**, we begin by determining I_z as:

$$I_z = I_{web} + I_{flange} \tag{a}$$

$$I_{web} = b_w h_w^3/12 \tag{b}$$

$$I_{flange} = 2\left[\frac{1}{12}bt_f^3 + bt_f\left(\frac{h_w + t_f}{2}\right)^2\right] \tag{c}$$

Using Eqs. (a), (b) and (c) to solve for Z gives:

$$Z = \frac{2I_z}{h} = \frac{1}{h}\left[\frac{b_w h_w^3}{6} + \frac{bt_f^3}{3} + bt_f(h_w + t_f)^2\right] \tag{d}$$

The area of the wide flange section is given by:

206 —Chapter 7 — Stresses in Beams

$$A = 2bt_f + b_w h_w$$

$$SE = \frac{1}{h(2bt_f + b_w h_w)} \left\{ \frac{b_w h_w^3}{6} + \frac{bt_f^3}{3} + bt_f(h_w + t_f)^2 \right\} \qquad (7.15)$$

The result in Eq. (7.15) involves many terms and the improvement in the effectiveness of the section is not apparent simply by inspecting this relation. To ascertain the benefits of removing the under-stressed material, let's consider Example 7.5.

EXAMPLE 7.5

If the wide flanged section is fabricated with the following geometry, determine the effectiveness of the section and compare it with the effectiveness of a rectangular section.

$b_w = 0.1\ b$; $h_w = 0.8\ h$;   then $t_f = 0.1\ h$

Solution: Substituting these quantities into Eq. (7.15) yields:

$$SE = \frac{1}{h(0.2bh + 0.08bh)} \left\{ \frac{0.1b(0.8h)^3}{6} + \frac{b(0.1h)^3}{3} + b(0.1h)[0.8h + 0.1h]^2 \right\}$$

Reducing this expression gives:

$$SE = 0.3209\ h$$

A comparison of this result with Eq. (7.14) indicates that the wide flanged section defined in this example with $SE = 0.3209\ h$ is nearly twice as efficient as the rectangular cross section with $SE = 0.1667\ h$.

Determining I_z and Z for the various non-rectangular cross sections is a time consuming task. Fortunately, others have resolved this problem for us. The American Institute of Steel Construction (AISC) has prepared tables showing the properties of the commercially available sections. An abridged version of these tables is presented in Appendix D.

7.4 BENDING OF BEAMS WITH TRANSVERSE FORCES

Moments and stresses are induced in beams when they are subjected to transverse forces. In structures, the most significant loads are usually due to gravitational forces although in machine components the loading is due to a number of different effects and gravitational forces are often negligible. It is important, when modeling a beam, to consider both the various types of loading and its supports. Typical examples of different loads and supports for the beams are illustrated in Fig. 7.11.

7.4.1 Shear and Bending Moments in Beams

Let's begin by considering the simply supported beam with a single concentrated force F as illustrated in Fig. 7.11. How does a concentrated force develop a moment? Where is the moment arm? We answer these questions by drawing several FBDs of the beam as shown in Fig. 7.12.

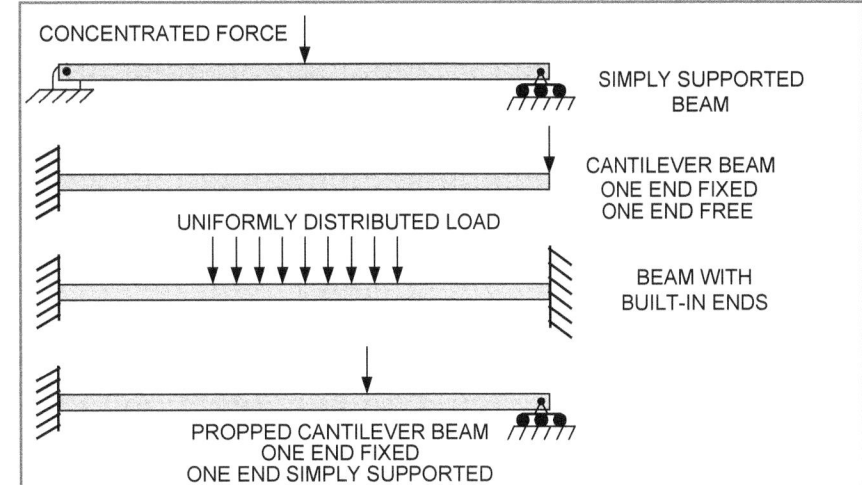

Fig. 7.11 Various types of transverse loading and supports for beams.

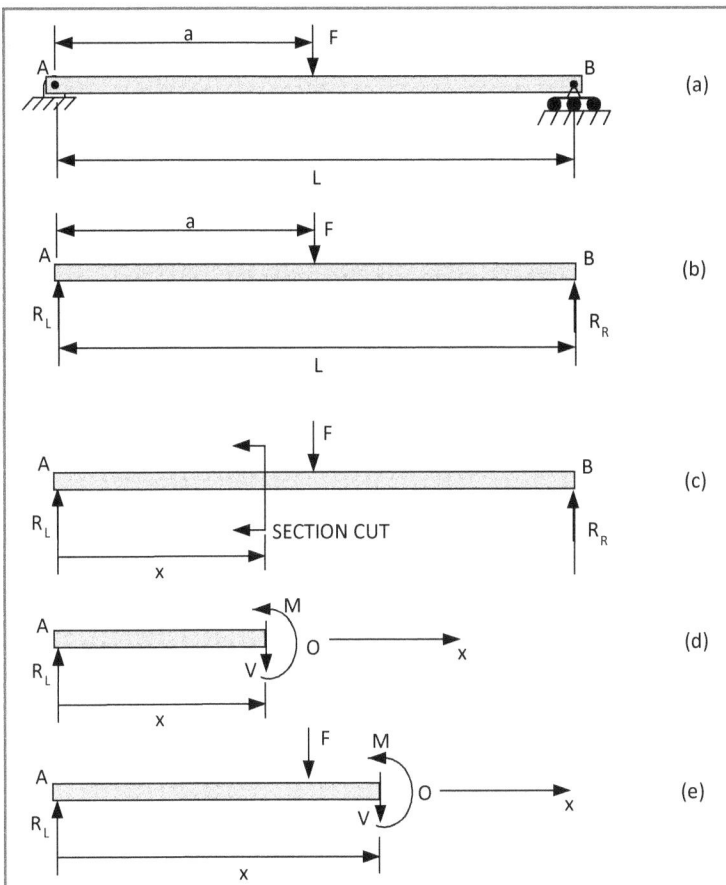

Fig. 7.12 FBDs of a simply supported beam with a concentrated load F.

In Fig. 7.12a, we show a simply supported beam of length L with a concentrated force F applied a distance, a, from the left end. A free body of the entire beam, in Fig. 7.12b, shows the supports replaced with reaction forces R_L and R_R. When the beam deforms, its ends are free to rotate over the simple supports (the pin at A or the roller at B); consequently the moment M = 0 at both ends of the beam.

We use the equilibrium equations to solve for the reaction forces R_L and R_R.

$$\Sigma M_A = R_R L - F a = 0$$

$$R_R = F\,a/L \qquad \text{(a)}$$

$$\Sigma F_y = R_L + R_R - F = 0$$

Substituting Eq. (a) into this relation gives:

$$R_L = F - R_R = F - F\,a/L = F[(L-a)/L] = F\,c/L \qquad \text{(b)}$$

where $c = L - a$

Next, we make a section cut a distance x from the left end of the beam where $0 < x < a$ as shown in Fig. 7.12c. The left side of the beam is removed as a free body. We apply a positive[3] internal shear force V and a positive internal moment M to the surface exposed by the section cut as indicated in Fig. 7.12d.

From the equilibrium equations, we write:

$$\Sigma F_y = R_L - V = Fc/L - V = 0$$

$$V = F\,c/L \qquad \Rightarrow \qquad 0 < x < a \qquad \text{(c)}$$

$$\Sigma M_O = M - R_L\,x = M - F\,x\,c/L = 0$$

$$M = F\,x\,c/L \qquad \Rightarrow \qquad 0 < x < a \qquad \text{(d)}$$

To determine the relations for the shear force V and the internal moment M for the remaining part of the beam, we consider the FBD shown in Fig. 7.12e and repeat the equilibrium analysis.

$$\Sigma F_y = R_L - F - V = F(L-a)/L - F - V = 0$$

$$V = -F\,a/L \qquad \Rightarrow \qquad a < x < L \qquad \text{(e)}$$

$$\Sigma M_O = M - R_L\,x + F(x-a) = M - F\,x(L-a)/L + F(x-a) = 0$$

$$M = F\,a(L-x)/L \qquad \Rightarrow \qquad a < x < L \qquad \text{(f)}$$

Note, Eqs. (e) and (f) are only valid for $a < x < L$, and Eqs. (c) and (d) only apply for $0 < x < a$.

Equations (c) and (e) give the shear force V and Eqs. (d) and (f) give the internal moment M as a function of position x along the length of the beam. Examination of Eqs (d) and (f) shows that the maximum moment occurs at the load point $x = a$. Accordingly, we write:

$$M_{max} = F\,a\,c/L \qquad (7.16)$$

The maximum stress σ is given by:

$$\sigma_{max} = F\,a\,c/(L\,Z) \qquad (7.17)$$

---

[3] We consider the internal shear force to be positive when it acts downward in the negative y direction on the FBD of the left end of the beam.

A shear stress τ also develops due to the shear force, V; however, we will delay the discussion of shear stress until Section 7.4.3. For long thin beams, the shear stress τ is small when compared to the normal stress σ and is usually ignored without compromising the safety of the structure. However, for short beams with significant height, shear stresses must be considered. Shear stresses are also important when the beams are fabricated from fibrous materials like wood, because the shear strength in the direction of the fibers is very low.

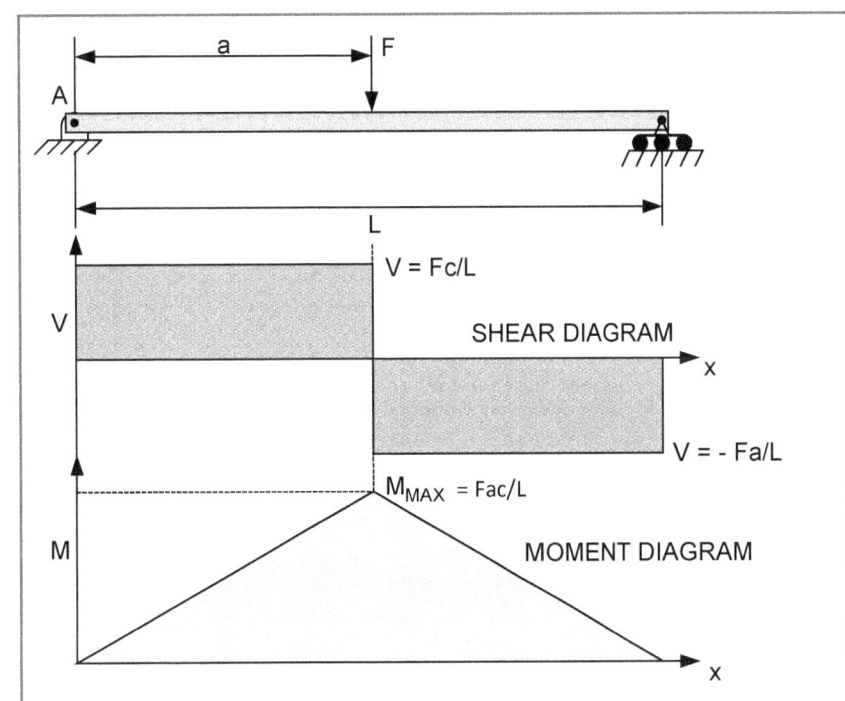

Fig. 7.13 Shear and bending moment diagrams show the distribution of V and M with x.

7.4.2 Shear and Bending Moment Diagrams

The equations for the shear force V and the bending moment M are often represented with diagrams that show the distribution of both of these quantities over the length of the beam. The diagrams are simply graphs of the equations developed by an analysis of the appropriate free bodies. An example of the shear and bending moment diagrams for the beam loaded with a single concentrated force F is presented Fig. 7.13.

Shear and bending moment diagrams are useful for two reasons. First, they aid in the visualization of the distribution of the shear force V and the bending moment M. Second, we use these diagrams to graphically determine the maximum moment and shear force. We have provided a table below to show the relationship among the loading condition and the lines or curves used in representing the shear and bending moment diagrams.

Table 7.1
Relationships among the loading conditions and the shear and bending moment diagrams

TYPE OF LOAD	SHEAR DIAGRAM	MOMENT DIAGRAM
NO LOAD	HORIZONTAL	OR / RAMP
CONCENTRATED LOAD	OR / VERTICAL JUMP	OR / SLOPE CHANGE
UNIFORMLY DISTRIBUTED LOAD	OR / RAMP	OR / PARABOLA
LINEARLY DISTRIBUTED LOAD	OR / 2nd DEGREE CURVE	OR / 3rd DEGREE CURVE
CONCENTRATED MOMENT	HORIZONTAL	CCW / CW OR / VERTICAL JUMP

Let's consider an example to demonstrate the graphical method of determining the shear and moment diagrams.

EXAMPLE E7.6

For the simply supported beam loaded as indicated in Fig. E7.6, construct the shear and bending moment diagrams using a graphical approach.

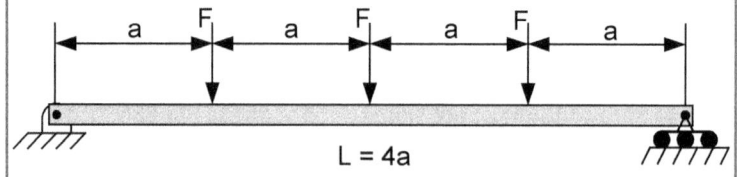

Fig. E7.6

Solution: The simply supported beam is symmetric; hence it is evident by inspection that:

$$R_L = R_R = 3F/2$$

We construct the FBD of the beam and then draw the shear diagram as shown in Fig. E7.6a. To construct the shear diagram, begin at the left end of the beam at the origin of the V – x axes. Because the left hand reaction is upward and equal to 3F/2, we draw a line along the positive V axis (upward) scaled to a length proportional to 3F/2. Move to the right on the beam (the positive x direction) and observe the beam is free of loads until x = a. This observation means the shear force is constant from 0 < x < a. At x = a, we encounter a concentrated force F which is directed downward. Because F is downward, we decrease the shear force by $\Delta V = -F$ so that $V = 3F/2 - F = F/2$ as shown in the shear diagram. below:

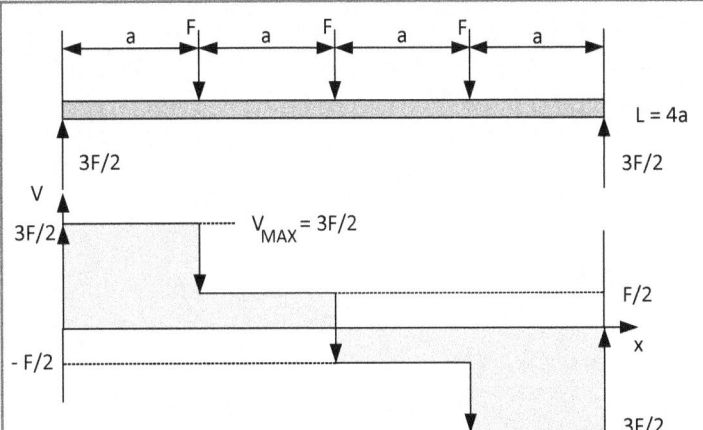

Fig. E10.6a

We proceed without change in the shear force until $x = 2a$, where we again encounter another concentrated force of magnitude F. As before, we decrease V by this amount so that $V = F/2 - F = -F/2$. We continue to increase x without change in the shear force until $x = 3a$, where another concentrated force of magnitude F is encountered. As before, we decrease V by this amount so that $V = -F/2 - F = -3F/2$.

We continue to increase x without change in the shear force until $x = 4a = L$, where we encounter the right hand reaction 3F/2 that is directed upward. Because the force is directed upward, we increase V by this amount so that $V = -3F/2 + 3F/2 = 0$. It is evident by the construction demonstrated that concentrated forces produce step changes in the shear diagram. Construction of the shear diagram then consists of drawing straight-line segments and adding or subtracting the concentrated forces to the shear force.

To construct the moment diagram, we make use of a relation between the shear forces and bending moments[4], namely:

$$dM/dx = V \qquad (7.18)$$

From Eq. (7.18), it is clear that the slope of the moment diagram at any position x is equal to the value of the shear force V at that same position. Integrating Eq. (7.18), we obtain:

$$M_{x_2} - M_{x_1} = \int_{x_1}^{x_2} V dx \qquad (7.19)$$

For the shear diagram in this example, the areas under the V – x graph are simple rectangles; hence, the integral $\int V dx$ between x_2 and x_1 is $V(x_2 - x_1)$. It is easy to draw the moment diagram shown in Fig. E7.6b using the shear diagram as a guide.

The area of the first rectangle between $x = 0$ and $x = a$ is $3Fa/2$. Because the area of the rectangle increases linearly with position x, we draw a straight line on the M – x diagram from the origin to the point with coordinates $x = a$, $M = 3Fa/2$. The area of the next rectangle from $a < x < 2a$ is $Fa/2$. We add this area to the existing moment to obtain $3Fa/2 + Fa/2 = 2Fa$. Again we draw a straight line on the M – x diagram from the point $(a, 3Fa/2)$ to the next point with coordinates $x = 2a$, $M = 2Fa$. We may construct the remainder of the moment diagram by repeating the procedure. Note that the next two areas on the V-x diagram lie below the axis and

---

[4] We will defer the derivation of Eq. (7.18) until Section 7.4.3.

are negative. These areas cause the M-x diagram to decrease and return to zero at x = 4a. The beam is symmetric; consequently, the moment diagram is also symmetric[5].

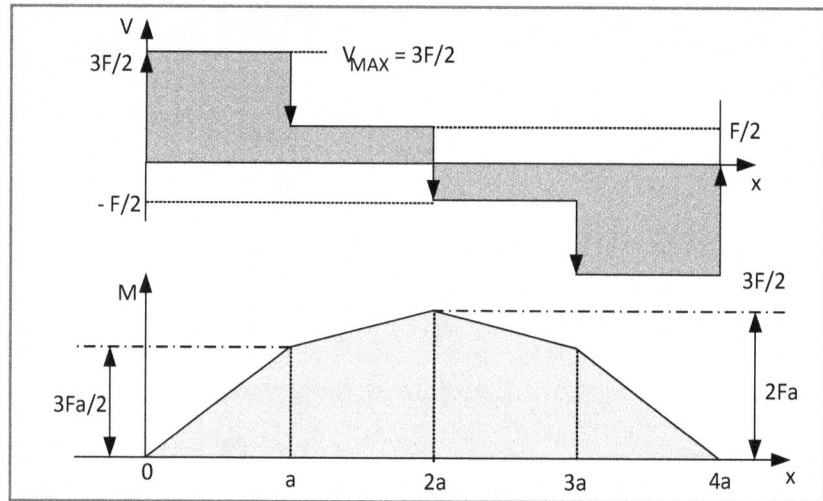

Fig. E7.6 b

EXAMPLE E7.7

For the simply supported beam with a uniformly distributed load of magnitude q, write the equations for the internal shear force V and the moment M as functions of position x. Also construct the shear and bending moment diagrams.

Fig. E7.7

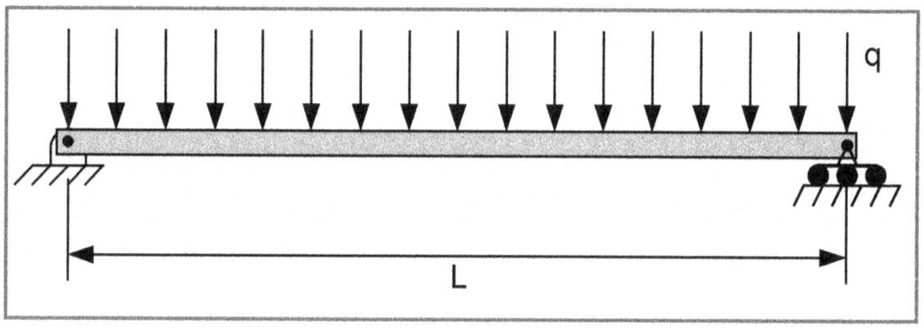

Solution:

The beam is symmetric so that the reactions R_L and R_R are equal. In addition, the total force in the negative y direction is F = qL; therefore, it is evident that

$$R_L = R_R = qL/2 \qquad (a)$$

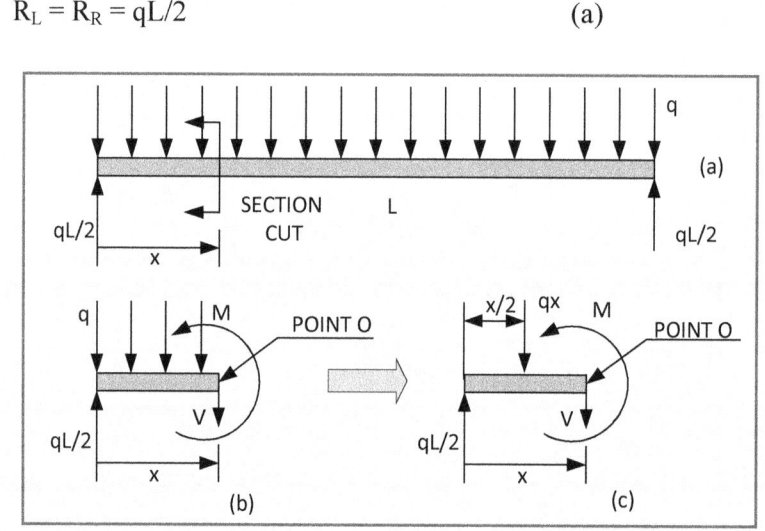

Fig. E7.7a

---

[5] The shear diagram is also symmetric with respect to the magnitude of V, but the signs of the shear forces are not symmetric.

Let's make a section cut at some arbitrary position x and construct a FBD of the left side of the beam as shown in Figs. E7.7a and inset (b).

The FBD of the left side of the beam is shown with a uniformly distributed load acting over a length x of the beam segment. We replace this uniformly distributed load with an equivalent concentrated force qx located at the center of the segment, as shown in Fig. E7.7a inset (c). Next, we apply the equilibrium equations and write:

$$\Sigma F_y = qL/2 - qx - V = 0$$

$$V = (q/2)(L - 2x) \quad \Rightarrow \quad 0 < x < L \quad (b)$$

$$\Sigma M_O = M + qx(x/2) - (qL/2)x = 0$$

$$M_x = (qx/2)(L - x) \quad \Rightarrow \quad 0 < x < L \quad (c)$$

Inspection of Eq. (b) shows that $V = 0$ at $x = L/2$. This zero crossing of the shear force on the V – x diagram locates the position of the maximum moment. Solving for M_{max} at $x = L/2$ using Eq. (c) gives:

$$M_{max} = qL^2/8 \quad (d)$$

We may construct graphs of Eqs. (b) and (c) showing V and M as a function of position along the length of the beam to produce the shear and moment diagrams. Alternatively we may construct them directly from the transverse forces applied to the beam as shown below.

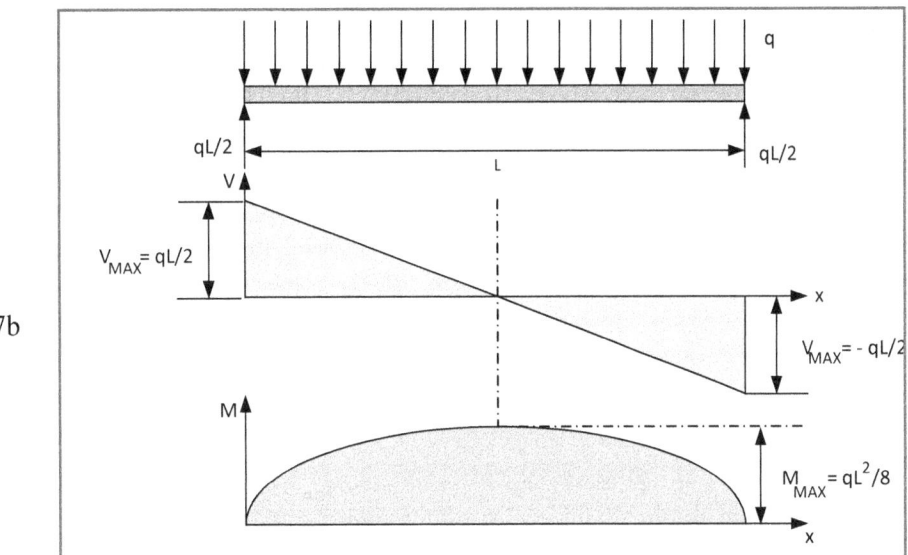

Fig. E7.7b

EXAMPLE E7.8

For the cantilever beam with a concentrated force F at its free end, write the equations for the internal shear force V and the moment M, and construct the shear and bending moment diagrams.

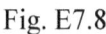
Fig. E7.8
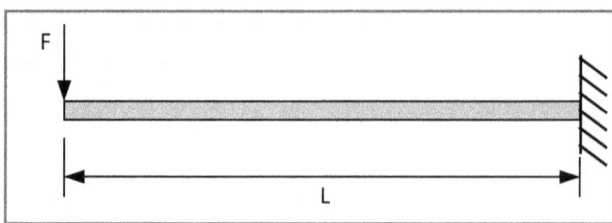

Solution: Let's begin by removing the built-in end and replacing it with a shear force V_{FE} and a moment M_{FE} when constructing the FBD of the cantilever beam as shown below:

Fig. E7.8a
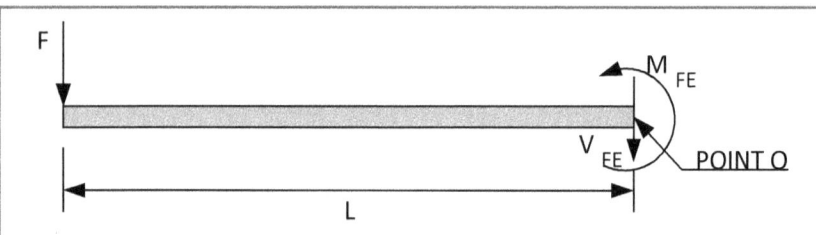

From the equilibrium equations, we write:

$$\Sigma F_y = -F - V_{FE} = 0$$

$$V_{FE} = -F \qquad (a)$$

$$\Sigma M_O = M_{FE} + FL = 0$$

$$M_{FE} = -FL \qquad (b)$$

The negative signs for V_{FE} and M_{FE} imply that the directions shown in the FBD are not correct. We change the directions in the FBD shown in Fig. E7.8 b.

Fig. E7.8b
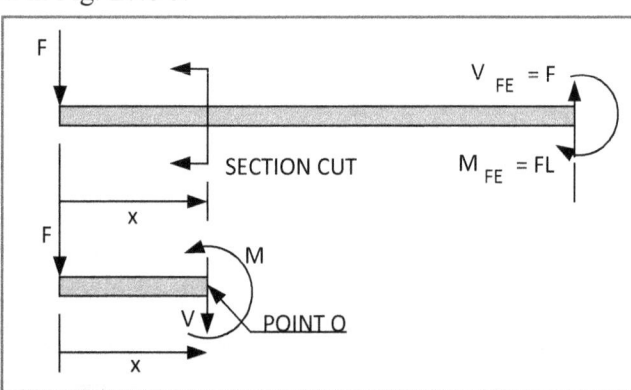

Next, we make a section cut at position x and remove the left side of the cantilever beam as a free body. Again, we write the equilibrium equations to obtain.

$$\Sigma F_y = -F - V = 0; \qquad V = -F \qquad (c)$$

$$\Sigma M = M + Fx = 0; \qquad M = -Fx \qquad (d)$$

The shear and moment diagrams are constructed by using either Eqs. (c) and (d) or by drawing the diagrams directly from the knowledge of the distribution of forces over the length of the beam, as indicated in Fig. E7.8 c. The shear force, V is a constant along the entire length of the beam. The bending moment M increases linearly in x—starting at zero at the free end and becoming a maximum $M_{max} = -FL$ at the fixed end.

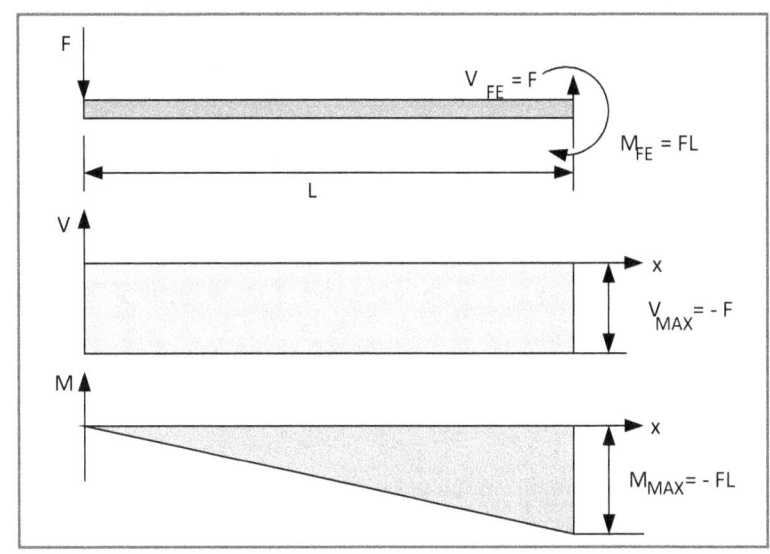

Fig. E7.8c

EXAMPLE E7.9

For the simply supported beam described in Fig. 7.13, determine the margin of safety if the beam and its loading are specified as:

- A wide flanged (WF) shape W 10×60
- F = 20 kip
- L = 24 ft and a = 10 ft.
- The yield strength of the hot rolled steel beam is 42 ksi.

Solution: The properties of the WF section are given in Appendix D as:

$$I_z = 341 \text{ in.}^4 \qquad Z = 66.7 \text{ in.}^3 \qquad \text{(a)}$$

The maximum moment occurs at x = a = 10 ft and is given by Eq. (7.16) as:

$$M_{max} = F\,a\,c/L = (20 \times 10^3)(10)(24-10)/24 = 116.7 \times 10^3 \text{ ft-lb} \qquad \text{(b)}$$

The maximum stress is given by Eq. (7.10) as:

$$\sigma_{max} = M_{max}/Z = [(116.7 \times 10^3)(12)]/66.7 = 21.0 \text{ ksi} \qquad \text{(c)}$$

Finally, the margin of safety is given by:

$$MOS = SF - 1 = S_y/\sigma - 1 = 42.0/21.0 - 1 = 100\% \qquad \text{(d)}$$

EXAMPLE E7.10

For the simply supported beam with a uniformly distributed load, described in Example E7.7, determine the margin of safety if the beam and its loading are specified as:

- A S8 × 23 American Standard shape
- q = 1 kip/ft
- L = 16 ft
- The yield strength of the steel from which the beam is hot rolled is 38 ksi.

Solution: The properties of the American Standard section are given in Appendix D as:

$$I_z = 64.9 \text{ in.}^4 \qquad Z = 16.2 \text{ in.}^3 \qquad (a)$$

The maximum moment occurs at x = L/2 = 8 ft and is given by:

$$M_{max} = qL^2/8 = (1 \times 10^3)(16)^2/8 = 32{,}000 \text{ ft-lb} \qquad (b)$$

The maximum stress is given by Eq. (7.10) as:

$$\sigma_{max} = M_{max}/Z = (32{,}000)(12)/16.2 = 23.7 \text{ ksi} \qquad (c)$$

Finally, the margin of safety is given by:

$$MOS = SF - 1 = S_y/\sigma - 1 = 38.0/23.7 - 1 = 60.3\% \qquad (d)$$

The margin of safety is probably somewhat low at 60.3%. While the beam is hot rolled to a standard section shape (its geometry is well established and not subject to error in fabrication), any significant over loading can endanger the stability of the structure. Consideration should be given to increasing the section size to an American Standard S10 × 25.4.

EXAMPLE E7.11

For the cantilever beam with a concentrated force F applied at its free end, described in Example E7.8, determine the margin of safety if the beam and its loading are specified as:

- A WT 5 × 30 structural tee
- F = 400 lb
- L = 15 ft
- The yield strength of the steel is 44 ksi.

Solution: The properties of the structural tee is given in Appendix D as:

$$I_z = 12.9 \text{ in.}^4 \qquad Z = 3.04 \text{ in.}^3 \qquad (a)$$

The maximum moment occurs at x = L = 15 ft and is given by:

> $M_{max} = FL = (400)(15) = 6{,}000$ ft-lb (b)
>
> The maximum stress is given by Eq. (7.10) as:
>
> $\sigma_{max} = M_{max}/Z = (6{,}000)(12)/(3.04) = 23.68$ ksi (c)
>
> Finally, the margin of safety is given by:
>
> $MOS = SF - 1 = S_y/\sigma - 1 = 44.0/23.68 - 1 = 85.81\%$ (d)
>
> It appears that the margin of safety is marginal given the specification for the beam and the loading. Consideration should be given to selection of a larger or heavier beam with a larger section modulus, such as WT5 × 44 or WT6 × 25.

7.4.3 Shear Stresses in Beams

The transverse forces applied to beams produce internal shear forces V in both horizontal and vertical directions (H and V respectively). These shear forces give rise to shear stresses τ_H and τ_V. Let's first establish the relationship between the internal bending moment M and the internal shear force V. Consider the simply supported beam with a uniformly distributed load q as illustrated in Fig. 7.14a. We make two section cuts and remove the segment of the beam between the section cuts with a length Δx. The FBD of this beam segment is presented in Fig. 7.14b. Writing the equilibrium equations yields:

$$\Sigma F_y = V - q\Delta x - (V + \Delta V) = 0$$

$$\Delta V/\Delta x = dV/dx = -q \quad (7.20)$$

$$\Sigma M_O = M + \Delta M - M + q\Delta x^2/2 - V\Delta x = 0$$

$$\Delta M/\Delta x = dM/dx = V \quad (7.18 \text{ bis})$$

Note, $q\Delta x^2/2$, a second order term, is neglected because it is small when compared to the other terms in the equation.

Next, let's sub-section this small segment of the beam with a horizontal cut some distance y above the neutral axis as shown in Fig. 7.14c. The FBD of the subsection of this beam segment is illustrated in Fig. 7.14d. We write the equilibrium equation $\Sigma F_x = 0$ for this free body to obtain:

$$\Sigma F_x = -\int(\sigma + \Delta\sigma)dA + \int\sigma dA + H = 0$$

$$H = \int\Delta\sigma \, dA \quad (a)$$

where H is a horizontal shear force.

From Eq. (7.7), we may write:

$$\Delta\sigma = -\Delta My/I_z \quad (b)$$

Substituting Eq. (7.18) into Eqs. (b) and (a) gives:

$$H = -(V\Delta x/I_z)\int y\,dA = -(V\Delta x/I_z)Q_1 \quad (c)$$

where Q_1 is defined by:

$$Q_1 = \int_{y_1}^{c} y\,dA \quad (7.21)$$

218 —Chapter 7 — Stresses in Beams

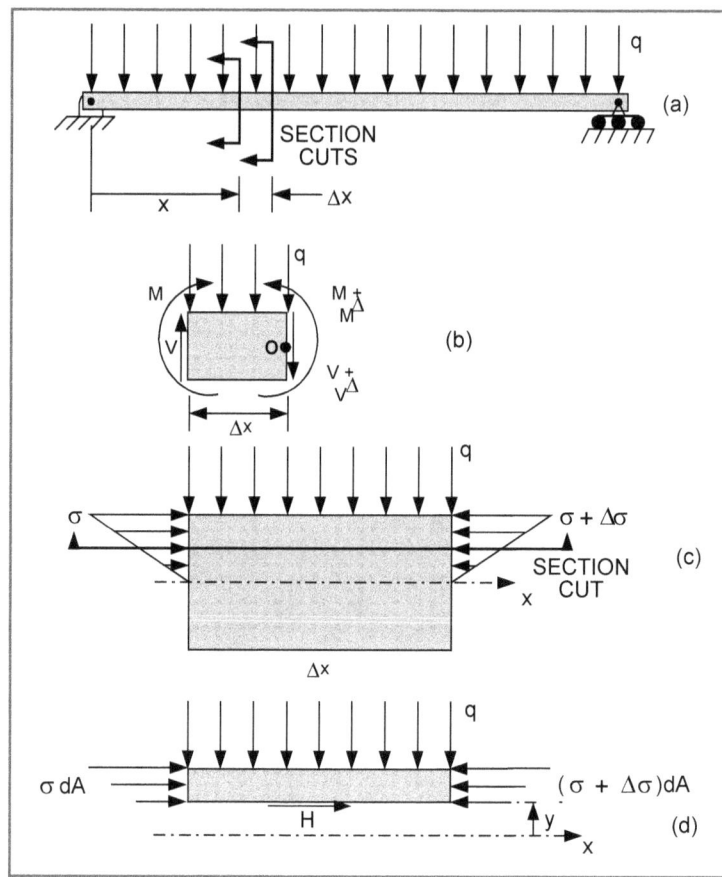

Fig. 7.14 Several FBDs that show the relation between M and V.

Q_1 is the first moment of the area from the section cut at position y_1 to the location of the outer fiber on the top surface of the beam at $y = c$.

The horizontal shear stress τ_H acting on the surface exposed by the section cut is given by:

$$\tau_H = H/A_H = - VQ_1/(bI_z) \qquad (7.22)$$

where $A_H = b\Delta x$ and b is the beam thickness.

We are not usually concerned with the sign in determining the shear stress. Failure of the material, from which structural components are fabricated, is independent of the direction of the shear stress. Finally, we show that the horizontal and vertical shear stresses[6] are equal.

$$\tau = \tau_V = \tau_H = VQ_1/(bI_z) \qquad (7.23)$$

The equality of the shear stresses ($\tau = \tau_V = \tau_H$) is established by taking moments about the point O for the element Δx by Δy shown in Fig. 7.15. This element may be removed from any plane body such as a beam. Also the location of point O is arbitrary.

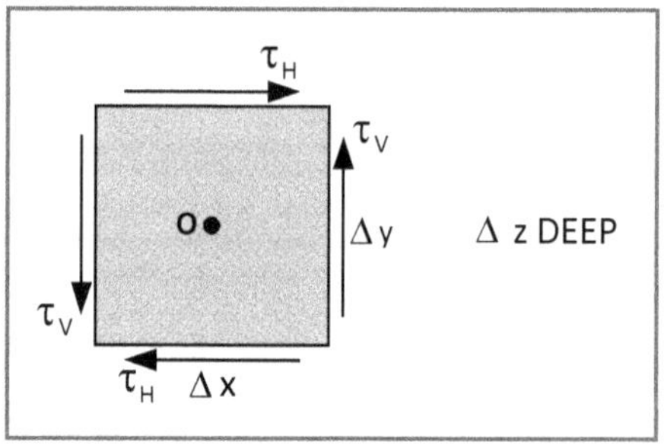

Fig. 7.15 An element Δx by Δy by Δz with shear stresses τ_V and τ_H acting on its faces.

---

[6] The vertical shear stresses τ_V and vertical shear force V act on the surface exposed by a vertical section cut.

$$\Sigma M_O = (\tau_V \Delta y\, \Delta z)\, \Delta x - (\tau_H \Delta x\, \Delta z)\, \Delta y = 0$$

$$\tau_V = \tau_H \tag{7.24}$$

Examination of Eq. (7.23) indicates that the shear stress τ_V is a maximum when Q_1 is a maximum. It is evident from Eq. (7.21) that Q_1 is a maximum for any given section when $y_1 = 0$. Therefore, the shear stress τ is a maximum at the neutral axis. As y_1 approaches the outer fiber at $y_1 = c$, Q_1 approaches zero. For this reason, the shear stresses τ are zero on the top and bottom surfaces of the beam, where the normal stresses σ are a maximum. We will demonstrate the method for solving for the shear stresses in a beam with a rectangular cross section in the following examples.

EXAMPLE E7.12

Consider a simply supported beam of length L with a concentrated force F of 10 kN applied at the center of the span. The beam cross section is rectangular with b = 50 mm and h = 125 mm and the beam length L = 5 m. Determine the maximum normal stress σ and the maximum shearing stress τ.

Solution: Because the simply supported beam is symmetric, it is clear that the reactions are equal to:

$$R_L = R_R = F/2 = 10/2 = 5.0 \text{ kN} \tag{a}$$

Next, we construct the shear and moment diagrams, as shown in Fig. E7.12. From the shear and bending moment diagrams, we determine that:

$$V_{max} = 5 \text{ kN} \quad \text{and} \quad M_{max} = 12.5 \text{ kN-m} \tag{b}$$

The maximum normal stress is determined from Eq. (7.12) as:

$$\sigma_{max} = 6M_{max}/(bh^2) = [(6)(12.5 \times 10^3 \text{ N-m})]/[0.05 \text{ m } (125)^2 \text{ mm}^2] = 96 \text{ MPa} \tag{c}$$

The maximum shear stress is given by Eq. (7.23) as:

$$\tau_{max} = V(Q_1)_{max}/(bI_z) \tag{d}$$

To evaluate Eq. (d), let's first determine $(Q_1)_{max}$ using Eq. (7.21)

$$Q_1 = \int_{y_1}^{c} y\, dA = b\int_{0}^{h/2} y\, dy = \frac{b}{2}\left[y^2\right]_0^{h/2} = \frac{bh^2}{8} \tag{e}$$

We may also determine Q_1 without integrating by recalling that the first moment of the area A_1 shown in Fig. E 7.12 a is given by:

$$Q_1 = A_1 \bar{y} \tag{f}$$

where A_1 is the area of the cross section above the section cut and $\bar{y}$ is the distance from the neutral axis to the centroid of the area A_1.

For a rectangular cross section the moment of inertia is:

$$I_z = bh^3/12 \tag{g}$$

Substituting Eqs. (e) and (g) into Eq. (d) yields:

$$\tau_{max} = (3/2)(V/bh) = (3/2)(V/A) \qquad (7.25)$$

$$\tau_{max} = (1.5)(5 \times 10^3)/[(50 \times 125)] = 1.20 \text{ MPa} \qquad (h)$$

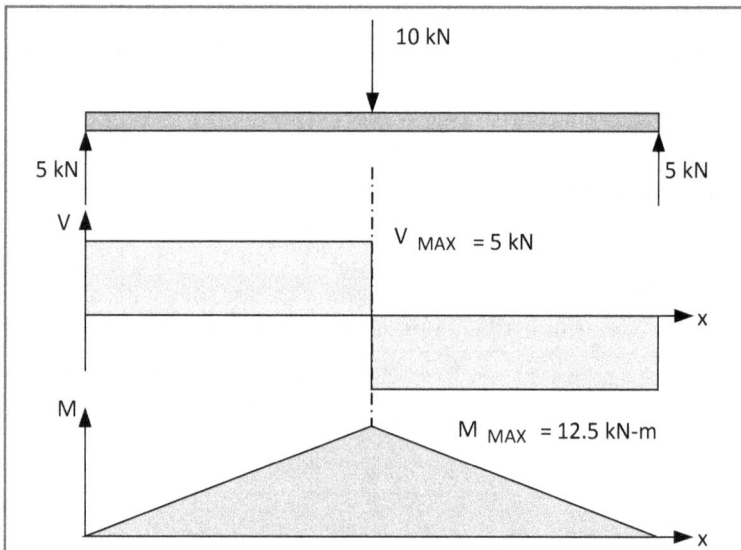

Fig. E7.12

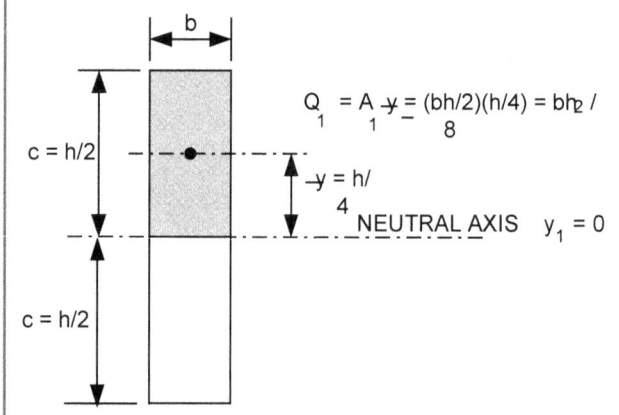

Fig. E7.12a

Comparison of the maximum normal stresses and the maximum shear stress for this example shows that the shear stresses are very small (1.25% of the maximum normal stresses). For long thin beams, the shear stresses are usually small enough to be neglected[7].

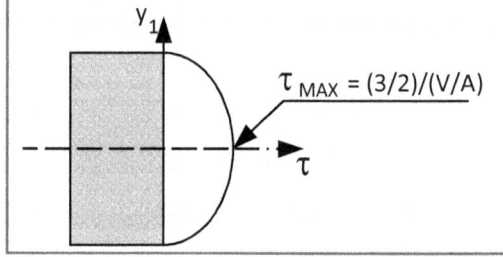

Fig. E7.12b

---

[7] The exception is when wood is used to fabricate the beam. The shear strength of wood is very low and even small shear stresses may be sufficient to split the wooden beam along its neutral axis.

Equation (7.25) indicates that the maximum shear stress is 1.5 times the average shear stress. The shear stress is not distributed uniformly over the cross sectional area, but varies as a parabolic function; the maximum occurs at the neutral axis as shown in Fig. E7.12b.

EXAMPLE E7.13

Let's again consider a simply supported beam of length L = 5 m with a concentrated force F = 10 kN applied at the center of the span. The beam cross section is circular with a diameter d = 100 mm. Determine the maximum normal stress σ and the maximum shearing stress τ.

Solution: The beam in this example is identical with that in the previous example except for its cross section. The reactions, shear and bending moment diagrams, maximum shear force and moment are the same:

$$R_L = R_R = F/2 = 10/2 = 5.0 \text{ kN} \tag{a}$$

$$V_{max} = 5 \text{ kN} \quad \text{and} \quad M_{max} = 12.5 \text{ kN-m} \tag{b}$$

Because the moment of inertia I_z for a circular area about its centerline is given by:

$$I_z = \pi d^4/64 = \pi r^4/4 \tag{c}$$

The maximum normal stress is given by:

$$\sigma_{max} = M_{max}d/(2I_z) = 64 M_{max}d/(2\pi d^4) = 32 M_{max}/(\pi d^3) \tag{7.26}$$

$$\sigma_{max} = (32)(12.5 \times 10^3)/[(\pi)(0.1)(100)^2] = 127.3 \text{ MPa} \tag{d}$$

The maximum shear stress is given by Eq. (7.23) as:

$$\tau_{max} = V(Q_1)_{max}/(bI_z) \tag{e}$$

We will determine Q_1 using the procedure described below. For the circular bar, the parameter b = 2r at the neutral axis, and $Q_1 = 2r^3/3$ as indicated in Fig. E7.13.

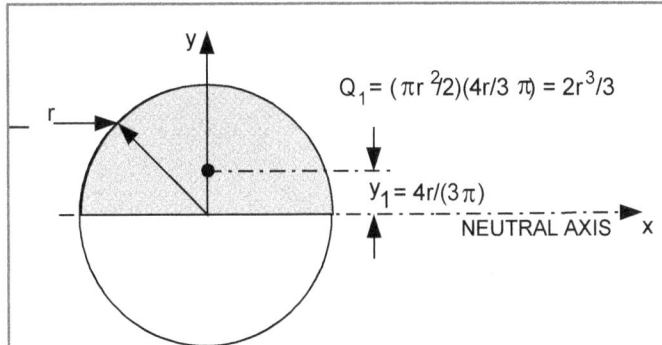

Fig. E7.13

Substituting these results and Eq. (c) into Eq. (e) gives:

$$\tau_{max} = (4/3)(V/\pi r^2) = (4/3)(V/A) \tag{7.27}$$

$$\tau_{max} = (4/3)(5 \times 10^3)/[\pi (50)^2] = 0.8488 \text{ MPa}$$

Again, it is evident that the maximum shear stress is very small (less than 1%) when compared to the normal stress.

7.5 BENDING OF COMPOSITE BEAMS

In our discussions of beams, we have considered only those fabricated from a single material, such as steel, aluminum or wood. However, in engineering design there is an increasing trend to employ beams fabricated from two or more materials. These are called **composite beams**; they offer the opportunity of using each of the materials employed in their construction to advantage. You may want to consider using them in the design of your structures.

7.5.1 Foam Core with Metal Cover Plates

Let's consider a composite beam fabricated with thin metal cover plates on the top and bottom with a plastic foam core, as shown by the cross sectional area depicted in Fig. 7.16.

The design concept for the foam core beam is to use light, low-strength foam to support the load-bearing metal plates positioned at the top and bottom, where they are the most efficient. To analyze the composite beam, we again assume that plane sections remain plane, before and after loading so the strain is linearly distributed as shown in Fig. 7.17. Note that the strain is continuous across the interface between the foam and the cover plates. The stress in the foam is given by:

$$\sigma_f = E_f \varepsilon \approx 0 \tag{a}$$

We set the stress in the foam equal to zero, because its modulus of elasticity E_f is small when compared to the modulus of elasticity of the metal.

The stress in the metal of the cover plates, σ_m is described by Hooke's law as:

$$\sigma_m = E_m \varepsilon = - E_m y/\rho \tag{b}$$

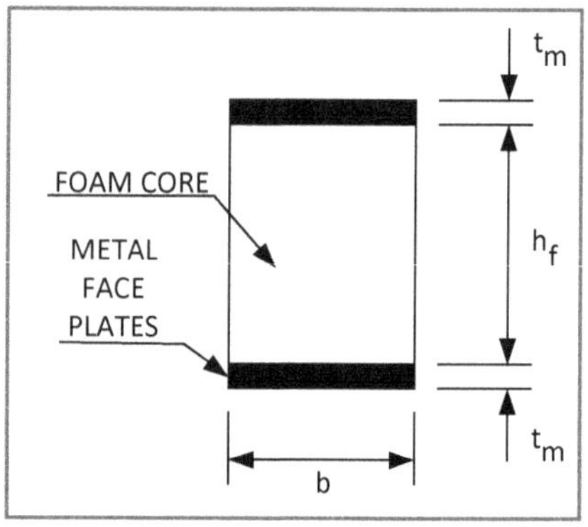

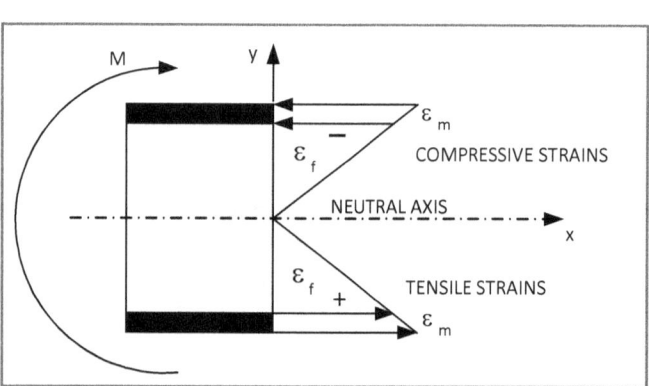

Fig. 7.16 Cross section of a composite beam with a foam core and metal cover plates.

Fig. 7.17 The strain distribution is continuous across the foam-metal interface.

We substitute Eq. (7.6) into Eq. (b) to obtain:

$$\sigma_m = - M y/I_z \tag{7.7 bis}$$

The relation for the stress is the same as that established previously in Section 7.2; however, the foam does not directly contribute to the load carrying capacity of the beam, because its modulus of elasticity is negligible. For this reason, we do not consider the foam in determining I_z. Instead, we consider only the metal plates when writing the relation for I_z as:

$$I_z \cong 2Ad^2 = 2\left[bt_m\left(\frac{h_f + t_m}{2}\right)^2\right] = \frac{bt_m}{2}(h_f + t_m)^2 \tag{7.28}$$

Combining Eq. (7.7) and (7.28) gives the maximum stress in the metal plates as:

$$\sigma_{max} = \frac{M(h_f + 2t_m)}{bt_m(h_f + t_m)^2} \tag{7.29}$$

Let's utilize this result to determine the effectiveness of foam core-metal plate composite beams.

EXAMPLE E7.14

Consider a simply-supported, foam core, metal cover plate composite beam with a uniformly distributed load of magnitude q. Aluminum cover plates 0.063 in. thick, 10 in. wide and 10 ft long are adhesively bonded to a polystyrene foam core. The foam is 10 in. wide, 6 in. high and 10 ft long. If the yield strength of the aluminum cover plates is 32 ksi, determine the distributed load q_Y that produces yielding.

Solution: From Example E7.7, we note the maximum moment at $x = L/2$ as:

$$M_{max} = qL^2/8 = q(10 \times 12)^2/8 = 1800\,q \tag{a}$$

where q is in units of lb/in.

When the composite beam yields, the stresses in the cover plates are:

$$\sigma_{max} = S_y = 32,000 \text{ psi.} \tag{b}$$

Substituting Eqs.(a) and (b) into Eq. (7.29) gives:

$$\sigma_{max} = \frac{M(h_f + 2t_m)}{bt_m(h_f + t_m)^2} = S_y = 32,000 = \frac{1800 q_Y(6 + 0.126)}{(10)(0.063)(6.063)^2}$$

Solving this relation for q_Y gives:

$$q_Y = 67.2 \text{ lb/in.} = 806 \text{ lb/ft} \tag{c}$$

The effectiveness of this beam may be judged by the ratio of its load carrying capacity to its weight per unit length. Its weight w per unit length is:

$$w = 2bt_m\gamma_m + bh_f\gamma_f \tag{d}$$

where $\gamma_m = 0.10$ lb/in.^{3} and $\gamma_f = 0.001$ lb/in.^{3} are the densities of the aluminum and foam respectively.

Substituting these quantities into Eq. (d) yields:

$$w = (2)(10)(0.063)(0.10) + (10)(6)(0.001) = 0.126 + 0.06 = 0.186 \text{ lb/in.}$$

The ratio of $q_Y/w = 67.2/0.186 = 361$

With the composite beam described, we can support a uniformly distributed load that is 360 times larger than the weight of the beam. That is a very good ratio. Composite structures with foam cores are used in aircraft structures, where minimizing structural weight is a critical design criterion.

EXAMPLE E7.15

A cantilever beam 5 m long is subjected to a concentrated force of 1 kN applied at its free end. The beam has a rectangular cross section with b = 100 mm. Determine the height of the foam and the thickness of the aluminum cover plates if the maximum stress is limited to 200 MPa. In this determination recall that the standard thickness of stock aluminum sheet from which the cover plates are cut is 0.016, 0.020, 0.025, 0.032, 0.040, 0.050, 0.063, and 0.090 in.

Solution: For a cantilever beam, the maximum moment is given by:

$$M_{max} = FL = (1 \times 10^3)(5) = 5.0 \times 10^3 \text{ N-m} = 5.0 \times 10^6 \text{ N-mm} \tag{a}$$

Because the maximum stress is limited to 200 MPa, we may write:

$$\sigma_{max} = \frac{M(h_f + 2t_m)}{bt_m(h_f + t_m)^2} = 200 \text{ N/mm}^2 \tag{b}$$

In composite beam construction, we often find that:

$$t_m \ll h_f; \quad \text{hence,} \quad h_f + 2t_m \approx h_f + t_m. \tag{c}$$

This approximation simplifies Eq. (b):

$$\sigma_{max} \approx \frac{M}{bt_m(h_f + t_m)} \tag{d}$$

Substituting known quantities into Eq. (d) and solving for h_f leads to:

$$h_f = \frac{M}{bt_m \sigma_{max}} - t_m = \frac{5.0 \times 10^6}{(100)t_m(200)} - t_m = \frac{250 \text{ mm}^2}{t_m} - t_m \tag{e}$$

In Eq. (e) we have expressed one unknown h_f in terms of another unknown t_m. However, the thickness of the metal cover plates is restricted to standard sizes. Let's evaluate Eq. (e) for all possible choices of t_m and then select the most suitable combination for the design of the beam.

The computation was performed with the aid of a spreadsheet and the results are shown in Table E7.15.

Examination of this table shows that we have many different solutions to consider in designing a beam; however, not all of them are realistic. If we use very thin cover plates (0.406 mm), the height of the foam is excessive (615 mm). If we use thick cover plates (2.29 mm), the height of the foam core (107 mm) is about the same as the width of the beam (100 mm). There is no unique solution, but the cross section with h_f = 307 mm and t_m = 0.8128 mm gives a h/b ratio of about 3/1 that appears reasonable.

Table E 7.15
Height of Foam as a Function of Cover Plate Thickness

COVER PLATE THICKNESS	COVER PLATE THICKNESS	HEIGHT OF FOAM
t_m (in.)	t_m (mm)	h_f (mm)
0.016	0.4064	614.7
0.020	0.5080	491.6
0.025	0.6350	393.0
0.032	0.8128	306.7
0.040	1.0160	245.0
0.050	1.2700	195.5
0.063	1.6002	154.6
0.090	2.2860	107.0

7.5.2 Reinforced Concrete

Another method of producing composite beams combines steel and concrete that utilizes both materials to advantage. **Concrete** is a mixture of sand, aggregate and cement that is mixed with water, poured into a mold and cured into the shape of some structural member. Concrete is a durable material that has replaced stone in many structural applications. Like stone, it exhibits a satisfactory compressive strength, but it fails when subjected to relatively low tensile stresses. Its compressive strength ranges from 3,000 to 10,000 psi depending on the constituents used in the mix. However, its tensile strength is only 100 to 200 psi. It is so weak in tension that it cannot be employed to fabricate beams without reinforcement.

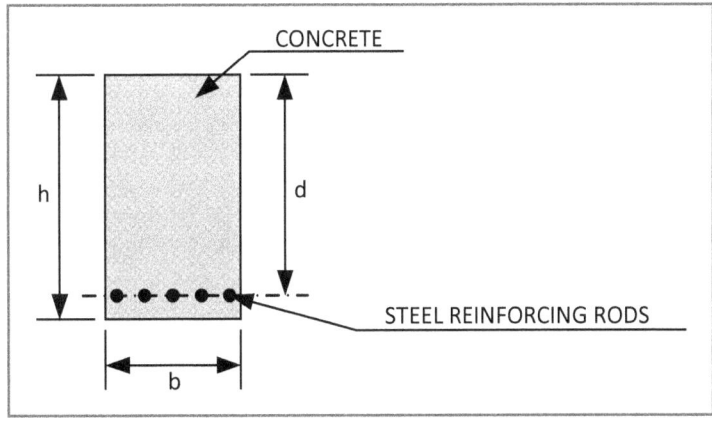

Fig. 7.18 Typical cross section of a reinforced concrete beam. Steel reinforcing rods are placed a short distance above the beam's bottom surface.

To reinforce concrete so that it may be used in beams, we embed steel rods in the concrete when it is cast into forms that define the shape of the beam. The concrete solidifies about the reinforcing steel rods, called **rebar**, producing a composite beam. A typical cross section of a concrete beam reinforced with steel rebar is illustrated in Fig. 7.18.

In preparing the drawing for Fig. 7.18, we assumed the beam was subjected to positive moments that produce tension in regions below the neutral axis. The positive bending moment produces tensile strains in both the concrete and the steel reinforcing rods, because plane sections before deformation remain plane after loading for composite beams as well as homogenous beams. These tensile strains produce tensile stresses that are sufficient to fail the concrete. When the concrete fails, micro-cracks initiate and grow into the tension region of the cross section. However, these cracks arrest near the neutral axis. The steel does not fail and the beam remains intact and rigid. However, the concrete on the tension side of the section does not contribute to the strength and rigidity, except to position the reinforcing rods some distance from the neutral axis of the beam. To accommodate for the lack of tensile strength of the concrete, we model the cross section of the beam as shown in Fig. 7.19.

The original cross section of the beam is shown in Fig. 7.19a. We remove all of the concrete located below the neutral axis from the model to account for the micro cracking in Fig. 7.19b. This leaves a rectangular section of concrete, s high by b wide, to resist the compressive forces acting above the neutral axis. We also indicate the total area A_s of the steel reinforcing rods that carry the tensile forces. The model in Fig. 7.19b is difficult to analyze because it involves two different materials, both of which contribute to the load carrying capacity of the beam.

To facilitate the analysis, we convert the composite (heterogeneous) model into a homogenous model by representing the steel area with an equivalent concrete area in Fig. 7.19c. The equivalent concrete area A_{eq} for the steel rods is:

$$A_{eq} = kA_s \qquad (7.30)$$

where

$$k = E_s/E_c \qquad (7.31)$$

Note that E_s and E_c are the elastic moduli of the steel and concrete, respectively.

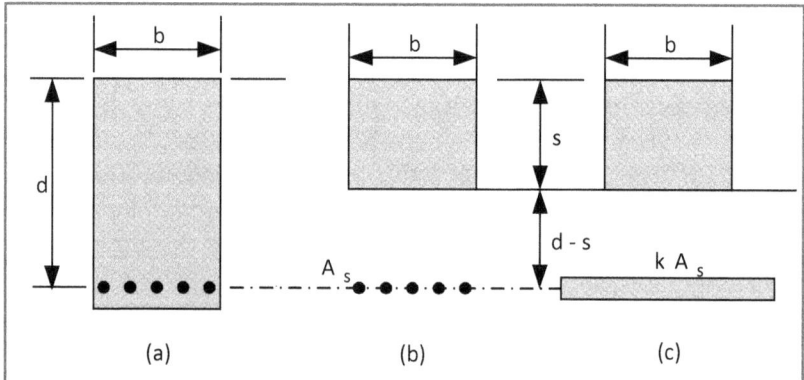

Fig. 7.19 Modeling a steel-bar reinforced concrete beam.

The analysis of the model shown in Fig. 7.19 requires that we first determine the position of the neutral axis. Recall that the neutral axis is coincident with the centroid of the section so that $\int y\,dA = 0$; hence, we may write:

$$(s)(b)(s/2) - kA_s(d-s) = 0$$

$$s^2 b/2 + kA_s s - kA_s d = 0 \qquad (7.32)$$

Solution of this quadratic equation gives the dimension s locating the neutral axis and establishing the dimensions of the concrete above the neutral axis.

The maximum stress σ_{con} in the concrete portion of the beam is given by:

$$\sigma_{con} = Ms/I_z \tag{7.33}$$

The stress in the steel reinforcing rods is given by:

$$\sigma_{steel} = kM(d-s)/I_z \tag{7.34}$$

Note in Eq. (7.34), we have incorporated the multiplier k in the relation to convert the results from an equivalent concrete beam to a real beam containing steel reinforcing rods. Finally, the second moment of the area, illustrated in Fig. 7.19c, is given by:

$$I_z = bs^3/3 + kA_s(d-s)^2 \tag{7.35}$$

The first term refers to the concrete above the neutral axis, and the second term is from the parallel axis theorem (Ad^2) for the steel rods. We have neglected the moment of inertia of the rods about their centroidal axis (I_z^2). This term is usually very small (much less than 1%) for reinforced concrete beams so it can be omitted without impairing safety.

EXAMPLE E7.16

A concrete floor slab is reinforced with four ½ in. diameter steel bars placed on 8 in. centers. The slab is 34 in. wide and 6 in. thick, and the centerline of the steel rebar is 1 in. above the bottom surface of the slab. The slab is 20 ft. long and is simply supported when installed. If a uniformly distributed load q is applied to the floor, determine the largest magnitude of q if the stress in the steel is limited to 15,000 psi and in the concrete to 1,200 psi. The modulus of elasticity of the steel and the concrete is 30×10^6 and 2.5×10^6 psi respectively.

Solution: Let's first determine the equivalent area of the model produced by the four steel rebars. From Eqs. (7.30) and (7.31), we write:

$$A_{eq} = kA_s = (E_s/E_c)(\pi d^2/4)N$$

where N is the number of rebars.

$$A_{eq} = (30/2.5)[\pi(0.5)^2/4]4 = (12)(\pi/16)(4) = 9.425 \text{ in}^2 \tag{a}$$

Next, let's use Eq. (7.32) to locate the position of the neutral axis

$$(b/2)s^2 + kA_s s - kA_s d = (b/2)s^2 + A_{eq}s - A_{eq}d = 0$$

$$(34/2)s^2 + 9.425 s - 9.425(6-1) = 0$$

Solving this quadratic equation gives:

$$s = 1.411 \text{ in.} \quad \text{and} \quad s = -1.965 \text{ in.} \tag{b}$$

We select the positive value because the negative result has no physical meaning.

Having established the position of the neutral axis we may determine I_z from Eq. (7.35) as:

$$I_z = bs^3/3 + kA_s(d-s)^2 = 34(1.411)^3/3 + 9.425(5 - 1.411)^2 = 153.2 \text{ in}^4. \quad (c)$$

The maximum stress σ_{con} in the concrete portion of the beam is given by Eq. (7.33) as:

$$\sigma_{con} = Ms/I_z = 1,200 = 1.411M/153.2 \quad (d)$$

Solving this relation for M yields:

$$M_{max} = 130,300 \text{ in-lb}$$

For a uniformly loaded beam, the maximum moment is:

$$M_{max} = qL^2/8 = 130,300 \quad (e)$$

$$q_{max} = (8)(130,300)/[(20)(12)^2] = 18.10 \text{ lb/in.} = 217.2 \text{ lb/ft} \quad (f)$$

We reach the limit of the stress in the concrete when the uniformly distributed load q is 217.2 lb/ft. However, it is necessary to determine if the limiting stress in the steel rebar is achieved with a higher or lower value of q. The stress in the steel reinforcing rods is given by Eq. (7.34) as:

$$\sigma_{steel} = kM(d-s)/I_z = 15,000 = (30/2.5)(5 - 1.411)M/153.2 \quad (g)$$

Solving this relation for M yields:

$$M_{max} = 53,360 \text{ in-lb}$$

And in a similar manner, we solve for the limiting value of q as:

$$q_{max} = 8 M_{max}/L^2 = (8)(53,360)/[(20)(12)^2] = 7.412 \text{ lb/in.} = 88.9 \text{ lb/ft} \quad (h)$$

Compare Eqs. (f) and (h) and select the smaller value to give $q_{max} = 88.9$ lb/in.

An examination of the results in Eqs. (f) and (h) indicates that the design of the reinforced slab is not balanced. The steel meets the limiting stress of 15,000 psi at loads less than one half that required for the concrete to reach its limit stress. To improve the design of the slab the amount of steel should be increased to better balance the load carrying capacity of the steel and concrete. Problem 7.35 provides you with the opportunity to increase the size or number of the rebars to increase the amount of steel and better balance the design of the reinforced concrete beam.

EXAMPLE E7.17

A concrete beam is reinforced with three 1 in. diameter steel bars placed on 6 in. centers. The section of the beam is rectangular—14 in. wide and 24 in. high with the centerline of the steel rebar 1.5 in. above the bottom surface of the beam. The beam is 24 ft. long and simply supported. If it supports three concentrated loads F = 10 kip, as shown in Fig. E7.17, determine the margin of safety for the beam. The yield strength of the steel is 42 ksi and the compressive strength of the concrete is 4,500 psi. The modulus of elasticity of the steel and the concrete is 30×10^6 and 3.0×10^6 psi respectively.

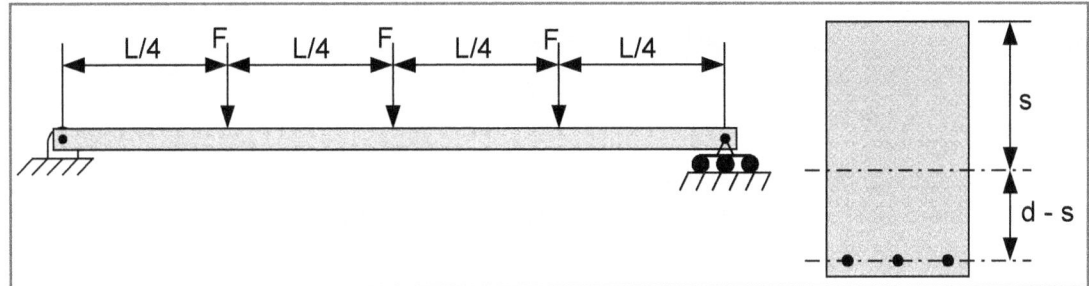

Fig. E7.17

Solution: The reactions for the symmetrically loaded beam are:

$$R = R_L = R_R = 3F/2 = 3(10)/2 = 15 \text{ kip} \quad (a)$$

The bending moment is a maximum at the center of the beam where $x = L/2$ and is:

$$M_{max} = R_L(L/2) - F(L/4) = FL/2 = (10,000)(24)(12)/2 = 1.44 \times 10^6 \text{ in-lb} \quad (b)$$

After the point of failure and the maximum bending moment for the composite beam have been determined, the location of the neutral axis is ascertained from Eq. (7.32).

$$s^2 b/2 + kA_s s - kA_s d = 0$$

$$7s^2 + 10(3\pi/4)s - 10(3\pi/4)(22.5) = 0$$

From the quadratic formula, we calculate that:

$$s = 7.18 \text{ in.} \quad \text{and } s = -10.55 \text{ in.} \quad (c)$$

Again, ignore the negative root ands select $s = 7.18$ in.

From Eq. (7.35), we establish that I_z is:

$$I_z = bs^3/3 + kA_s(d-s)^2$$

$$I_z = 14(7.18)^3/3 + 10(3\pi/4)(22.5 - 7.18)^2 = 7257 \text{ in}^4. \quad (d)$$

The maximum stress σ_{con} in the concrete portion of the beam is given by Eq. (7.33) as:

$$\sigma_{con} = Ms/I_z = (1.44 \times 10^6)(7.18)/7257 = 1425 \text{ psi} \quad (e)$$

The maximum stress in the steel reinforcing rods is given by Eq. (7.34) as:

$$\sigma_{steel} = kM(d-s)/I_z = (10)(1.44 \times 10^6)(22.5 - 7.18)/7257 = 30.40 \text{ ksi} \quad (f)$$

The safety factors and the margins of safety are given by:

$$SF_{con} = S_{con}/\sigma_{con} = 4500/1425 = 3.16 \qquad MOS = SF_{con} - 1 = 3.16 - 1 = 216\%$$

$SF_{steel} = S_{steel}/\sigma_{steel} = 42.0/30.4 = 1.38$ $MOS = SF_{steel} - 1 = 1.38 - 1 = 38\%$

A review of these results indicates that the margin of safety for the steel is too low, and for the concrete, it is more than adequate. Clearly, the design of the composite beam is deficient. The fact that one safety factor is high while the other is low indicates that the amount of steel relative to the size of section is not properly balanced. In this example, the amount of steel is inadequate. Adding more steel (increasing A_s) increases s, decreases (d − s), and increases I_z. All of these changes decrease the stresses in the steel rebar. Problem 7.36 is provided to enable you to design a reinforced concrete beam of the size described in Example E7.17 with a margin of safety for the steel that is more comparable with the margin of safety for the concrete.

7.6 PLASTIC BENDING

In our discussion of beams, we have considered only their elastic response and have limited our interest to situations, where the stresses exhibited magnitudes lower than the yield strength. This approach insures an adequate margin of safety based on the elastic behavior of the beam's material. However, it is important to realize that beams do not suddenly collapse when the applied bending moment exceeds that required to produce stresses equal to the yield strength.

The behavior of a beam subjected to moments larger than the yield moment depends on the post-yield response of the beam material. In this discussion, we consider a well-known class of materials: **elastic perfectly-plastic**. The stress-strain diagram for an elastic perfectly-plastic material is shown in Fig. 7.20. The material responds elastically until it yields, and then additional strain is imposed at constant stress $\sigma = S_y$.

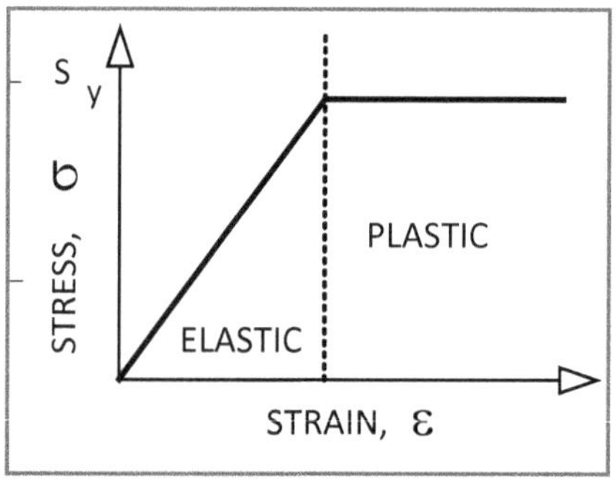

Fig. 7.20 The stress-strain diagram for an elastic perfectly-plastic material.

As we increase the moment applied to a beam, both the stress and the strain at the outer fibers of the beam increase linearly with the moment until $\sigma = S_y$, as illustrated in Fig. 7.21a. When the moment is increased by an additional increment, the strain continues to increase, and the distribution of strain from the neutral axis to the outside fibers remains linear with respect to the position y. As indicated in Fig. 7.21b, no discontinuity occurs at the elastic-plastic interface, as the curvature κ of the beam becomes larger.

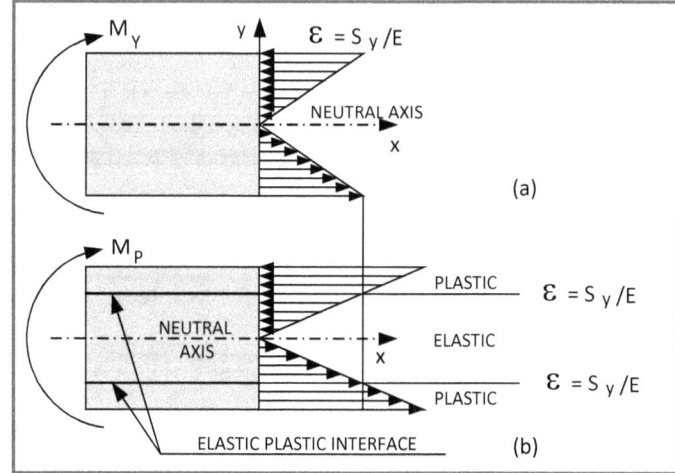

Fig. 7.21 Strain distribution due to pure bending: (a) at yield and (b) beyond yield.

The distribution of stress with position y across the section exhibits a discontinuity at the elastic-plastic interface as shown in Fig. 7.22. The maximum stress $\sigma_{MAX} = S_y$ is achieved at the outside fibers when M equals the yield moment M_Y. As the moment is increased beyond M_Y, the outer fibers yield, the stress in the plastic region remains constant at $\sigma = S_y$, and an elastic-plastic interface develops. This elastic-plastic interface separates the cross section into two regions—an elastic core near the neutral axis and plastic zones adjacent to the top and bottom surfaces.

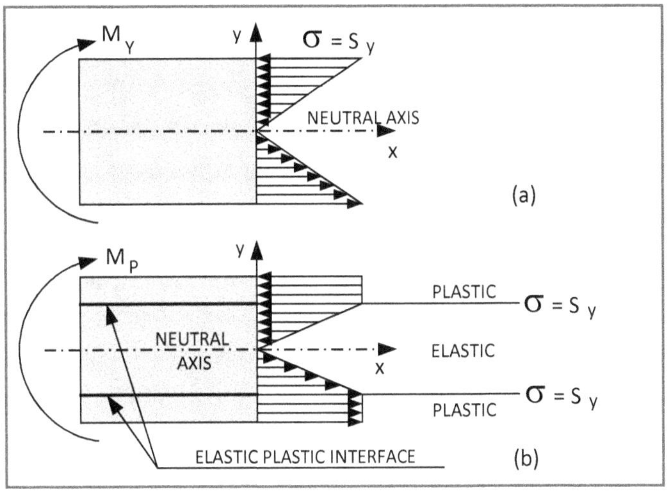

Fig. 7.22 Stress distributions (a) when $M = M_Y$; and (b) when $M_Y < M < M_L$.

The moment M_Y required to produce a stress $\sigma = S_y$ on the outside fibers of a beam with a rectangular cross section is given by:

$$M_Y = \frac{\sigma(2I_z)}{h} = \frac{S_y}{h}(2)\frac{1}{12}(bh^3) = S_y\frac{bh^2}{6} \qquad (7.36)$$

The moment M_P that is required to drive the elastic-plastic interface to some position y_p is determined from Fig. 7.23. The equal and opposite forces F_e and F_p form two couples—one associated with the elastic region ($F_e\, d_e$) and the other with the plastic region ($F_p\, d_p$).

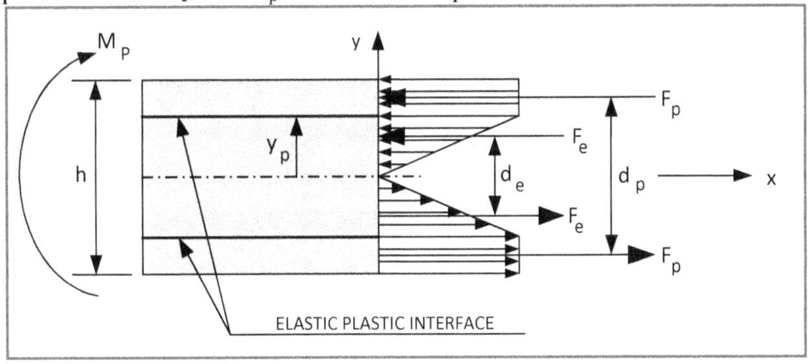

Fig. 7.23 Couples developed in the elastic and plastic regions of the cross section.

The forces, due to the stress σ, are given by:

$$F_p = (b/2)(h - 2y_p)S_y \qquad \text{and} \qquad F_e = (by_p/2)S_y \qquad (a)$$

And the distances d_p and d_e are given by:

$$d_p = (1/2)(h + 2y_p) \qquad \text{and} \qquad d_e = (4/3)y_p \qquad (b)$$

Summing moments about the origin of the x-y coordinate system gives:

$$\Sigma M_O = F_p\, d_p + F_e\, d_e - M_P = 0 \qquad (c)$$

Substituting Eqs. (a) and (b) into Eq. (c) and solving for M_P yields:

$$M_P = S_y b\left(\frac{1}{2}(h - 2y_p)\frac{1}{2}(h + 2y_p) + \frac{y_p}{2}\frac{4y_p}{3}\right) = \frac{S_y b}{12}(3h^2 - 4y_p^2) \tag{7.37}$$

Equation)7.37) indicates that beams are capable of supporting moments larger than the yield moment M_Y. The beam will be damaged in the processes, because it will be permanently deformed; however, it will not collapse. The question then arises as to how large of an overload can be placed on a beam before it collapses.

To answer this question, consider the behavior of the beam, when the elastic-plastic interface is driven into the neutral axis. The elastic core disappears and the entire cross section of the beam is in the plastic regime. At this point, the beam has become **fully plastic** reaching its **limit moment**, where additional moments cannot be supported.

It is easy to determine the limit moment, M_L from Eq. (7.37) if we set $y_p = 0$, thereby moving the position of the elastic-plastic interface to the neutral axis. Then, Eq. (7.37) reduces to:

$$M_L = S_y bh^2/4 \tag{7.38}$$

A comparison of the yield moment and the limit moment is possible from Eqs. (7.36) and (7.38). The ratio M_L/M_Y is given by:

$$\frac{M_L}{M_Y} = \frac{S_y bh^2}{4}\frac{6}{S_y bh^2} = \frac{3}{2} \tag{7.39}$$

Clearly for the rectangular cross section, the limit moment is 150% of the yield moment. This is a comfortable margin that adds to the safety factor, if collapse is an important criterion for design. However, the ratio of M_L/M_Y is a function of the shape of the cross section. For WF and standard I beam shapes the ratio is usually about 1.10 to 1.15.

EXAMPLE E7.18

Consider the simply supported beam of length L = 8 m with a concentrated force **F** located 3 m from the left end as shown in Fig. E7.18. Determine the force required to produce the yield moment and the force required for the beam to collapse. The cross section of the beam is rectangular with a height h = 100 mm and a width b = 50mm. The beam is fabricated from steel with a yield strength of 300 MPa. We will assume that the steel responds like an elastic-perfectly-plastic material.

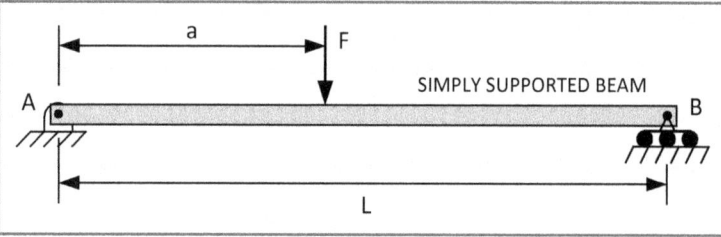

Fig. E7.18

Solution: Recall from Section 7.4.1 that the reaction forces R_R and R_L are given by:

$$R_R = Fa/L \quad \text{and} \quad R_L = Fc/L \qquad (a)$$

where $c = L - a$.

In Eq. (7.16), we showed that the moment was a maximum at $x = a$ (under the load point):

$$M_{max} = Fac/L \qquad (b)$$

Combining Eqs. (7.36) and (b) while setting $F = F_Y$ and $M_{MAX} = M_Y$ yields:

$$F_Y ac/L = S_y bh^2/6$$

$$F_Y = S_y bh^2 L/(6ac) \qquad (c)$$

Substituting the parameters describing the beam and the material into Eq. (c) yields:

$$F_Y = \frac{S_y bh^2 L}{6ac} = \frac{(300 \times 10^6)(0.05)(0.1)^2(8)}{(6)(3)(8-3)} = 13.33 \text{kN}$$

The force associated with the limit load is determined by substituting Eq. (7.38) into Eq. (b) or by recognizing that $M_L = 3M_Y/2$ as indicated in Eq. (7.39). Accordingly, we find that:

$$F_L = 3F_Y/2 = 20 \text{ kN}$$

When F_L produces the moment M_L, a **plastic hinge** forms under the load point and the beam collapses as indicated with the schematic illustration presented in Fig. E 7.18a.

Fig. E 7.18a

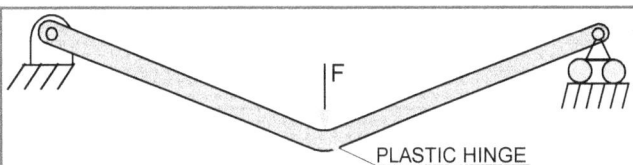

A plastic hinge has formed under the load when $M = M_L$, with both sides of the beam rotating about this point. The hinge is not frictionless, because work ($M_L \times \theta$) is produced by the force F_L as it deflects the beam and rotates the plastic hinge.

EXAMPLE E7.19

Consider a simply supported beam of length $L = 9$ m with a uniformly distributed load q acting over its entire the length. Determine the magnitude of the uniformly distributed load required to produce the yield and limit moments. The cross section of the beam is rectangular with $h = 120$ mm and $b = 40$ mm. The beam is fabricated from steel with a yield strength of 360 MPa. We will assume that the steel acts like an elastic-perfectly-plastic material.

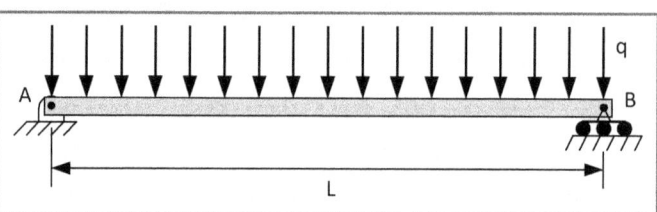

Fig. E7.19

Solution: Recall from Example E7.7 that the reaction forces R_R and R_L are given by:

$$R_R = R_L = qL/2 \qquad (a)$$

Also in Example E7.7, we showed that the moment was a maximum at $x = L/2$:

$$M_{MAX} = qL^2/8 \qquad (b)$$

Combining Eqs. (7.36) and (b) yields:

$$M_{MAX} = q_Y L^2/8 = M_y = S_y bh^2/6$$

$$q_Y = (4/3)S_y b(h/L)^2 \qquad (c)$$

Substituting the parameters describing the beam and its material into Eq. (c) yields:

$$q_Y = \frac{4S_y bh^2}{3L^2} = \frac{4(360\times 10^6)(0.04)(0.12)^2}{3(9)^2} = 3.413 \text{ kN/m}$$

The distributed load associated with the limit load is determined by substituting Eq. (7.38) into Eq. (b).

$$q_L = \frac{2S_y bh^2}{L^2} = \frac{2(360\times 10^6)(0.04)(0.12)^2}{(9)^2} = 5.120 \text{ kN/m}$$

When the distributed load $q = q_L$, a plastic hinge forms at the beam's center point and the beam collapses.

7.7 STRESS CONCENTRATIONS IN BEAMS

In our study of methods to analyze stresses in beams, we have assumed that the beams were continuous along their entire length. However, in design of structures it is common to cut notches in beams for one reason or another. In some instances, shouldered shafts or beams are employed and occasionally transverse holes are drilled in shafts to enable lubricant transfer. These geometric discontinuities produce stress concentrations in regions adjacent to the hole or notch. To determine the maximum stress, we account for these discontinuities by establishing the appropriate stress concentration factor. The stress concentration factor multiplies a nominal stress to give the maximum stress at the discontinuity. This procedure is identical to that described previously in Chapter 6.

7.7.1 Stress Concentration Factors

Due to Notches in Beams

Let's consider a beam with a rectangular cross section that is notched on its top and bottom surfaces, as shown in Fig. 7.24.

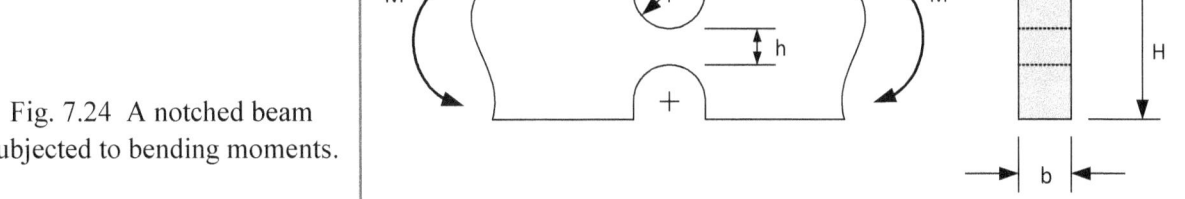

Fig. 7.24 A notched beam subjected to bending moments.

The stresses in the region adjacent to the notch are elevated due to the presence of the discontinuity. The maximum stress σ_{Max} occurs at the base of the notch, and is given by:

$$\sigma_{Max} = K\, \sigma_{Nom} \qquad (7.40)$$

where σ_{Nom} is the nominal stress and K is the stress concentration factor.

The nominal stress is the bending stress at the notched section that is given by:

$$\sigma_{Nom} = Mc/I_z = 6M/(bh^2) \qquad (7.41)$$

where b and h are defined in Fig. 7.24.

The stress concentration factor is a function of the geometry of the discontinuity and depends on the ratios H/h and r/h as shown in Fig. 7.25

EXAMPLE E7.20

A simply supported beam with a uniform load is shown in Fig. E7.20. If notches are cut on the top and bottom surfaces at the center span of the beam, determine the nominal stresses, the stress concentration factor and the maximum stresses at the notches. The beams dimensions are L= 20 ft, b = 2.5 in., H = 8 in. and h = 5 in. The notch radius r = 0.5 in. is centered on the top and bottom edges of the beam. The distributed load q = 0.8 kip/ft.

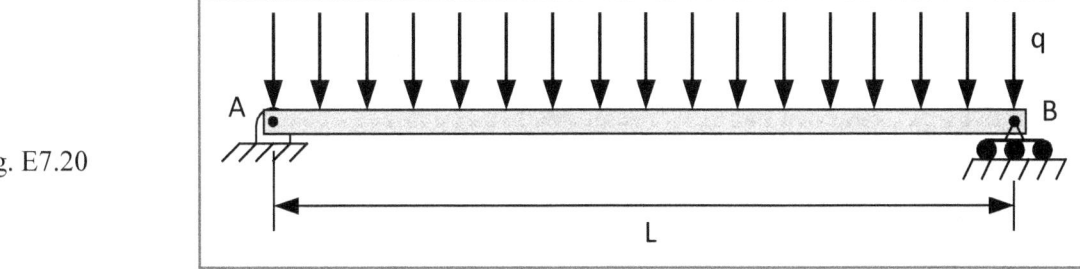

Fig. E7.20

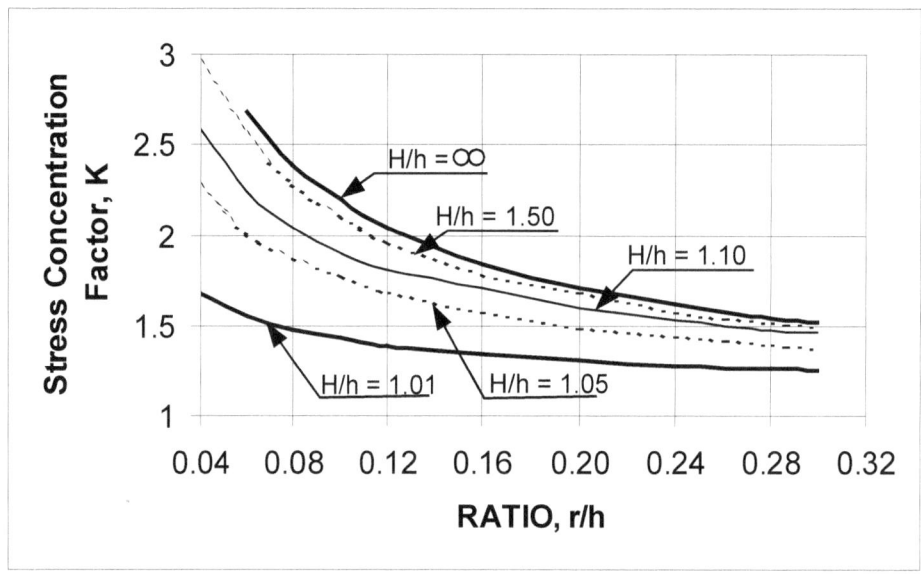

Fig. 7.25 Stress concentration factor for a notched beam in bending.

Solution: Let's begin by determining the moment at the midspan of the beam. In Example E7.7, we showed that the moment for a beam with a uniformly distributed load was a maximum at midspan given by:

$$M_{Max} = qL^2/8 = (800)(20)^2/8 = 40 \text{ kip-ft} = 480 \text{ kip-in.} \quad (a)$$

Next calculate the nominal stress σ_{Nom} from Eq. (7.41) as:

$$\sigma_{Nom} = Mc/I_z = 6M/(bh^2) = 6(480 \times 10^3)/[(2.5)(5)^2] = 46.08 \text{ ksi} \quad (b)$$

Reference Fig. 7.25 to find the stress concentration factor. The ratios giving the coordinates for K are $r/h = 0.50/5.0 = 0.10$ and $H/h = 8.0/5 = 1.60$. Then from Fig. 7.25, we establish that:

$$K = 2.11 \quad (c)$$

Substituting Eqs. (b) and (c) into Eq. (7.40) yields:

$$\sigma_{Max} = K \sigma_{Nom} = (2.11)(46.08) = 97.23 \text{ ksi} \quad (d)$$

Due to Grooves in Circular Shafts

Let's consider a shaft with a circular cross section that has a groove machined about its circumference as shown in Fig. 7.26. When the shaft is subjected to a bending moment M, the stresses in the region at the root of the groove are elevated due to the presence of the discontinuity. The procedure is the same as we described previously—we multiply a nominal value for the bending stresses by a suitable stress concentration factor to determine the maximum stresses in accordance with Eq. (7.40).

Fig. 7.26 A grooved shaft subjected to bending moments.

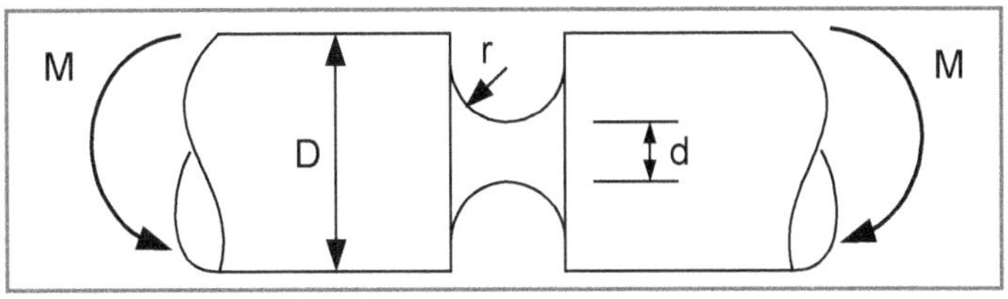

The nominal stress at the neck section where the groove is located is given by:

$$\sigma_{Nom} = 32M/\pi d^3 \tag{7.42}$$

The curves defining the stress concentration factors are presented in Fig. 7.27.

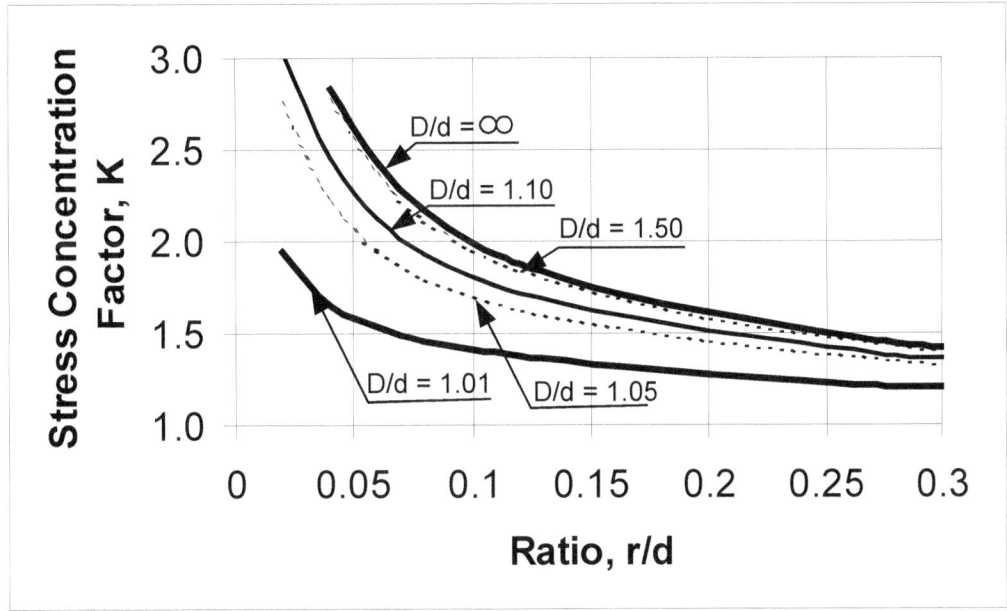

Fig. 7.27 Stress concentration factor for a grooved shaft.

EXAMPLE E7.21

A circular shaft is loaded in four-point bending producing a bending moment M = 5.0 kN-m over the central region of a circular shaft. If a groove is cut about the perimeter of the shaft at its midspan point, determine the nominal stresses, the stress concentration factor and the maximum stress at the base of the groove. The shaft's dimensions are D = 100 mm, d = 75 mm. The notch radius r = 15 mm is centered near the bottom of the grove.

Solution: Let's begin by determining the calculating the nominal stress at the location of the groove from Eq. (7.42) as:

$$\sigma_{Nom} = 32M/\pi d^3 = (32)(5.0 \times 10^3)/[\pi(0.075)^3] = 120.7 \text{ MPa} \tag{a}$$

Reference Fig. 7.27 to find the stress concentration factor. The ratios giving the coordinates for K are r/d = 15/75 = 0.20 and D/d = 100/75 = 1.3333. Then from Fig. 7.27, we establish that K = 1.53.

> Substituting the results for σ_{Nom} and K into Eq. (7.40) yields:
>
> $$\sigma_{Max} = K\, \sigma_{Nom} = (1.53)(120.7) = 184.7 \text{ MPa}$$

Due to Shoulder Fillets in a Stepped Beam

Some beams are not continuous and are of two different heights—H and h. We show such a stepped beam in Fig. 7.28, which is acted upon by a bending moment M. A circular fillet is placed at the transition to mitigate the concentration of stresses at the reentrant corner. When the beam is subjected to bending, the stresses in the region adjacent to the fillet are elevated due to the presence of the step. The procedure to determine the stresses at the fillet is the same as we described previously.

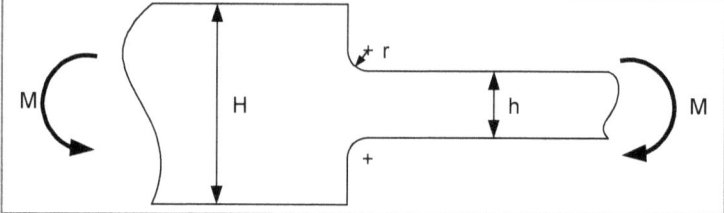

Fig. 7.28 A stepped beam subjected to bending.

The nominal stress σ_{Nom} for this case is determined from by Eq. (7.41), and the stress concentration factor is given in Fig. 7.29.

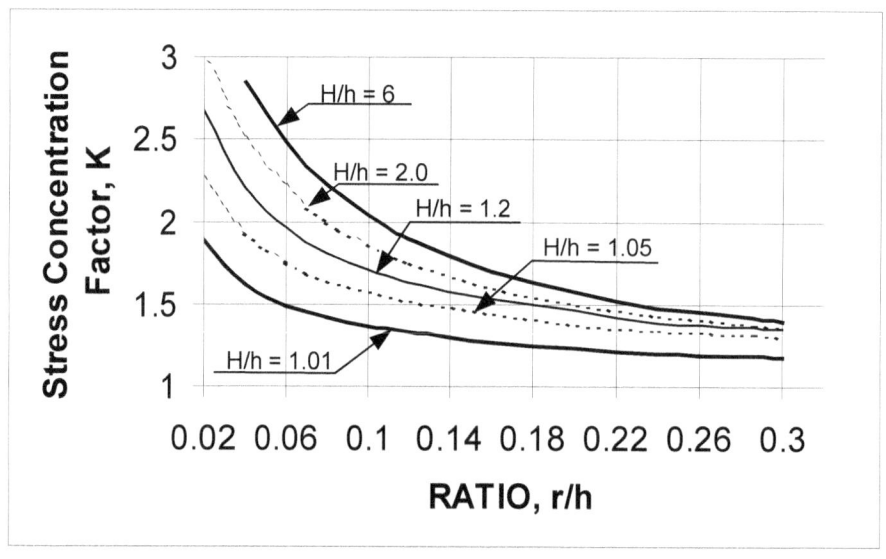

Fig. 7.29 Stress concentration factors for a stepped beam with shoulders.

Due to Shoulder Fillets in a Stepped Shaft

Many shafts are not continuous and are of two or more different diameters to accommodate the installation of gears, bearings, etc. We show such a stepped shaft in Fig. 7.30, which is acted upon by a bending moment M. A circular fillet is placed at the transition to mitigate the concentration of stresses at the reentrant corner. When the shaft is subjected to bending, the stresses in the region adjacent to the fillet are elevated due to the presence of the step. The procedure to determine the stresses at the fillet is the same as we described previously.

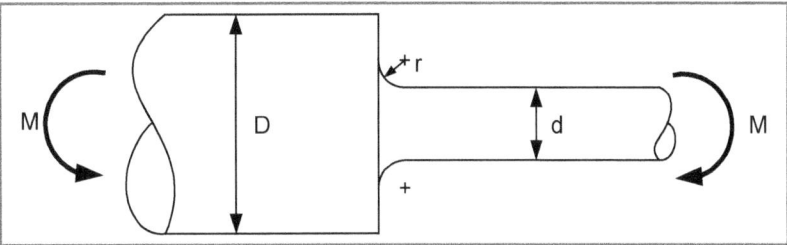

Fig. 7.30 Stepped shaft subjected to a bending moment M.

The nominal stress σ_{Nom} for this case is determined from by Eq. (7.42), and the stress concentration factor is given in Fig. 7.31.

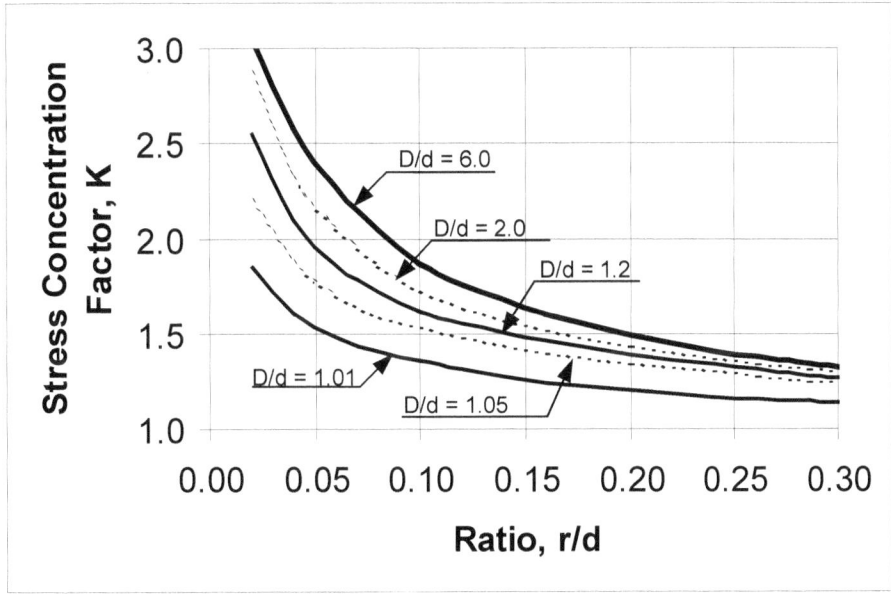

Fig. 7.31 Stress concentration factors for a stepped shaft.

EXAMPLE E7.22

A shaft with a circular cross section is designed with a step transition between two diameters—D = 2.0 in. and d = 1.0 in. The shaft is subjected to a bending moment of M = 150 ft-lb at the transition point. Determine the minimum radius of the fillet at the transition region, if the maximum stress is not to exceed 40.0 ksi.

Solution: Let's solve for the nominal stress at the transition region by using Eq. (7.42).

$$\sigma_{Nom} = 32M/(\pi d_3) = (32)(150)(12)/[\pi(1.0)^3] = 18.33 \text{ ksi} \qquad (a)$$

Next, determine the maximum allowable stress concentration factor from Eq. (7.40) as:

$$K_{Max} = \sigma_{Max}/\sigma_{Nom} = 40/18.33 = 2.182 \qquad (b)$$

Note that D/d = (2.0)/(1.0) = 2.0. Then select the dotted line curve for D/d = 2.0 from Fig. 7.31. Project a line horizontally from K = 2.182 and find the intercept with the dotted curve. Project a second line downward and read the intercept for r/d as 0.05. The radius of the fillet at the transition section must exceed r = (0.05)(1.0) = 0.05 in. so that $\sigma_{Max} \leq 40$ ksi.

Due to a Transverse Hole in a Circular Shaft

Many shafts are designed with transverse holes to enable the flow of lubricants or coolants to some other component in the system. We show a shaft with a small diameter transverse hole in Fig. 7.32. When the shaft is subjected to bending, the stresses in the region adjacent to the hole are elevated due to the presence of the discontinuity. The procedure to determine the maximum stresses at the hole is the same as we described previously.

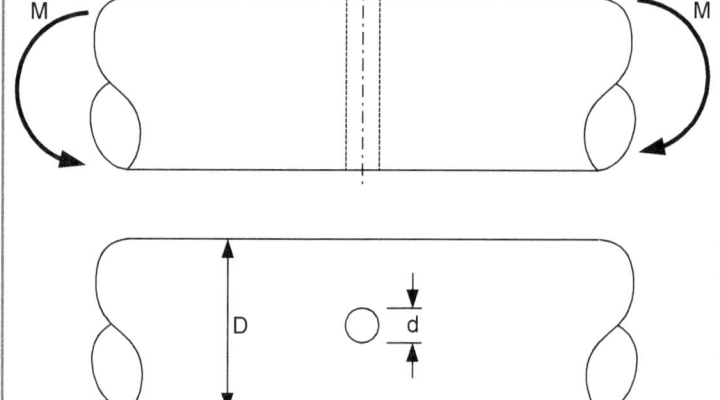

Fig. 7.32 A transverse hole in shaft with circular cross section that is subjected to a moment M.

The nominal stress σ_{Nom} at the location of the transverse hole is approximated by:

$$\sigma_{Nom} = \frac{M}{\dfrac{\pi D^3}{32} - \dfrac{D^2 d}{6}} \qquad (7.43)$$

The stress concentration factor, which is a function of the ratio of d/D, is presented in Fig. 7.33.

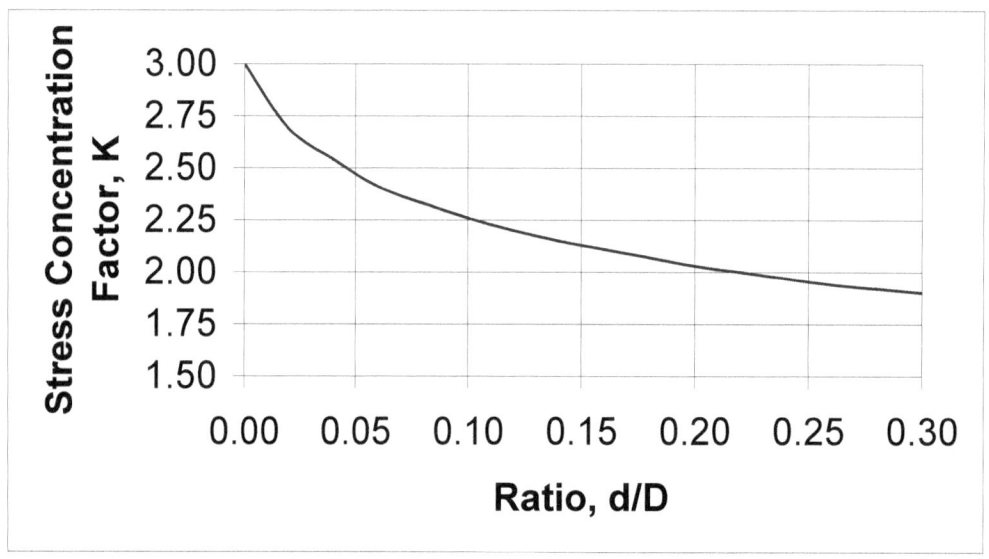

Fig. 7.33 Stress concentration factor for a shaft with a transverse hole.

EXAMPLE E7.23

A circular shaft with a diameter D = 125 mm is drilled with a transverse hole having a diameter d = 6 mm. If the shaft is subjected to a bending moment of 125 kN-m, determine the maximum stress adjacent the hole.

Solution: The nominal stress is determined from Eq. (7.43) as:

$$\sigma_{Nom} = \frac{M}{\frac{\pi D^3}{32} - \frac{D^2 d}{6}} = \frac{25 \times 10^6}{\frac{\pi (125)^3}{32} - \frac{(6)(125)^2}{6}} = 141.9 \text{ MPa} \quad (a)$$

Using the ratio D/d = 6.0/125 = 0.048 and the curve in Fig. 7.33, we find that K = 2.49. The from Eq. (7.40) we write:

$$\sigma_{Max} = K\, \sigma_{Nom} = (2.49)(141.9) = 353.3 \text{ MPa}$$

7.8 SUMMARY

Beams are long thin members subjected to transverse forces—either reactionary or applied. These transverse forces produce internal moments in the beam, which bend it into a circular arc. The curvature of the arc depends on the internal moment and may vary along the length of the beam.

In bending, we assume that plane sections of the beam remain plane as the beam deforms. This assumption permits us to determine the longitudinal strain in the beam if we know its radius of curvature, ρ.

$$\varepsilon = -y/\rho \quad (7.1)$$

The equations of equilibrium are employed to establish the position of the neutral axis of the beam at the centroid of its cross sectional area. We also showed the relation between the internal moment M and the radius of curvature as:

$$M = (EI_z/\rho) = \kappa EI_z \quad (7.5)$$

With a knowledge of the relation between curvature and the moment, we were able to establish the important relationship between the internal moment and normal stress:

$$\sigma = -My/I_z \quad (7.7)$$

Using Eq. (7.7), we consider beams with rectangular cross sections and showed that they were not efficient structural members, because they have extensive regions near their neutral axis, which are subjected to relatively small stresses.

We introduced more efficient designs of cross sections by considering American standard I and WF shapes. An effectiveness factor **SE** = Z/A was introduced to judge the adequacy of the design of various cross sections for beams.

$$\mathbf{SE} = \frac{1}{h(2bt_f + b_w h_w)} \left\{ \frac{b_w h_w^3}{6} + \frac{bt_f^3}{3} + bt_f (h_w + t_f)^2 \right\} \quad (7.15)$$

In most cases, the moment varies along the length of the beam and it is necessary to determine the shear forces and bending moments as a function of position. We introduced methods for constructing free bodies of the beam to determine the reactions, shear forces and bending moments. Also, the method for constructing shear and bending moment diagrams was introduced. It is important to recall the relations between loads, shear force and bending moment as indicated below, in constructing the shear and bending moment diagrams.

$$dV/dx = -q \quad (7.20)$$

$$dM/dx = V \quad (7.18)$$

$$M_{x_2} - M_{x_1} = \int_{x_1}^{x_2} V dx \quad (7.19)$$

Several examples are presented to demonstrate methods for calculating stresses, safety factors, and margins of safety for beams with various types of supports subjected to both concentrated forces and uniformly distributed forces.

We have briefly introduced the topic of shear stress in beams, and showed that:

$$\tau = \tau_V = \tau_H = VQ_1/(bI_z) \quad (7.23)$$

The shear stresses are not uniformly distributed over the cross section of the beam; they are a maximum at the neutral axis and zero at the top and bottom surfaces of the beam. In most cases, the shear stresses are small when compared to the normal stresses and they are often ignored. The exception is when the beams are fabricated from wood. The shear strength of wood is so small that even low shear stresses may be sufficient to split wooden beams at the neutral axis.

The bending of composite beams fabricated from two materials was discussed. The first composite section described consisted of a foam core with adhesively bonded metal cover plates. The maximum stress in the metal cover plates is given by:

$$\sigma_{max} = \frac{M(h_f + 2t_m)}{bt_m(h_f + t_m)^2} \quad (7.29)$$

We also developed methods for determining stresses in concrete reinforced with steel rods. This method involved the development of an equivalent section, where the steel was converted into an equivalent area of concrete. The beam was then analyzed using the equations listed below:

$$A_{eq} = kA_s \quad (7.30)$$

$$k = E_s/E_c \quad (7.31)$$

$$s^2 b/2 + kA_s s - kA_s d = 0 \quad (7.32)$$

$$\sigma_{con} = Ms/I_z \quad (7.33)$$

$$\sigma_{steel} = kM(d-s)/I_z \quad (7.34)$$

$$I_z = bs^3/3 + kA_s(d-s)^2 \quad (7.35)$$

Examples were presented to demonstrate the analytical procedure and to introduce the concept of balanced design of composite beams.

The plastic response of beams was introduced for an elastic-perfectly-plastic material. We showed that the strain increased linearly with increasing moment even in the plastic regime. However, the stresses were limited to the yield strength of the beam's material. As a consequence, a discontinuity formed in the distribution of the stresses across the section. This discontinuity defines the elastic-plastic interface. As the moment increases, the elastic-plastic interface is driven toward the neutral axis, a plastic hinge forms, and a simply supported beam collapses. Concepts of the yield moment, M_Y and the limit moment, M_L were introduced and equations were developed for computing these quantities for beams with rectangular cross sections.

$$M_Y = S_y \frac{bh^2}{6} \quad (7.36)$$

$$M_L = S_y bh^2/4 \quad (7.38)$$

Note that M_P is 1.5 times higher than M_L for beams with rectangular cross sections, but only 1.10 to 1.15 times higher for wide flange and standard I-beams.

Stress concentrations due to discontinuities in beams and circular shafts subjected to bending were presented. Stress concentration factors that account for the elevation of the stresses adjacent to the discontinuity were introduced for five different geometries. The maximum stresses are determined from Eq. (7.40).

$$\sigma_{Max} = K \sigma_{Nom} \quad (7.40)$$

PROBLEMS

7.1 Draw three different shapes of cross sectional areas that exhibit symmetry about the vertical or y axis.

7.2 For the beam shown in Fig. P7.2 construct the shear and bending moment diagrams and write the equations for the bending moment as a function of x as x varies from zero to L.

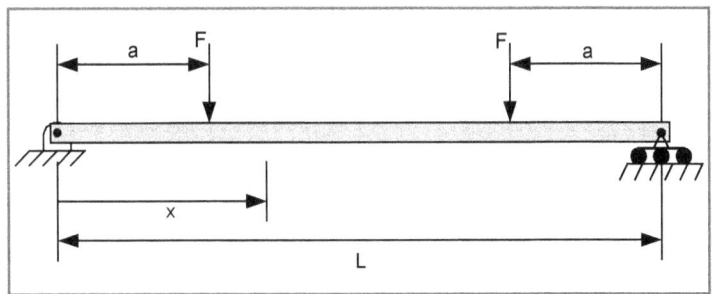

Fig. P7.2

7.3 In a manufacturing process, a metal sheet with a thickness h is passed around a roller with a diameter of D. Determine the maximum strain in the sheet. Also determine the maximum stress. Dimensions for h and D and the type of metal sheet are given in the table below.

Prob. No.	Material	h	D
7.3a	Aluminum	1.2 mm	450 mm
7.3b	Steel	0.8 mm	300 mm
7.3c	Brass	0.0625 in.	10.0 in.
7.3d	Titanium	0.040 in.	6.0 in.
7.3e	Magnesium	1.0 mm	250 mm

7.4 A beam with a rectangular cross section has dimensions b and h, as specified in the table below. If the yield strength of the beam's material is S_y, determine the maximum moment that the beam can support if the margin of safety is 100%.

Prob. No.	b	h	S_y
7.4a	2.0 in	7.5 in.	60 ksi
7.4b	2.2 in	6.4 in.	47 ksi
7.4c	80 mm	600 mm	200 MPa
7.4d	65 mm	500 mm	240 MPa
7.4e	40 mm	400 mm	500 MPa

7.5 Three planks with dimensions of 2.0 by 8.0 inches are adhesively bonded to form a timber I beam as shown in Fig. P7.5. Determine the stresses due to bending at points A, B, C and D. The bending moment applied to the beam is 27 ft-lb.

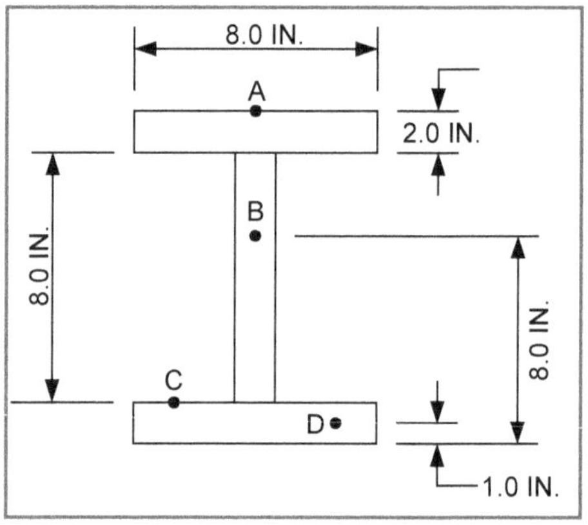

Fig. P7.5

7.6 A WT6 × 60 structural T section is employed as a cantilever beam. The beam is fabricated from steel that has a yield strength of 40 ksi in tension and 85 ksi in compression. Bending occurs about the z-axis and a safety factor of at least 2.8 is required. Determine the largest positive bending moment to which the beam can be subjected, if all of the requirements are satisfied. Also determine the maximum negative bending moment.

7.7 For the beam sections designated in the table below, verify the values of I_z and Z relative to the Z-Z axis that are given in Appendix D.

Prob. No.	Designation	Prob. No.	Designation
7.7a	W5 × 15	7.7d	S508 × 98
7.7b	W406 × 100	7.7e	S152 × 26
7.7c	S10 × 35	7.7f	WT6 × 60

7.8 Consider the flange and web section defined in Fig. P7.6. If $b_w = xb$ and $h_w = yh$, determine the effectiveness SE as x varies from 0.05 to 0.3 and y varies from 0.6 to 0.95. We suggest that you perform these calculations on a spreadsheet and prepare a graph displaying your results.

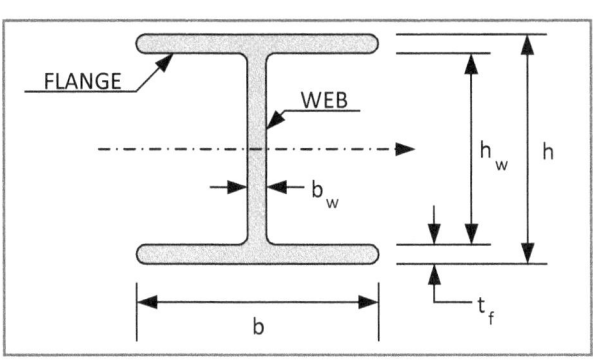
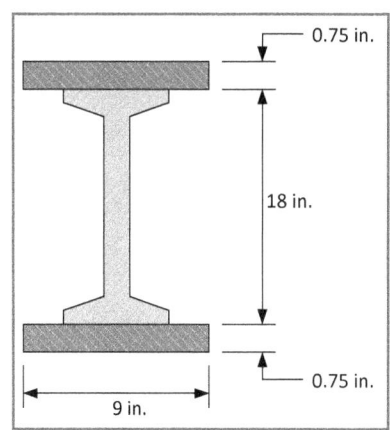

Fig. P7.8 Fig. P7.9

7.9 Two steel plates 9.0 in. by 0.75 in. are welded to the flanges of an S18 × 70 American Standard I-beam as shown in Fig. P7.9. Calculate the moment of inertia about each of its axes of symmetry.

7.10 If the thickness of the two plates were increased to 1.25 in. for the modified I-beam in Fig. 7.9, determine the moment of inertia about each of the axes of symmetry.

7.11 For the modified I-beam in Fig. P7.9, determine the added capacity of the beam in supporting bending moments.

7.12 Four different beams are shown in Fig. P7.12 with various end conditions. Construct FBDs of each beam showing the forces and/or moments at each support.

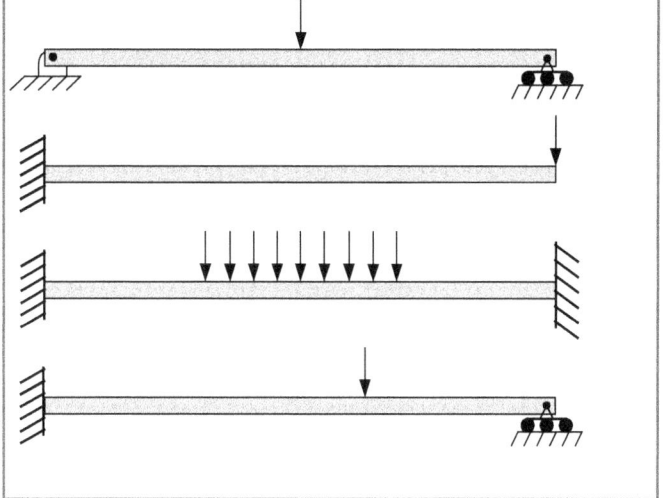

Fig. P7.12

7.13 Write equations for the shear and bending moments as a function of position across the beam shown in Fig. P7.13. Also construct the shear and bending moment diagrams.

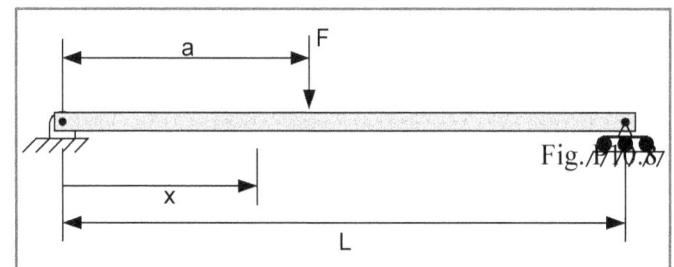

Prob. No.	a	L	F
7.13a	13 ft	22 ft	10 kip
7.13b	9 ft	18 ft	8.0 kip
7.13c	11 ft	20 ft	12 kip
7.13d	3.8 m	8.5 m	20 kN
7.13e	6.0 m	9.0 m	8 kN
7.13f	5.5 m	11 m	13 kN

Fig. P7.13

7.14 Write equations for the shear and bending moments as a function of position across the beam described in Fig. P7.14. Also construct the shear and bending moment diagrams. Dimensions and loads are specified in the table below.

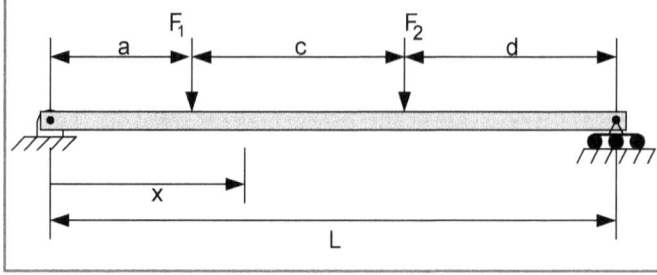

Fig P7.14

Prob. No.	a	c	L	F_1	F_2
7.14a	6 ft	5 ft	18 ft	5 kip	8 kip
7.14b	6 ft	8.0 ft	20 ft	7 kip	10 kip
7.14c	3 ft	10 ft	16 ft	10 kip	6 kip
7.14d	2 m	3.0 m	8 m	15 kN	18 kN
7.14e	3 m	2.0 m	7 m	10 kN	10 kN
7.14f	2.5 m	2.5 m	9 m	12 kN	16 kN

7.15 Write equations for the shear and bending moments as a function of position across the beam described Fig. P7.15. Also construct the shear and bending moment diagrams. Dimensions and loads are specified in the table below.

Fig. P7.15

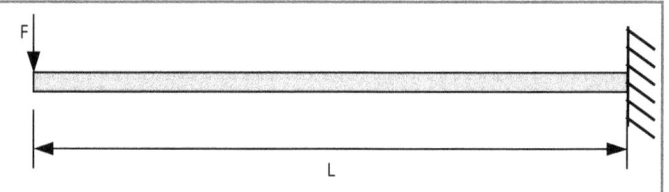

Prob. No.	L	F
7.15a	15 ft	7 kip
7.15b	18 ft	9 kip
7.15c	14 ft	7 kip
7.15d	6 m	14 kN
7.15e	7 m	15 kN
7.15f	8 m	10 kN

7.16 Write equations for the shear and bending moments, as a function of position across the beam described in Fig. P7.16. Also construct the shear and bending moment diagrams. Dimensions and loads are specified in the table below.

Fig. P7.16

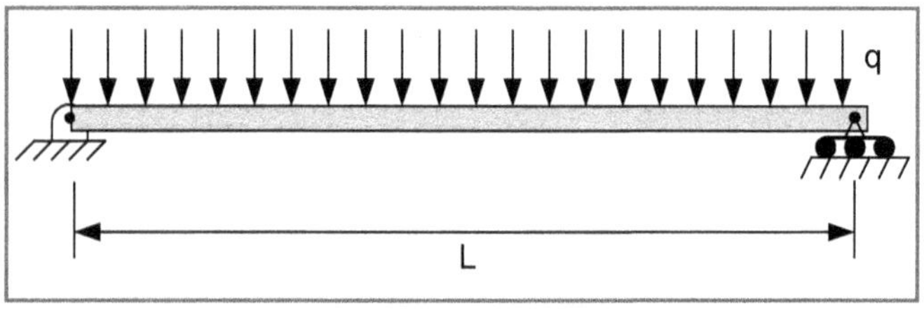

Prob. No.	L	q	Prob. No.	L	q
7.16a	15 ft	70 lb/ft	7.16d	6.3 m	1.0 kN/m
7.16b	18 ft	125 lb/ft	7.16e	7.4 m	0.8 kN/m
7.16c	14 ft	65 lb/ft	7.16f	5.9 m	1.2 kN/m

7.17 Write equations for the shear and bending moments as a function of position across the beam described in Fig. P7.17. Also construct the shear and bending moment diagrams. Dimensions and loads are specified in the table below.

Fig. P7.17

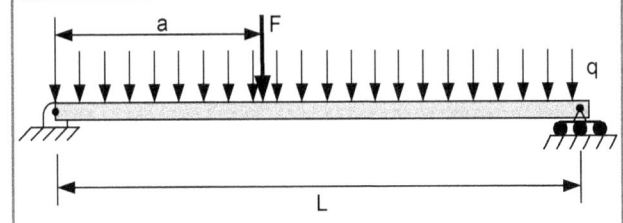

Prob. No.	a	L	F	q
7.17a	5.0 ft	15 ft	7 kip	70 lb/ft
7.17b	7.2 ft	18 ft	9 kip	125 lb/ft
7.17c	9.5 ft	14 ft	7 kip	65 lb/ft
7.17d	3.0 m	6 m	14 kN	1.0 kN/m
7.17e	4.6 m	7 m	15 kN	0.8 kN/m
7.17f	3.2 m	8 m	10 kN	1.2 kN/m

7.18 Write equations for the shear and bending moments as a function of position across the beam described in Fig. P7.18. Also construct the shear and bending moment diagrams. Dimensions and loads are specified in the table below.

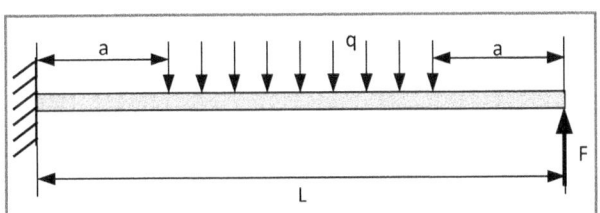

Prob. No.	a	L	F	q
7.18a	4.0 ft	15 ft	500 lb	100 lb/ft
7.18b	6.0 ft	18 ft	1200 lb	150 lb/ft
7.18c	2.5 ft	14 ft	1000 lb	75 lb/ft
7.18d	1.5 m	6 m	2 kN	1.2 kN/m
7.18e	2.0 m	7 m	3 kN	0.8 kN/m
7.18f	2.5 m	8 m	4.5 kN	1.5 kN/m

Fig. P7.18

7.19 Select a WF section for the steel beams described in the table below. The margin of safety (MOS) is and the yield strength of the material is also given in the table.

Prob. No.	Beam Configuration	MOS	S_y
7.19a	P7.13a	200%	42 ksi
7.19b	P7.14a	175%	36 ksi
7.19c	P7.15a	150%	48 ksi
7.19d	P7.16d	200%	300 MPa
7.19e	P7.17d	175%	320 MPa
7.19f	P7.18d	150%	350 MPa

7.20 Select an American Standard section for the steel beams described in the table below. The margin of safety (MOS) is and the yield strength of the material is also given in the table.

Prob. No.	Beam Configuration	MOS	S_y
7.20a	P7.13e	200%	300 MPa
7.20b	P7.14e	175%	320 MPa
7.20c	P7.15e	150%	350 MPa
7.20d	P7.16b	200%	42 ksi
7.20e	P7.17b	175%	36 ksi
7.20f	P7.18b	150%	48 ksi

7.21 Select a structural tee section for the steel beams described in the table below. The margin of safety (MOS) is and the yield strength of the material is also given in the table.

Prob. No.	Beam Configuration	MOS	S_y
7.21a	P7.13c	200%	42 ksi
7.21b	P7.14c	175%	36 ksi
7.21c	P7.15c	150%	48 ksi
7.21d	P7.16f	200%	300 MPa
7.21e	P7.17f	175%	320 MPa
7.21f	P7.18f	150%	350 MPa

7.22 For the beam described in the table below, determine the maximum shear stress τ if the cross section of the beam is rectangular. Dimensions for the beam are also given in the table.

Prob. No.	Beam Configuration	b	h
7.22a	P7.13b	1.0 in.	4.0 in.
7.22b	P7.15d	25 mm	100 mm
7.22c	P7.16e	35 mm	150 mm
7.22d	P7.17a	1.75 in.	6.0 in.

7.23 A circular shaft with a diameter of d is subjected to transverse loading from a gear set that produces a maximum shear force V_{Max} on the shaft. Determine the maximum shear stress τ_{Max} in the shaft due to this transverse load.

Prob. No.	d	V_{Max}
7.23a	50 mm	25 kN
7.23b	75 mm	50 kN
7.23c	1.50 in.	5.0 kip
7.23d	3.00 in.	15 kip

7.24 A timber company has designed a series of box beams with the geometry presented in Fig. P7.24. Determine the maximum shear stress τ_{Max} and the shear stress at the adhesive joints. The shear force V and the dimensions of the cross section are given in the table below.

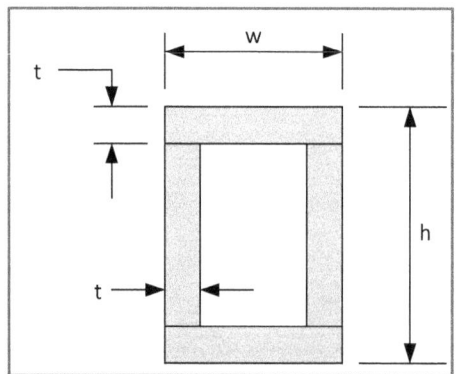

Prob. No.	V	t	h	w
7.24a	1200 lb	1.0 in.	4.0 in.	4.0 in
7.24b	5.0 kN	25 mm	100 mm	100 mm
7.24c	10.0 kN	35 mm	180 mm	100 mm
7.24d	2500 lb	1.75 in.	6.0 in.	4.0 in.
7.24e	4800 lb	2.0 in.	8.0 in.	5.0 in.
7.24f	14 kN	50 mm	250 mm	150 mm

Fig. P7.24

7.25 A timber company has designed a series of unsymmetrical T-sections to serve as beams as indicated in Fig. P7.25. Determine the maximum shear stress τ_{Max} and the shear stress at the adhesive joints. The shear force V and the dimensions of the cross section are given in the table below.

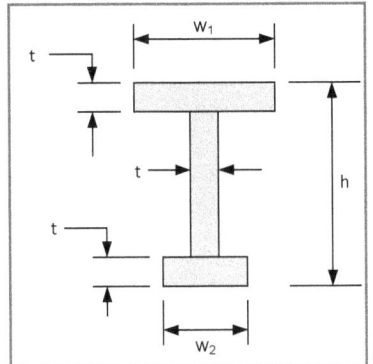

Fig. P7.25

Prob. No.	V	t	h	w_1	w_2
7.25a	1200 lb	1.0 in.	4.0 in.	4.0 in	2.0 in.
7.25b	5.0 kN	25 mm	100 mm	100 mm	50 mm
7.25c	10.0 kN	35 mm	180 mm	100 mm	75 mm
7.25d	2500 lb	1.75 in.	6.0 in.	4.0 in.	3.0 in.
7.25e	4800 lb	2.0 in.	8.0 in.	5.0 in.	4.0 in.
7.25f	14 kN	50 mm	250 mm	150 mm	100 mm

7.26 A timber company produces a series of I-beams fabricated from three pieces of timber, as shown in Fig. P7.26. The flange of the I-beam is fastened to the web with spikes that are each capable of resisting a shear force of 160 lb or 700 N. The beams are of length L and are simply supported. They carry a concentrated force F at its midspan. Determine the maximum spacing of the spikes needed to transmit the shear from the web to the flange of the beam for each of the conditions listed in the table below.

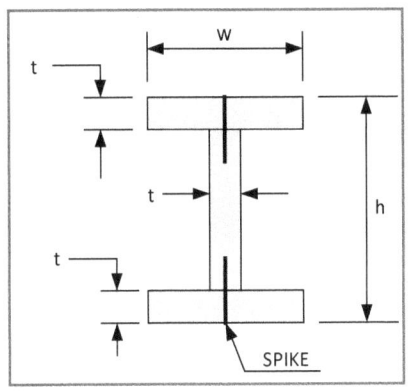

Fig. P7.26

Prob. No.	L	F	t	h	w
7.26a	12 ft	3500 lb	1.0 in.	4.0 in.	4.0 in
7.26b	4.0 m	10 kN	25 mm	100 mm	100 mm
7.26c	6.0 m	15 kN	35 mm	180 mm	100 mm
7.26d	15 ft	5000 lb	1.75 in.	6.0 in.	4.0 in.
7.26e	20 ft	8000 lb	2.0 in.	8.0 in.	5.0 in.
7.26f	5.0 m	30 kN	50 mm	250 mm	150 mm

7.27 Derive Eq. (7.29).

7.28 Consider a simply-supported, foam core, metal cover plate composite beam, as illustrated in Fig. P7.28. If the beam is loaded with a uniformly distributed load of magnitude q, determine the distributed load q_Y that produces yielding. Aluminum cover plates are t_m thick, b wide, and are adhesively bonded to a polystyrene foam core. The foam is b wide and h_f high. The beam is L long and the yield strength of the aluminum is S_y.

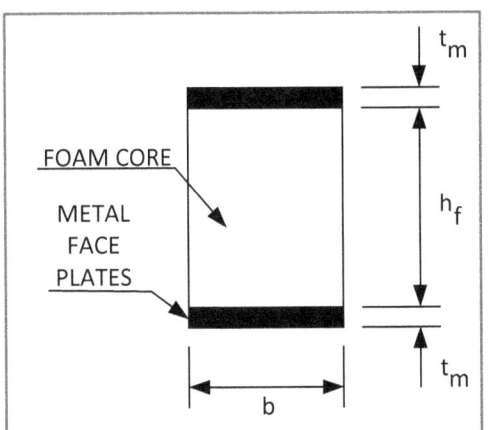

Fig. P7.28

Prob. No.	t_m	b	h_f	L	S_y
7.28a	1.0 mm	80 mm	200 mm	2.5 m	280 MPa
7.28b	0.8 mm	75 mm	180 mm	3.0 m	300 MPa
7.28c	1.2 mm	60 mm	150 mm	2.75 m	320 MPa
7.28d	0.040 in.	4.0 in.	10.0 in.	8.0 ft	47.0 ksi
7.28e	0.060 in.	3.0 in	7.0 in.	11.0 ft	40.0 ksi

7.29 Design a portable, one-way, bridge beam with a single structural member consisting of metal cover plates and foam core. The bridge is to support the weight of k students uniformly distributed over the span of the bridge beam. The average weight of a typical student is W. The bridge beam spans a ravine L wide.

Prob. No.	k	L	W
7.29a	2	4 m	700 N
7.29b	3	15 ft	155 lb
7.29c	4	5 m	700 N
7.29d	5	24 ft	155 lb

7.30 A cantilever beam L long is subjected to a concentrated force F applied at its free end. The beam has a rectangular cross section with a width b. Determine the height of the foam h_f and the thickness of the cover plates t_m, if the design stress in these plates is limited to σ_d. In this determination recall that the standard thickness of aluminum sheet stock from which the cover plates are cut is 0.016, 0.020, 0.025, 0.032, 0.040, 0.050, 0.063, and 0.090 in. Dimensions and loads are specified in the table below.

Prob. No.	F	L	b	σ_d
7.30a	450 lb	6.5 ft	4.0 in.	20 ksi
7.30b	380 lb	5.2 ft	5.6 in.	24 ksi
7.30c	1000 N	2.0 m	120 mm	200 MPa
7.30d	1200 N	1.5 m	90 mm	160 MPa

7.31 A beam, fabricated from western white pine, with dimensions shown in the table below, has been reinforced by bonding steel strips at on its top and bottom surfaces to form a composite beam. If the design stress is limited to 1.5 ksi in the wood and 30 ksi in the steel, determine the maximum bending moment, if the beam bends about its horizontal axis of symmetry.

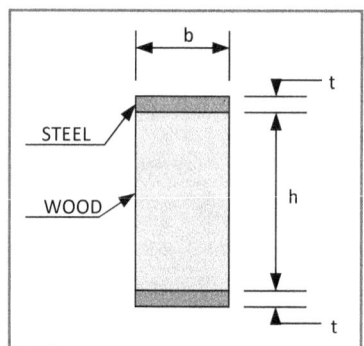

Prob. No.	h	b	t
7.31a	10 in.	4.0 in.	0.5 in.
7.31b	12 in.	6.0 in.	0.75 in.
7.31c	200 mm	100 mm	12 mm
7.31d	300 mm	125 mm	20 mm

Fig. P7.31

7.32 For the wood and steel composite beam described in Problem P7.31, determine the maximum bending moment if the beam is bent about its vertical axis of symmetry.

7.33 Derive Eqs. (7.32) and (7.35).

7.34 A concrete floor slab is reinforced with N steel bars each with a diameter D. The slab is b wide and h thick, and the centerline of the steel rebar is h/4 above the bottom surface of the slab. The slab is L long and is simply supported when installed. If a uniformly distributed load q is applied to the floor, determine its maximum magnitude, if the stress in the steel is limited to 20,000 psi and in the concrete to 1,500 psi. Note the modulus of elasticity of the steel and the concrete is 30×10^6 and 2.8×10^6 psi, respectively. Comment on the balance of the design of this reinforced concrete slab. Dimensions are specified in the table below.

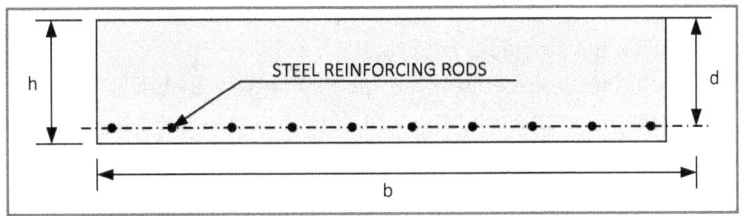

Problem No.	N	D	L	b	h
7.34a	12	½ in.	24 ft	48 in.	4 in.
7.34b	10	¾ in.	24 ft	36 in.	6 in.
7.34c	9	1.0 in.	20 ft	48 in.	8 in.
7.34d	14	½ in.	20 ft	60 in.	4 in.

7.35 A concrete beam is reinforced with N steel bars each with a diameter D. The section of the beam is rectangular—b wide and h high with the centerline of the steel rebar h/4 above the bottom surface of the beam. The beam is L long and simply supported. If it supports three concentrated forces F positioned at L/4, L/2 and 3L/4, determine the margin of safety for the beam. The yield strength of the rebar is 36 ksi and the compressive strength of concrete is 4 ksi. The modulus of elasticity of the steel and the concrete is 30×10^6 and 2.8×10^6 psi, respectively.

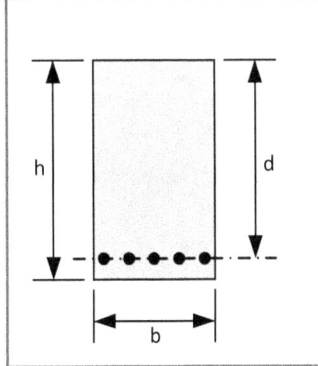

Prob. No.	N	D	L	b	h	F
7.35a	3	1.5 in.	22 ft	16 in.	24 in.	10 kip
7.35b	4	1.0 in.	24 ft	14 in.	18 in.	12 kip
7.35c	6	¾ in.	20 ft	17 in.	22 in.	8 kip
7.35d	8	½ in.	20 ft	12 in.	20 in.	6 kip

7.36 A concrete beam is reinforced with N steel rods each with an area A located at positions shown in Fig. P7.36. For an applied bending moment of M = 5 kN-m, determine the maximum stress in the steel rebars and the high-strength concrete.

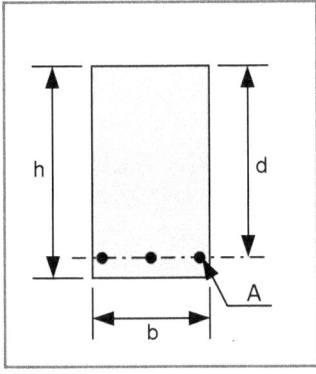

Prob. No.	N	A	h	b	d
7.36a	3	115 mm^2	150 mm	100 mm	110 mm
7.36b	2	300 mm^2	200 mm	100 mm	160 mm
7.36c	4	115 mm^2	250 mm	150 mm	210 mm
7.36d	3	300 mm^2	300 mm	150 mm	260 mm

7.37 Discuss the advantage of using steel bar reinforced concrete beams as structural components in a building in the event of a severe fire. In your discussion, discuss the effect of prolonged exposure to flames on the strength of steel.

7.38 Redesign the steel reinforced concrete beam in Example E7.36 to improve the balance between the stresses in the steel and concrete.

7.39 Redesign the steel reinforced concrete beam in Example E7.37 to improve the balance in the margin of safety between the steel and concrete.

7.40 A beam with a rectangular cross section, defined as shown in Fig. P7.40, is fabricated from steel with S_y = 40 ksi. Determine the yield moment M_Y. Dimensions for the cross section are given in the table below.

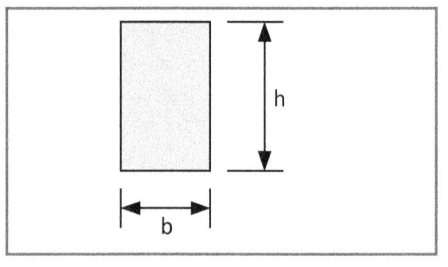

Fig. P7.40

Prob. No.	b	h
7.40a	2.0 in.	6 in.
7.40b	1.5 in.	5 in.
7.40c	1.25 in.	4 in.
7.40d	25 mm	125 mm
7.40e	35 mm	150 mm
7.40f	45 mm	180 mm

7.41 For a beam with a rectangular cross section, show that $M_P = M_Y$ when $y_p = h/2$. You may wish to examine Eq. (7.37) in developing this solution.

7.42 A beam with a rectangular cross section, defined in Fig. P7.40, is fabricated from steel with S_y = 320 MPa. Determine the moment M_p required to drive the elastic-plastic interface to the position $y_p = h/4$.

Prob. No.	b	h
7.42a	2.0 in.	6 in.
7.42b	1.5 in.	5 in.
7.42c	1.25 in.	4 in.
7.42d	25 mm	125 mm
7.42e	35 mm	150 mm
7.42f	45 mm	180 mm

7.43 Determine the ratio of M_L/M_Y for the web and flange section defined in Fig. P7.43, if $b_w = xb$ and $h_w = yh$. Let x vary from 0.05 to 0.3 and y vary from 0.6 to 0.95. We suggest that you perform these calculations on a spreadsheet and prepare a graph displaying your results.

Fig. P7.43

7.44 A cantilever beam illustrated in Figs. P7.44a and P7.44b is notched at position x. Determine the maximum tensile stress on the beam for the numerical parameters listed in the table below.

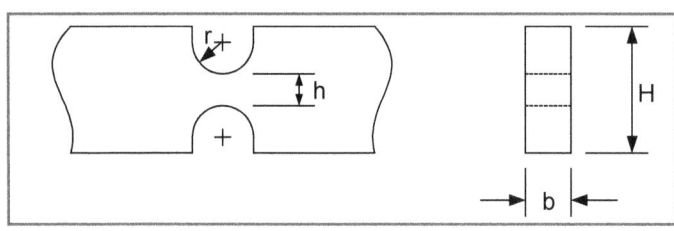

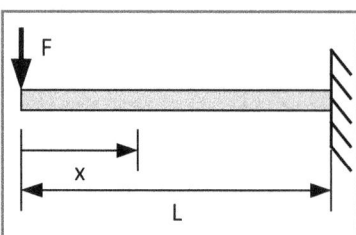

Fig. P7.44a   Fig. P7.44b

Prob. No.	F	L	x	b	H	h	r
7.44a	10 kip	11 ft	3.8 ft	2.0 in.	6 in.	4.0 in.	0.40 in.
7.44b	2.5 kN	3.2 m	1.9 m	60 mm	150 mm	100 mm	6.0 mm
7.44c	8.4 kip	14 ft	11 ft	1.75 in.	8 in.	4.0 in.	0.50 in.
7.44d	3.2 kN	2.8 m	1.8 m	75 mm	175 mm	100 mm	10 mm

7.45 An architect has proposed a series of stepped roof beams to provide a visual impact for a public building with high pedestrian traffic. A representative beam with its loading is presented in Figs. P7.45a and P7.45b. As the city's structural engineer you have the responsibility to approve or disapprove the design. Numerical parameters are provided in the table below.

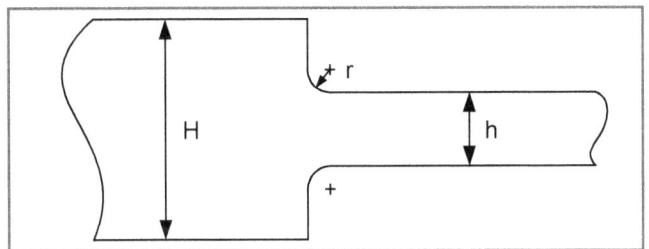

Fig. P7.45a   Fig. P7.45b

Prob. No.	q	L	x	b	H	h	r	S_y
7.45a	320 lb/ft	15 ft	7.5 ft	2.0 in.	6 in.	4.0 in.	0.40 in.	36 ksi
7.45b	2.5 kN/m	4.2 m	2.1 m	60 mm	150 mm	100 mm	6.0 mm	240 MPa
7.45c	480 lb/ft	18 ft	12 ft	1.75 in.	8 in.	4.0 in.	0.50 in.	36 ksi
7.45d	3.2 kN/m	5.0 m	3.0 m	75 mm	175 mm	100 mm	10 mm	240 MPa

7.46 A shaft in a drive train, illustrated in Figs. P7.46a and P7.46b, is subjected to bending moments due to a transverse force F imposed by a gear set. Determine the maximum stresses at the fillet due to the gear load. Numerical parameters are provided in the table below.

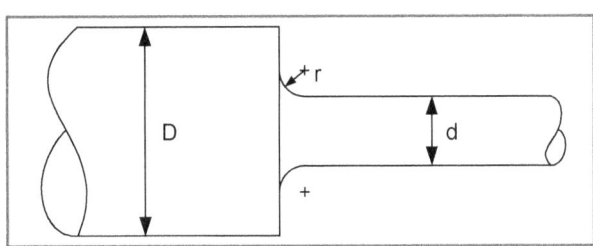
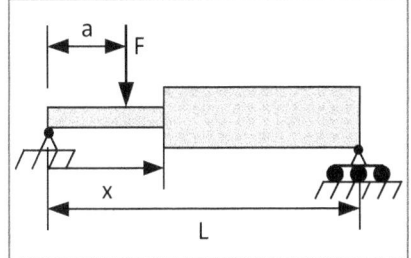

Fig. P7.46a   Fig. P7.46b

Prob. No.	F	L	a	x	D	d	r
7.46a	550 lb	2.5 ft	1.10 ft	1.25 ft	3.0 in.	2.25 in.	0.125 in.
7.46b	3.5 kN	1.2 m	0.50 m	0.60 m	75 mm	50 mm	6.0 mm
7.46c	800 lb	3.8 ft	1.60 ft	1.90 ft	4.0 in.	3.00 in.	0.150 in.
7.46d	5.2 kN	1.7 m	0.75 m	0.85 m	100 mm	60 mm	8.0 mm

CHAPTER 8

DEFLECTION OF BEAMS

8.1 INTRODUCTION

In Chapter 7, we studied the bending stresses produced in beams subjected to transverse loading. These uniaxial stresses σ_x were generated by the bending moments M(x) produced by the transverse loads. When a beam is subjected to a moment, as shown in Fig. 8.1, it bends into a circular arc and deflects. As the beam deforms, the neutral axis becomes a circular arc and plane sections remain plane.

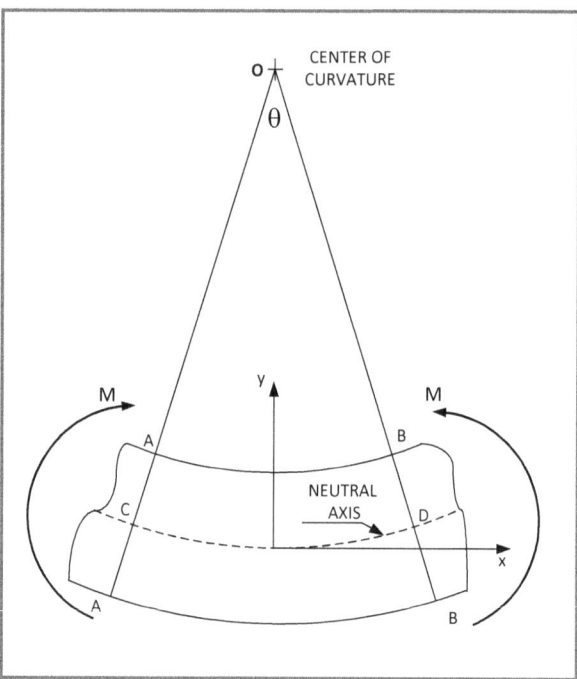

Fig. 8.1 A bending moment M causes the neutral axis of the beam to form a circular arc and deflect.

In Section 7.2, we used the geometry of the beam's deformation and equilibrium relations to position the neutral axis and to establish a relation between the bending moment and the radius of curvature ρ as:

$$M(x) = (EI_z/\rho) \qquad (7.5 \text{ bis})$$

$$\kappa = 1/\rho = M(x)/[EI_z] \qquad (7.6 \text{ bis})$$

where $\kappa = 1/\rho$ is the curvature of the beam and M(x) is the moment that varies with x.

8.2 THE ELASTIC CURVE FOR A BEAM

When a beam deflects under the action of a bending moment, the neutral axis becomes curved. Locally it is a circular arc with a radius of curvature ρ that depends on the bending moment M(x). As the local curvature varies with the bending moment, the neutral axis generates an elastic curve over the length of the beam. If we can write an equation for the elastic curve (the deformed shape of the neutral axis), the deflection of the beam will be apparent.

Let's rewrite Eq. (7.5) as:
$$1/\rho = (M/EI_z) \tag{8.1}$$

A well-known relation from calculus gives the curvature κ of any curve as a function of the derivatives of y as:

$$\kappa = \frac{1}{\rho} = \frac{\dfrac{d^2y}{dx^2}}{\left[1+\left(\dfrac{dy}{dx}\right)^2\right]^{3/2}} \tag{8.2}$$

Well-designed beams are stiff and consequently the deflections are very small compared to the length of the beam. Because of this fact, the slope of the elastic curve (dy/dx) is small when compared to 1 and $(dy/dx)^2$ is extremely small compared to 1. Hence, Eq. (8.2) can be closely approximated with:

$$\kappa = \frac{1}{\rho} = \frac{d^2y}{dx^2} \tag{8.3}$$

Substituting Eq. (8.1) into (8.3) yields:

$$\frac{d^2y}{dx^2} = \frac{M(x)}{EI_z} \tag{8.4}$$

If we recall Eq. (7.18), which states that shear force V(x) = dM(x)/dx, and substitute this relation into Eq. (8.4), we obtain:

$$\frac{d^3y}{dx^3} = \frac{V(x)}{EI_z} \tag{8.5}$$

Similarly recall Eq. (7.20) which states that the load q(x) = – dV(x)/dx and substitute this relation into Eq. (8.5) to obtain:

$$\frac{d^4y}{dx^4} = \frac{-q(x)}{EI_z} \tag{8.6}$$

Clearly the elastic curve y(x) can be determined by integrating any one of the three equations given above. We usually employ Eq. (8.4), because only two integrations are necessary to obtain the results for the elastic curve y(x) for the beam.

8.3 <u>ESTABLISHING THE ELASTIC CURVE BY INTEGRATION</u>

There are three different approaches for determining the equation describing the elastic curve, which include:

1. Integrating the equation for the bending moment M(x) with respect to x
2. Integrating the equation for the shear force V(x) with respect to x
3. Integrating the equation for the transverse loading q(x) with respect to x

We usually integrate Eq. (8.4) that involves M(x), because only two integrations are needed to convert d^2y/dx^2 into y(x). The mathematics give rise to two integration constants that are determined so as to satisfy constraints to the beam by its supports. In most examples, the integration is not difficult because M(x) is usually a polynomial in x. The difficulty arises in the form of the moment equation. In many instances, a single equation cannot be written that expresses M(x) over the entire length L of the beam. Instead, it is often necessary to write two or more different equations for M(x), each valid over a segment of the beam. Each of these moment equations must be integrated twice and the two constants of integration must be determined for each beam segment. The result is a time-consuming and tedious approach, in which there are many opportunities for mathematical errors.

It is possible to integrate Eq. (8.5) to obtain the equation for the elastic curve y(x). In this case, we integrate the function for the shear force V(x) with respect to x. Because the unknown y occurs as the third derivative in Eq. (8.5), it is clear that the solution for the differential equation requires three integrations giving rise to three constants of integration. Because all of the mathematical difficulties mentioned in the previous paragraph occur with this approach, there is little merit in pursuing the topic.

It is also possible to integrate Eq. (8.6) to obtain the equation for the elastic curve y(x). In this case, we integrate the function for the load q(x) with respect to x. As the unknown y occurs as the fourth derivative in Eq. (8.6), it is clear that the solution for the differential equation requires four integrations giving rise to four constants of integration. Because all of the mathematical difficulties mentioned in the previous two paragraphs become more severe with this approach, there is very little merit in pursuing the topic. Regardless of the approach employed to determine the equation of the elastic curve by integration we make several assumptions, which include:

- The deflections are small compared to the length of the beam so that the slope (dy/dx) is small and the square of the slope $(dy/dx)^2$ is negligible when compared to unity.
- The stresses in the beam are lower than the yield strength so that the modulus of elasticity E can be treated as a constant.
- The beam's cross-section does not change along the length of the beam so that the second moment of inertia I_z is a constant with respect to x.
- The distortion due to the shearing stress τ_{xy} is sufficiently small that the out of plane deformations of the beam's section can be ignored. (Plane sections remain plane.)

Let's consider six examples that demonstrate the integration method, describe the effect of beam constraints (boundary conditions), and make use of compatibility and symmetry conditions. We will test the small slope approximation and illustrate some of the difficulties encountered with this when using the integration method.

EXAMPLE E8.1

Determine the elastic curve for the simply supported beam subjected to a uniformly distributed load q as shown in Fig. E8.1.

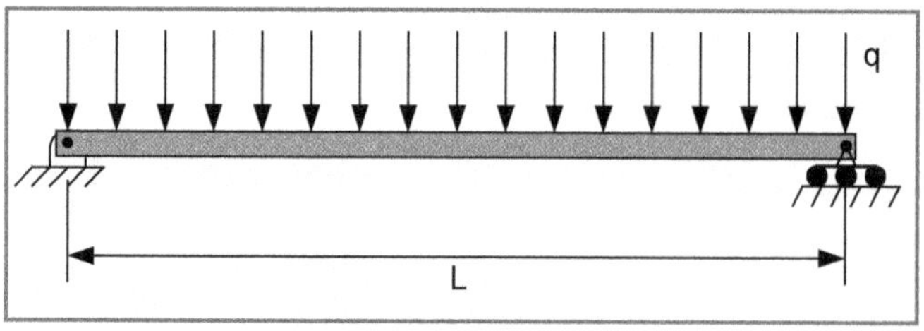

Fig. E8.1

Solution:

Step 1: Prepare a FBD of the entire beam and employ the equilibrium relations to determine the reactions at the beam's simple supports. We have performed this task in Example 7.7 and obtained the results shown below for the reaction forces R_L and R_R as:

$$R_L = R_R = qL/2 \qquad (a)$$

Step 2: Prepare a FBD of a segment removed from the left side of the beam, and solve for the internal shear force $V(x)$ and the internal moment $M(x)$. See Fig. E7.7a.

$$V(x) = (q/2)(L - 2x) \quad \Rightarrow \quad 0 < x < L \qquad (b)$$

$$M(x) = (qx/2)(L - x) \quad \Rightarrow \quad 0 < x < L \qquad (c)$$

Step 3: Write the differential equation for the elastic curve by substituting Eq. (c) into Eq. (8.4) to obtain:

$$EI_Z \frac{d^2 y}{dx^2} = \frac{qx}{2}(L - x) \qquad (d)$$

Step 4: Integrate both sides of Eq. (d) and simplify to obtain:

$$EI_Z \frac{dy}{dx} = \frac{q}{2}\left(\frac{x^2 L}{2} - \frac{x^3}{3}\right) + C_1 = \frac{qx^2}{12}(3L - 2x) + C_1 \qquad (e)$$

$$EI_Z y = \frac{q}{12}\left(x^3 L - \frac{x^4}{2}\right) + C_1 x + C_2 = \frac{qx^3}{24}(2L - x) + C_1 x + C_2 \qquad (f)$$

Step 5: The integration process has generated two constants of integration — C_1 and C_2. To determine these constants, we examine the constraints placed on the beam because these constraints control the shape of the elastic curve. The beam is constrained at its supports. On the left end, where $x = 0$, the beam is constrained by a pin and clevis arrangement, which provides the following constraint:

$$y(0) = 0 \qquad (g)$$

At the left end, where $x = L$, the beam is constrained by a set of rollers, which provides the following constraint:

$$y(L) = 0 \qquad (h)$$

Step 6: The relations presented in Eqs. (g) and (h) are known as boundary conditions. Let's substitute the boundary requirement of Eq. (g) into Eq. (f) to obtain:

$$y(0) = 0 \quad \Rightarrow \quad C_2 = 0$$

Now substitute Eq. (h) into Eq. (f) to obtain:

$$EI_Z y = \frac{qL^3}{24}(2L - L) + C_1 L = 0 \qquad (i)$$

Solving for C_1 yields:
$$C_1 = -\frac{qL^3}{24} \tag{j}$$

Substituting Eq. (j) into Eq. (f) gives the relation for the elastic curve from $0 < x < L$ as:

$$y = \frac{qx}{24EI_Z}\left[x^2(2L-x)-L^3\right] \tag{8.7}$$

Following a similar process, we can write the expression for the slope (dy/dx) as:

$$\frac{dy}{dx} = \frac{q}{24EI_Z}\left(6x^2L - 4x^3 - L^3\right) \tag{8.8}$$

We have evaluated Eq. (8.7) to determine $24EI_Zy/q$ as a function of x where x varied from 0 to L with L = 1.0 units. The result is shown in Fig. E8.1a.

Fig. E8.1a

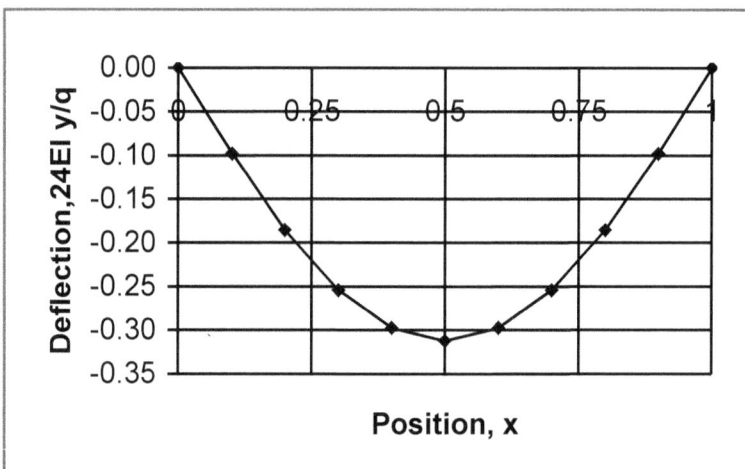

The normalized value of y associated with the elastic curve is a negative quantity because we defined y positive upward previously in Chapter 7. The elastic curve presented in Fig. E8.1a indicates that the deflection y is zero at x = 0 and x = L, as required by the constraints of the simply supported beam. The deflection is a maximum at the center of the span of the beam.

EXAMPLE E8.2

The simply supported beam in Example 8.1 is fabricated from a wide flange rolled steel shape. The architect designing the building has indicated a W8 × 40 beam from 1020 HR steel. The beam is 20 ft long and the uniformly distributed load q = 800 lb/ft. Determine the maximum stresses in the beam and the safety factor against failure by yielding, the maximum deflection and the maximum slope. Interpret the results.

Solution:

Step 1: The maximum bending stress σ_{Max} in the beam is calculated from Eq. (7.10) as:

$$\sigma_{Max} = M_{Max}/Z = qL^2/(8Z) \tag{a}$$

where $M_{Max} = qL^2/8$ and Z is the section modulus.

Reference to Appendix D indicates that the section modulus Z and moment of inertia I_z for a W8 × 40 beam are:

$$Z = 35.5 \text{ in.}^3 \qquad I_z = 146 \text{ in.}^4 \tag{b}$$

From Eq. (a) and (b), we write:

$$\sigma_{Max} = qL^2/(8Z) = (800)(20)^2(12)/[(8)(35.5)] = 13.52 \text{ ksi} \tag{c}$$

From Appendix B, we find $S_y = 42$ ksi for 1020 HR steel. Then we can determine the safety factor as:

$$SF = S_y/\sigma_{Max} = 42/13.52 = 3.11 \tag{d}$$

This is a reasonable safety factor considering that the loads in a building are well known and not subject to significant changes over the anticipated service life of the structure.

Step 2: The maximum deflection y is calculated by setting $x = L/2$ in Eq. (8.7) to obtain:

$$y_{Max} = -\frac{5qL^4}{384EI_Z} \tag{8.9}$$

Substituting numerical parameters into Eq. (8.9) gives:

$$y_{Max} = -\frac{5qL^4}{384EI_Z} = -\frac{(5)(800)(20)^4(12)^4}{(384)(12)(30 \times 10^6)(146)} = -0.6575 \text{ in.} \tag{e}$$

While this deflection is small compared to the length of the beam it is relatively large compared to the 8.25 in. height of the beam. A suggestion to the architect to consider a W16 × 40 section is in order. This beam has the same weight of 40 lb/ft, but much higher values for Z and I_z. The increase in cost would be modest for significant improvements in both strength and rigidity. The only disadvantage is the increased height of the structure that would be required to accommodate the increased height of the beam.

Step 3: It is evident from an examination of Fig. E8.1a that the maximum slope occurs at the supports for the beam where $x = 0$ and L. At $x = 0$, Eq. (8.8) reduces to:

$$\left(\frac{dy}{dx}\right)_{Max} = -\frac{qL^3}{24EI_Z} = -\frac{(800)(20)^3(12)^3}{(24)(12)(30 \times 10^6)(146)} = -8.767 \times 10^{-3} \tag{f}$$

The negative sign in Eq. (f) indicates that the deflection y is becoming more negative as x is increasing. Let's square Eq. (f) and compare the result to unity.

$$\left(\frac{dy}{dx}\right)^2_{Max} = \left(-8.767 \times 10^{-3}\right)^2 = 7.686 \times 10^{-5} \tag{g}$$

Clearly, the square of the maximum slope for this practical example is extremely small when compared with unity. This fact verifies the assumption made in reducing the form of Eq. (8.2) to the much simpler version presented as Eq. (8.3).

EXAMPLE E8.3

Derive the equation for the elastic curve for the simply supported beam loaded with a single concentrated force F as shown in Fig. E8.3.

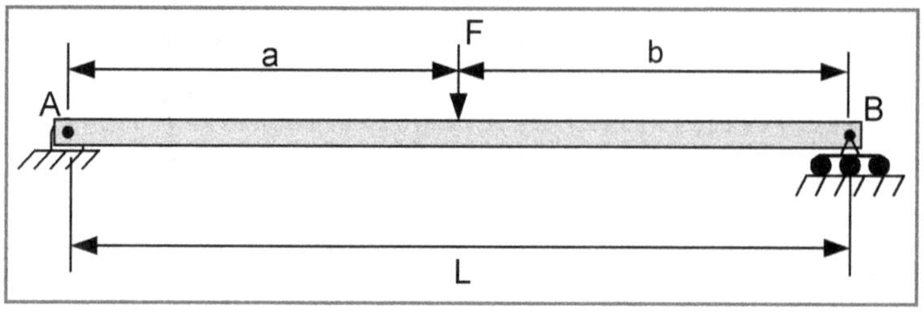

Fig. E8.3

Solution:

Step 1: Determine the reaction forces at the simple supports. Reference to Section 7.4.1 shows the approach and gives the results as:

$$R_L = Fb/L \qquad R_R = Fa/L \qquad (a)$$

Step 2: Write equations for the bending moment M as a function of x covering the range in x from zero to L. The procedure followed is illustrated in Section 7.4.1.

$$M(x) = Fbx/L \qquad 0 \leq x < a \qquad (b)$$

$$M(x) = Fbx/L - F(x-a) \qquad a \leq x \leq L \qquad (c)$$

These results differ from those presented in Example 8.1, where a single relation was written to describe the distribution of the bending moments over the entire length of the beam. The fact that it is necessary to write two equations to express M(x), complicates the solution.

Step 3: Write the differential equations for the elastic curve by substituting Eqs. (b) and (c) into Eq. (8.4) to obtain:

$$EI_z \frac{d^2 y_1}{dx^2} = M(x) = \frac{Fbx}{L} \qquad 0 \leq x \leq a \qquad (d)$$

$$EI_z \frac{d^2 y_2}{dx^2} = M(x) = \frac{Fbx}{L} - F(x-a) \qquad a \leq x \leq L \qquad (e)$$

Step 4: Integrate these second order differential equations once to obtain the equation for the slopes of the elastic curve as:

$$EI_z \frac{dy_1}{dx} = \frac{Fbx^2}{2L} + C_1 \qquad 0 \leq x \leq a \qquad (f)$$

$$EI_z \frac{dy_2}{dx} = \frac{Fbx^2}{2L} - \frac{F(x-a)^2}{2} + C_2 \qquad a \leq x \leq L \qquad (g)$$

Step 5: Before proceeding, let's use one of the compatibility conditions to evaluate the constants of integration in Eqs. (f) and (g). The elastic curve is continuous at the point of load application. Hence, the slope of the elastic curve dy_1/dx is the same as the slope dy_2/dx at $x = a$ where we transition from one solution for the slope to another. Accordingly, we equate Eq. (f) and Eq. (g) and set $x = a$ to obtain:

$$\frac{Fba^2}{2L} + C_1 = \frac{Fba^2}{2L} - 0 + C_2 \qquad \text{at } x = a$$

$$C_1 = C_2 \qquad \qquad (h)$$

Step 6: Integrate Eqs. (f) and (g) to obtain the equations for the elastic curve as:

$$EI_z y_1 = \frac{Fbx^3}{6L} + C_1 x + C_3 \qquad 0 \leq x \leq a \qquad (i)$$

$$EI_z y_2 = \frac{Fbx^3}{6L} - \frac{F(x-a)^3}{6} + C_1 x + C_4 \qquad a \leq x \leq L \qquad (j)$$

Step 7: Let's again recognize the compatibility conditions that exist at $x = a$. Because the elastic curve is continuous, it is clear that $y_1 = y_2$ at $x = a$. Hence, we can equate Eq. (i) and Eq. (j) and set $x = a$ to obtain:

$$\frac{Fba^3}{6L} + C_1 a + C_3 = \frac{Fba^3}{6L} - \frac{F(a-a)^3}{6} + C_1 a + C_4 \qquad \text{at } x = a$$

$$C_3 = C_4 \qquad \qquad (k)$$

We have used the two compatibility conditions to establish that $C_1 = C_2$ and $C_3 = C_4$, which reduces the four constants of integration to two.

Step 8: Let's consider the constraints placed on the elastic curve by the simple supports at each end of the beam. Clearly, the simple supports constrain the deflection at both ends of the beam to zero. Hence, we can write:

$$y_1(0) = 0 \qquad \qquad (l)$$

$$y_2(L) = 0 \qquad \qquad (m)$$

Substituting Eq. (l) into Eq. (i) gives:

$$C_3 = 0 = C_4 \qquad \qquad (n)$$

Substituting Eq. (m) into Eq. (j) gives:

$$0 = \frac{FbL^3}{6L} - \frac{F(L-a)^3}{6} + C_1 L + 0$$

$$C_1 = -\frac{Fb}{6L}(L^2 - b^2) \qquad \qquad (o)$$

Step 5: With the four constants of integration established, we write the solutions for the elastic curve as:

$$EI_z y_1 = -\frac{Fbx}{6L}(L^2 - x^2 - b^2) \qquad 0 \leq x \leq a \qquad (8.10)$$

$$EI_z y_2 = -\frac{Fbx}{6L}(L^2 - x^2 - b^2) - \frac{F(x-a)^3}{6} \qquad a \leq x \leq L \qquad (8.11)$$

We will determine the maximum deflection of a simply supported beam with a concentrated load in the next example. However, you should be aware that the double integration method often gives rise to four or even six constants of integration. As such, it is tedious and time consuming to execute solutions to even simple problems. We will introduce a method involving singularity functions in Section 8.4 that involves the determination of only two constants of integration.

EXAMPLE E8.4

For the simply supported beam, described in Example 8.3, determine the maximum deflection y_{Max}. The beam spans 6 m and is made from an American Standard shape with a designation of S203 × 34. The beam is fabricated from 1020 HR steel. A concentrated force of F = 10 kN is applied at a = 4 m.

Solution:

Step 1: Let's begin by deriving the equation for the maximum deflection by considering the slope dy_1/dx. At the position x where the deflection is a maximum[1], $dy_1/dx = 0$. Differentiating Eq. (8.10) and setting the result to zero gives the location of the maximum deflection as:

$$EI_z \frac{dy_1}{dx} = -\frac{Fb}{6L}(L^2 - 3x^2 - b^2) = 0$$

$$x = \sqrt{\frac{L^2 - b^2}{3}} \qquad (a)$$

Evaluating Eq. (a) gives:

$$x = \sqrt{\frac{L^2 - b^2}{3}} = \sqrt{\frac{36 - 4}{3}} = 3.266 \text{ m} \qquad (b)$$

Step 2: Determine the bending stresses in the beam to insure that the beam is elastic. The moment is a maximum at x = a = 4.0 m. Hence, we write:

$$M_{Max} = Fba/L = (10 \times 10^3)(2.0)(4.0)/6.0 = 13.33 \text{ kN-m} \qquad (c)$$

And the bending stress is given by:

$$\sigma_{Max} = M_{Max}/Z = [13.33 \times 10^3 \text{ (N-m)}(10^3 \text{ mm/m})/(265 \times 10^3 \text{ mm}^3) = 50.31 \text{ MPa}$$

where the section modulus $Z = 265 \times 10^3 \text{ mm}^3$ is listed in Appendix D.

Because σ_{Max} is less than the yield strength of 1020 HR steel (290 MPa), we conclude that the beam under load is elastic and Eq. (8.10) is valid.

---

[1] Because a > b the maximum deflection will occur on the left portion of the beam where x < a.

Step 3: To determine the expression for the maximum deflection, substitute Eq. (a) into Eq. (8.10) to determine the expression for y_{Max} as:

$$y_{Max} = -\frac{Fb}{9\sqrt{3}LEI_z}(L^2 - b^2)^{3/2} \qquad (8.12)$$

Evaluating Eq. (8.12) with the numerical parameters listed in the problem statement and in Appendices B and D gives:

$$y_{Max} = -\frac{(10 \times 10^3)(2000)}{9\sqrt{3}(6000)(207 \times 10^3)(27 \times 10^6)}\left[(6000)^2 - (2000)^2\right]^{3/2}$$

$$y_{Max} = -6.926 \text{ mm} \qquad (d)$$

Again we note that the deflection is small compared to both the length and the height of the beam.

EXAMPLE E8.5

Determine the equation for the elastic curve for the cantilever beam presented in Fig. E8.5. Also write the equation for the maximum deflection and the maximum slope.

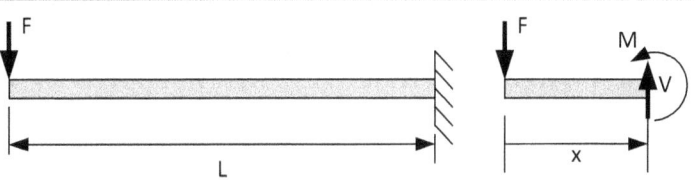

Fig. E8.5

Solution:

Step 1: Prepare a FBD of a segment of the beam where $0 \leq x \leq L$ and write an equation for $M(x)$ as:

$$M(x) = -Fx \qquad (a)$$

Step 3: Write the differential equation for the elastic curve by substituting Eq. (a) into Eq. (8.4) to obtain:

$$EI_z \frac{d^2y}{dx^2} = M(x) = -Fx \qquad 0 \leq x \leq L \qquad (b)$$

In this problem a single moment equation is valid over the entire length of the beam indicating that only two constants of integration will be required in the integration process.

Step 4: Integrating Eq. (b), we obtain:

$$EI_z \frac{dy}{dx} = -\frac{Fx^2}{2} + C_1 \qquad 0 \leq x \leq L \qquad (c)$$

Step 5: The constraint condition (a built-in end) at the right support of the beam, where $x = L$, forces the slope at the wall to be zero. Hence, Eq. (d) evaluated at $x = 0$ reduces to:

$$EI_z \frac{dy}{dx} = 0 = -\frac{FL^2}{2} + C_1 \quad \Rightarrow \quad C_1 = FL^2/2 \qquad (d)$$

Substituting the value for C_1 into Eq. (c) gives the equation for the slope of the elastic curve at any location along its length as:

$$\frac{dy}{dx} = \frac{F(L^2 - x^2)}{2EI_z} \qquad 0 \le x \le L \qquad (8.13)$$

Step 6: Integrate Eq. (c) to obtain the expression for the elastic curve as:

$$EI_z y = -\frac{Fx^3}{6} + \frac{FL^2 x}{2} + C_2 \qquad (e)$$

Step 7: The constraint condition on the right end of the beam due to the built-in support is that the deflection at the wall is zero. Evaluating Eq. (e) with $y = 0$ at $x = L$ gives:

$$EI_z y = 0 = -\frac{FL^3}{6} + \frac{FL^3}{2} + C_2 \quad \Rightarrow \quad C_2 = -\frac{FL^3}{3} \qquad (f)$$

Substituting for the constant C_2 into Eq. (e) gives:

$$EI_z y = -\frac{Fx^3}{6} + \frac{FL^2 x}{2} - \frac{FL^3}{3} \qquad (8.14)$$

Step 8: Inspection of Eq. (8.14) shows that the deflection is a maximum at the free end where $x = 0$.

$$y_{Max} = -\frac{FL^3}{3EI_z} \qquad (8.15)$$

The maximum slope also occurs at the free end. Setting $x = 0$ in Eq. (8.13) gives the relation for the maximum slope as:

$$\left(\frac{dy}{dx}\right)_{Max} = \frac{FL^2}{2EI_z} \qquad (8.16)$$

EXAMPLE E8.6

For the simply supported beam, shown in Fig. E8.6, determine the equation for the elastic curve. Also determine the maximum deflection and the maximum slope of the beam.

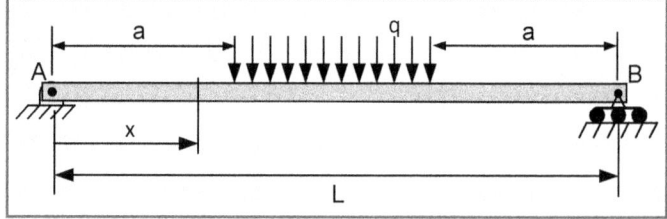

Fig. E8.6

Solution:

Step 1: Write the equations for the reaction forces at the beam's supports and the moments as a function of position x. FBDs of the entire beam and segments of the beam and the application of the equations of equilibrium lead to:

$$R_L = R_R = qb/2 \qquad \text{with } b = (L - 2a) \qquad (a)$$

$$M(x) = qbx/2 \qquad 0 \leq x \leq a \qquad (b)$$

$$M(x) = qbx/2 - q(x - a)^2/2 \qquad a \leq x \leq a + b \qquad (c)$$

$$M(x) = qbx/2 - qb(2x - L)/2 \qquad a + b \leq x \leq L \qquad (d)$$

We note that three different equations are necessary for describing the bending moments over the entire length of this beam. If we double integrate all of these equations, six constants of integration will be generated. Clearly, this approach will be long and tedious and should be avoided if possible. Fortunately, we recognize that the loading and the supports of the beam are symmetric. Symmetry enables us to solve for the equation of the elastic curve for the left half of the beam and to infer the solution for the right half of the beam. With this approach, we will integrate only the first two of the three moment equations.

Step 2: Substituting Eqs (b) and (c) into Eq. (8.4) gives a second order differential equation for the elastic curve as:

$$EI_z \frac{d^2 y_1}{dx^2} = M(x) = \frac{qbx}{2} \qquad 0 \leq x \leq a \qquad (e)$$

$$EI_z \frac{d^2 y_2}{dx^2} = M(x) = \frac{qbx}{2} - \frac{q(x-a)^2}{2} \qquad a \leq x \leq a + b \qquad (f)$$

Step 3: Integrating Eq. (e) twice gives:

$$EI_z \frac{dy_1}{dx} = \frac{qbx^2}{4} + C_1 \qquad 0 \leq x \leq a \qquad (g)$$

$$EI_z y_1 = \frac{qbx^3}{12} + C_1 x + C_2 \qquad 0 \leq x \leq a \qquad (h)$$

To satisfy the boundary condition at the left support where y = 0 at x = 0, it is evident that:

$$C_2 = 0 \qquad (i)$$

Step 4: Integrating Eq. (f) twice gives:

$$EI_z \frac{dy_2}{dx} = \frac{qbx^2}{4} - \frac{q(x-a)^3}{6} + C_3 \qquad a \leq x \leq a + b \qquad (j)$$

$$EI_z y_2 = \frac{qbx^3}{12} - \frac{q(x-a)^4}{24} + C_3 x + C_4 \qquad a \leq x \leq a+b \qquad (k)$$

Step 5: Determine C_3 by noting that the beam is symmetric; hence, $(dy_2/dx) = 0$ at $x = L/2$. Substituting this information into Eq. (j) and simplifying the result yields:

$$C_3 = -\frac{qb(3L^2 - b^2)}{48} \qquad (l)$$

To evaluate the remaining two constants of integration, we use the compatibility conditions, which indicate that $(dy_1/dx) = (dy_2/dx)$ and $y_1 = y_2$ at $x = a$. From Eqs. (g) and (j) and the compatibility condition $(dy_1/dx) = (dy_2/dx)$ at $x = a$, it is apparent that:

$$C_1 = C_3 = -\frac{qb(3L^2 - b^2)}{48} \qquad (m)$$

From Eqs. (h) and (k) and the compatibility condition $y_1 = y_2$ at $x = a$, we find that:

$$C_2 = C_4 = 0 \qquad (n)$$

Step 6: Substituting the results for the four constants into Eqs. (h) and (k) yields:

$$y_1 = \frac{qbx}{48EI_z}\left[4x^2 - 3L^2 + b^2\right] \qquad 0 \leq x \leq a \qquad (8.17)$$

$$y_2 = \frac{q}{48EI_z}\left[4bx^3 - 2(x-a)^4 - bx(3L^2 - b^2)\right] \qquad a \leq x \leq a+b \qquad (8.18)$$

Step 6: We write the relation for the maximum deflection by observing that $y_{Max} = y_2$ at $x = L/2$. Substituting $x = L/2$ into Eq. (8.18) gives:

$$y_{Max} = -\frac{qb}{384EI_z}\left[8L^3 + b^3 - 4b^2 L\right] \qquad (8.19)$$

Substituting the results for the constants C_1 and C_3 into Eqs. (g) and (j) gives the equations for the slopes of the elastic curve as:

$$\frac{dy_1}{dx} = \frac{qb}{48EI_z}\left[12x^2 - 3L^2 + b^2\right] \qquad 0 \leq x \leq a \qquad (8.20)$$

$$\frac{dy_2}{dx} = \frac{q}{48EI_z}\left[12bx^2 - 8(x-a)^3 - 3bL^2 + b^3\right] \qquad a \leq x \leq a+b \qquad (8.21)$$

The maximum slope occurs at the supports. We write the relation for the maximum slope by setting $x = 0$ in Eq. (8.20) to obtain:

$$\left(\frac{dy_1}{dx}\right)_{Max} = -\frac{qb}{48EI_z}\left[3L^2 - b^2\right] \qquad (8.22)$$

These six examples have shown that the double integration of the moment equation enables one to establish the equation of the elastic curve and the equation for the slope over the entire length of the beam. The difficulty that often arises is in determining the constants of integration that are generated during the integration process. When several different moment equations must be written to cover the span of the beam, the number of constants of integration increases and the work involved in the solution becomes excessive. We will introduce an approach in Section 8.4 that avoids this difficulty.

8.3.1 Boundary and Compatibility Conditions

In the solutions to the previous examples, we used information describing the beam's constraints in deriving the equation for the elastic curve of a beam in bending. Beams are supported (constrained) by simple supports or by built-in supports. If simple supports are used, a minimum of two supports is necessary for stability. Usually the supports are placed at the ends of the beam, but this is not required if overhangs provide some design advantage. If the beam has a built-in support at one end, the other end may be free. Fixed and simple supports may both be employed in constraining the deformation of a beam. However, this practice produces a statically indeterminate beam.

Simple supports are designed with several different arrangements that permit free rotation of the beam over the support (M = 0), while constraining the deflection so that y = 0. Several of the designs for simple supports are illustrated in Fig. 8.2. The roller and the step configurations are the most common, because they are less expensive to implement in construction.

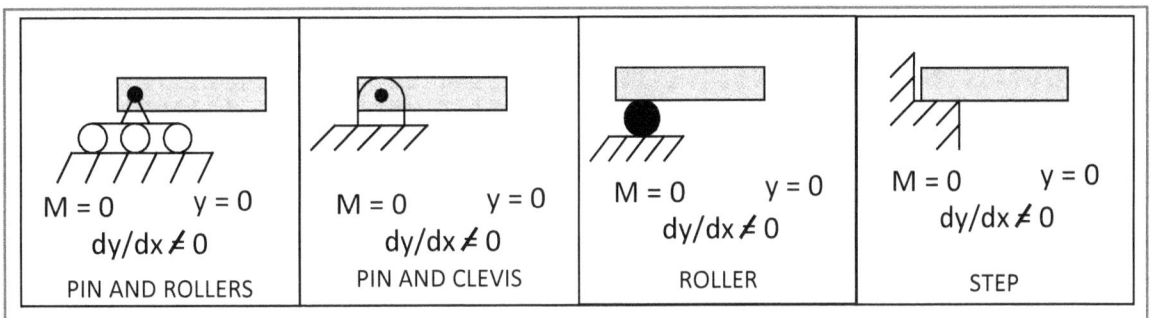

Fig. 8.2 Illustrations of different simple supports for a beam that yield the same boundary conditions.

Fixed or built-in supports are more difficult to achieve in practice. Two methods are illustrated in Fig. 8.3 together with the boundary conditions. Both the slope and the deflection are constrained so that (dy/dx) = 0 and y = 0. However, the moment at the support is not equal to zero.

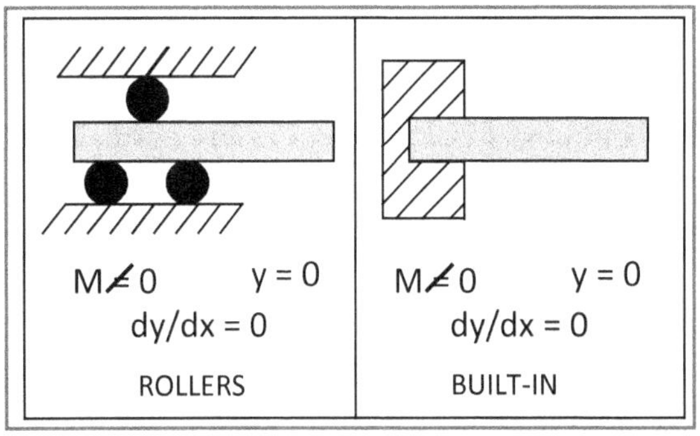

Fig. 8.3 Illustration of fixed or built-in supports for a beam.

With a built-in or fixed support at one end of a beam, the other end may be free, as is the case for a cantilever beam. The free end will deflect and rotate so that (dy/dx) ≠ 0 and y ≠ 0. However, the moment M and the shear force V both are zero at this point.

Compatibility and Symmetry Conditions

When it is necessary to write two or more moment equations to describe M(x) over the entire length of the beam, the integration method produces two or more equations for the elastic curve. Each of these equations is valid over some segment of the beam. For instance in Example 8.3, the expression for y_1 was valid for $0 \le x \le a$, and the relation for y_2 was valid for $a \le x \le L$. However at $x = a$, the results for both y_1 and y_2 must be identical. In addition the slopes dy_1/dx and dy_2/dx must also be identical at $x = a$. The deflection given by both solutions must be continuous as we transition from one solution to another. We illustrate the concept of continuity of the slope and the deflection by observing the elastic curve in Fig. 8.4

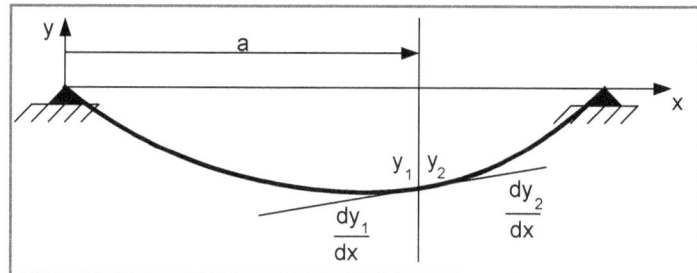

Fig. 8.4 For the elastic curve to be continuous, $y_1 = y_2$ and $(dy_1/dx) = (dy_2/dx)$ at $x = a$.

When the loading on a uniform beam is symmetric, the elastic curve is also symmetric as illustrated in Fig. 8.5. Because of this symmetry, the slope at the center of the beam where $x = L/2$ is zero. We will employ boundary conditions, compatibility conditions, and symmetry to establish constants of integration, when establishing the equation for the elastic curve.

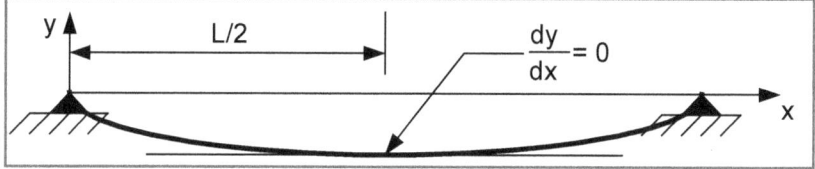

Fig. 8.5 The slope is zero at the mid point of a symmetrically loaded beam.

8.3.2 Integrating the Shear Force Relation and the Load Relation

It is possible to write one or more equations describing the variation of the shear force V, as a function of position x over the span of the beam. Substituting V(x) into Eq. (8.5) yields a third order differential equation in terms of the deflection y. By integrating this differential equation three times, we can obtain the equation for the elastic curve. The difficulty with this approach is the number of constants of integration that are generated during the integration process. For a simply supported beam with a single concentrated force, two equations are needed to express the shear force over the length of the beam. Integrating these equations three times produces six constants of integration that can be determined from boundary and compatibility conditions. However, the time and effort required is excessive.

Similarly it is possible to write equations describing the forces q applied to the beam as a function of x. Substituting q(x) into Eq. (8.6) yields a fourth order differential equation in terms of the deflection y. By integrating this differential equation four times, we can obtain the equation for the elastic curve. We encounter the same difficulty with this approach — the number of constants of integration that are generated require an excessive effort to determine. A much better approach for determining the equation for the elastic curve is presented in the next section.

8.4 DEFLECTIONS AND SLOPES WITH SINGULARITY FUNCTIONS

8.4.1 Singularity Functions

In Section 8.3, we found that the integration method was tedious and time consuming, when two or more equations were required to express the bending moment over the entire length of the beam. To avoid this difficulty, we will consider a family of singularity functions that will enable us to express the moment over the length of the beam in a single equation regardless of the complexity of the loading. Macauley first introduced singularity functions in 1919 and they were adapted by the mechanics community in the 1960-70 time frame. To begin, consider a family of singularity functions given by:

$$f_n(x) = \langle x - a \rangle^n \qquad (8.23)$$

When $n \geq 0$ the function f_n is evaluated with the following rule:

$$f_n(x) = 0 \qquad \text{if } x < a$$
$$f_n(x) = (x - a)^n \qquad \text{if } x \geq a \qquad (8.24)$$

Clearly, the pointed brackets are identical to ordinary brackets if the quantity within the pointed brackets is zero or positive. However, if the quantity within the pointed brackets is negative, we set f_n equal to zero. Think of the pointed brackets as a switch that turns off a term, when the quantity within the pointed brackets is negative.

Let's determine $f_n(x)$ if $n = 0$, 1 and 2 as a function of x. The three results are presented as graphs in Fig. 8.6.

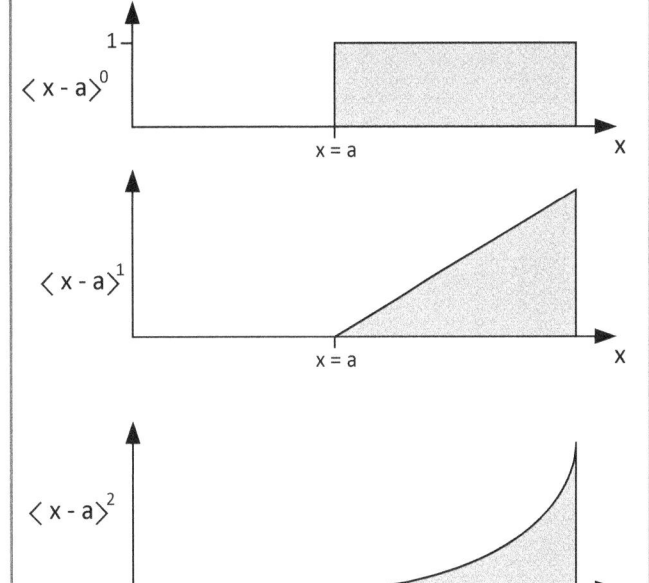

Fig. 8.6 Singularity functions for $n = 0$, 1 and 2.

The function $\langle x - a \rangle^0 = 1$ when $x \geq a$ is called the unit step function, and $\langle x - a \rangle^1$ is called the ramp function, which begins at $x = a$. We integrate these functions using the following rule:

$$\int_{-\infty}^{x} \langle x - a \rangle^n \, dx = \frac{\langle x - a \rangle^{n+1}}{n+1} \qquad n \geq 0 \qquad (8.25)$$

Different rules apply for $f_n(x)$ when n is either -1 or -2.

$$f_n(x) = 0 \quad \text{if } x < a$$
$$f_n(x) = 0 \quad \text{if } x > a \qquad (8.26)$$
$$f_n(x) \Rightarrow \infty \quad \text{if } x = a$$

The singularity in f_n at $x = a$ does not lead to mathematical difficulty, because when $<x-a>^{-1}$ and $<x-a>^{-2}$ are integrated, the result is finite as indicated below:

$$\int_{-\infty}^{x} <x-a>^{-2} dx = <x-a>^{-1} \qquad (8.27)$$

$$\int_{-\infty}^{x} <x-a>^{-1} dx = <x-a>^{0} \qquad (8.28)$$

The function $<x-a>^{-1}$ gives rise to a concentrated load and the function $<x-a>^{-2}$ produces a concentrated moment at $x = a$, as shown in Fig. 8.7.

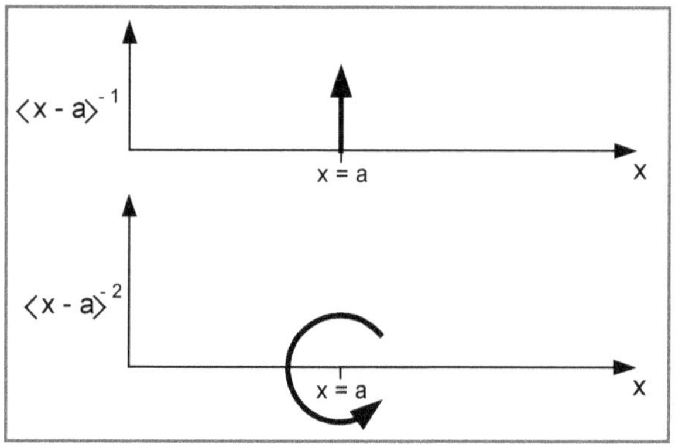

Fig. 8.7 Singularities produced by the function $<x-a>^{-1}$ and $<x-a>^{-2}$ at $x = a$.

Next, let's show the application of the new singularity function by writing equations for the load, shear force and bending moments, as a function of position over the entire length of a beam.

8.4.2 Writing Equations with Singularity Functions

Typical loading $q(x)$ applied to beams includes the four cases listed below. The equation for the load $q(x)$, for each of these different loads, can be written using singularity functions as:

1. A concentrated moment at some location $x = a$.

$$q(x) = M_o <x-a>^{-2} \qquad (8.29)$$

2. A concentrated force F at some location $x = a$.

$$q(x) = F<x-a>^{-1} \qquad (8.30)$$

3. A uniformly distributed load q_o beginning at $x = a$.

$$q(x) = q_o <x-a>^{0} \qquad (8.31)$$

4. A linearly increasing (ramp) load q(x) beginning at x = a.

$$q(x) = (q_o/b)\langle x - a \rangle^1 \tag{8.32}$$

where q_o/b is the slope of the ramp function.

Equations for the shear force V(x) can be written by substituting Eq. (8.29) into Eq. (7.20) and integrating using Eq. (8.27) to obtain:

$$V(x) = -\int_{-\infty}^{x} q(x)dx = -M_o \int_{-\infty}^{x} \langle x-a \rangle^{-2} dx = -M_o \langle x-a \rangle^{-1} \tag{8.33}$$

Similarly, equations for the bending moment can be obtained by substituting V(x) into Eq. (7.18) and integrating using Eq. (8.28) to obtain:

$$M(x) = \int_{-\infty}^{x} V(x)dx = -M_o \int_{-\infty}^{x} \langle x-a \rangle^{-1} dx = -M_o \langle x-a \rangle^{0} \tag{8.34}$$

Using this same procedure, the equations for V(x) and M(x) were derived for each of the four types of beam loading functions. The summary of the results obtained is presented in Fig. 8.8.

BEAM LOADING	LOAD EQUATION	SHEAR FORCE EQ.	MOMENT EQUATION
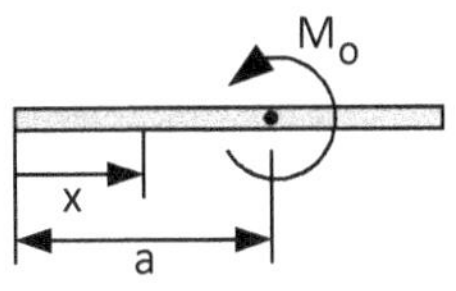 M_o applied moment at a	$q(x) = M_o \langle x-a \rangle^{-2}$	$V(x) = -M_o \langle x-a \rangle^{-1}$	$M(x) = -M_o \langle x-a \rangle^{0}$
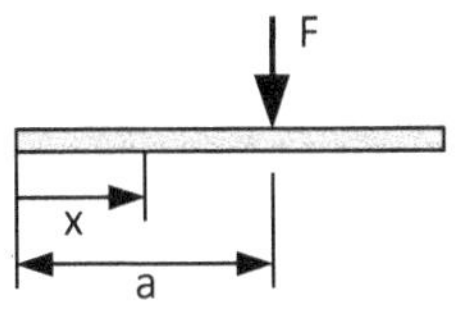 Concentrated force F at a	$q(x) = F \langle x-a \rangle^{-1}$	$V(x) = -F \langle x-a \rangle^{0}$	$M(x) = -F \langle x-a \rangle^{1}$
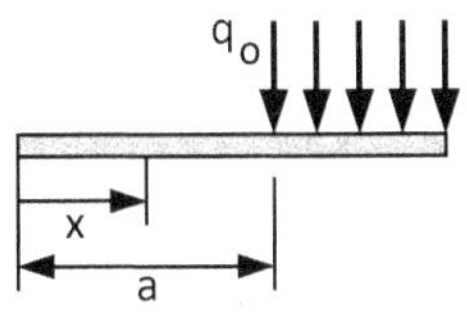 Uniform distributed load q_o from a	$q(x) = q_o \langle x-a \rangle^{0}$	$V(x) = -q_o \langle x-a \rangle^{1}$	$M(x) = -\dfrac{q_o}{2} \langle x-a \rangle^{2}$
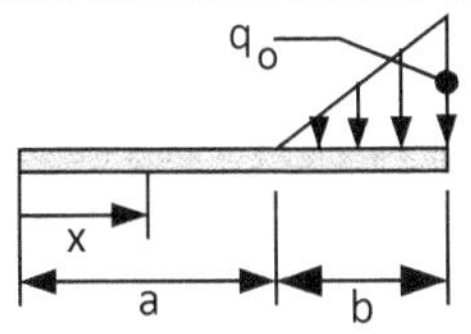 Ramp load q_o beginning at a over span b	$q(x) = \dfrac{q_o}{b} \langle x-a \rangle^{1}$	$V(x) = -\dfrac{q_o}{2b} \langle x-a \rangle^{2}$	$M(x) = -\dfrac{q_o}{6b} \langle x-a \rangle^{3}$

Fig. 8.8 Equations for q(x), V(x) and M(x) for four different types of beam loading.

8.4.3 Applying Singularity Functions to Deflection of Beams

Let's consider five examples to demonstrate the economy of effort in solving for the equation of the elastic curve of a beam, when using singularity functions. The examples will illustrate several concepts including:

1. Writing shear and bending moment equations for different beam loadings using singularity functions.
2. Writing the equations for the elastic curve and its slope for different beam loadings using singularity functions.
3. Solving for constants of integration that are generated in the integration process.
4. Integrating the pointed brackets that are inherent when using singularity functions.

EXAMPLE E8.7

For the simply supported beam shown in Fig. E8.7, write the equation for the loading function q(x). Then integrate this function to determine V(x) and M(x). Solve for the integration constants that arise in the integration process.

Fig. E8.7

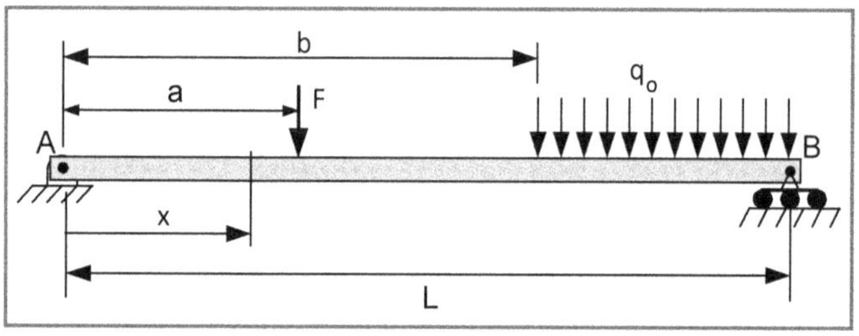

Solution:

Replace the pin and clevis support with a concentrated force R_L directed in the positive y direction. Beginning at $x = 0$, we must accommodate three loadings in writing q(x), R_L, F and q_o. Using the results from Fig. 8.8, we write:

$$q(x) = -R_L\langle x - 0 \rangle^{-1} + F\langle x - a \rangle^{-1} + q_o\langle x - b \rangle^0 \qquad (a)$$

where the sign of the term containing R_L is negative because it is in the positive y direction.

Integrate Eq. (a) using Eq. (7.20) and the integration rules for pointed brackets to obtain the shear force V(x) as:

$$V(x) = R_L\langle x - 0 \rangle^0 - F\langle x - a \rangle^0 - q_o\langle x - b \rangle^1 + C_1 \qquad (b)$$

Integrate Eq. (b) using Eq. (7.18) and the integration rules to obtain the bending moment M(x) as:

$$M(x) = R_L\langle x - 0 \rangle^1 - F\langle x - a \rangle^1 - (q_o/2)\langle x - b \rangle^2 + C_1 x + C_2 \qquad (c)$$

The constants of integration C_1 and C_2 both are zero because $M(0) = 0$ and $V(0) = R_L$. In addition, because R_L occurs at the left end of the beam where $x = 0$, the pointed brackets for the first term in Eqs. (b) and (c) can be omitted. Hence, we write:

$$V(x) = R_L - F\langle x-a\rangle^0 - q_o\langle x-b\rangle^1 \tag{d}$$

$$M(x) = R_L x - F\langle x-a\rangle^1 - (q_o/2)\langle x-b\rangle^2 \tag{e}$$

EXAMPLE E8.8

For the simply supported beam shown in Fig. E8.7, write the equation for the elastic curve and the slope of this curve as a function of x. Locate the position of the maximum deflection and determine y_{Max}. Assume EI_z is a constant with respect to x and the dimensions in Fig. E8.7 are a = 2 m, b = 4 m and L = 6m. The force F = 30 kN and the distributed load q_o = 12 kN/m. The beam is fabricated from a steel American Standard S381×64 section.

Solution: Prepare a FBD of the entire beam and solve for the reaction load R_L at x = 0, by writing a moment equation to obtain:

$$R_L = 2F/3 + 2q_o/6 = (2)(30)/3 + (2)(12)/6 = 24 \text{ kN} \tag{a}$$

Substitute the result for M(x) from Example 8.7 into Eq. (8.4) and integrate to obtain:

$$EI_z \frac{d^2y}{dx^2} = 24x - 30\langle x-2\rangle^1 - 6\langle x-4\rangle^2 \tag{b}$$

$$EI_z \frac{dy}{dx} = 12x^2 - 15\langle x-2\rangle^2 - 2\langle x-4\rangle^3 + C_1 \tag{c}$$

$$EI_z y = 4x^3 - 5\langle x-2\rangle^3 - \frac{1}{2}\langle x-4\rangle^4 + C_1 x + C_2 \tag{d}$$

The constants of integration are determined from the boundary conditions, which are:

$$y(0) = 0 \qquad y(6) = 0 \tag{e}$$

From the boundary condition y(0) = 0 it is evident that C_2 = 0. From the boundary condition y(6) = 0, we determine C_1 from Eq. (d) as:

$$EI_z y = 4(6)^3 - 5(4)^3 - \frac{1}{2}(2)^4 + C_1(6) = 0 \quad \Rightarrow \quad C_1 = -89.33 \text{ kN-m}^2 \tag{f}$$

We can then substitute Eq. (f) into Eqs. (c) and (d) and write the equations for the elastic curve and the slope of the elastic curve as:

$$EI_z \frac{dy}{dx} = 12x^2 - 15\langle x-2\rangle^2 - 2\langle x-4\rangle^3 - 89.33 \tag{g}$$

$$EI_z y = 4x^3 - 5<x-2>^3 - \frac{1}{2}<x-4>^4 - 89.33x \qquad (h)$$

To locate the position of the maximum deflection, we use the fact that (dy/dx) = 0, when y is a maximum. Also, the position of the force F and its magnitude compared to q_o leads us to conclude that the maximum deflection will occur in the region 2 m < x < 4 m. Substituting this information into Eq. (g), we obtain:

$$EI_z \frac{dy}{dx} = 12x^2 - 15(x-2)^2 - 89.33 = 0 \qquad (i)$$

Note that the term $<x-4>^3 = 0$ for $x \leq 4$ m.

Solving the quadratic equation gives the position for the maximum deflection as:

$$x = 2.913 \text{ m} \qquad (j)$$

This result verifies the conclusion that the maximum location[2] would occur from 2 m < x < 4 m. The other root of Eq. (i)—x = 17.087 m—is neglected, because it not realistic as the beam is only 6 m long.

Substituting Eq. (j) into Eq. (h) gives $EI_z y_{Max}$ as:

$$EI_z y_{Max} = \frac{4}{6}(2.913)^3 - 5(0.913)^3 - 0 - (89.33)(2.913) = -165.1 \text{ kN-m}^3 \qquad (k)$$

The negative sign is to be expected because the beam deflects downward, while the y-axis is upward. Let E = 207 GPa for steel and reference Appendix D to find $I_z = 186 \times 10^6$ mm^4. Substituting these values into Eq. (k) yields:

$$y_{Max} = \frac{-(165.1 \times 10^3)(10^3)}{(207 \times 10^3)(186 \times 10^6)} = -4.288 \text{ mm} \qquad (l)$$

This deflection is very small compared to the dimensions of the beam (height h = 381 mm and length L = 6000mm).

EXAMPLE 8.9

A simply supported beam, with a constant cross sectional area over its entire length, is subjected to loads F and q as shown in Fig. E8.9. Determine the equation for the elastic curve and its slope. Also determine the maximum deflection y_{Max}. The dimension in Fig. E8.9 are a = 6 ft and L = 24 ft. The loads are F = 20 kip and q_o = 6 kip/ft. Assume the beam is fabricated from a steel wide flange W16×67 section.

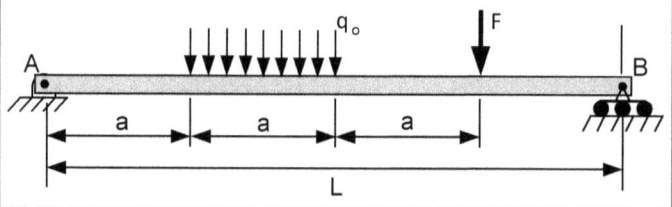

Fig. E8.9

---

[2] The maximum deflection is usually near the center of the span of the beam, even when the loading is not symmetric.

Solution:

As we examine the uniformly distributed load in Fig. E8.9, we are faced with a problem. We can use a singularity function to activate q_o at $x = a$, but how do we turn it off at $x = 2a$? The answer is to introduce a new loading equivalent to the loading shown in Fig. E8.9 that is more amendable to using singularity functions. This new loading involves a downward acting q_o from $a \leq x \leq L$ and an upward acting q_o from $2a \leq x \leq L$. The two distributed loads cancel each other for $x \geq 2a$ producing the loading shown in Fig. E8.9a.

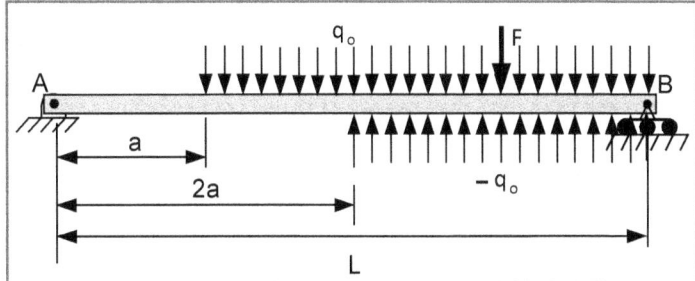

Fig. E8.9

Let's write the equation for the loading function q(x) by using the results in Fig. 8.8 as:

$$q(x) = -R_L <x-0>^{-1} + q_o <x-a>^0 - q_o <x-2a>^0 + F<x-3a>^{-1} \tag{a}$$

where the sign of the term containing R_L is negative because it is in the positive y direction.

Integrate Eq. (a) using Eq. (7.20) and the integration rules for pointed brackets to obtain the shear force V(x) as:

$$V(x) = R_L - q_o <x-a>^1 + q_o <x-2a>^1 - F<x-3a>^0 + C_1 \tag{b}$$

Integrate Eq. (b) using Eq. (7.18) and the integration rules to obtain the bending moment M(x) as:

$$M(x) = R_L x - (q_o/2)<x-a>^2 + (q_o/2)<x-2a>^2 - F<x-3a>^1 + C_1 x + C_2 \tag{c}$$

The constants of integration C_1 and C_2 both are zero, because $M(0) = 0$ and $V(0) = R_L$. Hence, we write:

$$V(x) = R_L - 6<x-6>^1 + 6<x-12>^1 - 20<x-18>^0 \tag{d}$$

$$M(x) = R_L x - 3<x-6>^2 + 3<x-12>^2 - 20<x-18>^1 \tag{e}$$

Prepare a FBD of the entire beam and solve for the reaction load R_L at $x = 0$, by writing a moment equation to obtain:

$$R_L = (q_o/4)(15) + F/4 = (6/4)(15) + 20/4 = 27.5 \text{ kip} \tag{f}$$

Substitute the result for M(x) from Eq. (e) into Eq. (8.4) and integrate to obtain:

$$EI_z \frac{d^2 y}{dx^2} = 27.5x - 3<x-6>^2 + 3<x-12>^2 - 20<x-18>^1 \tag{g}$$

$$EI_z \frac{dy}{dx} = 13.75x^2 - <x-6>^3 + <x-12>^3 - 10<x-18>^2 + C_1 \qquad (h)$$

$$EI_z y = 4.583x^3 - 0.250<x-6>^4 + 0.250<x-12>^4 - 3.333<x-18>^3 + C_1 x + C_2 \qquad (i)$$

The constants of integration are determined from the boundary conditions, which are:

$$y(0) = 0 \qquad\qquad y(24) = 0 \qquad (j)$$

From the boundary condition $y(0) = 0$ it is evident that $C_2 = 0$. From the boundary condition $y(24) = 0$, we determine C_1 as:

$$C_1 = -1,732 \text{ kip-ft}^2 \qquad (k)$$

We can then substitute Eq. (k) into Eqs. (h) and (i) and write the equations for the elastic curve and the slope of the elastic curve as:

$$EI_z \frac{dy}{dx} = 13.75x^2 - <x-6>^3 + <x-12>^3 - 10<x-18>^2 - 1732 \qquad (l)$$

$$EI_z y = 4.583x^3 - 0.250<x-6>^4 + 0.250<x-12>^4 - 3.333<x-18>^3 - 1732x \qquad (m)$$

To locate the position of the maximum deflection we use the fact that $(dy/dx) = 0$, when y is a maximum. Also, the position of the force F and its magnitude compared to q_o leads us to conclude that the maximum deflection will occur in the region 6 ft < x < 12 ft. Substituting this information into Eq. (l), we obtain:

$$EI_z \frac{dy}{dx} = 13.75(x)^2 - (x-6)^3 - 1732 = 0 \qquad (n)$$

Note the terms $<x-12>^3$ and $<x-18>^2$ go to zero for $x \leq 12$ ft.

Solving the cubic equation obtained by expanding Eq. (n) gives the position for the maximum deflection as:

$$x = 11.86 \text{ ft} \qquad (o)$$

The two other roots — x = 25.01 ft and – 5.113 ft — are impossible, because they do not lie on the beam. This result verifies the conclusion that the maximum location would occur slightly to the left of the center of the beam, where 6 ft < x < 12 ft.

Substituting Eq. (o) into Eq. (m) gives $EI_z y_{Max}$ as:

$$EI_z y_{Max} = 4.583(11.86)^3 - 0.250(11.86-6)^4 - (1732)(11.86) = -13,191 \text{ kip-ft}^3 \qquad (p)$$

The negative sign is to be expected, because the beam deflects downward, while the y-axis is upward. Let $E = 30 \times 10^6$ psi for steel and reference Appendix D to find $I_z = 954$ in^4. for the wide flange section. Substituting these values into Eq. (p) yields:

$$y_{Max} = \frac{-(13{,}191)(12^3)}{(30 \times 10^3)(954)} = -0.7964 \text{ in.} \qquad (q)$$

Again it is noted that the magnitude of the deflection of the beam is small compared to its height.

EXAMPLE E8.10

A cantilever beam, with a constant cross sectional area over its entire length, is subjected to a concentrated force F and a ramp load with a maximum distributed force of q_o, as shown in Fig. E8.10. Determine the equation for the elastic curve and its slope. Also determine the maximum deflection y_{Max}. The dimensions are a = 4 m and L = 12 m and the loads are F = 24 kN and q_q = 8 kN/m. Assume the beam is aluminum with a W686 × 217 wide flange section.

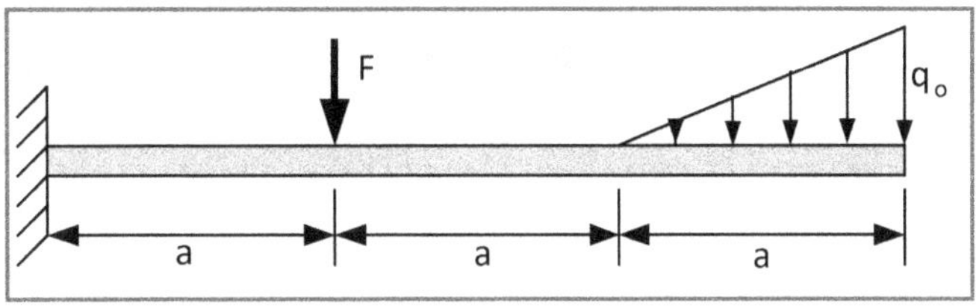

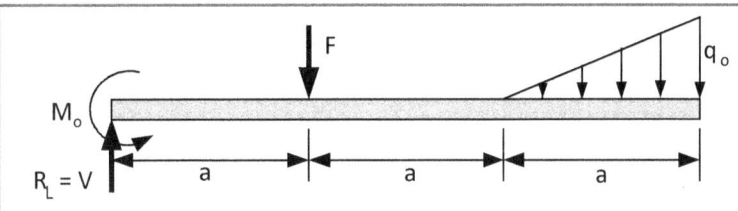

Fig. E8.10

Solution:

Let's begin by drawing a FBD of the entire beam to show the moment and shear force at the built-in support as shown in Fig. E8.10a.

Fig. E8.10a

At the support on the left side of the beam $\Sigma F = 0$ and $\Sigma M = 0$ yields:

$$R_L = V(0) = F + q_o a/2 \qquad (a)$$

$$M_o = Fa + (4q_o a^2)/3 \qquad (b)$$

Let's write the equation for the loading function q(x) by using the results in Fig. 8.4 as:

$$q(x) = M_o \langle x - 0 \rangle^{-2} - R_L \langle x - 0 \rangle^{-1} + F \langle x - a \rangle^{-1} + (q_o/a) \langle x - 2a \rangle^1 \qquad (c)$$

Integrate Eq. (c) using Eq. (7.20) and the integration rules for pointed brackets to obtain the shear force V(x) as:

$$V(x) = -M_o \langle x - 0 \rangle^{-1} + R_L \langle x - 0 \rangle^0 - F \langle x - a \rangle^0 - (q_o/2a) \langle x - 2a \rangle^2 + C_1 \qquad (d)$$

Integrate Eq. (b) using Eq. (7.18) and the integration rules to obtain the bending moment M(x) as:

$$M(x) = -M_o \langle x - 0 \rangle^0 + R_L \langle x - 0 \rangle^1 - F \langle x - a \rangle^1 - (q_o/6a) \langle x - 2a \rangle^3 + C_1 x + C_2 \qquad (e)$$

The constants of integration C_1 and C_2 both are zero because $M(0) = -M_o$ and $V(0) = R_L$. Hence, we write:

$$V(x) = -M_o/x + R_L - F<x-a>^0 - (q_o/2a)<x-2a>^2 \qquad (f)$$

$$M(x) = -M_o + R_Lx - F<x-a>^1 - (q_o/6a)<x-2a>^3 \qquad (g)$$

Substitute numerical values for F and q_o in Eqs. (a) and (b) to obtain:

$$R_L = V(0) = F + q_oa/2 = 24 + (8)(4)/2 = 40 \text{ kN} \qquad (h)$$

$$M_o = Fa + (4q_oa^2)/3 = (24)(4) + (4)(8)(4)^2/3 = 266.7 \text{ kN-m} \qquad (i)$$

Substitute the result for M(x) from Eq. (g) into Eq. (8.4) and simplify using Eqs. (h) and (i). Then integrate the result to obtain:

$$EI_z \frac{d^2y}{dx^2} = -266.7 + 40x - 24<x-4>^1 - \frac{1}{3}<x-8>^3 \qquad (j)$$

$$EI_z \frac{dy}{dx} = -266.7x + 20x^2 - 12<x-4>^2 - \frac{1}{12}<x-8>^4 + C_1 \qquad (k)$$

$$EI_z y = -133.4x^2 + 6.667x^3 - 4<x-4>^3 - \frac{1}{60}<x-8>^5 + C_1x + C_2 \qquad (l)$$

The constants of integration are determined from the boundary conditions, which are:

$$y(0) = 0 \qquad (dy/dx)(0) = 0 \qquad (m)$$

From the boundary condition $y(0) = 0$, it is evident that $C_2 = 0$ and from the boundary condition $(dy/dx)(0) = 0$, we determine $C_1 = 0$:

We can set $C_1 = C_2 = 0$ in Eqs. (l) and (k) and write the equations for the elastic curve and the slope of the elastic curve as:

$$EI_z \frac{dy}{dx} = -266.7x + 20x^2 - 12<x-4>^2 - \frac{1}{12}<x-8>^4 \qquad (n)$$

$$EI_z y = -133.4x^2 + 6.667x^3 - 4<x-4>^3 - \frac{1}{60}<x-8>^5 \qquad (o)$$

It is evident from inspection that the maximum deflection of the cantilever beam is at its free end, where x = L = 12 m. Accordingly, we write:

$$EI_z y_{Max} = -133.4(12)^2 + 6.667(12)^3 - 4(8)^3 - \frac{1}{60}(4)^5 = -9,754 \text{ kN-m}^3 \qquad (p)$$

Let E = 72 GPa for aluminum and reference Appendix D to find $I_z = 2{,}345 \times 10^6$ mm^4 for the wide flange section. Substituting these values into Eq. (p) yields:

$$y_{Max} = \frac{-(9{,}754 \times 10^3)(10^3)^3}{(72 \times 10^3)(2345 \times 10^6)} = -57.77 \text{ mm} \tag{q}$$

EXAMPLE E8.11

A simply supported beam with a constant cross sectional area over its entire length is designed with an overhang, as shown in Fig. E8.11. The beam carries a uniformly distributed load q_o and a concentrated force F. Determine the equation for the elastic curve and its slope. Also determine the maximum deflection y_{Max}. The dimensions are a = 5 ft and L = 20 ft and the loads are F = 15 kip and q_q = 1.2 kip/ft. Assume an aluminum beam fabricated from a 12 × 52 structural tee section.

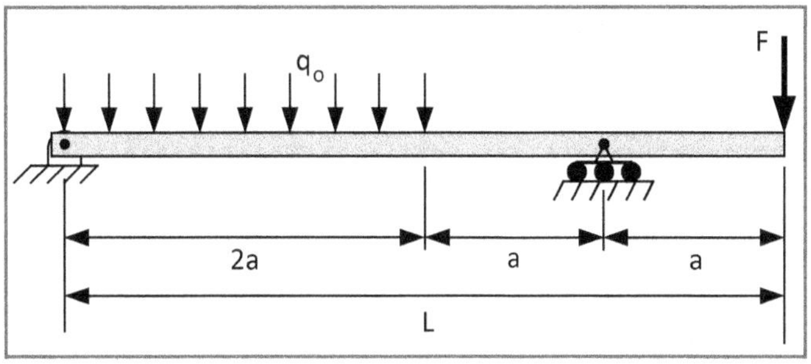

Fig. E8.11

Solution:

Let's begin by determining the reactions R_L and R_R at the two supports. Prepare a FBD and take moments about the right hand support to obtain:

$$R_L = [4aq_o - F]/3 = [(4)(5)(1.2) - 15]/3 = 3 \text{ kip} \tag{a}$$

$$R_R = 2aq_o + F - R_L = 12 + 15 - 3 = 24 \text{ kip} \tag{b}$$

When we attempt to write the equation for the loading function q(x) by using the results in Fig. 8.4, we note a difficulty. If we accommodate the uniform load with the term $q_o <x - 0>^0$ it applies that loading over the entire length of the beam. To remove the uniform loading from the right half of the beam, we must introduce an equal and opposite loading that begins at x = 2a and continues to the right side of the beam, as shown in Fig. E8.11a.

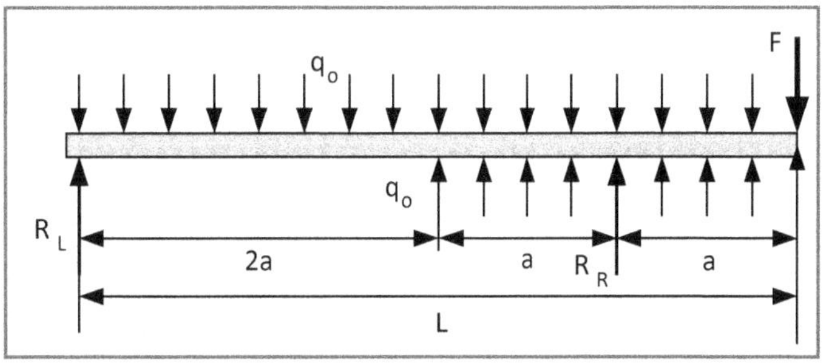

Fig. E8.11a

Writing the relation for the loading shown in Fig. E8.11a yields:

$$q(x) = -R_L <x - 0>^{-1} + q_o <x - 0>^0 - q_o <x - 2a>^0 - R_R <x - 3a>^{-1} \tag{c}$$

Integrate Eq. (c) using Eqs. (7.20) and the integration rules for pointed brackets to obtain the shear force V(x). Also substitute numerical values for a, F, q_o, R_L and R_R into Eq. (c).

$$V(x) = +3 <x-0>^0 - 1.2 <x-0>^1 + 1.2 <x-10>^1 + 24 <x-15>^0 + C_1 \quad (d)$$

Integrate Eq. (d) using Eqs. (7.18) and the integration rules to obtain the bending moment M(x) as:

$$M(x) = +3 <x-0>^1 - 0.6 <x-0>^2 + 0.6 <x-10>^2 + 24 <x-15>^1 + C_1 x + C_2 \quad (e)$$

The constants of integration C_1 and C_2 both are zero because $M(0) = 0$ and $V(0) = R_L$. Hence, we write:

$$V(x) = +3 - 1.2x + 1.2 <x-10>^1 + 24 <x-15>^0 \quad (f)$$

$$M(x) = +3x - 0.6x^2 + 0.6 <x-10>^2 + 24 <x-15>^1 \quad (g)$$

Substitute the result for M(x) from Eq. (g) into Eq. (8.4) and then integrate to obtain:

$$EI_z \frac{d^2y}{dx^2} = 3x - 0.6x^2 + 0.6 <x-10>^2 + 24 <x-15>^1 \quad (h)$$

$$EI_z \frac{dy}{dx} = 1.5x^2 - 0.2x^3 + 0.2 <x-10>^3 + 12 <x-15>^2 + C_1 \quad (i)$$

$$EI_z y = 0.5x^3 - 0.05x^4 + 0.05 <x-10>^4 + 4 <x-15>^3 + C_1 x + C_2 \quad (j)$$

To solve for C_1 and C_2, we use the boundary conditions that $y(0) = 0$ and $y(15) = 0$. The first of these boundary conditions yields $C_2 = 0$. The second boundary condition gives:

$$0 = 0.5(15)^3 - 0.05(15)^4 + 0.05(5)^4 + 4(0) + C_1(15) \Rightarrow \quad C_1 = 54.16 \text{ kip-ft}^2 \quad (k)$$

We can then substitute Eq. (k) into Eqs. (i) and (j) and write the equations for the elastic curve and the slope of the elastic curve as:

$$EI_z \frac{dy}{dx} = 1.5x^2 - 0.2x^3 + 0.2 <x-10>^3 + 12 <x-15>^2 + 54.16 \quad (l)$$

$$EI_z y = 0.5x^3 - 0.05x^4 + 0.05 <x-10>^4 + 4 <x-15>^3 + 54.16x \quad (m)$$

There are two possibilities for the location of the maximum deflection of the beam — between the supports or at the free end of the overhanging beam. If y_{Max} occurs between the supports, then $(dy/dx) = 0$ when y is a maximum. Let's first assume that the maximum will be near the center of the beam with $x \leq 15$ so that we can write Eq. (l) as:

$$EI_z \frac{dy}{dx} = 1.5x^2 - 0.2x^3 + 0.2(x-10)^3 + (0) + 54.16 = 0 \quad (n)$$

Expanding and solving Eq. (n) gives the location of y_{Max} as:

$$x = 10.14 \text{ ft} \quad \text{or} \quad 3.196 \text{ ft} \qquad (o)$$

Both roots of Eq. (n) must be checked to determine the one which is correct. Substituting Eq. (o) into Eq. (m) gives $EI_z y_{Max}$ as:

$$EI_z y_{Max} = 0.5(10.14)^3 - 0.05(10.14)^4 + 0.05(0.14)^4 + 0 + 54.16(10.14) = 541.9 \text{ kip-ft}^3 \qquad (p)$$

$$EI_z y_{Max} = 0.5(3.196)^3 - 0.05(3.196)^4 + 0 + 0 + 54.16(3.196) = 184.2 \text{ kip-ft}^3 \qquad (q)$$

The larger of the two values is given by Eq. (p) as 541.9 kip-ft^3.

In this problem the maximum deflection is positive indicating the beam deflects in the positive y direction with a deflection curve of the shape shown in Fig. E8.11a. The result given by Eq. (p) is the maximum deflection between the supports, but it may not be the maximum for the beam. It is necessary to check the deflection at the free end of the beam.

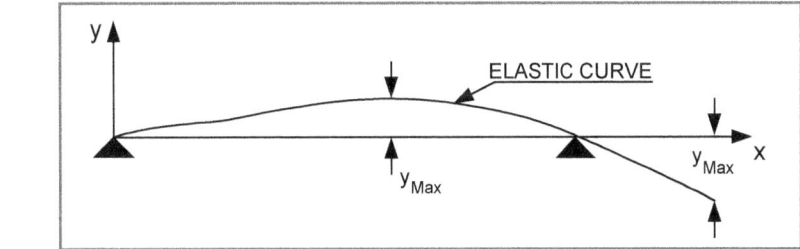

Fig. E8.11b

To determine the deflection at the end of the beam, where x = 20 ft, we employ Eq. (m), set x = 20 ft and write:

$$EI_z y_{Max} = 0.5(20)^3 - 0.05(20)^4 + 0.05(10)^4 + 4(5)^3 + 54.16(20) = -1917 \text{ kip-ft}^3 \qquad (r)$$

Comparing the results from Eq. (p) and (r) indicates that the maximum deflection occurs at the free end of the overhanging beam and is given by Eq. (r).

Let $E = 10.4 \times 10^3$ ksi for aluminum and reference Appendix D to find $I_z = 189$ in.^{4} for the structural tee section. Substituting these values into Eq. (r) yields:

$$y_{Max} = \frac{-(1917)(12)^3}{(10.4 \times 10^3)(189)} = -1.685 \text{ in.} \qquad (s)$$

8.5 <u>SUPERPOSITION CONCEPTS</u>

The principle of superposition is useful in solving for the deflection of beams that are loaded with two or more forcing functions. We have presented the integration method that is effective, when the bending moment equation M(x) is valid from $0 \leq x \leq L$. This method becomes cumbersome when two or more bending moment equations are necessary to describe M(x) over the entire length of the beam. We introduced singularity functions that were employed with the integration method to significantly reduce the mathematical complexities involved in solving for constants of integration. Superposition represents a third method for determining the equation of the elastic curve.

Superposition involves combining together the results of the deflection from two or more separate solutions to yield a new solution containing all of the loading functions involved in the separate solutions. This concept of superposition of displacement functions is demonstrated better by the graphic shown in Fig. 8.9. Three concentrated forces — F1, F2 and F3 — load the beam, shown on the left in Fig. 8.9. These forces are located a distance a, b and c from the left side of the beam, respectively. If the solution is known for the elastic curve for a beam loaded with a single concentrated force, we can use superposition to add together the solutions for the displacement y(x) associated with the three separate forces.

One of the advantages of the method of superposition is that we make use of the many solutions that already exist. Mechanics of materials is a subject with a long history, and the classic problems of bending of beams were solved many years ago. We have provided a catalog of several of these solutions for your convenience in Appendix E.

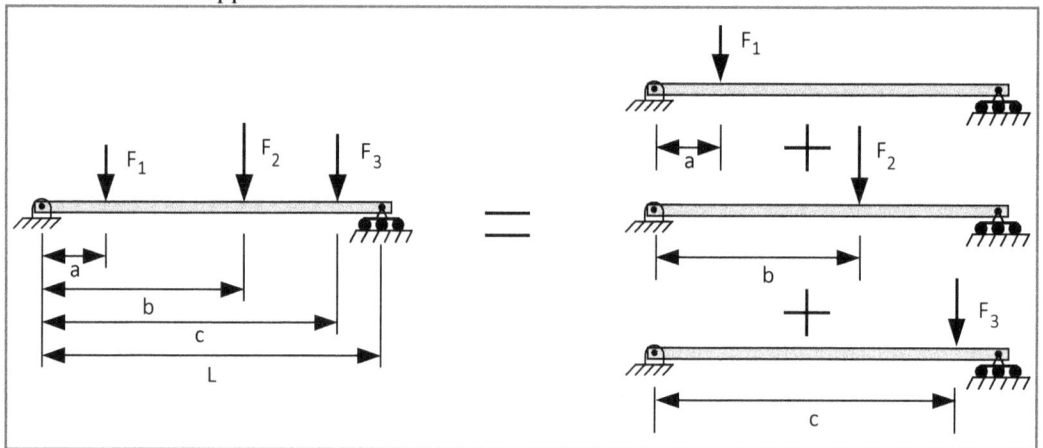

Fig. 8.9 Superposition is adding solutions for y(x) for a single loading function to obtain solutions for two or more loading functions.

There are two requirements that must be satisfied before using the principle of superposition. The first is the linearity requirement. The solution for y(x) must be linear with respect to the loading functions. We have solved several different problems determining y(x) for several different loading functions including concentrated forces, concentrated moments, uniformly distributed loads and ramp functions. In all cases we found that y(x) was linear with respect to the loading function. The second requirement pertains to the integrity of the geometry of the beam during the loading. Does the beam change its dimensions? The beam on the left in Fig. 8.9 must have essentially the same geometry as each of the three beams on the right of Fig. 8.9. If the beam deformations are small, we can conclude that the geometry does not change significantly as the beam is loaded. We have solved for the maximum deflection in several problems and noted that the deflection of the beam was always small when compared to its length. Hence, we conclude that the principle of superposition is valid and can be employed as another method for determining the deflection of a beam in bending.

We will demonstrate the methods of superposition with several examples that show combinations of different loading functions applied to beams supported in different ways.

EXAMPLE E8.12

Consider the simply supported beam subjected to a uniformly distributed load and a concentrated force, as illustrated in Fig. E8.12. Determine the deflection at mid span and the slope at x = L. The beam is fabricated from a steel wide flange section W203×60 with a length L = 6 m. The force F = 50 kN is applied at a = 4 m and q_o = 10 kN/m = 10 N/mm.

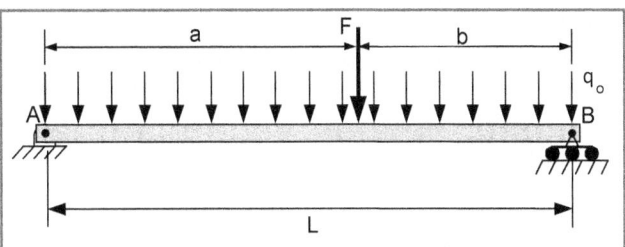

Fig. E8.12

Solution:

Let's employ the principle of superposition to solve this problem. Two identical simply supported beams each with a single load, as illustrated in Fig. E8.12a, represent the simply supported beam with the two different loadings.

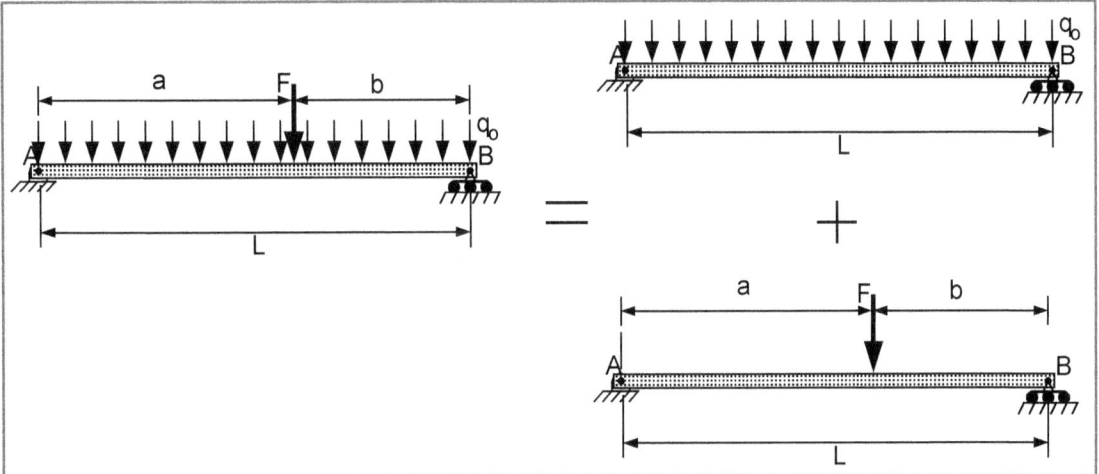

Fig. E8.12a

Consider first the simply supported beam with the uniformly distributed load. Reference to the solutions presented in Appendix E gives the equations for the deflection and slope as:

$$y = -\frac{q_0 x}{24EI_z}\left(x^3 - 2Lx^2 + L^3\right) \qquad (a)$$

$$\left(\frac{dy}{dx}\right)_{x=0,L} = \pm\frac{q_0 L^3}{24EI_z} \qquad (b)$$

Note that $I_z = 60.8 \times 10^6$ mm^4 is given in Appendix D for the specified wide flange section. Substituting numerical parameters into Eq. (a) and Eq. (b) gives

$$y_{x=L/2} = -\frac{q_0 L}{48EI_z}\left(\frac{L^3}{8} - \frac{L^3}{2} + L^3\right) = -\frac{5q_0 L^4}{384EI_z} \qquad (c)$$

$$y_{x=L/2} = -\frac{5q_0 L^4}{384EI_z} = -\frac{(5)(10)(6\times 10^3)^4}{(384)(207\times 10^3)(60.8\times 10^6)} = -13.41 \text{ mm} \qquad (d)$$

$$\left(\frac{dy}{dx}\right)_{x=L} = \frac{q_0 L^3}{24 EI_z} = \frac{(10)(6 \times 10^3)^3}{(24)(207 \times 10^3)(60.8 \times 10^6)} = 7.150 \times 10^{-3} \text{ Radians} \quad (e)$$

Next, reference the solution for the simply supported beam subjected to a concentrated load at position x= a in Appendix E and write the relations for the displacement and the slope as:

$$y = -\frac{Fbx}{6EI_z L}\left(L^2 - b^2 - x^2\right) \quad (f)$$

$$\left(\frac{dy}{dx}\right)_{x=L} = +\frac{Fab(L+a)}{6EI_z L} \quad (g)$$

Substituting x = L/2 and numerical parameters into Eq. (f) and solving gives:

$$y_{x=L/2} = -\frac{Fb}{48 EI_z}\left(3L^2 - 4b^2\right) = -\frac{(50 \times 10^3)(2 \times 10^3)}{(48)(207 \times 10^3)(60.8 \times 10^6)}\left[3(36) - 4(4)\right] \times 10^6 = -15.23 \text{ mm} \quad (h)$$

Substituting numerical parameters into Eq. (g) gives:

$$\left(\frac{dy}{dx}\right)_{x=L} = \frac{Fab(L+a)}{6EI_z L} = \frac{(50 \times 10^3)(4 \times 10^3)(2 \times 10^3)(10 \times 10^3)}{6(207 \times 10^3)(60.8 \times 10^6)(6 \times 10^3)} = 8.828 \times 10^{-3} \text{ Radians} \quad (i)$$

The final step in this process is to superimpose the results from the two solutions. Adding the results from Eqs. (d) and (h) gives the total displacement at mid span as:

$$y_{x=L/2} = -(13.41 + 15.23) = -28.64 \text{ mm} \quad (j)$$

Similarly, the slope at x = L is given by adding the results from Eqs. (e) and (i) to obtain:

$$(dy/dx)_{x=L} = (7.150 + 8.828) \times 10^{-3} = 0.01598 \text{ Radians} = 0.9155° \quad (k)$$

The use of the superposition principle enables one to solve for deflections in beams subjected to multiple loads by considering identical beams subjected to a single load function. The solutions for the beams with single loadings are then added to obtain the results for the beam with multiple loads. It is a simple technique that makes use of the existing solutions that are tabulated in Appendix E.

EXAMPLE E8.13

Consider the simply supported beam subjected to a ramp type distributed load and a concentrated moment, as illustrated in Fig. E8.13. Determine the deflection at midspan and the slope at both ends of the beam. The steel beam is fabricated from an American Standard flange section S10×35 with a length L = 24 ft. The concentrated moment M_o = 2,400 ft-lb is applied at x = 0. The ramp load is specified with q_o = 1.44 kip/ft = 120 lb/in.

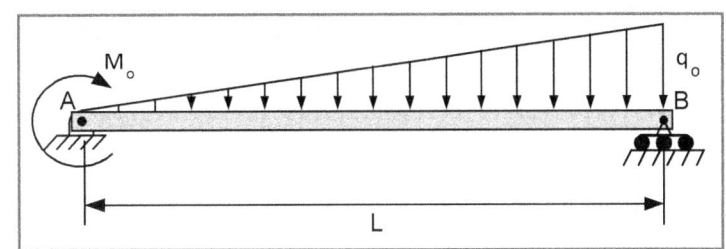

Fig. E8.13

Solution:

We again use the principle of superposition in solving this problem. Two identical simply supported beams each with a single load, as illustrated in Fig. E8.13a, represent the simply supported beam with the ramp distribution and the concentrated moment loads.

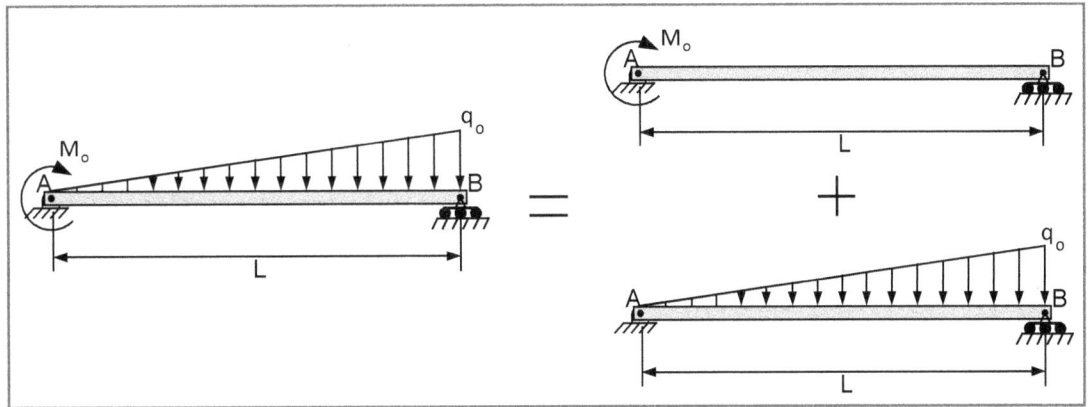

Fig. E8.13a

First, consider the solution for the simply supported beam subjected to a concentrated moment at x = 0 that is given in Appendix E. Write the relations for the displacement and the slope as:

$$y = -\frac{M_0 x}{6EI_z L}\left(x^2 - 3Lx + 2L^2\right) \qquad (a)$$

$$\left(\frac{dy}{dx}\right)_{x=0} = -\frac{ML}{3EI_z} \qquad (b)$$

$$\left(\frac{dy}{dx}\right)_{x=L} = \frac{ML}{6EI_z} \qquad (c)$$

At midspan x = L/2 and Eq. (a) reduces to:

$$y = -\frac{3M_0 L^2}{48EI_z} \qquad (d)$$

Note that I_z = 147 in.^{4} is given in Appendix D for the specified American Standard section. Substituting numerical parameters into Eqs. (b), (c) and (d) gives:

$$y = -\frac{3M_0L^2}{48EI_z} = -\frac{3(2400)(12)(24)^2(12)^2}{48(30\times10^6)(147)} = -0.03385 \text{ in.} \quad (e)$$

$$\left(\frac{dy}{dx}\right)_{x=0} = -\frac{ML}{3EI_z} = -\frac{(2400)(12)(24)(12)}{3(30\times10^6)(147)} = -6.269\times10^{-4} \text{ Radians} \quad (f)$$

$$\left(\frac{dy}{dx}\right)_{x=L} = \frac{ML}{6EI_z} = 3.135\times10^{-4} \text{ Radians} \quad (g)$$

Consider next the simply supported beam with the ramp type distributed load. Reference to the solutions presented in Appendix E gives the equations for the deflection and slope as:

$$y = -\frac{q_0 x}{360EI_z L}\left(3x^4 - 10L^2x^2 + 7L^4\right) \quad (h)$$

$$\left(\frac{dy}{dx}\right)_{x=0} = -\frac{7q_0L^3}{360EI_z} \quad (i)$$

$$\left(\frac{dy}{dx}\right)_{x=L} = \frac{q_0L^3}{45EI_z} \quad (j)$$

Substitute $x = L/2$ into Eq. (h) to obtain:

$$y_{x=L/2} = -\frac{q_0}{(2)(360)EI_z}\left(\frac{3L^4}{16} - \frac{10L^4}{4} + 7L^4\right) = -\frac{75q_0L^4}{11,520EI_z} \quad (k)$$

Substituting numerical parameters into Eqs. (i), (j) and (k) gives:

$$y_{x=L/2} = -\frac{75q_0L^4}{11,520EI_z} = -\frac{(75)(120)(24)^4(12)^4}{11,520(30\times10^6)(147)} = -1.219 \text{ in.} \quad (l)$$

$$\left(\frac{dy}{dx}\right)_{x=0} = -\frac{7q_0L^3}{360EI_z} = -\frac{7(120)(24)^3(12)^3}{360(30\times10^6)(147)} = -0.01264 \text{ Radians} \quad (m)$$

$$\left(\frac{dy}{dx}\right)_{x=L} = \frac{q_0L^3}{45EI_z} = \frac{(120)(24)^3(12)^3}{45(30\times10^6)(147)} = 0.01444 \text{ Radians} \quad (n)$$

The final step in this process is to superimpose the results from the two solutions. Adding the results from Eqs. (e) and (l) gives the total displacement at mid span as:

$$y_{x=L/2} = -(0.03385 + 1.219) = -1.253 \text{ in.} \quad (o)$$

Similarly the slope at $x = 0$ and $x = L$ is obtained by adding the results to obtain:

$$(dy/dx)_{x=0} = -(6.269\times10^{-4} + 0.01264) = -0.01327 \text{ Radian} = 0.7601° \quad (p)$$

$$(dy/dx)_{x=L} = (0.0003135 + 0.01444) = 0.01475 \text{ Radian} = 0.8453° \quad (q)$$

The use of the superposition principle enables one to solve for deflections in beams subjected to multiple loads by considering identical beams subjected to a single load function. The solutions for the beams with single loadings are then added to obtain the results for the beam with multiple loads.

EXAMPLE E8.14

Consider the cantilever beam subjected to a concentrated force applied at its free end and a uniformly distributed load over the left half of the beam as shown in Fig. E8.14. Determine the deflection and the slope at the free end of the beam. The steel beam is fabricated from a structural tee section WT178 × 61 with a length L = 5.0 m. The concentrated force is F = 2 kN and the uniformly distributed load is specified with q_o = 2.0 kN/m = 2.0 N/mm.

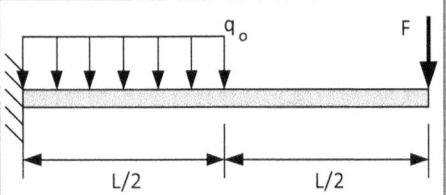

Fig. E8.14

Solution:

Apply the principle of superposition by dividing the loading into two separate parts as shown in Fig. E8.14a.

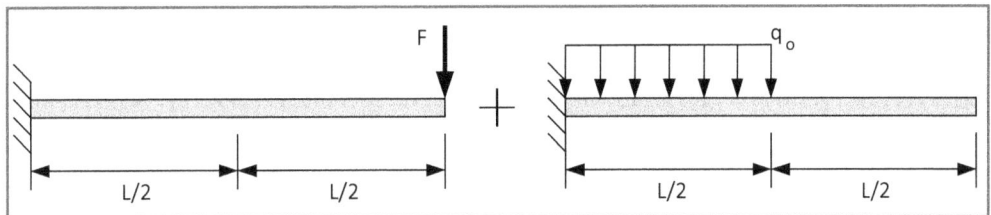

Fig. E8.14a

Consider first the solution for the cantilever beam, with a concentrated load applied at its free end that is presented in Appendix E. We write the expressions for the deflection and the slope at the free end and substitute numerical parameters to obtain:

$$y_{x=L} = -\frac{FL^3}{3EI_z} = -\frac{2000(5000)^3}{3(207 \times 10^3)(17.1 \times 10^6)} = -23.54 \text{ mm} \quad (a)$$

$$\left(\frac{dy}{dx}\right)_{x=L} = -\frac{FL^2}{2EI_z} = -\frac{2000(5000)^2}{2(207 \times 10^3)(17.1 \times 10^6)} = -7.063 \times 10^{-3} \text{ Radians} \quad (b)$$

The value of $I_z = 17.1 \times 10^6$ mm^4 was determined from Appendix D.

Next consider the solution for the cantilever beam with a uniformly distributed load applied over its left half. Using the solution presented in Appendix E, we write the expressions for the deflection and the slope at the free end and substitute numerical parameters to obtain:

$$y_{x=L} = -\frac{7q_0L^4}{384EI_z} = -\frac{7(2)(5000)^4}{384(207\times10^3)(17.1\times10^6)} = -6.437 \text{ mm} \qquad (c)$$

$$\left(\frac{dy}{dx}\right)_{x=L} = -\frac{q_0L^3}{48EI_z} = -\frac{2(5000)^3}{48(207\times10^3)(17.1\times10^6)} = -1.471\times10^{-3} \text{ Radians} \qquad (d)$$

Finally we add the results for the deflections and the slopes to obtain:

$$y_{x=L} = -(23.54 + 6.437) - 29.98 \text{ mm} \qquad (e)$$

$$\left(\frac{dy}{dx}\right)_{x=L} = -(7.063 + 1.471)\times10^{-3} = 0.008534 \text{ Radian} = 0.4890° \qquad (f)$$

8.6 STATICALLY INDETERMINATE BEAMS

In most instances beams are statically determinate. The reactions at the supports may be determined from the three applicable equations of equilibrium — $\Sigma F_x = 0$, $\Sigma F_y = 0$ and $\Sigma M_o = 0$. However, in some cases beams are designed with additional supports that produce additional reaction forces making the beam statically indeterminate. Two common examples of statically indeterminate beams are illustrated in Fig. 8.10 — the propped cantilever beam and the continuously supported beam.

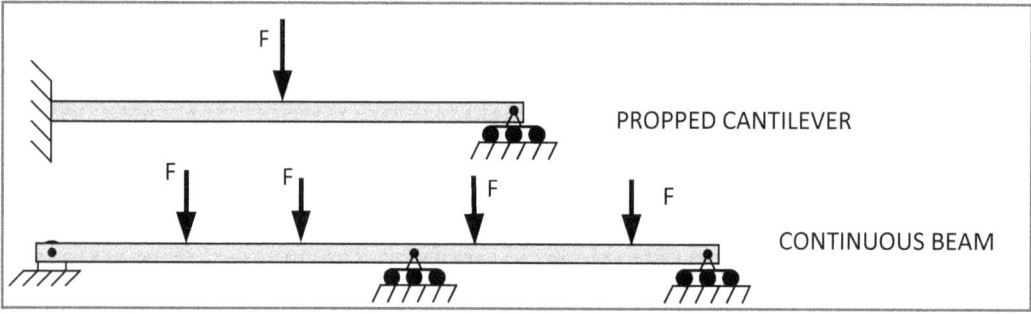

Fig. 8.10 Two examples of statically indeterminate beams.

The propped cantilever has one reaction force R_{By} at the prop and three possible reaction forces R_{Ax}, R_{Ay} and M_A at the built-in end. Three possible equations and four unknowns, yield a statically indeterminate beam. For the continuous beam, two reaction forces are possible at the clevis and one at each of the roller supports at locations B and C. Clearly, the continuous beam is statically indeterminate.

There are several methods to solve for the additional reaction force; however, we supplement the equilibrium relations with a deformation equation to enable a solution. We will demonstrate two different methods to determine the reaction forces for statically indeterminate beams including the method of superposition and the method of integration with singularity functions.

8.6.1 Superposition Methods

In applying the method of superposition to the solution of statically indeterminate beams, we supplement the equilibrium equations by writing additional equations based on the deformation of the beam. This is accomplished by representing the statically indeterminate beam with two beams, each of which is statically determinate. The deflections of these two beams are added to yield the boundary conditions imposed on the statically indeterminate beam. This process yields the reaction force at one of the supports. Let's consider the propped cantilever beam, as an example to demonstrate this approach.

EXAMPLE E8.15

Determine the reaction force at the prop of the cantilever beam shown in Fig. E8.15. The force F = 3,000 lb. The steel beam, 18 ft long, is fabricated from a wide flange section specified as W6 × 25. After solving for the reaction force, solve for the deflection at mid span.

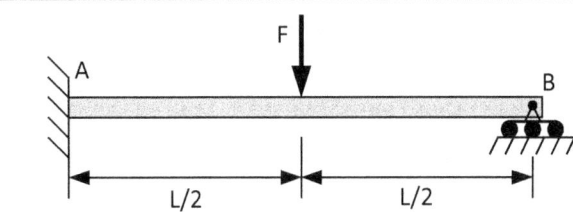

Fig. E8.15

Solution:

The prop at B causes the cantilever beam to be statically indeterminate. It may have been added to reduce the deflection or the stress of the beam; however, the prop complicates the analysis.

We begin the analysis by replacing the prop with a reaction force R and noting the boundary condition at x = L as:

$$y(L) = 0 \quad \text{(a)}$$

We then apply the principle of superposition and represent the beam in Fig. E8.15 with the two statically determinate beams shown in Fig. E8.15a.

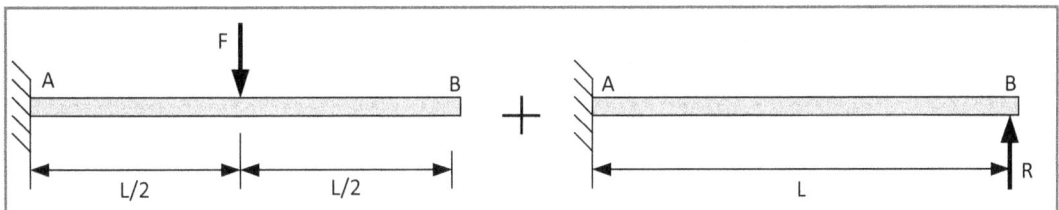

Fig. E8.15a

Let's determine the deflection at the free end for each of the two beams using the solutions tabulated in Appendix E. For the cantilever beam with the force F applied at midspan, we write:

$$y_{x=L} = -\frac{5FL^3}{48EI_z} \quad \text{(b)}$$

For the cantilever beam with the upward directed reaction force, we write:

$$y_{x=L} = \frac{RL^3}{3EI_z} \quad \text{(c)}$$

However, the end of the cantilever beam is propped and y(L) = 0. Hence, it is clear from Eqs. (b) and (c) that:

$$y_{x=L} = \frac{RL^3}{3EI_z} - \frac{5FL^3}{48EI_z} = 0 \quad \text{(d)}$$

Solving Eq. (d) and inserting numerical parameters gives:

$$R = 5F/16 = 5(3000)/16 = 937.5 \text{ lb} \qquad (e)$$

It is interesting that the partitioning of the force F between the built-in support and the prop is independent of the length of the beam and its stiffness (EI_z).

Let's continue the analysis using the result for the reaction force R and the method of superposition to determine the deflection at mid span. From Appendix E, we write the expressions for the deflection at $x = L/2$ for the two beams illustrated in Fig. 8.15a as:

$$y_{x=L/2} = -\frac{FL^3}{24EI_z} \qquad (f)$$

$$y_{x=L/2} = \frac{5RL^3}{48EI_z} \qquad (g)$$

Substituting Eq. (e) into Eq. (g) yields:

$$y_{x=L/2} = \frac{25FL^3}{768EI_z} \qquad (h)$$

Superimposing the results from Eqs. (f) and (h) and substituting numerical parameter yields:

$$y_{x=L/2} = \frac{FL^3}{EI_z}\left(\frac{25}{768} - \frac{1}{24}\right) = -\frac{7FL^3}{768EI_z} = -\frac{7(3000)(18)^3(12)^3}{768(30\times10^6)(53.4)} = -0.1720 \text{ in.} \qquad (i)$$

In this example, we employed the principle of superposition twice — first to find the reaction force and then to determine the mid span deflection. In both applications, the solutions tabulated in Appendix E greatly reduced the computational effort required.

8.6.2 Integration Method with Singularity Functions

Singularity functions may also be applied to establish the deformation equation needed to solve for the additional reaction force in statically indeterminate beams. With this approach, we remove one of the supports (usually a roller) and replace the support with an unknown reaction force R. This substitution enables us to write expressions for $q(x)$, $V(x)$ and $M(x)$ over the entire length of the beam. We then integrate twice the $M(x)$ relation to obtain $y(x)$. Of course, the equation $y(x)$ contains the unknown force R and two integration constants. These three unknowns are determined from three boundary conditions.

Let's consider an example of a continuous beam to demonstrate the process followed with this approach.

EXAMPLE E8.16

Determine the reaction force at the center support of the continuous beam shown in Fig. E8.16. The forces are both equal to 12 kN. The steel beam is 2L long, where L = 6 m and a = b = 3 m. The beam is fabricated from a wide flange section specified as W203 × 36. After solving for the reaction force solve for the deflection at mid span (x = L/2).

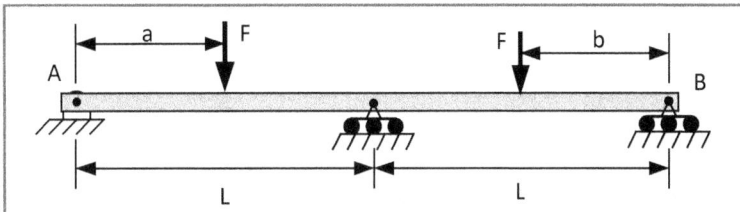

Fig. E8.16

Solution:

Let's remove the support and replace it with a reaction force R directed upwards, as illustrated in Fig. E8.16a.

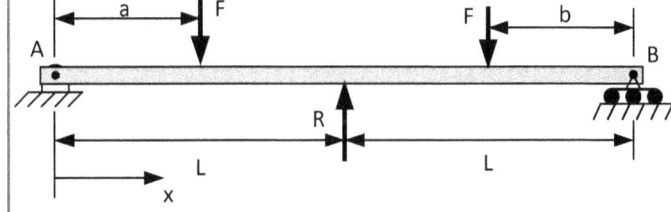

Fig. E8.16a

Next write the expression for q(x) with the singularity functions described in Fig. 8.8.

$$q(x) = -R_A \langle x-0 \rangle^{-1} + F\langle x-a \rangle^{-1} - R\langle x-L \rangle^{-1} + F\langle x-3a \rangle^{-1} \quad (a)$$

where R_A is the reaction force at the left side support.

Integrate Eq. (a) to obtain the equation for the shear force V(x) as:

$$V(x) = R_A \langle x-0 \rangle^0 - F\langle x-a \rangle^0 + R\langle x-L \rangle^0 - F\langle x-3a \rangle^0 + C_1 \quad (b)$$

Integrate Eq. (b) to obtain the equation for the bending moment M(x) as:

$$M(x) = R_A \langle x-0 \rangle^1 - F\langle x-a \rangle^1 + R\langle x-L \rangle^1 - F\langle x-3a \rangle^1 + C_1 x + C_2 \quad (c)$$

The constants of integration C_1 and C_2 both are zero because M(0) = 0 and V(0) = R_A. Hence, we write:

$$M(x) = R_A x - F\langle x-a \rangle^1 + R\langle x-L \rangle^1 - F\langle x-3a \rangle^1 \quad (d)$$

Substitute Eq. (d) into Eq. (8.4) and integrate to obtain:

$$EI_z \frac{d^2y}{dx^2} = R_A x - F\langle x-a \rangle^1 + R\langle x-L \rangle^1 - F\langle x-3a \rangle^1 \quad (e)$$

$$EI_z \frac{dy}{dx} = \frac{R_A}{2} x^2 - \frac{F}{2}\langle x-a \rangle^2 + \frac{R}{2}\langle x-L \rangle^2 - \frac{F}{2}\langle x-3a \rangle^2 + C_1 \quad (f)$$

$$EI_z y = \frac{R_A}{6}x^3 - \frac{F}{6}<x-a>^3 + \frac{R}{6}<x-L>^3 - \frac{F}{6}<x-3a>^3 + C_1 x + C_2 \qquad (g)$$

Examination of Eq. (g) indicates that we have four unknowns — R_A, R, C_1 and C_2. Boundary conditions for the deflection y listed below enable us to write three independent equations.

$$y(0) = 0 \qquad y(L) = 0 \qquad y(2L) = 0 \qquad (h)$$

An equilibrium relation $\Sigma M_B = 0$ gives the fourth equation as:

$$2R_A + R = 2F \qquad (i)$$

From the boundary condition $y(0) = 0$, it is evident that $C_2 = 0$. From the boundary condition $y(L) = 0$, we determine C_1 as:

$$EI_z y_{x=L} = \frac{R_A}{6}(L)^3 - \frac{F}{6}\left(L - \frac{L}{2}\right)^3 + 0 - 0 + C_1 L = 0$$

$$C_1 = (L^2/48)[F - 8 R_A] \qquad (j)$$

From the boundary condition $y(2L) = 0$, we can determine another independent equation in terms of the remaining unknowns R_A and R as:

$$EI_z y_{x=2L} = \frac{R_A}{6}(2L)^3 - \frac{F}{6}\left(\frac{3L}{2}\right)^3 + \frac{R}{6}(L)^3 - \frac{F}{6}\left(\frac{L}{2}\right)^3 + \frac{2L^3}{48}(F - 8R_A) = 0$$

$$6R_A + R = (13/4)F \qquad (k)$$

From Eqs. (i) and (k), it is clear that:

$$R_A = (5/16)F \qquad \Rightarrow \qquad R = (11/8)F = (11/8)(12)\ 16.5 \text{ kN} \qquad (l)$$

It is interesting that the load partitioning between the supports is independent of the stiffness of the beam and the length of the beam. The partitioning of the loads among the three supports depends only on the placement of the two concentrated forces F, as indicated by $a = b = L/2$.

Let's continue the analysis and determine the deflection at mid span between the first and second supports located by $x = a = L/2 = 3$ m. Begin by substituting the values for R and R_A from Eq. (l) into Eq. (g) to give the expression for y(x) as:

$$EI_z y = \frac{5F}{96}x^3 - \frac{F}{6}<x-a>^3 + \frac{11F}{48}<x-L>^3 - \frac{F}{6}<x-3a>^3 - \frac{FL^2}{32}x \qquad (m)$$

Evaluating Eq. (m) at mid span where $x = a = L/2$ gives:

$$EI_z y_{x=L/2} = -\frac{7FL^3}{768} \quad (n)$$

Reference to Appendix D gives $I_z = 34.5 \times 10^6$ mm^4 for the W203 × 36 wide flange section. Then substituting numerical parameters into Eq. (n) gives:

$$y_{x=L/2} = -\frac{7FL^3}{768 EI_z} = -\frac{7(12{,}000)(6000)^3}{768(207 \times 10^3)(34.5 \times 10^6)} = -3.308 \text{ mm} \quad (o)$$

8.7 SUMMARY

In this chapter, we returned to study beams subjected to transverse forces, with an emphasis on the deflection. The geometry of the deformation of the beam (plane sections remain plane) enabled writing an expression relating curvature κ to the bending moment.

$$\kappa = 1/\rho = M(x)/[EI_z] \quad (7.6 \text{ bis})$$

We referred to an equation derived in analytical geometry that gives the curvature as a function of the derivatives of the elastic curve. Then we simplified this equation by assuming the square of the slope of the beam was small compared to unity.

$$\kappa = \frac{1}{\rho} = \frac{\dfrac{d^2 y}{dx^2}}{\left[1 + \left(\dfrac{dy}{dx}\right)^2\right]^{3/2}} \cong \frac{d^2 y}{dx^2} \quad (8.2)$$

We showed that derivatives of the elastic curve y(x) were related to M(x), V(x) and q(x) as:

$$\frac{d^2 y}{dx^2} = \frac{M(x)}{EI_z} \quad (8.4)$$

$$\frac{d^3 y}{dx^3} = \frac{V(x)}{EI_z} \quad (8.5)$$

$$\frac{d^4 y}{dx^4} = -\frac{q(x)}{EI_z} \quad (8.6)$$

There are three different approaches for determining the equation for the elastic curve y(x) which include:

1. Integrating Eq. (8.4) with the moment M(x) with respect to x
2. Integrating Eq. (8.5) with the shear force V(x) with respect to x
3. Integrating Eq. (8.6) with the transverse loading q(x) with respect to x

Several examples were presented to show the application of Eq. (8.4) to determine the equation of the elastic curve. To determine the constants of integration, we made use of boundary conditions, symmetry and compatibility conditions. It was clear for beams with multiple loadings that the double integration method was time consuming and tedious because of the large number of constants of integration that developed during the integration process. To reduce the number of constants of integration, we introduced the concept of singularity functions $f_n(x)$ given by:

$$f_n(x) = <x - a>^n \quad (8.23)$$

When the exponent $n \geq 0$, the pointed brackets were interpreted either as zero or as regular brackets depending upon the value of the constant a as indicated below:

$$\begin{aligned} f_n(x) &= 0 & \text{if } x < a \\ f_n(x) &= (x - a)^n & \text{if } x \leq a \end{aligned} \quad (8.24)$$

When integrating pointed brackets with $n \geq 0$, the integration rule is:

$$\int_{-\infty}^{x} <x-a>^n \, dx = \frac{<x-a>^{n+1}}{n+1} \quad n \geq 0 \quad (8.25)$$

However, when n is negative (– 1 or –2) a singularity exists and we define $f_n(x)$ as:

$$\begin{aligned} f_n(x) &= 0 & \text{if } x < a \\ f_n(x) &= 0 & \text{if } x > a \\ f_n(x) &\Rightarrow \infty & \text{if } x = a \end{aligned} \quad (8.26)$$

The singularity in f_n at $x = a$ does not lead to mathematical difficulty, because when $f_n(x) = <x - a>^{-1}$ and $f_n(x) = <x - a>^{-2}$ are integrated, the result is finite as indicated below:

$$\int_{-\infty}^{x} <x-a>^{-2} \, dx = <x-a>^{-1} \quad (8.27)$$

$$\int_{-\infty}^{x} <x-a>^{-1} \, dx = <x-a>^{0} \quad (8.28)$$

We then described the use of singularity functions for writing the expression for four common loading functions as:

1. A concentrated moment at $x = a$: $\Rightarrow q(x) = M_o<x - a>^{-2}$ \quad (8.29)
2. A concentrated force F at $x = a$: $\Rightarrow q(x) = F<x - a>^{-1}$ \quad (8.30)
3. A uniformly distributed load q_o beginning at $x = a$: $\Rightarrow q(x) = q_o<x - a>^0$ \quad (8.31)
4. A linearly increasing (ramp) load $q(x)$ beginning at $x = a$:

$$q(x) = (q_o/b)<x - a>^1 \quad (8.32)$$

where q_o/b is the slope of the ramp function.

A graphic in Fig. 8.8 is presented to aid you in writing expressions for loads, shear forces and moments in terms of singularity functions. Several examples are presented to demonstrate the techniques used in applying singularity methods to solve for beam deflections.

We introduced another technique for solving for beam deflections called superposition. Superposition involves combining together the results of the deflection from two or more separate solutions to yield a new solution containing all of the loading functions involved in the separate solutions. One of the advantages of the method of superposition is that we make use of many solutions that already exist. Mechanics of Materials is a subject with a long history, and the classic problems in bending of beams were solved many years ago. We have provided a catalog of several of these solutions for your convenience in Appendix E.

There are two requirements that must be satisfied before using the principle of superposition. First is a linearity requirement. The solution for y(x) must be linear with respect to the loading functions, which is satisfied. The second requirement pertains to the integrity of the geometry of the beam during the loading. The beam does not undergo significant changes in its dimensions, and we conclude that the principle of superposition is valid. The methods of superposition were demonstrated in several examples.

In most instances beams are statically determinate. The reactions at the supports may be determined from the three applicable equations of equilibrium. However, in some cases beams are designed with additional supports that produce additional reaction forces making the beam statically indeterminate. There are several methods to solve for the additional reaction forces; however, we often supplement the equilibrium relations with a deformation equation to enable a solution. Two different methods to determine the reaction forces for statically indeterminate beams were demonstrated, namely, superposition and integration with singularity functions.

PROBLEMS

8.1 For the cantilever beams shown in Fig. P8.1, write the equation for the elastic curves. Also determine the deflection and slope at the free end.

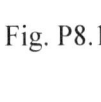

Fig. P8.1

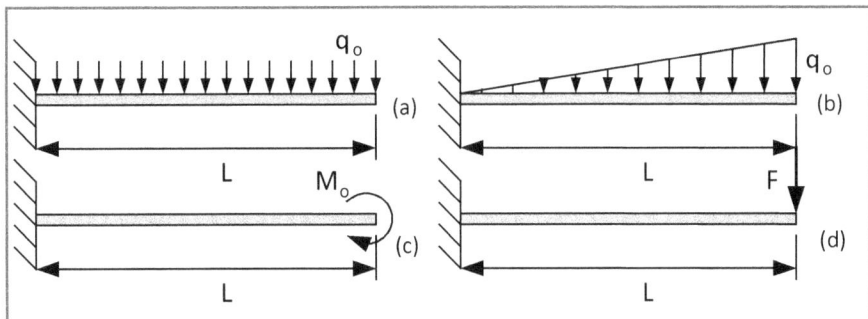

8.2 For the steel cantilever beam, shown in Fig. P 8.1a, determine the deflection and slope at the free end for the numerical parameters defined in the table below:

Prob. No.	Section	q_o	L
8.2a	W10 × 60	600 lb/ft	16 ft
8.2b	W305 × 74	10 kN/m	5.0 m
8.2c	S10 × 25.4	150 lb/ft	20 ft
8.2d	S254 × 38	2.5 kN/m	6.0 m

8.3 For the steel cantilever beam, shown in Fig. P 8.1b, determine the deflection and slope at the free end for the numerical parameters defined in the table below:

Prob. No.	Section	q_o	L
8.3a	W10 × 60	500 lb/ft	20 ft
8.3b	W305 × 74	12 kN/m	5.5 m
8.3c	S10 × 25.4	250 lb/ft	18 ft
8.3d	S254 × 38	3.6 kN/m	4.8 m

8.4 For the steel cantilever beam, shown in Fig. P 8.1c, determine the deflection and slope at the free end for the numerical parameters defined in the table below:

Prob. No.	Section	M_o	L
8.4a	W14 × 82	150 kip-ft	20 ft
8.4b	W457 × 89	180 kN-m	5.5 m
8.4c	W18 × 60	120 kip-ft	18 ft
8.4d	W254 × 45	60 kN-m	4.8 m

8.5 For the steel cantilever beam, shown in Fig. P 8.1d, determine the deflection and slope at the free end for the numerical parameters defined in the table below:

Prob. No.	Section	F	L
8.5a	S18 × 70	8.0 kip	16 ft
8.5b	S381 × 64	20 kN	5.0 m
8.5c	S12 × 35	2.5 kip	20 ft
8.5d	S203 × 27	1.0 kN	6.0 m

8.6 For the simply supported beams, shown in Fig. P8.6, write the equation for the elastic curves. Also determine the deflection at mid span and the slope at the both ends.

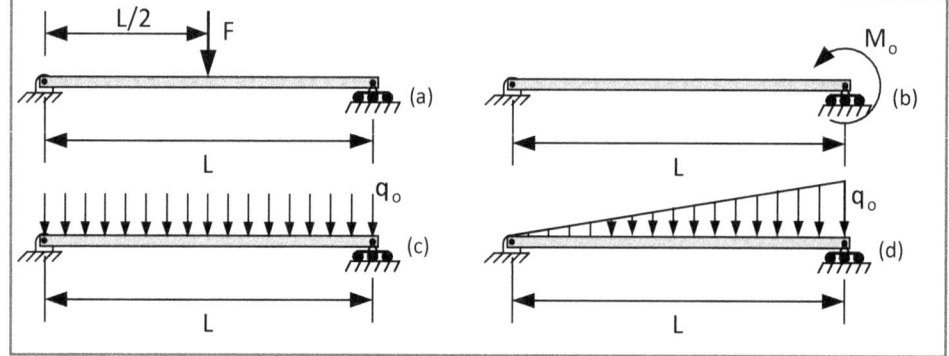

Fig. P8.6

8.7 For the steel simply supported beam shown in Fig. P8.6a, determine the mid span deflection and slope at both ends for the numerical parameters defined in the table below:

Prob. No.	Section	F	L
8.7a	S18 × 70	1.6 kip	16 ft
8.7b	S381 × 64	4.0 kN	5.0 m
8.7c	S12 × 35	0.5 kip	20 ft
8.7d	S203 × 27	0.25 kN	6.0 m

8.8 For the steel simply supported beam, shown in Fig. P8.6b, determine the mid span deflection and slope at both ends for the numerical parameters defined in the table below:

Prob. No.	Section	M_o	L
8.8a	W14 × 82	22 kip-ft	20 ft
8.8b	W457 × 89	25 kN-m	5.5 m
8.8c	W18 × 60	150 kip-ft	18 ft
8.8d	W254 × 45	10 kN-m	4.8 m

8.9 For the steel simply supported beam, shown in Fig. P8.6c, determine the mid span deflection and slope at both ends for the numerical parameters defined in the table below:

Prob. No.	Section	q_o	L
8.9a	W10 × 60	1500 lb/ft	20 ft
8.9b	W305 × 74	30 kN/m	5.5 m
8.9c	S10 × 25.4	700 lb/ft	18 ft
8.9d	S254 × 38	10 kN/m	4.8 m

8.10 For the steel simply supported beam, shown in Fig. P8.6d, determine the mid span deflection and slope at both ends for the numerical parameters defined in the table below:

Prob. No.	Section	q_o	L
8.10a	W10 × 60	700 lb/ft	16 ft
8.10b	W305 × 74	12 kN/m	5.0 m
8.10c	S10 × 25.4	200 lb/ft	20 ft
8.10d	S254 × 38	3.0 kN/m	6.0 m

8.11 For the simply supported beam, shown in Fig. P8.11, determine the equation of the elastic curve and the slope at both ends.

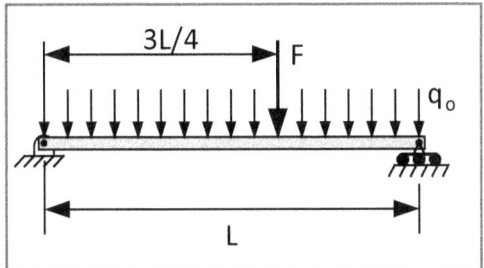

Prob. No.	Section	q_o	F	L
8.12a	W10 × 60	500 lb/ft	1.0 kip	12 ft
8.12b	W305 × 74	10 kN/m	2.0 kN	3.5 m
8.12c	S10 × 25.4	100 lb/ft	0.25 kip	15 ft
8.12d	S254 × 38	2.0 kN/m	0.250 N	4.0 m

Fig. P8.11

8.12 For the steel simply supported beam, shown in Fig. P8.11, determine the mid span deflection and the slope at both ends for the numerical parameters listed in the table above.

8.13 For the simply supported beam, shown in Fig. P8.13, determine the equation of the elastic curve and the slope at both ends.

Prob. No.	Section	q_o	L
8.14a	S18 × 70	1.2 kip/ft	14 ft
8.14b	S381 × 64	3.0 kN/m	4.5 m
8.14c	S12 × 35	0.5 kip/ft	18 ft
8.14d	S203 × 27	0.5 kN/m	5.0 m

Fig. P8.13

8.14 For the steel simply supported beam, shown in Fig. P8.13, determine the mid span deflection and the slope at both ends for the numerical parameters listed in the table above.

8.15 For the simply supported beam, shown in Fig. P8.15, determine the equation of the elastic curve and the slope at both ends.

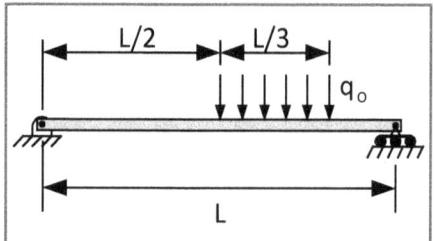

Prob. No.	Section	q_o	L
8.16a	W14 × 82	1.0 kip/ft	15 ft
8.16b	W457 × 89	1.5 kN/m	6.0 m
8.16c	W18 × 60	1.0 kip/ft	18 ft
8.16d	W254 × 45	0.75 kN/m	4.5 m

Fig. P8.15

8.16 For the aluminum simply supported beam, shown in Fig. P8.15, determine the mid span deflection and the slope at both ends for the numerical parameters listed in the table above.

8.17 For the simply supported beam, shown in Fig. P8.17, determine the equation of the elastic curve and the slope at both ends.

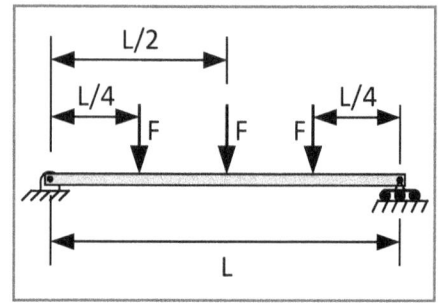

Prob. No.	Section	F	L
8.18a	S20 × 86	2.0 kip	15 ft
8.18b	S457 × 81	4.0 kN	4.0 m
8.18c	S15 × 50	1.5 kip	24 ft
8.18d	S508 × 112	2.5 kN	5.0 m

Fig. P8.17

8.18 For the steel simply supported beam, shown in Fig. P8.17, determine the mid span deflection and the slope at both ends for the numerical parameters listed in the table above.

8.19 For the simply supported beam, shown in Fig. P8.19, determine the equation of the elastic curve and the slope at both ends.

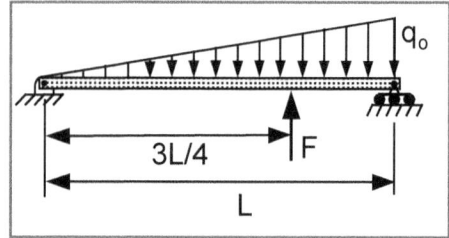

Prob. No.	Section	q_o	F	L
8.20a	W21 × 83	150 lb/ft	1.2 kip	25 ft
8.20b	W533 × 92	1.0 kN/m	1.5 kN	4.0 m
8.20c	S12 × 50	500 lb/ft	4.5 kip	24 ft
8.20d	S305 × 74	2.5 kN/m	2.5 kN	5.0 m

Fig. P8.19

8.20 For the steel simply supported beam, shown in Fig. P8.19, determine the mid span deflection and the slope at both ends for the numerical parameters listed in the table above.

8.21 For the cantilever beam, shown in Fig. P8.21, write the equation for the elastic curve. Also determine the equations for deflection and slope at the free end.

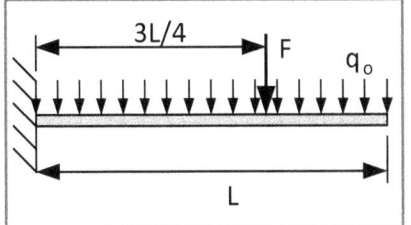

Fig. P8.21

Prob. No.	Section	q_o	F	L
8.22a	W21 × 83	250 lb/ft	1.5 kip	15 ft
8.22b	W533 × 92	1.4 kN/m	1.0 kN	3.0 m
8.22c	S12 × 50	750 lb/ft	2.5 kip	16 ft
8.22d	S305 × 74	2.4 kN/m	1.8 kN	4.5 m

8.22 For the steel cantilever beam, shown in Fig. P8.21, determine the deflection and the slope at the free end for the numerical parameters listed in the table above.

8.23 For the cantilever beam, shown in Fig. P8.23, write the equation for the elastic curve. Also determine the equations for the deflection and slope at the free end.

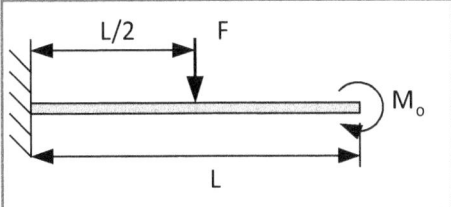

Fig. P8.23

Prob. No.	δ	M_o	F	L
8.24a	1.4 in.	1250 ft-lb	1.5 kip	18 ft
8.24b	16 mm	1.4 kN-m	2.0 kN	4.0 m
8.24c	0.86 in.	2500 ft-lb	2.5 kip	22 ft
8.24d	15 mm	2.4 kN-m	3.5 kN	5.5 m

8.24 For the steel cantilever beam, shown in Fig. P8.23, determine the minimum weight wide flange section if the deflection δ at the free end is limited. Numerical parameters for the deflection, geometry and loading are listed in the table above.

8.25 For the cantilever beam, shown in Fig. P8.25, write the equation for the elastic curve. Also determine the equations for the deflection and slope at the free end.

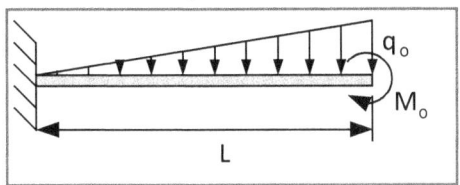

Fig. P8.25

Prob. No.	δ	M_o	q_o	L
8.26a	1.8 in.	1500 ft-lb	1.0 kip/ft	16 ft
8.26b	26 mm	1.8 kN-m	2.4 kN/m	3.5 m
8.26c	0.70 in.	2100 ft-lb	2.2 kip/ft	21 ft
8.26d	19 mm	3.2 kN-m	3.1 kN/m	4.4 m

8.26 For the steel cantilever beam, shown in Fig. P8.25, determine the minimum weight American Standard section if the deflection δ at the free end is limited as specified in the table above.

8.27 For the cantilever beam, shown in Fig. P8.27, write the equation for the elastic curve. Also determine the deflection and slope at the free end.

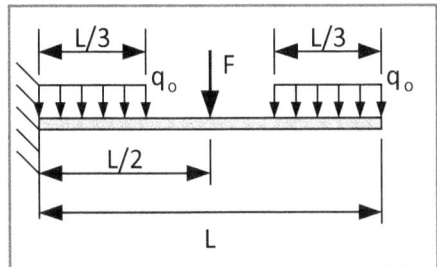

Fig. P8.27

Prob. No.	δ	F	q_o	L
8.28a	1.1 in.	1500 lb	1.2 kip/ft	15 ft
8.28b	20 mm	1.6 kN	2.0 kN/m	3.0 m
8.28c	0.85 in.	2000 lb	2.0 kip/ft	21 ft
8.28d	26 mm	3.0 kN	2.5 kN/m	4.5 m

8.28 For the steel cantilever beam, shown in Fig. P8.27, determine the minimum weight structural tee section if the deflection δ at the free end is limited as specified in the table above.

8.29 For the cantilever beam, shown in Fig. P8.29, write the equation for the elastic curve. Also determine the deflection and slope at the free end.

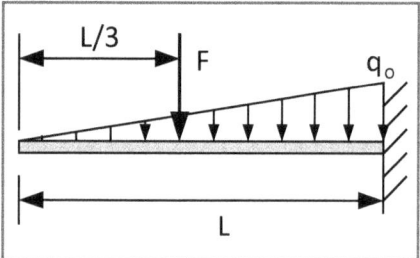

Prob. No.	δ	F	q_o	L
8.30a	1.3 in.	2500 lb	1.0 kip/ft	18 ft
8.30b	15 mm	3.6 kN	2.6 kN/m	4.0 m
8.30c	0.50 in.	4000 lb	2.4 kip/ft	24 ft
8.30d	12 mm	5.0 kN	1.5 kN/m	5.5 m

Fig. P8.29

8.30 For the steel cantilever beam, shown in Fig. P8.29, determine the minimum weight wide flange section if the deflection δ at the free end is limited as specified in the table above.

8.31 Using the method of superposition, determine the equation for the elastic curve for the simply supported beam, shown in Fig. P8.31. Also, determine the deflection at mid span and the slope at the left end of the beam.

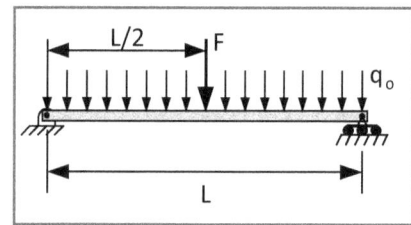 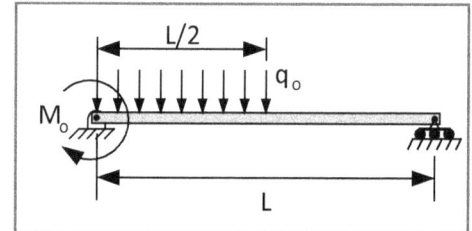

Fig. P8.31   Fig. P8.32

8.32 Using the method of superposition, determine the equation for the elastic curve for the simply supported beam, shown in Fig. P8.32. Also, determine the deflection at mid span and the slope at the left end of the beam.

8.33 Using the method of superposition, determine the equation for the elastic curve for the simply supported beam, shown in Fig. P8.33. Also, determine the deflection at mid span and the slope at the right end of the beam.

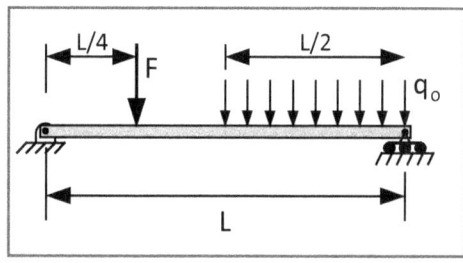 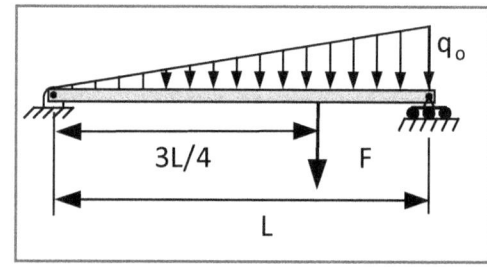

Fig. P8.33   Fig. P8.34

8.34 Using the method of superposition, determine the equation for the elastic curve for the simply supported beam, shown in Fig. P8.34. Also, determine the deflection at mid span and the slope at the right end of the beam.

8.35 Using the method of superposition, determine the equation for the elastic curve for the cantilever beam, shown in Fig. P8.35. Also, determine the deflection at mid span and the slope at the free end of the beam.

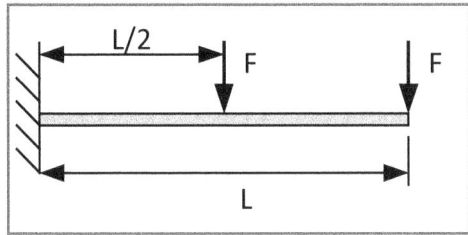
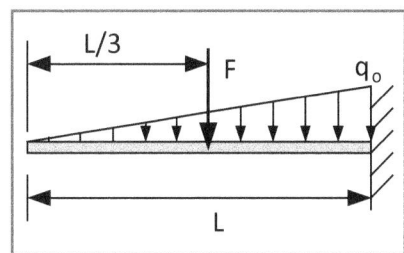

Fig. P8.35 Fig. P8.36

8.36 Using the method of superposition, determine the equation for the elastic curve for the cantilever beam, shown in Fig. P8.36. Also, determine the deflection at mid span and the slope at the free end of the beam.

8.37 Using the method of superposition, determine the equation for the elastic curve for the cantilever beam, shown in Fig. P8.37. Also, determine the deflection at mid span and the slope at the free end of the beam.

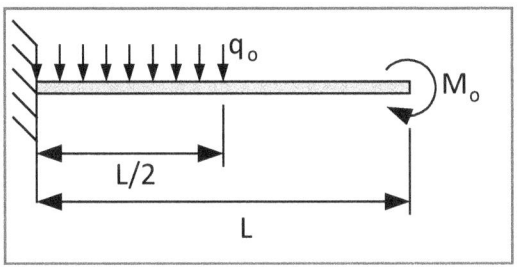
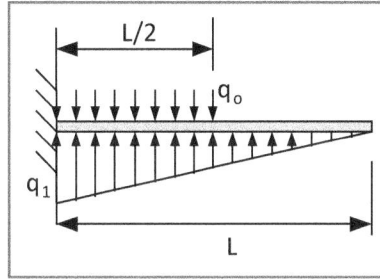

Fig. P8.37 Fig. P8.38

8.38 Using the method of superposition, determine the equation for the elastic curve for the cantilever beam, shown in Fig. P8.38. Also, determine the deflection at mid span and the slope at the free end of the beam.

8.39 Determine the reactions at the prop and the wall on the cantilever beam, shown in Fig. P8.39.

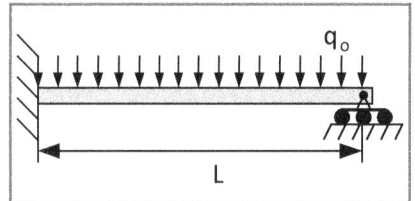

Fig. P.39

8.40 Determine the reactions at the prop and the wall on the cantilever beam, shown in Fig. P8.40.

Fig. P8.40

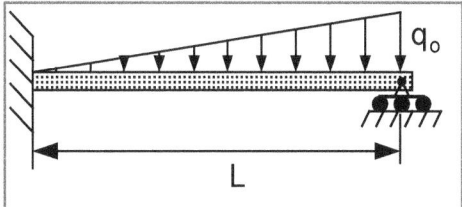

8.41 Determine the reactions at the prop and the wall on the cantilever beam, shown in Fig. P8.41.

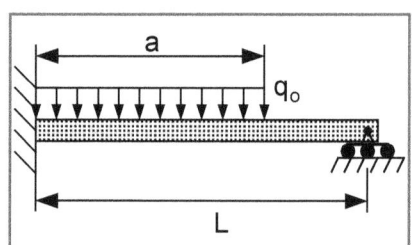

Fig. P8.41

Prob. No	a
8.41a	L/5
8.41b	L/4
8.41c	L/3
8.41d	L/2
8.41e	3L/4
8.41f	4L/5

8.42 Determine the reactions at the prop and the wall on the cantilever beam, shown in Fig. P8.42.

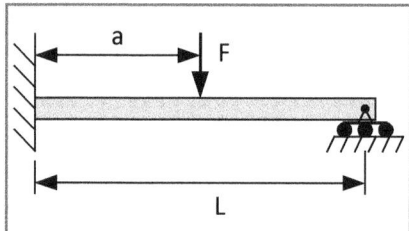

Fig. P8.42

Prob. No	a
8.42a	L/5
8.42b	L/4
8.42c	L/3
8.42d	L/2
8.42e	3L/4
8.42f	4L/5

8.43 Determine the reaction at the prop on the cantilever beam, shown in Fig. P8.43.

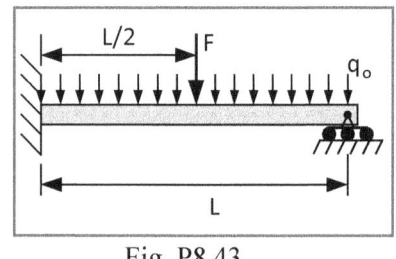
Fig. P8.43

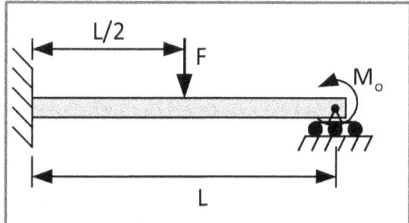
Fig. P8.44

8.44 Determine the reactions at the prop and the wall on the cantilever beam, shown in Fig. P8.44.

8.45 Determine the reactions at the supports of the continuous beam, shown in Fig. P8.45.

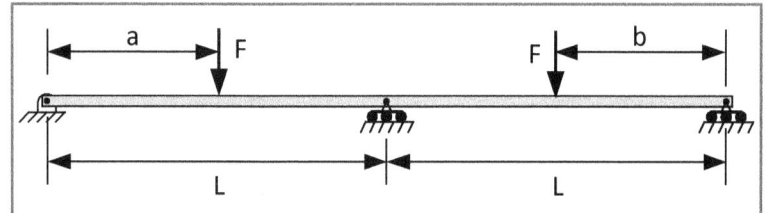

Prob. No.	a	b
8.45a	L/4	L/3
8.45b	L/3	L/4
8.45c	L/2	L/2
8.45d	L/4	L/2

Fig. P8.45

8.46 Determine the reactions at the supports of the continuous beam, shown in Fig. P8.46.

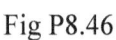
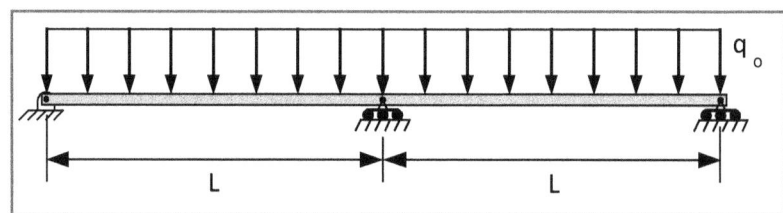

Fig P8.46

8.47 Determine the reactions at the supports of the continuous beam, shown in Fig. P8.47.

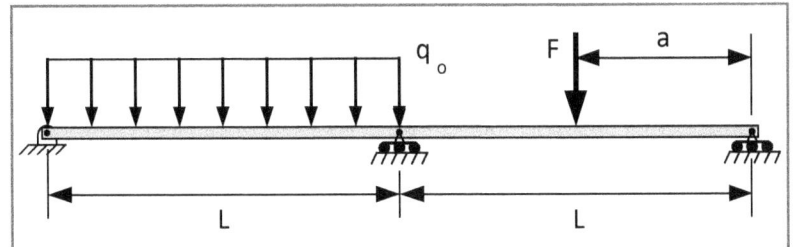

Prob. No.	a
8.47a	L/2
8.47b	L/4
8.47c	3L/4

Fig. P8.47

8.48 Determine the reactions at the supports of the continuous beam, shown in Fig. P8.48.

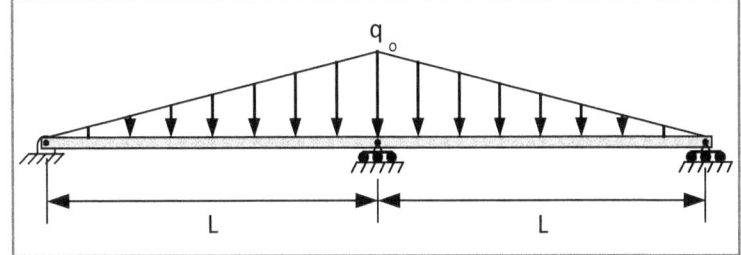

Fig. P8.48

8.49 Determine the reactions at the supports of the continuous beam, shown in Fig. P8.49 if a = L/2.

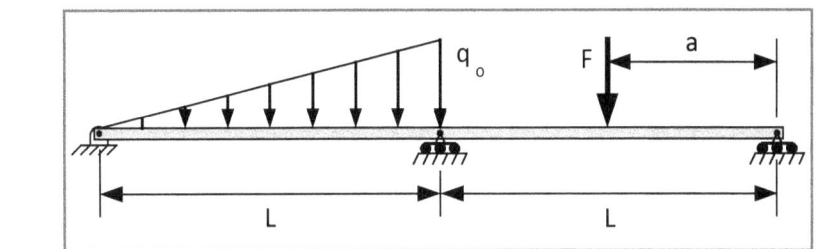

Fig. P8.49

8.50 Determine the reactions at the supports of the built-in beam, shown in Fig. P8.50.

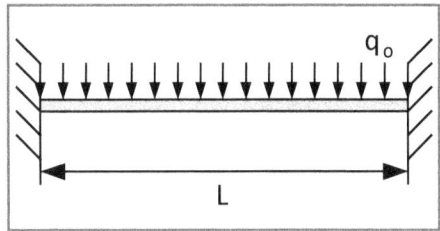

Fig. P8.50

8.51 Determine the reactions at the supports of the built-in beam, shown in Fig. P8.51.

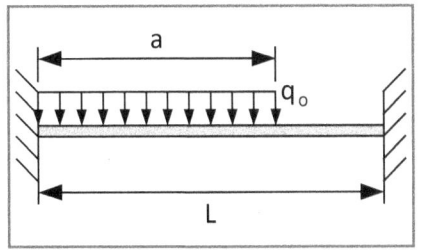

Fig. P8.51

Prob. No.	a
8.51a	L/4
8.51b	L/2
8.51c	3L/5

8.52 Determine the reactions at the supports of the built-in beam, shown in Fig. P8.52.

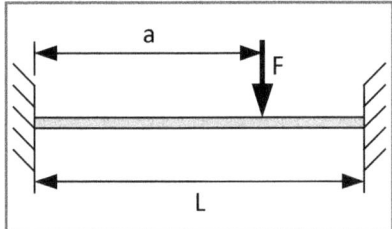

Prob. No.	a
8.52a	L/4
8.52b	L/2
8.52c	3L/5

Fig. P8.52

8.53 Determine the reactions at the supports of the built-in beam, shown in Fig. P8.53.

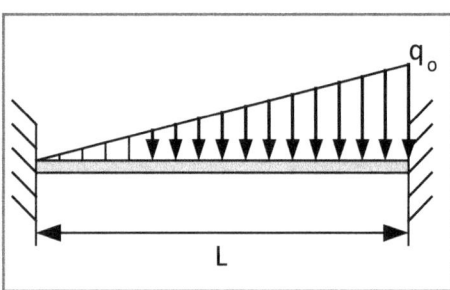

Fig. P8.53

CHAPTER 9

STRESSES IN THIN WALLED PRESSURE VESSELS

9.1 INTRODUCTION

Thin walled[1] pressure vessels (shells) are widely employed as containers for liquids and gasses. They are usually subjected to internal pressure, which generates membrane stresses in the walls of the pressure vessels. Bending stresses do not occur unless the shell is designed with discontinuities, such as nozzles or skirts. Because the state of stress is primarily membrane, the pressure vessels are very efficient structures with stresses that are uniformly distributed through the thickness of the wall.

Applications involving pressure vessels vary widely. Pressure vessels are fabricated in various sizes and shapes. They can be small such as a can for soda, or large such as a storage tank for natural gas. While the most common shape is cylindrical, they may be designed with spherical, conical or toroidal shapes.

9.2 SPHERICAL PRESSURE VESSELS

Let's consider a portion of a thin spherical shell, to illustrate the concept of membrane stresses. The portion of a spherical shell with a radius r and wall thickness t subjected to an internal pressure p is illustrated in Fig. 9.1. Writing the equilibrium equation $\Sigma F_y = 0$ yields:

$$p\pi r_x^2 - (N \sin \phi)(2\pi r_x) = 0 \qquad (9.1)$$

where N is the membrane force per unit length and r_x is the radius defined in Fig. 9.1

Solving Eq. (9.1) for N and noting that $r_x/\sin \phi = r$ gives:

$$N = pr_x/(2 \sin \phi) = pr/2 \qquad (9.2)$$

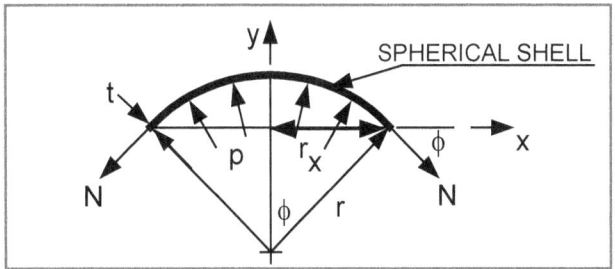

Fig. 9.1 A portion of a spherical shell subjected to an internal pressure p.

---

[1] A pressure vessel is considered to be thin walled if the ratio of r/t > 10. The membrane stresses in thin walled pressure vessels are uniformly distributed through the thickness of the wall.

It is evident from Eq. (9.2) that the value of N is independent of the angle ϕ; hence, the membrane force per unit length is independent of position. The membrane stresses σ_m for a spherical pressure vessel are obtained from Eq. (9.2) as:

$$\sigma_m = N/t = pr/2t \tag{9.3}$$

Because the spherical pressure vessel is symmetric, it is evident that the membrane stresses in the meridian σ_{my} and tangential σ_{mx} directions are the same and the stress state is biaxial with:

$$\sigma_1 = \sigma_2 = \sigma_{my} = \sigma_{mx} = pr/2t \tag{9.4}$$

These stresses are also principal stresses, because axes of symmetry always coincide with principal planes.

EXAMPLE E9.1

A spherical pressure vessel with a radius r = 20 ft and a wall thickness of 3/8 in. is subjected to a pressure of 75 psi. Determine the membrane stresses in the longitudinal and meridian directions. Also determine the maximum shear stress.

Solution: From Eq. (9.4), we write:

$$\sigma_{my} = \sigma_{mx} = pr/2t = (75)(20)(12)/[(2)(3/8)] = 24,000 \text{ psi} \tag{a}$$

Also $\tau_{xy} = 0$, because the meridian and longitudinal membrane stresses are principal stresses.

From Eq. (2.21), the maximum shear stress is given by:

$$\tau_{Max} = \sqrt{\left(\frac{\sigma_{xx} - \sigma_{yy}}{2}\right)^2 + \tau_{xy}^2} = \sqrt{\left(\frac{24000 - 24000}{2}\right)^2 + 0} = 0$$

The maximum shear stress is zero for in-plane stresses. For out of plane stresses τ are given by:

$$\tau_{Max} = (\sigma_1 - \sigma_3)/3 = \sigma_1/2 = pr/(4t) \tag{9.4a}$$

9.3 CYLINDRICAL PRESSURE VESSELS

Cylindrical pressure vessels are the most common type employed, because the cylindrical shape is easy to fabricate. Thin metal plates are formed into rounds and the longitudinal (axial) seam is welded to produce a thin walled tube. End caps are welded on this tube to produce a thin walled vessel capable of maintaining internal pressures.

To determine the membrane stresses σ_h (hoop or circumferential direction) and σ_a (axial or longitudinal direction) in a thin walled cylindrical pressure vessel, consider a cylinder with a radius r and a wall thickness t, as illustrated in Fig. 9.2. Next remove a semi-circular segment of the cylinder, as a FBD. The forces due to pressure p and the membrane forces acting on this FBD are shown in Fig. 9.3.

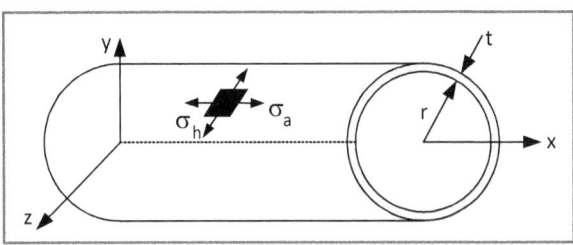

Fig. 9.2 A cylindrical pressure vessel with a radius r and a wall thickness of t.

To solve for the hoop stress σ_h, write $\Sigma F_z = 0$:

$$2(\sigma_h t\Delta x) - 2pr\Delta x = 0 \quad (a)$$

$$\sigma_h = pr/t \quad (9.5)$$

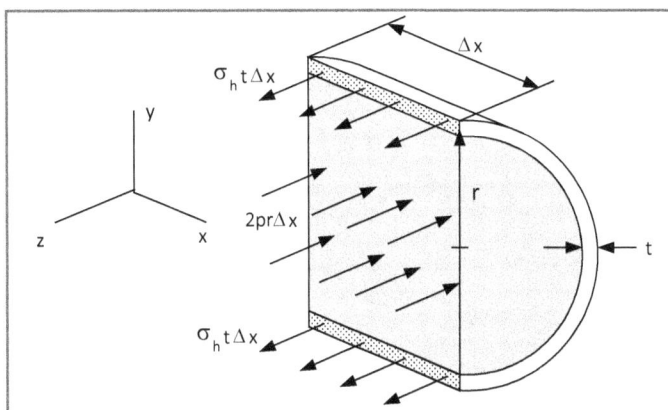

Fig. 9.3 A FBD of a semi-circular segment from the thin walled cylindrical pressure vessel.

The axial stresses are determined, by considering a segment of the cylinder formed by a section cut perpendicular to the axis of the cylinder, as shown in Fig. 9.4. This FBD shows the forces due to pressure and the stresses in the axial direction. Writing $\Sigma F_x = 0$ yields:

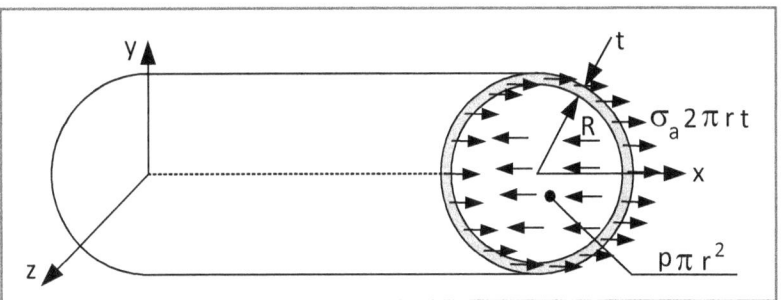

Fig. 9.4 FBD of a cylindrical pressure vessel showing the forces in the axial direction.

$$\sigma_a (2\pi rt) - p(\pi r^2) = 0 \quad (b)$$

Solving Eq. (b) for σ_a gives:

$$\sigma_a = pr/2t \quad (9.6)$$

Comparing Eq. (9.5) and Eq. (9.6) indicates that the hoop stress is twice as large as the axial stress in a thin walled circular pressure vessel. Also, from symmetry it is evident that σ_h and σ_a are principal stresses with:

$$\sigma_1 = \sigma_h \quad \sigma_2 = \sigma_a \quad \tau_{ha} = 0 \quad (9.7)$$

The maximum shear stress τ_{Max} is determined from Eq. (2.21) as:

$$\tau_{Max} = \sqrt{\left(\frac{\sigma_{xx} - \sigma_{yy}}{2}\right)^2 + \tau_{xy}^2} = \sqrt{\left(\frac{\sigma_1 - \sigma_2}{2}\right)^2} = \frac{pr}{4t} \qquad (9.8)$$

EXAMPLE E9.2

A cylindrical pressure vessel, shown in Fig. E9.2, is employed as a water storage tank for a small municipality. Determine the height of the water h that can be maintained in the tank if the city's engineering director specifies a safety factor of 3.0. The tank is fabricated from 1020 HR steel, with a diameter of D = 40 ft., a height H = 50 ft. and a thickness t = 0.375 in. The engineering director also indicates that the safety factor should be based on the maximum shear stress theory of yielding.

Fig. E9.2

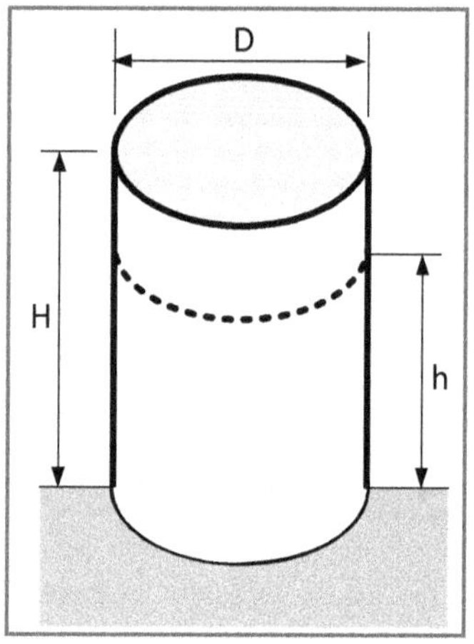

Solution: The pressure in the water tank varies from zero at the water level to a maximum at the base. The maximum pressure is given by:

$$p_{Max} = \gamma h \qquad (9.9)$$

where $\gamma = 62.4$ lb/ft^3 is the density of fresh water.

At the base of the tank, the pressure is:

$$p_{Max} = (62.4)h \quad (\text{lb/ft}^2) \qquad (a)$$

The hoop stress is given by Eq. (9.5) as:

$$\sigma_h = \sigma_1 = pR/t = [(62.4)(20)(12)/(0.375)]h = 39{,}940\,h \quad (\text{lb/ft}^2) \qquad (b)$$

where h is measured in ft.

From the failure diagram for yielding under maximum shear stress, shown previously, we can write:

$$(\sigma_1)_{Design} = S_y/SF \qquad (c)$$

Reference to Appendix B-2 for the yield strength of 1020 HR steel gives:

$$(\sigma_1)_{Design} = S_y/SF = 42.0/3.0 = 14.0 \text{ ksi} \qquad (d)$$

Converting the units of Eq. (d) from ksi to lb/ft² gives:

$$(\sigma_1)_{Design} = (14 \times 10^3 \text{ lb/in.}^2)(144 \text{ in.}^2/\text{ft}^2) = 2,016 \times 10^3 \text{ lb/ft}^2 \qquad (e)$$

From Eq. (b) and Eq. (e), we solve for h as:

$$h = (2,016 \times 10^3)/39,940 = 50.5 \text{ ft} \qquad (f)$$

Reference to Fig. E9.2 shows that the tank is 50 ft high and if filled to the brim, the stresses would exceed the allowable design limit of 14.0 ksi by about one percent. You may avoid even this slight excess by placing an overflow line at a location 49 ft above the tank's base.

Fig. E9.2a

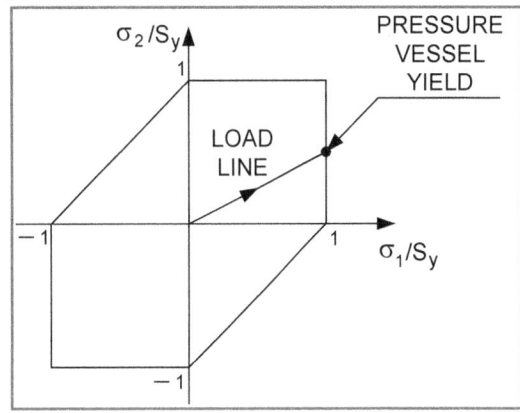

9.4 SUMMARY

Equations for the stresses in thin-walled pressure vessels were derived. For spherical pressure vessels the stress state is isotropic with $\sigma_1 = \sigma_2$.

$$\boxed{\sigma_1 = \sigma_2 = \sigma_{my} = \sigma_{mx} = pr/2t} \qquad (9.4)$$

For cylindrical pressure vessels, we showed that the hoop stress σ_h and axial stress σ_a are given by:

$$\boxed{\sigma_h = pr/t} \qquad (9.5)$$

$$\boxed{\sigma_a = pr/2t} \qquad (9.6)$$

PROBLEMS

9.1 The stresses on the external surface of a cylindrical pressure vessel in the hoop and axial direction are predicted from a simple analysis to be σ_h and σ_a, when the vessel is pressurized to 125 psi. If two strain gages are placed on the pressure vessel to measure the strains in the hoop and axial directions, predict the strains that are expected to occur when the pressure of 125 psi is applied. The pressure vessel is fabricated from the material specified in the table below together with the prediction for σ_h and σ_a.

Prob. No.	σ_h	σ_a	Material
9.1a	35 ksi	17.5 ksi	Steel
9.1b	120 MPa	60 MPa	Aluminum
9.1c	22 ksi	11 ksi	Brass
9.1d	98 MPa	49 MPa	Titanium
9.1e	25 ksi	12.5 ksi	Stainless Steel

9.2 A single strain gage oriented in the hoop direction on a cylindrical pressure vessel gives the measurement listed in the table below. Predict the hoop and axial stresses if the pressure vessel is fabricated from the material specified in the table below.

Prob. No.	$\varepsilon_h \times 10^{-6}$	Material
9.2a	1200	Steel
9.2b	2200	Aluminum
9.2c	900	Brass
9.2d	1300	Titanium
9.2e	1100	Stainless Steel

9.3 Determine the pressure p required to cause yielding in a spherical pressure vessel with the parameters listed in the table below:

Problem No.	r	t	Material	Failure Theory
9.3a	5.0 ft	0.25 in.	1020 HR Steel	Max Normal Stress
9.3b	48 in.	0.375 in.	2024 T-4 Aluminum	Max Shear Stress
9.3c	1.5 m	8 mm	304 A Stainless Steel	von Mises
9.3d	800 mm	6 mm	80-20 Brass	Max Shear Stress
9.3e	36 in.	0.25 in.	1045 HR Steel	von Mises

9.4 Determine the pressure p required to cause yielding in a cylindrical pressure vessel, with the parameters listed in the table below:

Problem No.	r	t	Material	Failure Theory
9.4a	4.0 ft	1/8 in.	1020 HR Steel	Max Normal Stress
9.4b	42 in.	1/4 in.	2024 T-4 Aluminum	Max Shear Stress
9.4c	1.2 m	6 mm	304 A Stainless Steel	von Mises
9.4d	600 mm	4 mm	80-20 Brass	Max Shear Stress
9.4e	32 in.	3/16 in.	1045 HR Steel	von Mises

A pressure vessel, shown in Fig. P9.5, is employed as a water storage tank for a small municipality. Determine the height of the water, h that can be maintained in the tank if the city's engineering director specifies a safety factor of SF. The tank is fabricated from steel with a diameter D, a height H and a thickness t. The city's engineering director also indicates that the safety factor should be based on different theories of yielding. Numerical parameters for the problem are listed in the table below.

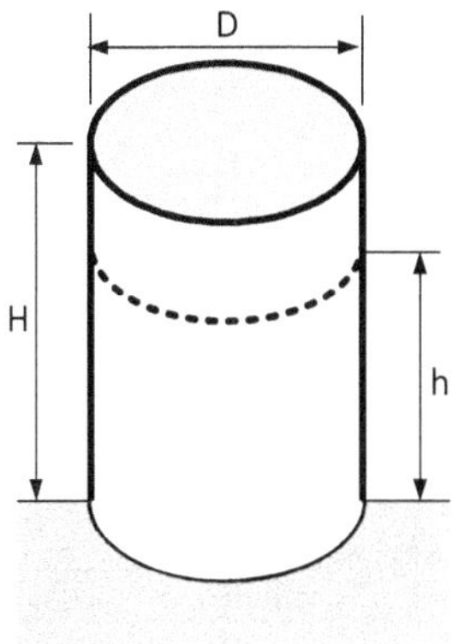

Fig. P9.5

Problem No.	D	t	H	Material	Failure Theory	SF
9.5a	30.0 ft	1/8 in.	60 ft	1020 HR Steel	Max Normal Stress	2.5
9.5b	36 ft	1/4 in.	90 ft	1010 A Steel	Max Shear Stress	3.0
9.5c	12 m	6 mm	25 m	1018 A Steel	von Mises	2.8
9.5d	10 m	4 mm	22 m	1020 HR Steel	Max Shear Stress	3.4
9.5e	42 ft	3/16 in.	75 ft	1018 A Steel	von Mises	2.2

9.5 A cylindrical pressure vessel is fabricated by wrapping thin steel plate in a spiral about a mandrel of diameter D and butt-welding the seam, as illustrated in Fig. P9.6. The weld seam forms an angle of θ with the axis of the cylinder. Determine the normal stress σ and the shearing stress τ acting on the weld, when the vessel is subjected to a pressure p and the plate thickness is t.

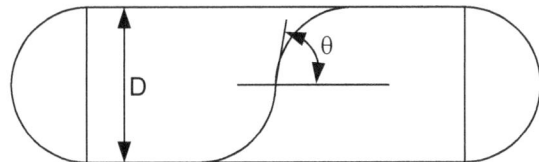

Fig. P9.6

Problem No.	D	t	θ	p
9.6a	10 ft	1/8 in.	60°	120 psi
9.6b	15 ft	1/4 in.	50°	150 psi
9.6c	3.0 m	6 mm	45°	700 kPa
9.6d	4.0 m	4 mm	65°	1.2 MPa
9.6e	12 ft	3/16 in.	55°	200 psi

9.6 A 36 in. diameter pipeline is to be designed to transmit a variety of fluids and to operate at a maximum pressure of 500 psi. Determine the thickness of the steel plate used in its fabrication if the pipeline is designed with a safety factor of SF. Information for the solution of this problem is given in the table below.

Problem No.	SF	Material	Theory of Yieldingθ
9.7a	2.5	1020 HR Steel	Max Normal Stress
9.7b	3.5	1010 A Steel	Max Shear Stress
9.7c	3.0	1018 A Steel	von Mises
9.7d	4.0	1020 HR Steel	Max Shear Stress
9.7e	3.3	1018 A Steel	von Mises

9.7 A pair of orthogonal strain gages is mounted on a cylindrical pressure vessel, with the orientation indicated in Fig. P9.8. For the parameters indicated in the table below, determine the hoop and axial stresses in the cylinder, the internal pressure applied when the strains were measured and the factor of safety based on the theory for yielding defined in the table.

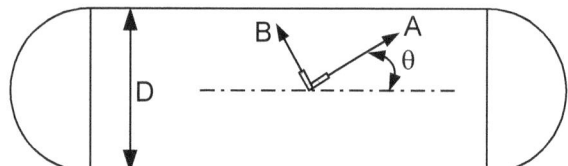

Fig. P9.8

Prob. No.	$\varepsilon_A \times 10^{-6}$	$\varepsilon_B \times 10^{-6}$	θ	D	t	Material	Theory of Yieldingθ
9.8a	420	210	5°	5.6 ft	1/8 in.	1020 HR Steel	Max Normal Stress
9.8b	642	355	8°	8.5 ft	1/4 in.	1010 A Steel	Max Shear Stress
9.8c	538	275	10°	3.0 m	6 mm	1018 A Steel	von Mises
9.8d	680	261	15°	2.3 m	4 mm	1020 HR Steel	Max Shear Stress
9.8e	610	340	12°	6.5 ft	3/16 in.	1018 A Steel	von Mises

9.8 A thin walled spherical water tank, presented in Fig. P9.9, is filled with water and vented to the atmosphere. Determine the meridional σ_m and tangential σ_t stresses anywhere along the equator of the sphere. Your result should be written as an equation in terms of the water density γ, tank diameter D and its wall thickness t.

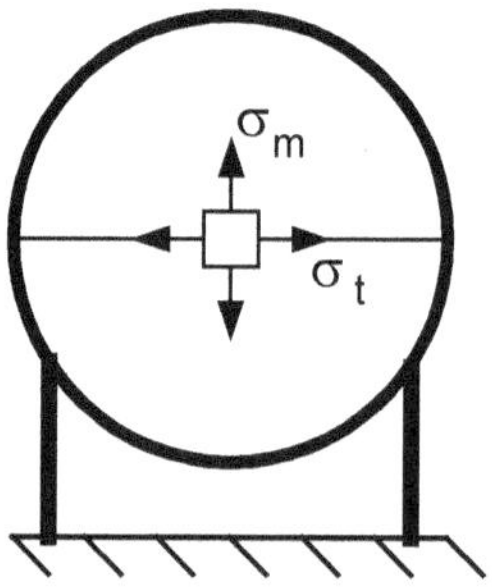

Fig. P9.9

CHAPTER 10

COMBINED LOADING

10.1 INTRODUCTION

In most structures fabricated from beams, struts and columns, a single type of load such as an axial or transverse force or a bending moment produces the stresses. However, in the design of vehicles or machines, it is common to encounter two or more different types of loads acting on a machine component. For example, shafts that transmit power are subjected to torsion and transverse forces that produce internal bending moments and shear forces. When dealing with components subjected to two or three different types of loading, locating the point of maximum normal or shearing stresses is often difficult. In addition when stresses are due to two or more different types of load, determining the combined stress at a given point must be done with care. A guide to assist you in dealing with both of these difficulties is given below:

1. The bending stress is a maximum at a point located the maximum distance from the neutral axis of the beam.
2. The bending stress is a maximum at a point where the bending moment M is a maximum, which coincides with the location where the shear force V is zero.
3. The transverse and longitudinal shear stresses are equal and a maximum at a point located at the neutral axis of the beam, where Q is also a maximum.
4. The transverse and longitudinal shear stresses are a maximum at a point where the shear force V is a maximum, which occurs at a point of application of a transverse load.
5. The shear stress due to torsion is a maximum at the point, where the torsion T is a maximum.
6. The maximum shear stress on a shaft occurs on its outside surface.

Even with these guides, the location of the maximum stress is not apparent. In many analyses, it is necessary to evaluate the stresses at several points to discover the most critical location. When two or more normal stresses or shear stresses act on the same plane, they can be added together using the method of superposition providing the combined value does not exceed the yield strength of the material. However, when the stresses act on mutually orthogonal planes, appropriate stress equations of transformation must be used to determine the maximum principal and shear stresses.

Let's consider three examples to demonstrate the procedure for performing this type of analysis.

EXAMPLE E10.1

A U shaped machine component, illustrated in Fig. E10.1, is subjected to opposing forces of 20 kN. If the component is fabricated from 1045 HR steel, determine the safety factor based on yield strength using the maximum normal stress theory. Comment on the appropriateness of this theory to predict the onset of yielding for this example.

314 Combined Loading; Chapter 10

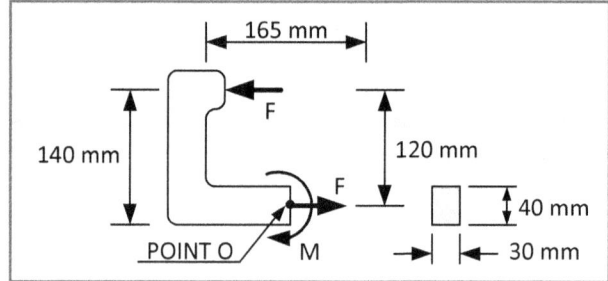

Fig. E10.1

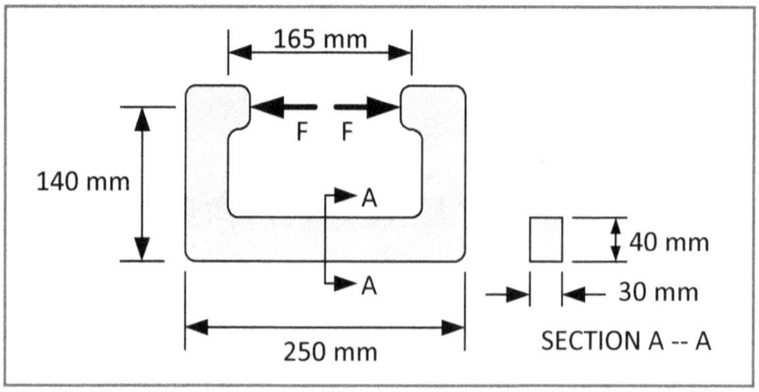

Solution: Let's begin the analysis by constructing the FBD of the left side of the U shaped member as shown in Fig. E 10.1a.

Fig E10.1a

It is evident from the equilibrium relation $\Sigma F_x = 0$, that $P = F = 20$ kN. This internal force produces a uniformly distributed axial stress σ_{xx} that is given by:

$$\sigma_{xx} = P/A = (20 \times 10^3)/1200 = 16.67 \text{ MPa} \tag{a}$$

The equilibrium relation for the moments about the point O gives:

$$\Sigma M_o = 0 \quad\quad 120F - M = 0 \quad\quad M = (120)(20 \times 10^3) = 240 \times 10^4 \text{ N-mm} \tag{b}$$

The moment M produces a bending stress σ_{xx} that is linearly distributed over the cross section of the component at section A-A. The maximum and minimum bending stress is given by:

$$\sigma_{xx} = \pm Mc/I = \pm 6M/bh^2 \tag{c}$$

Substituting the results from Eq. (b) into Eq. (c) yields:

$$\sigma_{xx} = \pm \frac{(6)(240 \times 10^4)}{(30)(40)^2} = \pm 300 \text{ MPa} \tag{d}$$

The stresses due to P and M both act on the same plane and are in the x direction; hence, the results can be superimposed to give:

$$\sigma_{Max} = 16.67 + 300 = 316.7 \text{ MPa} \quad\quad \sigma_{Min} = 16.67 - 300 = -283.3 \text{ MPa} \tag{e}$$

The superposition process and the resulting distribution of stresses across the section A-A is presented in Fig. 10.1b.

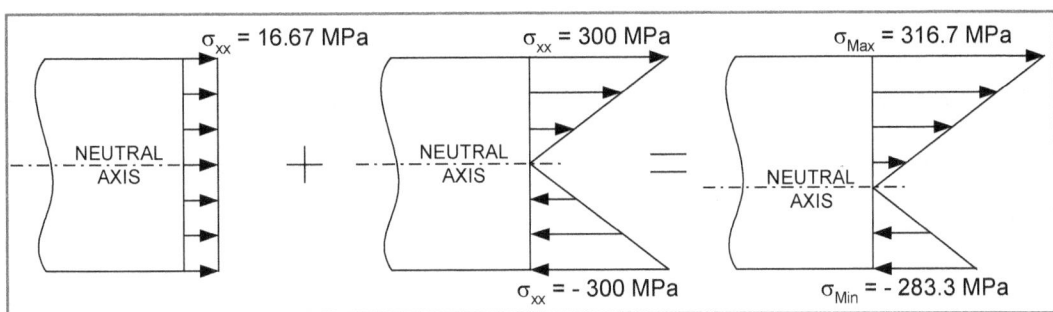

Fig. E10.1b

From Appendix B-2, we determine that S_y = 414 MPa for 1045 HR steel. The maximum normal stress theory for yielding indicates that the onset for yielding occurs when $\sigma_{Max} \geq S_y$. Accordingly the safety factor is determined from:

$$SF = S_y/\sigma_{Max} = 414/316.7 = 1.307$$

The maximum normal stress theory is appropriate in this case because $\sigma_2 = \sigma_3 = 0$. For a uniaxial state of stress, all three of the failure theories described in this chapter predict yielding to occur when $\sigma_{Max} \geq S_y$. We note that the safety factor is relatively low and warn the design engineer of this fact.

EXAMPLE E10.2

The design engineer responsible for the U shaped machine component, described in Example 10.1, must drill a hole 6 mm in diameter through the member at Section A-A to permit passage of a signal wire. She knows that the hole should pass through the neutral axis to minimize its effects on the maximum stresses and the safety factor. She also is aware that the neutral axis has shifted and requests that you locate it.

Solution: We locate the position of the neutral axis by using the similar triangles depicted in Fig. E10.1b. Let's redraw the two triangles and define the location of the neutral axis relative to the bottom edge of the U shaped member, as illustrated in Fig. E10.2.

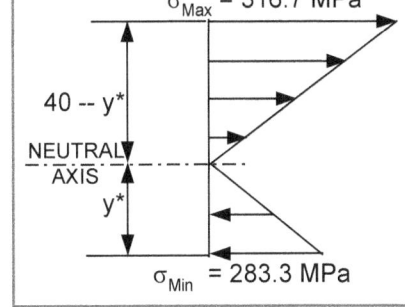

Fig. E10.2

From the proportionality of similar triangles, we write:

$$\frac{y^*}{283.3} = \frac{40 - y^*}{316.7} \qquad (a)$$

$$y^* = 18.89 \text{ mm}$$

You also indicate that drilling a 6 mm hole at the neutral axis will elevate the stresses slightly and reduce the safety factor.

EXAMPLE E10.3

A traffic sign along a city street is cantilevered from a pole, as illustrated in Fig. E10.3. The pressure due to the wind impinging on the sign is uniformly distributed over its area and equal to 12 lb/ft² in the x direction. The pole is fabricated from a tube with a 10 in outside diameter and a 0.25 in. wall thickness. Determine the stresses at points A and B located at the base of the pole.

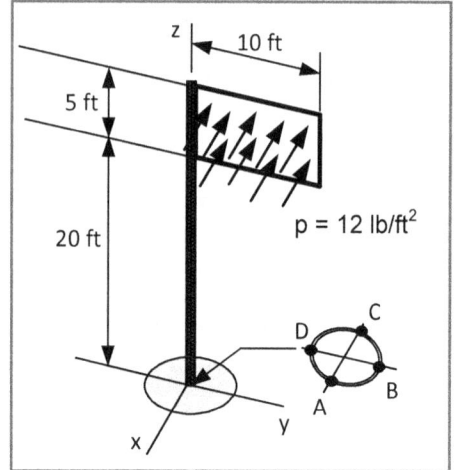

Fig. E10.3

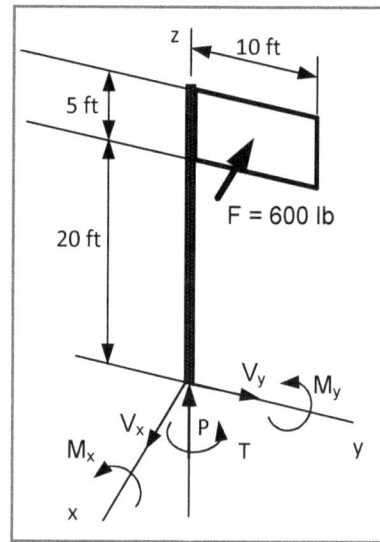

Fig. 10.3a FBD of pole showing all reaction forces and moments.

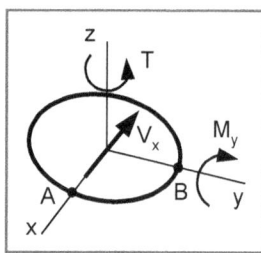

Fig. E10.3b Force and moment directions at ground level.

Solution: The effective force due to the wind pressure acts at the center of the sign and is given by:

$$F_x = pA = (12)(5)(10) = 600 \text{ lb} \quad (a)$$

Next let's prepare a FBD showing the forces at the base of the pole, as indicated in Fig. E10.3a and Fig. 10.3b. Then we write the six equations of equilibrium to obtain:

$$\Sigma F_x = 0 \qquad V_x = 600 \text{ lb} \quad (b)$$

$$\Sigma F_y = 0 \qquad V_y = 0 \quad (c)$$

$$\Sigma F_z = 0 \qquad P = 0 \quad (d)$$

$$\Sigma M_x = 0 \qquad M_x = 0 \quad (e)$$

$$\Sigma M_y = 0 \qquad M_y = (600)(22.5) = 13{,}500 \text{ ft-lb} \quad (f)$$

$$\Sigma M_z = 0 \qquad T = -(600)(5) = -3{,}000 \text{ ft-lb} \quad (g)$$

The geometric parameters that are needed to determine the stresses at the base of the pole are:

$$A = (\pi/4)(d_o^2 - d_i^2) = (\pi/4)[(10)^2 - (9.5)^2] = 7.658 \text{ in}^2. \tag{h}$$

$$I = I_{xx} = I_{yy} = (\pi/64)(d_o^4 - d_i^4) = (\pi/64)[(10)^4 - (9.5)^4] = 91.05 \text{ in}^4 \tag{i}$$

$$J = 2I = 182.1 \text{ in}^4 \tag{j}$$

$$r = c = 5.0 \text{ in.} \tag{k}$$

For circular shafts or tubes it can be shown that Q is given by:

$$Q = (2/3)/(r_o^3 - r_i^3) \tag{m}$$

$$Q = (2/3)/[(5.0)^3 - (4.75)^3] = 11.885 \text{ in}^3 \tag{n}$$

Let's determine the stresses due to the three internal forces M_y, V_x and T and, in the process, use the geometric parameters listed above.

$$\sigma_{zz} = M_y c/I = (13,500)(12)(5)/91.05 = 8,896 \text{ psi} \tag{o}$$

$$\tau_T = Tr/J = (3,000)(12)(5)/182.1 = 988.4 \text{ psi} \tag{p}$$

$$\tau_{Vx} = (V_x Q)/(It) = [(600)(11.885)]/[(91.05)(2)(0.25)] = 156.6 \text{ psi} \tag{q}$$

Consider point A, which is on the tension side of the pole, and show the stresses on an elemental volume removed from the annular ring at the base of the pole.

$$(\sigma_{zz})_A = \sigma_{zz} = 8,896 \text{ psi} \tag{r}$$

$$(\tau_T)_A = \tau_T = 988.4 \text{ psi} \tag{s}$$

$$(\tau_{Vx})_A = 0 \tag{t}$$

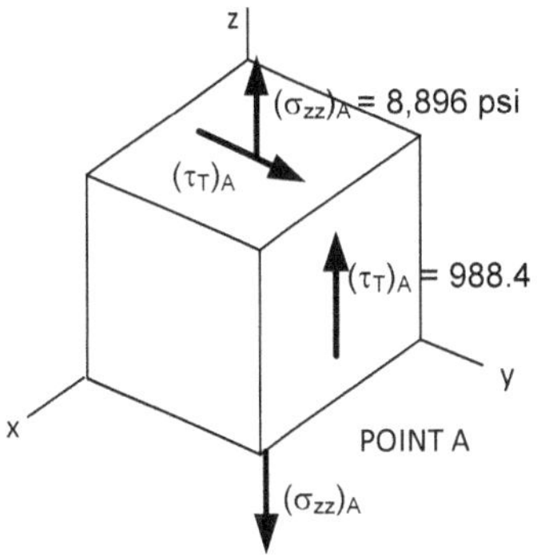

Fig. E10.3c

For point A, only the shear stress (τ_T) due to the torque appears and it acts in the + y direction. The shear stress (τ_{Vx}) due to the shear force V_x does not appear, because of the location of point A. Recall from the analysis of beams that the shear stress due to transverse loadings was a maximum at the neutral axis and zero at the top and bottom edges of the beam. For the shear force V_X, point A is at the top edge of the cross section; hence, $\tau_{Vx})_A = 0$.

Finally consider point B, which is on the neutral axis of the pole and show the stresses on an elemental volume removed from the annular ring at the base of the pole.

$$(\sigma_{zz})_B = 0 \quad \text{and} \quad (\tau)_B = -\tau_T - \tau_{Vx} = -1,145 \text{ psi} \tag{u}$$

318 Combined Loading; Chapter 10

For point B, the normal stress component $(\sigma_{zz})_A$ vanishes, because point B is located on the y-axis, which acts as the neutral axis for M_y. Also both shear stress components are present and algebraically sum because they both act in the same direction (– x axis).

EXAMPLE E10.4

A machine component is fabricated from a tube that has been formed into a 90° bend, as illustrated in Fig. E10.4. The outside diameter of the tube is 60 mm and its inside diameter is 50 mm. A design engineer wants to place an access hole in the tube at a location 150 mm from its base and is concerned with the possibility of failure by doing so. He requests you to determine the maximum principal stress and the maximum shear stress at point A, at this elevation on the tube. An exercise is included in the list of problems to determine the same quantities at point B.

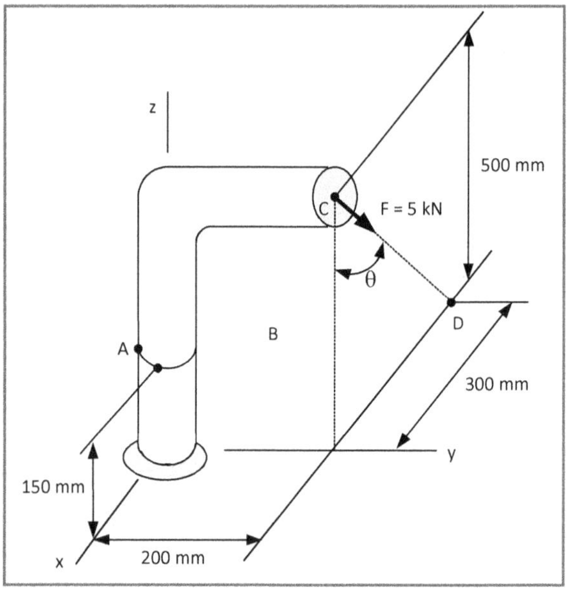

Fig. E10.4

Solution: Let's first determine the angle θ and resolve the force **F** into its components F_x and F_z.

$$\theta = \tan^{-1}(300/500) = 30.96° \quad (a)$$

$$F_x = F \sin\theta = 5 \sin(30.96°) = 2.572 \text{ kN} \quad (b)$$

$$F_z = F \cos\theta = 5 \cos(30.96°) = 4.287 \text{ kN} \quad (c)$$

Next prepare a FBD of the segment of the tube with a section cut at the 150 mm elevation, as shown in Fig. E10.4a.

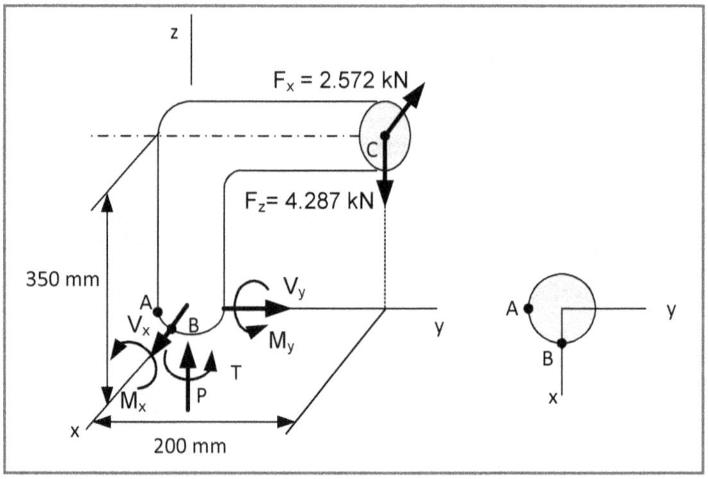

Fig. E10.4a

Let's use this FBD together with the equilibrium relations to determine the internal forces at the section cut.

$\Sigma F_x = 0$ $\qquad V_x = 2.572$ kN $\qquad$ (d)

$\Sigma F_y = 0$ $\qquad V_y = 0$ $\qquad$ (e)

$\Sigma F_z = 0$ $\qquad P = 4.287$ kN $\qquad$ (f)

$\Sigma M_x = 0$ $\qquad M_x = (4.287 \times 10^3)(0.200) = 857.4$ N-m $\qquad$ (g)

$\Sigma M_y = 0$ $\qquad M_y = (2.572 \times 10^3)(0.350) = 900.2$ N-m $\qquad$ (h)

$\Sigma M_z = 0$ $\qquad T = -(2.572 \times 10^3)(0.200) = -514.4$ N-m $\qquad$ (i)

The geometric parameters that are needed to determine the stresses at points A and B are:

$A = (\pi/4)(d_o^2 - d_i^2) = (\pi/4)[(60)^2 - (50)^2] = 863.9$ mm^2. $\qquad$ (j)

$I = I_{xx} = I_{yy} = (\pi/64)(d_o^4 - d_i^4) = (\pi/64)[(60)^4 - (50)^4] = 329.4 \times 10^3$ mm^4 $\qquad$ (k)

$J = 2I = 658.8 \times 10^3$ mm^4 $\qquad$ (l)

$r = c = 30$ mm. $\qquad$ (m)

$Q = (2/3)(r_o^3 - r_i^3) = (2/3)[(30)^3 - (25)^3] = 7{,}583$ mm^3 $\qquad$ (n)

Let's determine the stresses at point A due to the five internal forces P, M_x, M_y, V_x and T. In the process, we will use the geometric parameters listed above.

$\sigma_{zz} = P/A = (4{,}287)/863.9 = 4.962$ MPa $\qquad$ (compressive) $\qquad$ (o)

$\sigma_{zz} = M_x c/I = (857.4 \times 10^3)(30)/(329.4 \times 10^3) = 78.09$ MPa $\qquad$ (p)

$\sigma_{zz} = M_y y/I = 0$ $\qquad$ (q)

$\sigma_{xx} = \sigma_{yy} = 0$ $\qquad$ (r)

$(\sigma_{zz})_{Total} = 78.09 - 4.962 = 73.13$ MPa $\qquad$ (s)

$\tau_T = Tr/J = (514.4 \times 10^3)(30)/(658.8 \times 10^3) = 23.42$ MPa $\qquad$ (t)

$\tau_{Vx} = (V_x Q)/(It) = [(2.572 \times 10^3)(7{,}583)]/[(329.4 \times 10^3)(10)] = 5.921$ MPa $\qquad$ (u)

$(\tau)_{Total} = \tau_T - \tau_{Vx} = 23.42 - 5.921 = 17.50$ MPa $\qquad$ (v)

Before proceeding to determine the principal stresses and the maximum shear stress, let's draw an element removed from the tube at point A, as shown in Fig. E10.4b. Let's also examine a two dimensional element representing the x-z plane at point A.

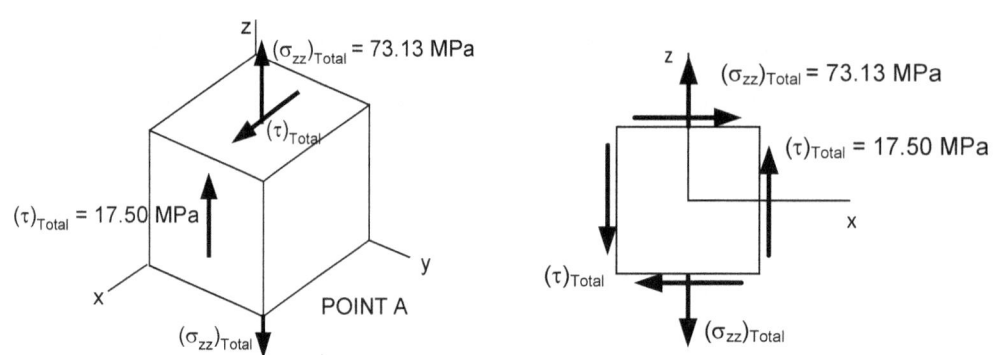

Fig. E10.4b

The principal stresses are given by:

$$\sigma_{Max, Min} = \frac{\sigma_{xx}+\sigma_{zz}}{2} \pm \sqrt{\left(\frac{\sigma_{xx}-\sigma_{zz}}{2}\right)^2 + \tau_{xz}^2} \quad (w)$$

$$\sigma_{Max, Min} = \frac{0+73.13}{2} \pm \sqrt{\left(\frac{0-73.13}{2}\right)^2 + (17.50)^2} = 36.56 \pm 40.54 \text{ MPa} \quad (x)$$

$$\sigma_{Max} = 77.10 \text{ MPa} \qquad \sigma_{Min} = -3.980 \text{ MPa} \quad (y)$$

The maximum shear stress is given by:

$$\tau_{Max} = \sqrt{\left(\frac{\sigma_{xx}-\sigma_{zz}}{2}\right)^2 + \tau_{xz}^2} = \sqrt{\left(\frac{0-73.13}{2}\right)^2 + (17.50)^2} = 40.54 \text{ MPa} \quad (z)$$

EXAMPLE E10.5

Determine the maximum normal stress on the vertical section A -- A of the cantilever beam, shown in the Fig. E10.5.

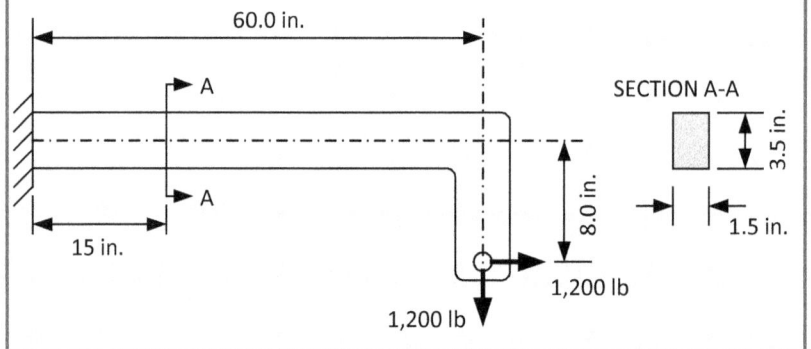

Solution:

First prepare a FBD of the right hand portion of the cantilever beam at section cut, as shown in Fig. E10.5a.

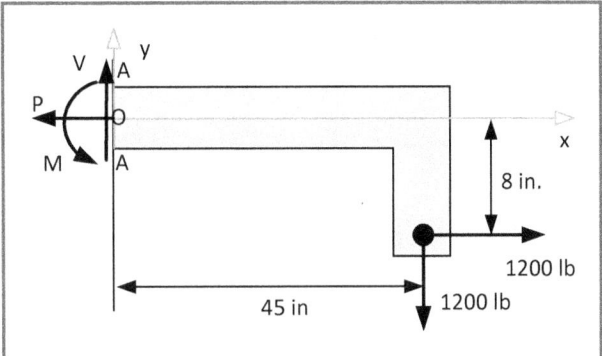

Fig. E10.5a

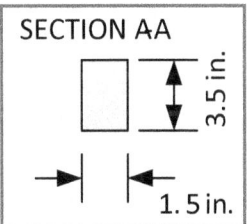

From $\Sigma F_x = 0$, we find that:
$$P = 1,200 \text{ lb} \quad (a)$$

From $\Sigma F_y = 0$, we find that:
$$V = 1,200 \text{ lb} \quad (b)$$

Because $\Sigma M_0 = 0$, we write:
$$M = (1,200)(45) - (1,200)(8) = 44,000 \text{ in-lb} \quad (c)$$

The area of the section is: $A = (1.5)(3.5) = 5.25 \text{ in}^2$. $\quad (d)$

The moment of the area is: $1/12 \, b \, h^3 = (1.5)(3.5)^3/12 = 5.395 \text{ in}^4 \quad (e)$

The distance from the centerline to the outer fiber is:
$$c = 3.5/2 = 1.75 \text{ in} \quad (f)$$

The normal stress due to P is:
$$\sigma_P = P/A = 1,200/5.25 = 228.6 \text{ psi} = 0.2286 \text{ ksi} \quad (g)$$

The bending stress at the top edge of the beam is:
$$\sigma_M = Mc/I = (44,000)(1.75)/5.359 = 14.50 \text{ ksi} \quad (h)$$

The maximum stress occurs at the top edge of the beam and is:
$$\sigma_{Max} = \sigma_P + \sigma_M = 0.2286 + 14.50 = 14.73 \text{ ksi}$$

EXAMPLE E10.6

The large bracket, shown in Fig. E10.6, is loaded in its plane of symmetry with a force F = 4 kN inclined by an angle θ = 10°. Determine the principal stresses and the maximum shearing stresses at points A, B and C. Points A and C are in the web adjacent to the flange of the bracket.

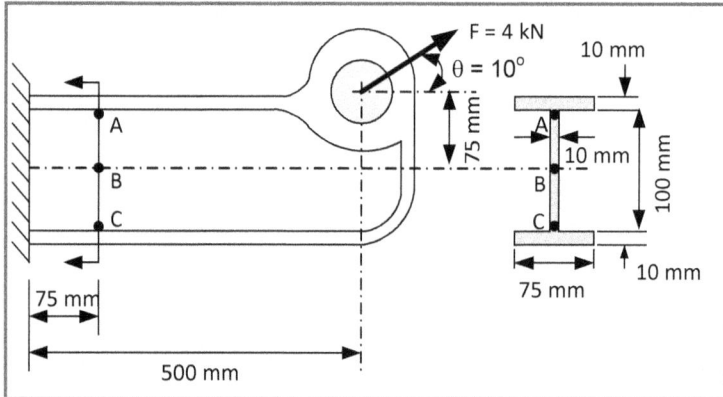

Fig. E10.6

Solution: First prepare a FBD of the right hand portion of the large bucket at the section cut, as shown in Fig. E10.6a.

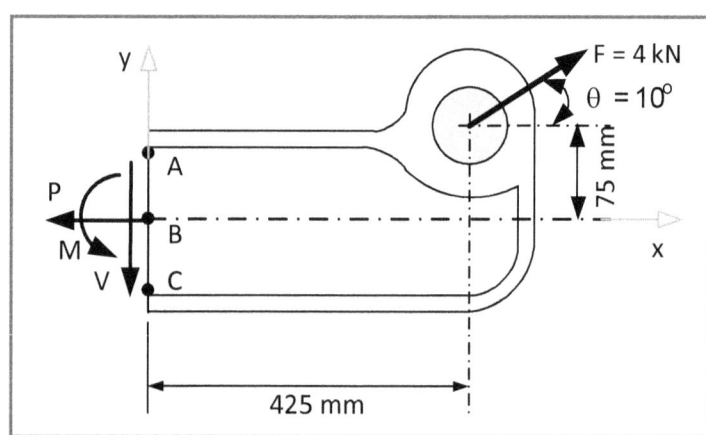

Fig. E10.6a

From $\Sigma F_x = 0$, we find that:
$$P = F \cos \theta = (4) \cos (10°) = 3.939 \text{ kN} \qquad (a)$$

From $\Sigma F_y = 0$, we find that:
$$V = F \sin \theta = (4) \sin (10°) = 0.6946 \text{ kN} \qquad (b)$$

Because $\Sigma M_B = 0$, we write:
$$M + (4) \sin (10°)(425) - (4) \cos (10°)(75) = 0 \qquad (c)$$

Solving Eq. (c) for the moment M gives:
$$M = 0.2404 \text{ kN-mm} \qquad (d)$$

Referencing the drawing of the cross section in Fig. E10.6, enables us to determine the area of the section as:

$$A = 2(75)(10) + (100)(10) = 2,500 \text{ mm}^2. \tag{e}$$

The second moment of the area is given by:

$$I = (1/12)(75)(120)^3 + (1/12)(65)(100)^3 = 5.383 \times 10^6 \text{ mm}^4 \tag{f}$$

The distance from the centerline to points A and C is:

$$y_A = y_C = 50 \text{ mm} \tag{g}$$

The first moment of the area at points A and C is:

$$Q_A = Q_C = (75)(10)(55) = 41.25 \times 10^3 \text{ mm}^3 \tag{h}$$

The first moment of the area at point B is:

$$Q_B = Q_A + (50)(10)(25) = 53.75 \times 10^3 \text{ mm}^3 \tag{i}$$

The normal stress σ due to the axial force P is:

$$\sigma = P/A = 3,939/2,500 = 1.576 \text{ MPa} \tag{j}$$

The bending stress at point A is:

$$\sigma_{MA} = M\, y_A/I = (240.4)(50)/5.383 \times 10^6 = 0.002233 \text{ MPa} \tag{k}$$

Note that:
$$\sigma_{MA} = -\sigma_{MC} \tag{l}$$

The shear stress due to the shear force V at points A and C is:

$$\tau_{VA} = \tau_{VC} = V\, Q_A/I\, b = (694.6)(41.25 \times 10^3)/(5.383 \times 10^6)(10) = 0.5323 \text{ MPa} \tag{m}$$

The shear stress due to the shear force V at point B is:

$$\tau_{VB} = V\, Q_B/I\, b = (694.6)(53.75 \times 10^3)/(5.383 \times 10^6)(10) = 0.6936 \text{ MPa} \tag{n}$$

Superimposing the normal stresses at point A, B and C gives:

$$\sigma_A = \sigma_P + \sigma_M = 1.576 + 0.002233 = 1.578 \text{ MPa}$$

$$\sigma_B = \sigma_P = 1.576 \text{ MPa} \tag{o}$$

$$\sigma_C = \sigma_P - \sigma_M = 1.576 - 0.002233 = 1.574 \text{ MPa}$$

The shear stresses at points A, B and C are given by:

$$\tau_A = \tau_{VA} = 0.5323 \text{ MPa}$$

$$\tau_B = \tau_{VB} = 0.6936 \text{ MPa} \tag{p}$$

$$\tau_C = \tau_{VC} = 0.5323 \text{ MPa}$$

A graphic depicting the stress states at elements A, B and C is presented in Fig. E10.6b:

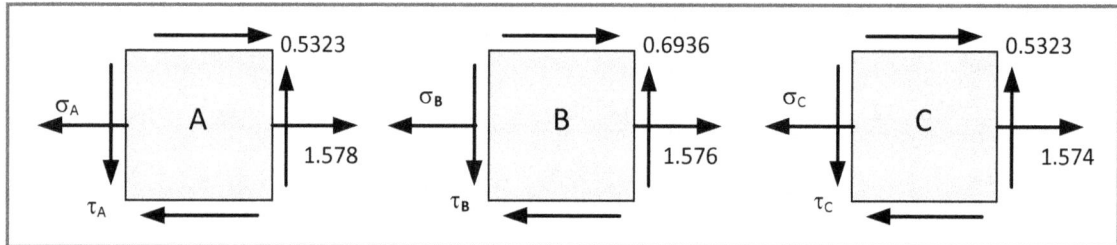

Fig. E10.6b

Consider point A and determine the maximum and minimum normal stresses and the maximum shear stress using Eqs. (9.15) and (9.18).

$$\sigma_{P1, P2} = \frac{\sigma_{xx} + \sigma_{yy}}{2} \pm \sqrt{\left(\frac{\sigma_{xx} - \sigma_{yy}}{2}\right)^2 + \tau_{xy}^2}$$

$$\sigma_{P1, P2} = \frac{1.578 + 0}{2} \pm \sqrt{\left(\frac{1.578 - 0}{2}\right)^2 + (0.5323)^2} \tag{q}$$

$$\sigma_{P1, P2} = 0.789 \pm 0.9519 \text{ MPa}$$

$$\sigma_{P1} = 1.741 \text{ MPa} \quad \text{and} \quad \sigma_{P2} = -0.1628 \text{ MPa}$$

The maximum shear stress at point A is:

$$\tau_{Max} = (\sigma_{P1} - \sigma_{P2})/2 = (1.741 + 0.1628)/2 = 0.9519 \text{ MPa} \tag{r}$$

Similarly for point B:

$\sigma_{P1} = 1.838$ MPa $\sigma_{P2} = -0.2618$ MPa $\tau_{Max} = 1.050$ MPa   (s)

And for point C:

$\sigma_{P1} = 1.737$ MPa $\sigma_{P2} = -0.1631$ MPa $\tau_{Max} = 0.9501$ MPa   (s)

EXAMPLE E10.7

A thin walled cylindrical pressure vessel with an inner diameter 600 mm and a wall thickness of 10 mm, shown in Fig. E10.7, is subjected to an internal pressure of 1.0 MPa. It is also subjected to a torque of 50 kN-m and an axial force of 300 kN, which act through heavy plates attached to the ends of the cylinder. Determine the principal stresses and the maximum shear stresses at point A located on the outer surface of the cylinder.

Fig. E10.7

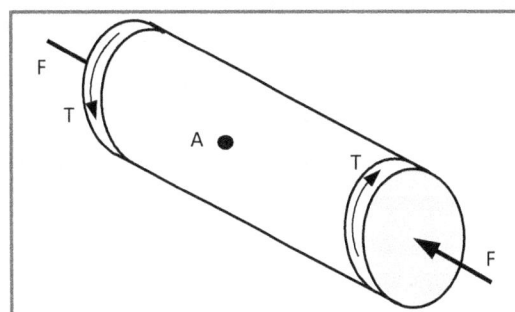

Solution:

Prepare a FBD as shown in Fig. E10.7a.

Fig. E10.7a

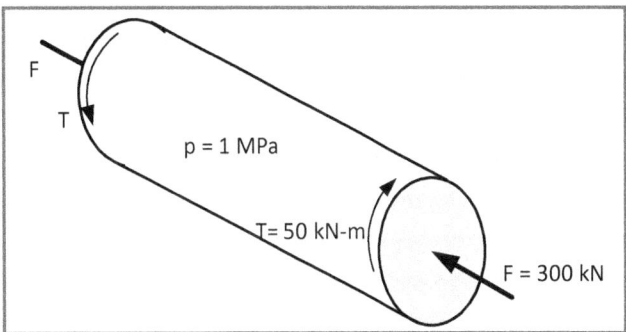

We will determine stresses due to the three loadings — p, F and T. Let's solve for the stresses due to the internal pressure p in the pressure vessel.

$$\sigma_h = p\, r_i/t = (1)(300)/(10) = 30 \text{ MPa}$$

$$\sigma_a = p\, r_i/2t = \sigma_h/2 = 15 \text{ MPa}$$

(a)

The area A of a section is:

$$A = \pi\, (r_o^2 - r_i^2) = \pi\, [(310)^2 - (300)^2] = 19{,}164 \text{ mm}^2 \qquad (b)$$

The polar moment of inertia is:

$$J = (\pi/2)\, (r_o^4 - r_i^4) = (\pi/2)\, [(310)^4 - (300)^4] = 1.783 \times 10^9 \text{ mm}^4 \qquad (c)$$

The stress due to the axial force F is:

$$\sigma_F = F/A = (300 \times 10^3)/(19{,}164) = 15.65 \text{ MPa} \qquad (d)$$

The stress due to the torque T is:

$$\tau_T = T\, r_o/J = (50 \times 10^3)(10^3)(310)/(1.783 \times 10^9) = 8.693 \text{ MPa} \qquad (e)$$

Consider any point on the outer surface of the pressure vessel. The stress occurring an element cut from the surface are presented in Fig. E10.7b.

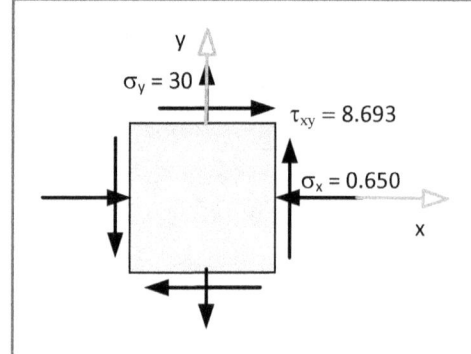

Fig. E10.7b

Next determine the principal stresses as:

$$\sigma_{P1,P2} = \frac{\sigma_{xx}+\sigma_{yy}}{2} \pm \sqrt{\left(\frac{\sigma_{xx}-\sigma_{yy}}{2}\right)^2 + \tau_{xy}^2} = \frac{-0.65+30}{2} \pm \sqrt{\left(\frac{0.65-30}{2}\right)^2 + (8.693)^2}$$

$\sigma_{P1,P2} = 14.675 \pm 17.619$ MPa (f)

$\sigma_{P1} = 32.29$ MPa   and $\sigma_{P2} = -2.944$ MPa

The maximum shear stresses are:

$$\tau_{Max} = (\sigma_{P1} - \sigma_{P2})/2 = (32.29 + 2.944)/2 = 17.62 \text{ MPa} \qquad (g)$$

EXAMPLE E10.8

An angled strut 6 in. thick is subjected to a force F = 40 kip and a uniformly distributed load q = 10 kip/ft, as illustrated in Fig. E10.8. Determine the principal stresses at points A, B and C.

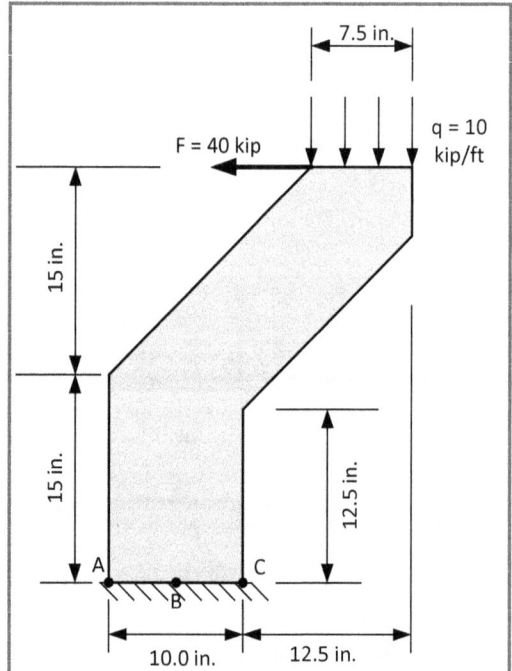

Fig. E10.8

Solution: Prepare a FBD as shown in Fig. E10.8a.

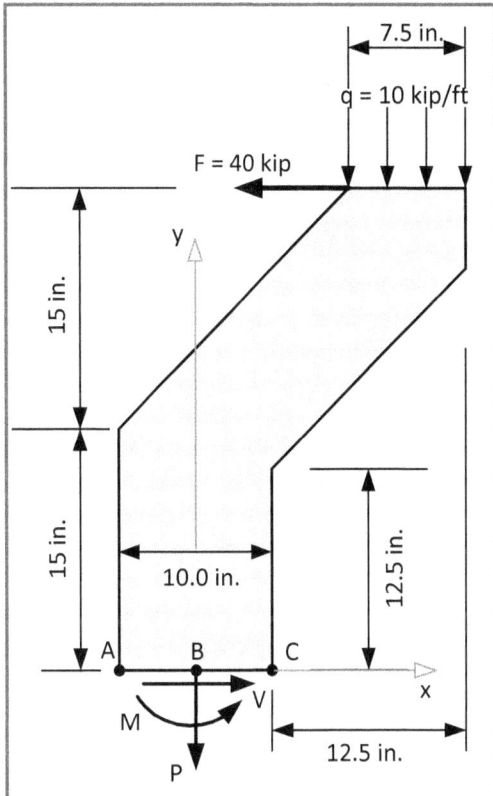

Fig. E10.8a

From $\Sigma F_y = 0$, we find that:

$$P = -(10 \text{ kip/ft})(7.5 \text{ in})(\text{ft}/12 \text{ in}) = -6.25 \text{ kip} \quad (a)$$

From $\Sigma F_x = 0$, we find that $V = 40$ kip $\quad$ (b)

Because $\Sigma M_B = 0$, we write:

$$M + (40)(30) - (6.25)(17.5 - 3.75) = 0$$

$$M = -1,114 \text{ kip-in} \quad (c)$$

The area at the base of the member is: $A = (6)(10) = 60 \text{ in}^2$. $\quad$ (d)

The moment of the area is: $1/12 \, b \, h^3 = (6)(10)^3/12 = 500 \text{ in}^4$ $\quad$ (e)

The distance from the centerline to the outer fiber is:

$$c = 10/2 = 5 \text{ in} \qquad b = 6 \text{ in} \quad (f)$$

The first moment of the area at point B is:

$$Q_B = (6)(5)(2.5) = 75 \text{ in}^3 \quad (g)$$

The normal stress due to P is:

$$\sigma_P = P/A = -6.25/60 = -0.1042 \text{ ksi} \quad (h)$$

The bending stress due to M is:

$$\sigma_M = Mc/I = (1114)(5)/500 = 11.14 \text{ ksi} \qquad (i)$$

The shear stress at point B is:

$$\tau_{VB} = V Q_B/I b = (40)(75)/(500)(6) = 1 \text{ ksi} \qquad (j)$$

Consider point A and note:

$$\sigma_A = \sigma_P - \sigma_M = -0.1042 - 11.14 = -11.24 \text{ ksi} \qquad (k)$$

and $\qquad \tau_A = 0 \qquad (l)$

At point B the normal stress $\sigma_B = \sigma_P = -0.1042$ ksi $\qquad (m)$

At point C the normal stress $\sigma_B = \sigma_P + \sigma_M = -0.1042 + 11.14 = 11.04$ ksi $\qquad (n)$

and $\qquad \tau_C = 0 \qquad (l)$

We show the state of stress at the three points graphically in Fig. 10.8b.

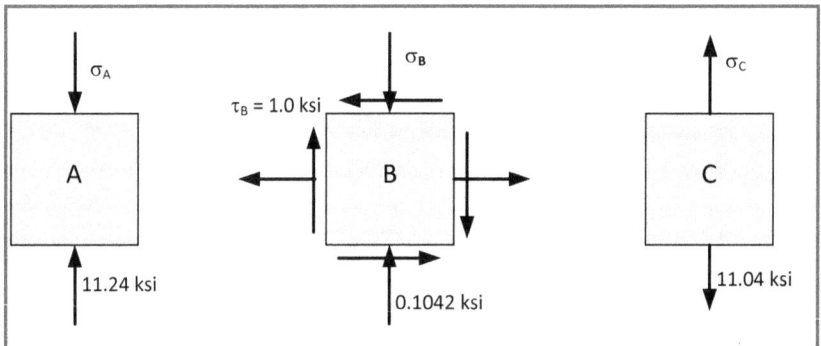

Fig. 10.8b

At point A it is evident that $\sigma_{P1} = 0 \qquad$ and $\qquad \sigma_{P2} = -11.24$ ksi $\qquad (o)$

At point B we write:

$$\sigma_{P1, P2} = \frac{-0.1042 + 0}{2} \pm \sqrt{\left(\frac{-0.1042 - 0}{2}\right)^2 + (1)^2}$$

which yields:
$$\sigma_{P1} = 0.9493 \text{ ksi} \qquad \text{and} \qquad \sigma_{P2} = -1.053 \text{ ksi} \qquad (p)$$

At point C it is evident that $\sigma_{P1} = 11.04$ ksi and $\sigma_{P2} = 0 \qquad (q)$

10.2 SUMMARY

This chapter covered stresses due to combined loading of machine components. While no new equations were derived in this section, several examples were included to demonstrate the methods used in combining shear and normal stresses.

PROBLEMS

10.1 An axle of an all-terrain vehicle is subjected to the forces and torque, shown in the figure below. If the axle diameter is d, determine the principal planes and the principal stresses at point A on the shaft. Also determine the maximum shearing strains at point B.

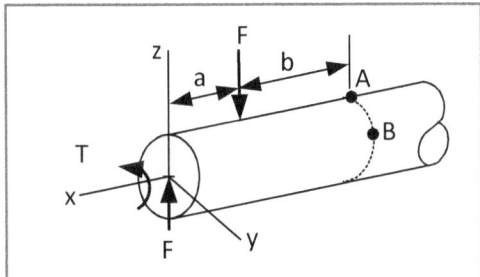

Prob. No.	d	a	b	F	T
10.1a	15 mm	100 mm	150 mm	2.00 kN	200 N-m
10.1b	18 mm	120 mm	175 mm	2.25 kN	250 N-m
10.1c	20 mm	140 mm	200 mm	2.50 kN	300 N-m
10.1d	22 mm	160 mm	225 mm	3.00 kN	400 N-m
10.1e	25 mm	200 mm	250 mm	4.00 kN	500 N-m

10.2 An advertising sign located along a city street is cantilevered from a pole as illustrated in The figure below. The pressure due to the wind impinging on the sign is uniformly distributed over its area and equal to p in the x direction. The pole is fabricated from a tube with an outside diameter d and a wall thickness t. Determine the stresses at points A and B located at the base of the pole.

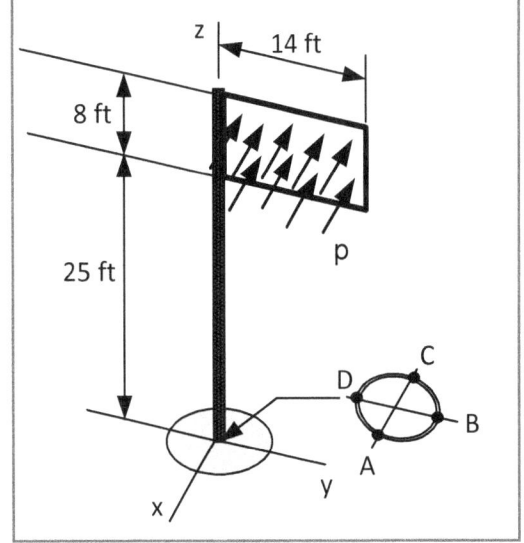

Prob. No.	d	t	p
10.2a	8 in.	0.20 in.	10 lb/ft^2
10.2b	10 in.	0.25 in.	12 lb/ft^2
10.2c	6 in	0.125 in.	15 lb/ft^2
10.2d	12 in	0.312 in.	18 lb/ft^2
10.2e	14 in	0.375 in.	20 lb/ft^2

330 Combined Loading; Chapter 10

10.3 Repeat Problem 10.2 for points C and D, located at the base of the pole.

10.4 A machine component is fabricated from a tube that has been formed into a 90° bend, as illustrated in the figure below. The outside diameter of the tube is d_o and its inside diameter is d_i. Determine the maximum principal stress and the maximum shear stress at the elevation L at points A and B.

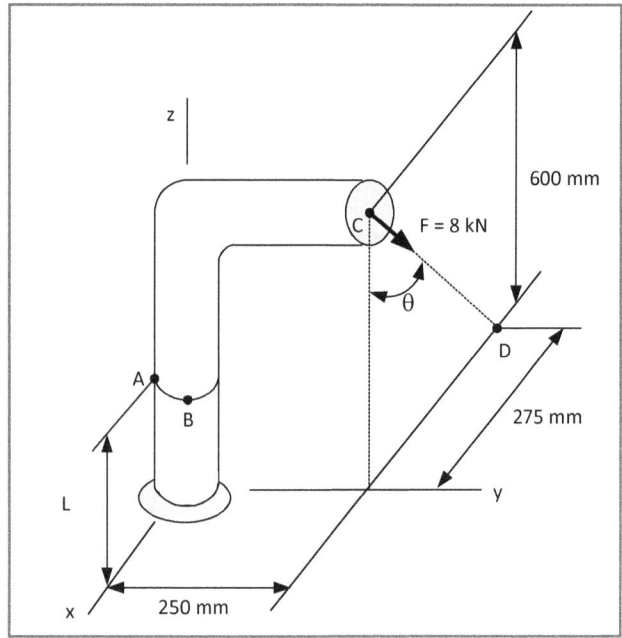

Prob. No.	d_o	d_i	L
10.4a	100 mm	88 mm	100 mm
10.4b	150 mm	140 mm	120 mm
10.4c	125 mm	110 mm	180 mm
10.4d	200 mm	180 mm	200 mm
10.4e	90 mm	75 mm	150 mm

10.5 A U shaped machine component, illustrated in the figure below, is subjected to opposing forces of magnitude F. If the component is fabricated from the type of steel specified in the table below, determine the safety factor based on yield strength using the maximum normal stress theory.

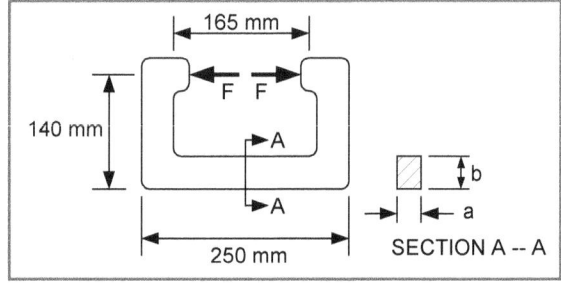

Prob. No.	F	a	b	Steel
10.5a	10 kN	20 mm	40 mm	1018 A
10.5b	20 kN	20 mm	50 mm	1212 HR
10.5c	25 kN	30 mm	60 mm	1045 HR
10.5d	20 kN	30 mm	50 mm	4340HR
10.5e	50 kN	25 mm	75 mm	52100 A

10.6 Determine the maximum normal stress on the vertical section A -- A of the cantilever beam shown in the figure below. The numerical parameters for this problem are given in the table below.

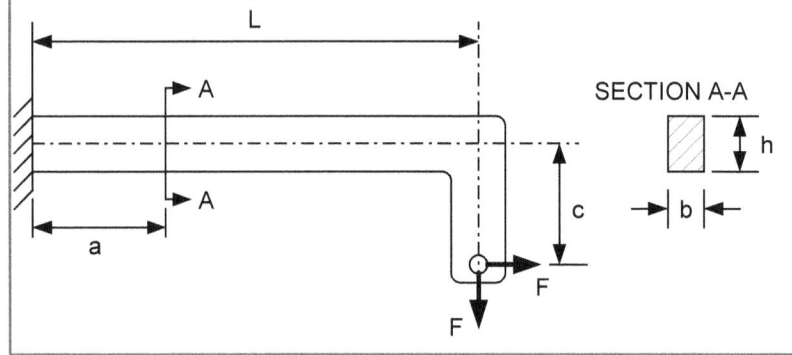

Prob. No.	F	a	b	c	h	L
10.6a	4.0 kN	300 mm	40 mm	300 mm	75 mm	1.5 m
10.6b	1200 lb	15 in.	1.5 in.	8.0 in.	3.5 in.	60 in.
10.6c	5.0 kN	500 mm	30 mm	250 mm	60 mm	2.0 m
10.6d	1000 lb	32 in.	1.25 in.	10.0 in.	3.0 in.	71 in.
10.6e	8.0 kN	150 mm	45 mm	325 mm	90 mm	1.75 m

10.7 The large bracket, shown in the figure below, is loaded in its plane of symmetry with a force F inclined by an angle θ. Determine the principal stresses and the maximum shearing stresses at points A, B and C. Points A and C are in the web adjacent to the flange of the bracket. The numerical parameters for this problem are given in the table below.

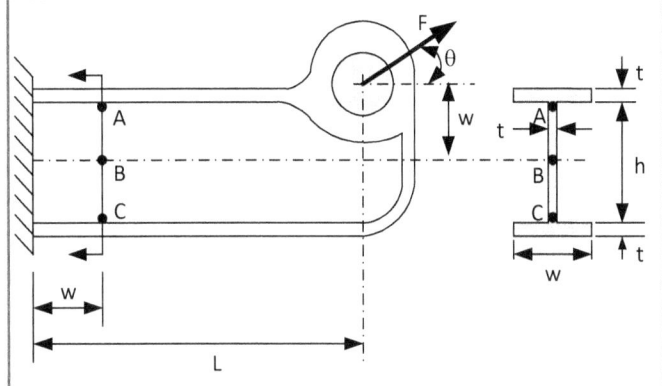

Prob. No.	F	θ	t	w	h	L
10.7a	4.0 kN	10°	10 mm	75 mm	100 mm	500 mm
10.7b	1200 lb	20°	0.25 in.	4.0 in.	4.0 in.	10 in.
10.7c	5.0 kN	30°	5 mm	50 mm	80 mm	400 mm
10.7d	1000 lb	45°	0.15 in.	3.0 in.	3.0 in.	20 in.
10.7e	8.0 kN	60°	8 mm	60 mm	90 mm	300 mm

10.8 A thin walled cylindrical pressure vessel with an inner diameter d and a wall thickness t, shown in the table below, is subjected to an internal pressure p. It is also subjected to a torque T and an axial force F, which act through heavy plates attached to the ends of the cylinder. Determine the principal stresses and the maximum shear stresses of a point located on the outside of the cylinder.

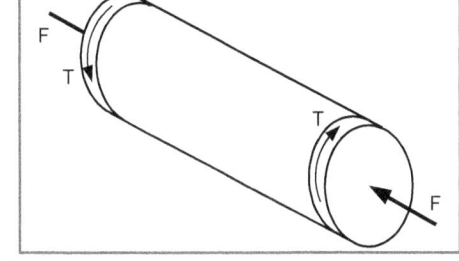

Prob. No.	F	T	p	d	t
10.8a	300 kN	50 kN-m	1.0 MPa	600 mm	10 mm
10.8b	120 kip	50 kip-ft	100 psi	20.0 in.	0.40 in.
10.8c	200 kN	40 kN-m	1.2 MPa	500 mm	15 mm
10.8d	100 kip	65 kip-ft	150 psi.	30.0 in.	0.375 in.
10.8e	250 kN	30 kN-m	0.8 MPa	400 mm	18 mm

10.9 An angled strut 8 in. (200 mm) thick is fabricated from concrete (no reinforcement) and subjected to a force F and a uniformly distributed load q, as illustrated in the figure below. Determine the principal stresses at points A, B and C. Note that a = 0.75 L, b = 1.5 L and c = 1.25 L.

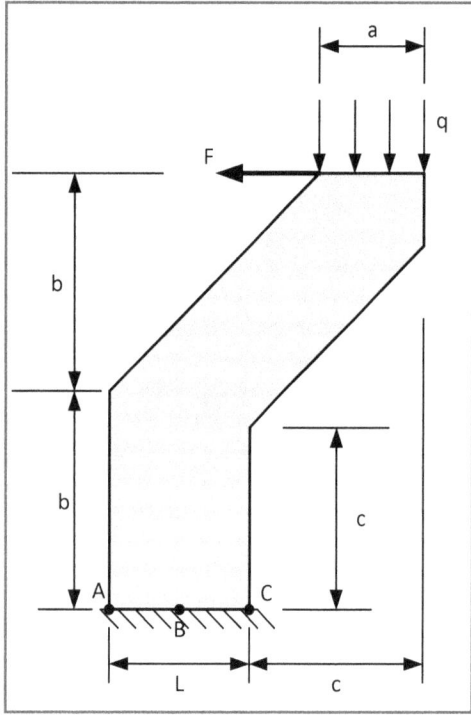

Prob. No.	F	q	L
10.9a	100 kN	20 kN/m	200 mm
10.9b	40 kip	10 kip/ft	10 in.
10.9c	80 kN	25 kN/m	240 mm
10.9d	25 kip	15 kip/ft	12 in..
10.9e	125 kN	30 kN/m	280 mm

CHAPTER 11

BUCKLING OF COLUMNS

11.1 INTRODUCTION

In this chapter, the concept of instability of columns under the action of a compressive force is introduced. This is a new concept that leads to a markedly different mode of failure of many structural elements, such as columns, plates and shells. Even the analysis employed differs from our normal approach. In the previous chapters, we were concerned with stresses and deflections. If the stresses were less than the material's strength, the structural element was safe. If the deflections did not exceed some specified limit, the structural element was satisfactory. The concept of stability requires a designer to determine a critical load to insure that the structural element will not become unstable and collapse prior to either yield or fracture.

When the compressive force on a column exceeds some critical value P_{CR}, the column becomes unstable and collapses. This mode of failure is extremely dangerous, because the collapse is catastrophic. When a column buckles, without shedding load, it initially deforms in the shape of a half sine wave and then simply folds when a plastic hinge forms at its center, as illustrated in Fig. 11.1.

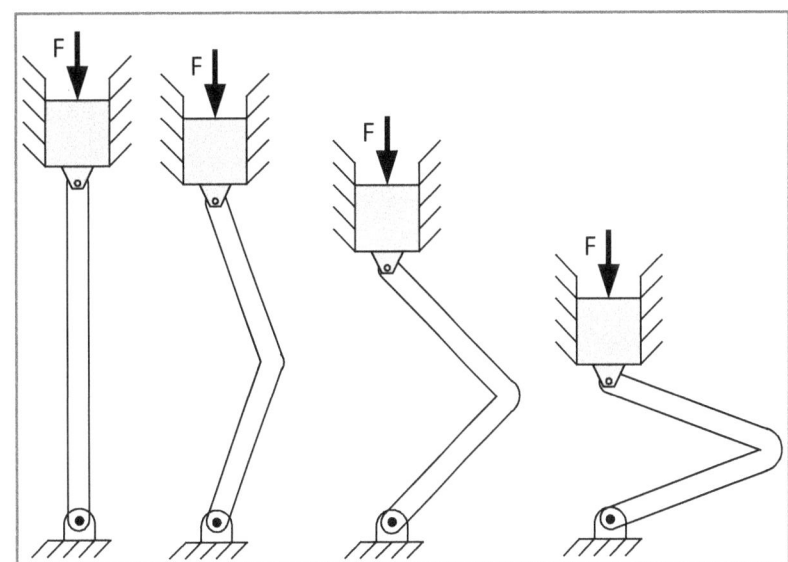

Fig. 11.1 Catastrophic collapse of a column with pinned ends during a buckling event.

In developing the equations for predicting the buckling of a column, we deviate from our previous practice of neglecting the small deformations of a structural element under the action of a load. When computing stress from $\sigma = P/A$, we do not take into account the change in A due to the strains induced by the force P. However, the entire concept of elastic instability is based on load-induced deformations. In analyzing columns, we consider the moment produced by the axial force P times the deflection y. To show this new analytical approach, consider a column with both ends pinned subjected to a centric force P in the next section.

11.2 BUCKLING OF COLUMNS WITH BOTH ENDS PINNED

Consider a column of length L with both ends pinned subjected to a centric force P, as illustrated in Fig. 11.2a. As the magnitude of the centric force increases, the column will bow slightly, as indicated in Fig 11.2b. The column is at this stage of loading elastic and stable. To perform the analysis, we model the pinned end supports and the column, as shown in Fig. 11.2c. We have selected the x-axis in the vertical direction and the y-axis in the horizontal direction to adapt the column coordinate system to that previously used for a beam. The origin is placed at the top of the column.

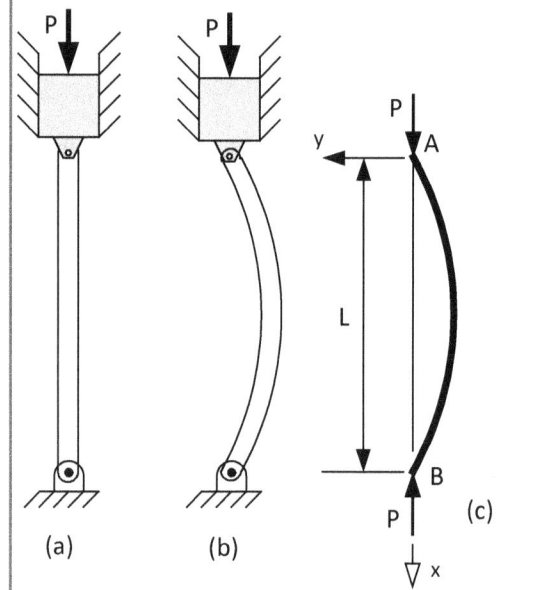

Fig. 11.2 Modeling a column with pinned ends.

Next, let's make a section cut and remove a portion of the top of the column as a FBD, as indicated in Fig. 11.3.

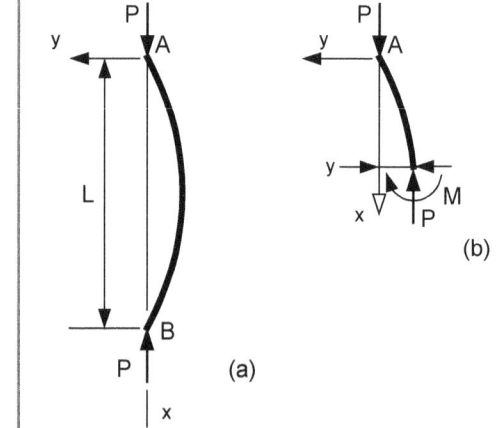

Fig. 11.3 FBD of the top portion of the column.

An examination of the FBD shows a couple, which produces a moment $M = -Py$. This is a bending moment and Eq. (8.4) is valid. Hence, we write:

$$\frac{d^2y}{dx^2} = \frac{M(x)}{EI} = -\frac{Py}{EI} \qquad (11.1)$$

Rearranging Eq. (11.1) to write the differential equation in standard format yields:

$$\frac{d^2y}{dx^2} + p^2 y = 0 \tag{11.2}$$

where $p = \sqrt{\dfrac{P}{EI}}$ (11.3)

Equation (11.2) is a linear, second-order differential equation with constant coefficients, which can be solved for y to obtain:

$$y = C_1 \sin(px) + C_2 \cos(px) \tag{11.4}[1]$$

where C_1 and C_2 are constants of integration.

We follow the procedure established in Chapter 8 and solve for the constants of integration using boundary conditions. Because both ends of the column are pinned, we write:

$$y(0) = 0 \qquad y(L) = 0 \tag{11.5}$$

It is clear from the first of the boundary conditions listed in Eq. (11.5) that $C_2 = 0$. Substituting the second boundary condition (y = 0 at x = L) into Eq. (11.4) yields:

$$C_1 \sin pL = 0 \tag{11.6}$$

There are two interpretations for Eq. (11.6). First, the equation is satisfied if $C_1 = 0$, which gives y = 0 for all of x. This is not realistic, because it implies that the column remains straight and will never buckle. Because buckling has been demonstrated in the laboratory, this solution is discarded. The remaining solution to Eq. (11.6) is obtained by seeking values of pL that permit sin pL = 0.

If $\quad pL = n\pi \quad$ then $\quad \sin pL = 0$

Then from Eq. (11.3), we write:

$$p = \sqrt{\frac{P}{EI}} = \frac{n\pi}{L}$$

$$P = \frac{n^2 EI \pi^2}{L^2} \tag{11.7}$$

Again, it is necessary to interpret Eq. (11.7), because it contains a counter n, which can be 0, 1, 2, etc. We seek the lowest meaningful value of n. Clearly n = 0 does not produce a realistic result, because this value gives P = 0. We seek a value for P > 0. If we select n = 1, then:

$$P_{CR} = \frac{\pi^2 EI}{L^2} \tag{11.8}$$

where P_{CR} is the critical centric compression load that initiates buckling and catastrophic collapse of the column. Note that I is the minimum value of the moment of inertia for the cross sectional area of the column.

---

[1] This solution can be verified by differentiating Eq. (11.4) twice with respect to x to obtain d^2y/dx^2. Then substitute d^2y/dx^2 and y into Eq.(11.2).

Equation (11.8) is known as the Euler relation for elastic buckling. It is valid for long slender columns[2] with both ends pinned.

Returning to Eq. (11.4), we may use Eq. (11.7) and express the deflection of the column as:

$$y = C_1 \sin \frac{\pi x}{L} \qquad (11.9)$$

This result indicates that the column deflects as a sinusoid, when the load $P = P_{CR}$. The fact that we cannot determine the amplitude (C_1) of the sin wave is not important, because the column has become unstable and will quickly collapse.

Let's apply the Euler buckling theory to determine the critical load for two different columns both with pinned ends.

EXAMPLE E11.1

A column 18 feet long is fabricated from an aluminum tube with a 6 in. outside diameter and a 5.6 in. inside diameter. Determine the critical buckling force if the tube is pinned at both ends.

Solution: We employ Eq. (11.8) to write the expression for the critical buckling force as:

$$P_{CR} = \frac{\pi^2 EI}{L^2} \qquad (a)$$

The modulus of elasticity $E = 10.4 \times 10^6$ psi for an aluminum alloy is from Appendix B-1.

The minimum value of the moment of inertia for a tube is:

$$I = (\pi/64)[d_o^4 - d_i^4] = (\pi/64)[(6)^4 - (5.6)^4] = 15.34 \text{ in.}^4 \qquad (b)$$

Substituting Eq. (b) into Eq. (a) yields:

$$P_{CR} = \frac{\pi^2 EI}{L^2} = \frac{\pi^2 (10.4 \times 10^6) 15.34}{(18)^2 (12)^2} = 33.75 \text{ kip}$$

EXAMPLE E11.2

A column 10 m long is fabricated from a square tube (1020 HR steel) shown in Fig. E11.2. Determine the critical buckling force if the tube is pinned at both ends. Also determine the compressive stress in the tube and comment on its magnitude.

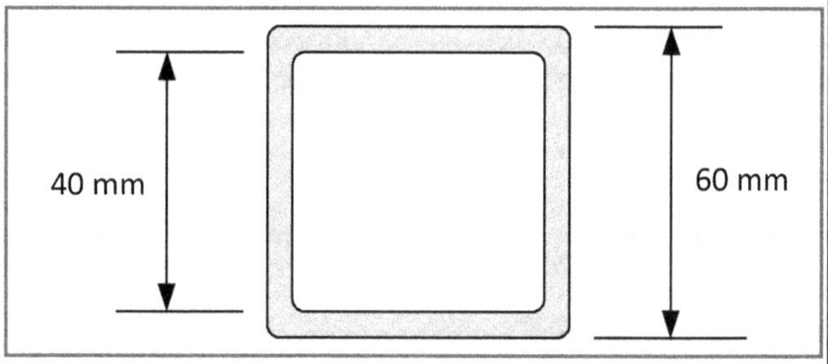

Fig. E11.2

---

[2] We will discuss the distinction among long slender columns, intermediate columns and short columns later in this chapter.

Solution: We employ Eq. (11.8) to write the expression for the critical buckling force as:

$$P_{CR} = \frac{\pi^2 EI}{L^2} \quad \text{(a)}$$

The modulus of elasticity E = 207 GPa for the steel tube is from Appendix B-1.

The minimum value of the moment of inertia for the square tube is:

$$I = (1/12)[h_o^4 - h_i^4] = (1/12)[(60)^4 - (40)^4] = 0.8667 \times 10^6 \text{ mm}^4 \quad \text{(b)}$$

Substituting Eq. (b) into Eq. (a) yields:

$$P_{CR} = \frac{\pi^2 EI}{L^2} = \frac{\pi^2 (207 \times 10^3)(0.8667 \times 10^6)}{(10 \times 10^3)^2} = 17.71 \text{ kN} \quad \text{(c)}$$

The axial stress produced by the critical force P_{CR} is:

$$\sigma_{CR} = P_{CR}/A = 17.71 \times 10^3/[(60)^2 - (40)^2] = 8.855 \text{ MPa} \quad \text{(d)}$$

The yield strength of 1020 HR steel is listed in Appendix B-2 as 290 MPa; hence, the stress at the critical load that produces buckling is only 3.0% of the yield strength. This example illustrates the danger of collapse due to buckling when columns are long and slender.

11.3 INFLUENCE OF END CONDITIONS

The critical buckling load is profoundly affected by the end conditions of the column. Increasing the constraint at either end markedly increases the critical load required for the onset of elastic instability. We will consider three additional sets of boundary conditions to show the influence of constraint on buckling load.

1. Top — pinned and bottom — built-in.
2. Top — free and bottom — built-in.
3. Top — built-in and bottom — built-in.

11.3.1 Column Buckling with One Pinned and One Built-in End.

Consider the column with its bottom built-in and its top pinned, as shown in Fig. 11.4. It is subjected to an axial load P, as shown in Fig. 11.5. To begin the analysis, we draw the deformed shape of the column in the initial stages of instability. In preparing this drawing, we recognize that the slope of the elastic curve at the built-in end and the deflection at both ends must be zero.

Next, let's make a section cut and remove a portion of the top of the column, as a FBD as indicated in Fig. 11.6.

338 — Buckling of Columns — Chapter 11

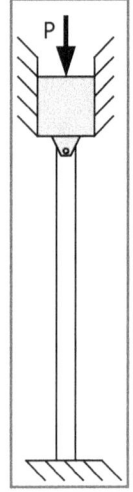

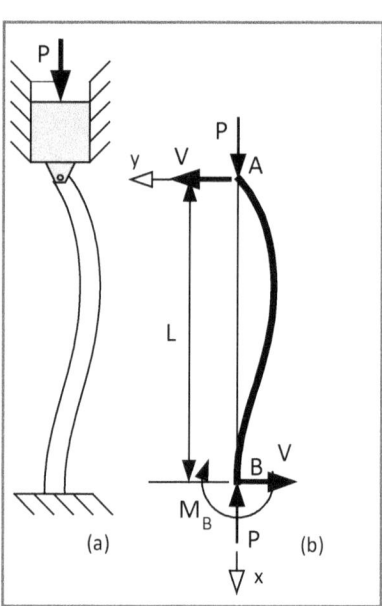

Fig. 11.4 Column with one end fixed and the other end pinned.

Fig. 11.5 Deformation of the column at initiation of elastic instability.

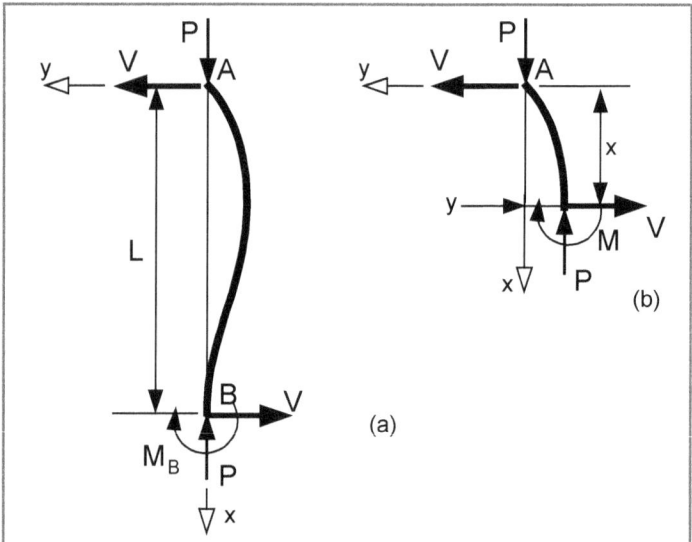

Fig. 11.6 FBD of the top portion of the column.

An examination of the FBD in Fig. 11.6b, shows two couples produced by the equal and opposite forces V and P. These couples result in a bending moment M given by:

$$M = -Py - Vx \qquad (11.10)$$

Substituting Eq. (11.10) into Eq. (8.4) yields:

$$\frac{d^2y}{dx^2} = \frac{M(x)}{EI} = -\frac{1}{EI}(Py + Vx) \qquad (11.11)$$

Rearranging Eq. (11.11) to write the differential equation in standard format gives:

$$\frac{d^2y}{dx^2} + p^2 y = -\frac{Vx}{EI} \qquad (11.12)$$

where $p = \sqrt{\dfrac{P}{EI}}$

Equation (11.12) is a linear, second-order differential equation with constant coefficients; however, it is not homogeneous. We obtain the homogeneous solution by setting the right hand side of Eq. (11.12) equal to zero and solving for y_h to obtain:

$$y_h = C_1 \sin(px) + C_2 \cos(px) \qquad (11.13)$$

where C_1 and C_2 are constants of integration.

The particular solution is an expression for y_p that satisfies Eq. (11.12). Consider the expression for y_p shown below:

$$y_p = -Vx/P \qquad (11.14)$$

The expression in Eq. (11.14) satisfies Eq. (11.12) and hence, it represents the particular solution. The general solution is given by:

$$y = y_h + y_p = C_1 \sin(px) + C_2 \cos(px) - Vx/P \qquad (11.15)$$

We follow the usual procedure to solve for the constants of integration using boundary conditions. Because the top of the column is pinned, we write:

$$y(0) = 0 \qquad (11.16)$$

The boundary conditions for the built-in end are:

$$y(L) = 0 \qquad dy/dx\,(L) = 0 \qquad (11.17)$$

It is clear from the boundary conditions listed in Eq. (11.16) that $C_2 = 0$. Substituting the first boundary condition ($y = 0$ at $x = L$) from Eq. (11.17) into Eq. (11.15) yields:

$$C_1 \sin pL = VL/P \qquad (11.18)$$

Next differentiate Eq. (11.15) with respect to x to obtain:

$$dy/dx = pC_1 \cos(px) - V/P \qquad (11.19)$$

Substituting the boundary condition ($dy/dx = 0$ at $x = L$) from Eq. (11.17) into Eq. (11.19) gives:

$$pC_1 \cos pL = V/P \qquad (11.20)$$

Dividing Eq. (11.18) by Eq. (11.20) leads to the transcendental equation:

$$\tan pL = pL \qquad (11.21)$$

Solving this equation by trial and error leads to the first meaningful solution for pL as:

$$pL = 4.4934 \qquad (11.22)$$

Then substituting Eq. (11.3) into Eq. (11.22) and solving, we obtain:

$$P_{CR} = \frac{20.19 EI}{L^2} \approx \frac{\pi^2 EI}{(0.7L)^2} \qquad (11.23)$$

Comparing the result in Eq. (11.23) with that in Eq. (11.8) shows that they differ only by the coefficient multiplying the term (EI/L^2). Recognizing this similarity, let's write a general equation for the buckling of columns as:

$$P_{CR} = \frac{\pi^2 EI}{(kL)^2} \qquad (11.24)$$

For the column with two pinned ends $k = 1$, and for the column with one end pinned and the other built-in $k = 0.7$. Clearly, building-in one end of the column increased the critical load required for buckling by a factor of more than two.

11.3.2 Column Buckling with One Free and One Built-in End.

This case deals with a column with a free end, where the load is applied, and a built-in end, as depicted in Fig. 11.7

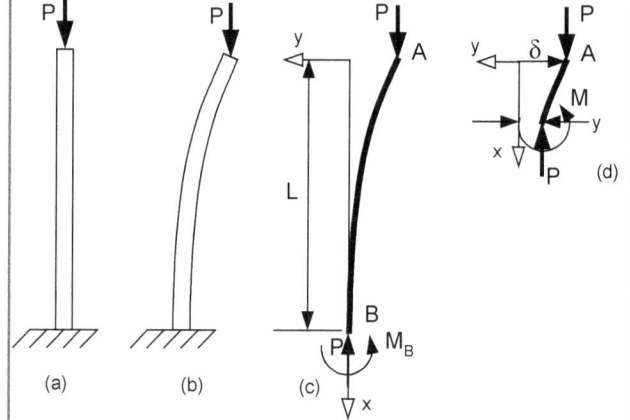

Fig. 11.7 A cantilever type column with one end free and the other built-in.

A FBD of the free end of the cantilever column is shown in Fig. 11.7d. From this FBD we write the moment equation as:

$$M = P(\delta - y) \qquad (11.25)$$

Substituting Eq. (11.25) into Eq. (8.4) and rearranging to write the differential equation in standard format gives:

$$\frac{d^2 y}{dx^2} + p^2 y = p^2 \delta \qquad (11.26)$$

Using similar procedures to solve Eq. (11.26), we obtain:

$$y = y_h + y_p = C_1 \sin(px) + C_2 \cos(px) + \delta \qquad (11.27)$$

We solve for the constants of integration using boundary conditions. Because the top of the column undergoes a deflection δ, we write:

$$y(0) = \delta \qquad (11.28)$$

The boundary conditions for the built-in end are:

$$y(L) = 0 \qquad dy/dx\,(L) = 0 \qquad (11.29)$$

It is clear from the boundary condition listed in Eq. (11.28) that $C_2 = 0$. Next differentiate Eq. (11.27) with respect to x to obtain:

$$dy/dx = pC_1 \cos(px) \qquad (11.30)$$

Substituting the boundary condition (dy/dx = 0 at x = L) into Eq. (11.30) gives:

$$C_1 \cos pL = 0 \qquad (11.31)$$

It is evident that Eq. (11.31) is satisfied when $pL = n\pi/2$. The first bucking mode occurs when n = 1; hence, $pL = \pi/2$. Substituting this value into Eq. (11.3) gives the critical buckling load as:

$$P_{CR} = \frac{\pi^2 EI}{4L^2} \qquad (11.32)$$

Comparison of Eq. (11.32) with Eq. (11.23) shows a remarkable reduction in the critical buckling load when the pin at the top of the column is removed and the axial compression load is applied to the column's free end. The reduction is a factor of about eight.

11.3.3 Column Buckling with Two Built-in Ends.

The final case considered is the column with both ends built-in as shown in Fig. 11.8. The procedure for deriving the Euler equation for the critical load for buckling is similar to that described previously. We begin with the FBD as shown in Fig. 11.8d and write the moment equation as:

$$M = -Py + M_A \qquad (11.33)$$

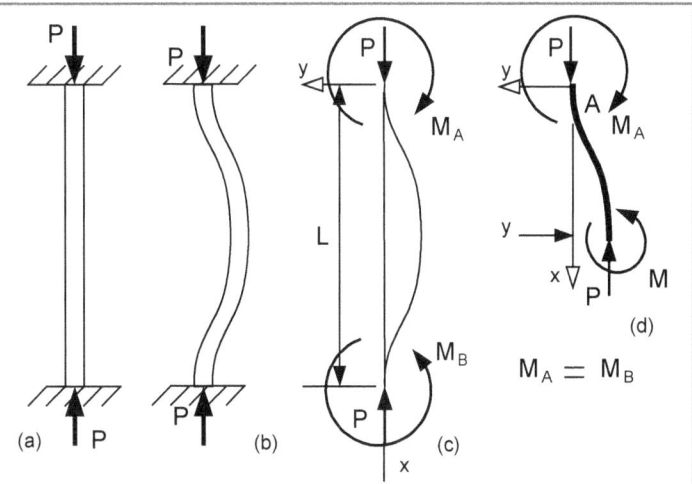

Fig. 11.8 A column with both ends built-in under axial compression loading.

Substituting Eq. (11.33) into Eq. (8.4) and rearranging gives the differential equation in standard format as:

$$\frac{d^2y}{dx^2} + p^2 y = \frac{M_A}{EI} \qquad (11.34)$$

Solving Eq. (11.34) for the homogeneous and particular solutions gives the general solution as:

$$y = y_h + y_p = C_1 \sin(px) + C_2 \cos(px) + M_A/P \qquad (11.35)$$

We follow the usual procedure to solve for the constants of integration using boundary conditions. At the top of the column (x = 0), y = 0 and dy/dx = 0. At the middle of the column, symmetry exists so that dy/dx = 0 at x L/2. At the bottom of the column, (x = L), y = 0 and dy/dx = 0. It is clear from the boundary condition (y = 0 at x = 0) and Eq. (11.35) that:

$$C_2 = -M_A/P \qquad (11.36)$$

Differentiating Eq. (11.35) with respect to x and substituting Eq. (11.36) yields:

$$dy/dx = C_1 p \cos(px) + (M_A/P) p \sin(px) \qquad (11.37)$$

Substituting the boundary condition (dy/dx = 0 at x = 0) into Eq. (11.37) gives $C_1 = 0$. We can now write Eq. (11.34) as:

$$y = C_2[\cos(px) - 1] \qquad (11.38)$$

From the boundary condition (y = 0 at x = L) and Eq. (11.38), we can write:

$$\cos(pL) - 1 = 0 \qquad (11.39)$$

Equation (11.39) is satisfied when $pL = 2n\pi$. The first mode of buckling occurs with n = 1, which leads to:

$$pL = 2\pi \qquad (11.40)$$

Then substituting Eq. (11.3) into Eq. (11.40), we obtain:

$$P_{CR} = \frac{4\pi^2 EI}{L^2} \qquad (11.41)$$

Let's apply these new relations to determine the critical load for two different columns with various end conditions.

EXAMPLE E11.3

A column 15 feet long is fabricated from a steel tube with the rectangular cross section, presented in Fig. E11.3. The column is pinned at one end and built-in at the other. Determine the critical buckling force.

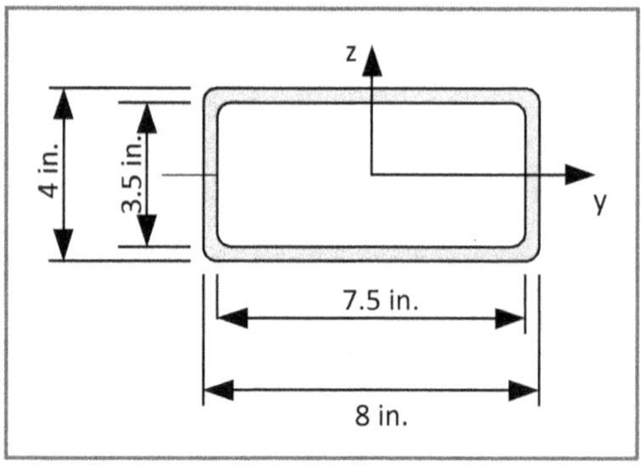

Fig. E11.3

Solution: We employ Eq. (11.23) to write the expression for the critical buckling force for this column as:

$$P_{CR} = \frac{20.19 EI}{L^2} \tag{a}$$

The modulus of elasticity $E = 30 \times 10^6$ psi for the steel tube is from Appendix B-1.

The cross section is rectangular; hence the moment of inertia is a function of the choice of axis. For this reason, we will determine both I_y and I_z below:

$$I_y = (1/12)[b_o h_o^3 - b_i h_i^3] = (1/12)[(8)(4)^3 - (7.5)(3.5)^3] = 15.87 \text{ in.}^4 \tag{b}$$

$$I_z = (1/12)[b_o h_o^3 - b_i h_i^3] = (1/12)[(4)(8)^3 - (3.5)(7.5)^3] = 47.62 \text{ in.}^4 \tag{c}$$

Substituting the lower value of the moment of inertia from Eq. (b) into Eq. (a) yields:

$$(P_{CR})_y = \frac{20.19 EI_y}{L^2} = \frac{20.19(30 \times 10^6)15.87}{(15)^2(12)^2} = 296.7 \text{ kip} \tag{d}$$

Examination of the results indicates the importance of the choice of axis when determining the moment of inertia when the cross section is not circular or square. In this case, the difference in the inertia is about a factor of three.

EXAMPLE E11.4

A column 6.4 m long is fabricated from a steel wide flange section. with a designation of W152 × 24. Determine the critical buckling force for the following two cases: (a) the column is pinned at one end and built-in at the other, and (b) the column is built-in at both ends.

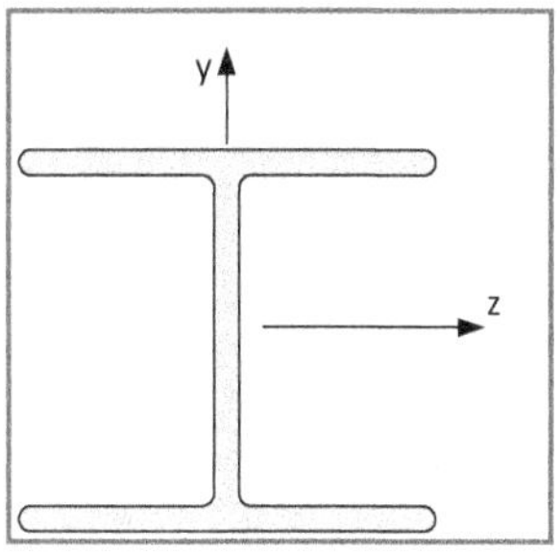

Fig. E11.4

Solution: Employ Eq. (11.32) to write the expression for the critical buckling force for the first column as:

$$P_{CR} = \frac{\pi^2 EI}{4L^2} \tag{a}$$

The modulus of elasticity (207 GPa) for the steel column is from Appendix B-1. The moment of inertia is a function of the choice of axis. Both I_y and I_z are given in Appendix D as:

$$I_y = 1.84 \times 10^6 \text{ mm}^4 \qquad I_z = 13.4 \times 10^6 \text{ mm}^4 \tag{b}$$

We select the minimum value of the moment of inertia I_y given in Eq. (b) and substitute this value into Eq. (a) to obtain:

$$(P_{CR})_y = \frac{\pi^2 EI_y}{4L^2} = \frac{\pi^2 (207 \times 10^3)(1.84 \times 10^6)}{(4)(6400)^2} = 22.94 \text{ kN} \qquad (c)$$

Examination of the results indicates the importance of the choice of the end conditions for column buckling. In this case, the difference in the critical buckling load is in excess of a factor of sixteen.

Next, employ Eq. (11.41) to obtain the expression for the critical buckling force for the second column:

$$(P_{CR})_y = \frac{4\pi^2 EI_y}{L^2} = \frac{4\pi^2 (207 \times 10^3)(1.84 \times 10^6)}{(6400)^2} = 367.1 \text{ kN} \qquad (d)$$

Again, we have selected the minimum value of the moment of inertia I_y for evaluating the critical buckling force.

11.3.4 Summary of Equations For Critical Column Buckling Loads

We have derived four equations for the buckling of columns under the action of centric compressive loads. These equations all are of the form $P_{CR} = \pi^2 EI/(kL)^2$. We summarize the results for k as a function of the various end constraints imposed on the column in Table 11.1

Table 11.1 Constant k as a function of column end constraints.

Case No.	End Constraints	k
1	Pinned—Pinned	1
2	Pinned—Built-in	0.7
3	Free—Built-in	2
4	Built-in—Built-in	0.5

The determination of the critical buckling load for columns subjected to centric compressive forces involves identifying the end constraints and substituting the correct value of k into Eq. (11.24). Care must also be exercised in selecting the correct axis about which to determine the inertia I, as two choices exist in most cases. In most instances the minimum inertia is employed in the determination. Four examples have been described to demonstrate the method for computing the Euler buckling loads.

We have not emphasized the critical buckling stress σ_{CR} in this section, as we were more concerned with determining the critical buckling force. The topic of stresses produced in columns is discussed in much more detail in the next section.

11.4 <u>COLUMN STRESSES AND LIMITATIONS OF EULER'S THEORY</u>

As a centric load is applied to a column, an axial compressive stress develops that increases until the column becomes unstable. At the inception of buckling, the stress is called the critical stress and is given by:

$$\sigma_{CR} = P_{CR}/A \qquad (11.42)$$

If we consider the column pinned at both ends, then $P_{CR} = \pi^2 EI/L^2$. Substituting this value into Eq. (11.42) and rearranging symbols yields:

$$\sigma_{CR} = \frac{\pi^2 EI}{L^2 A} = \frac{\pi^2 EAr^2}{L^2 A} = \frac{\pi^2 E}{(L/r)^2} \qquad (11.43)$$

Note $I = Ar^2$ in the expansion of Eq. (11.43), where A is the cross sectional area of the column and r is the radius of gyration of the area.

The term (L/r) is known as the slenderness ratio of the column. As (L/r) increases, the column becomes more flexible and will buckle at lower and lower stress levels. Let's consider an example to illustrate the reduction in the critical stress σ with increasing (L/r).

EXAMPLE E11.5

Consider a steel column with E = 207 GPa and S_y = 300 MPa. Prepare a graph showing σ_{CR} as a function of the slenderness ratio (L/r).

Solution: The solution involves evaluating Eq. (11.43) as (L/r) is varied from 10 to 200 in steps of 10.

$$\sigma_{CR} = \frac{\pi^2 E}{(L/r)^2} = S_y \qquad (a)$$

The results obtained from a spreadsheet calculation using Eq. (a) are presented in Fig. E11.5.

Examination of Fig. E11.5 indicates that the critical stress σ_{CR} exceeds the yield strength S_y of the steel specified in the example statement for (L/r) < 82.5. This fact means that the Euler theory for buckling is valid only when (L/r) ≥ 82.5. For slenderness ratios less than this value, failure of the column will occur by compressive yielding.

The limiting value of (L/r) depends on the column constraint and both the modulus of elasticity and the yield strength of the material, from which the column is fabricated. We will explore this dependency in the next example.

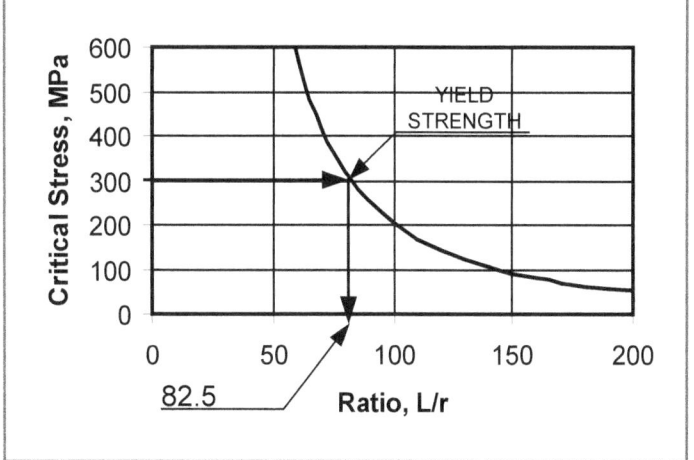

Fig. E11.5 Critical stress σ_{CR} as a function of slenderness ratio (L/r).

EXAMPLE E11.6

Consider a column with both ends pinned. Determine the limiting value of (L/r) as a function of yield strength for columns fabricated from both steel and aluminum. Consider the yield strength varying from 20 to 200 ksi.

Solution: Set $\sigma_{CR} = S_y$ in Eq. (11.43) and solve for (L/r) to obtain:

$$L/r = \sqrt{\frac{\pi^2 E}{S_y}} \qquad (a)$$

Let $E = 30 \times 10^6$ pi and 10.4×10^6 psi for steel and aluminum, respectively. Using a spreadsheet to evaluate Eq. (a) and varying S_y from 20 to 200 ksi in steps of 10, we obtain the results presented in Fig. E11.6.

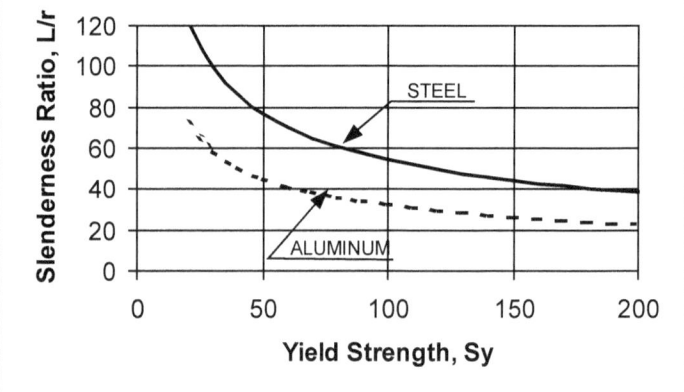

Fig. E11.6 The limiting value of the slenderness ratio as a function of yield strength for a column with both ends pinned.

The marked difference in the limiting value of the slenderness ratio for steel and aluminum alloy is due to the fact that aluminum is less resistant to buckling, because of its much lower modulus of elasticity.

These two examples have illustrated that the Euler theory of buckling has limits of applicability. Long slender columns with high values of (L/r) will buckle under the action of centric compressive loads and Euler's equations are valid. However, if (L/r) for the column is less than the limit value, as is the case for shorter columns with larger radii of gyration, then the Euler theory is not valid. In these cases, the column does not buckle. The failure mode is one of yielding due to excessive compressive stress. The different modes of failure are illustrated in Fig. 11.9.

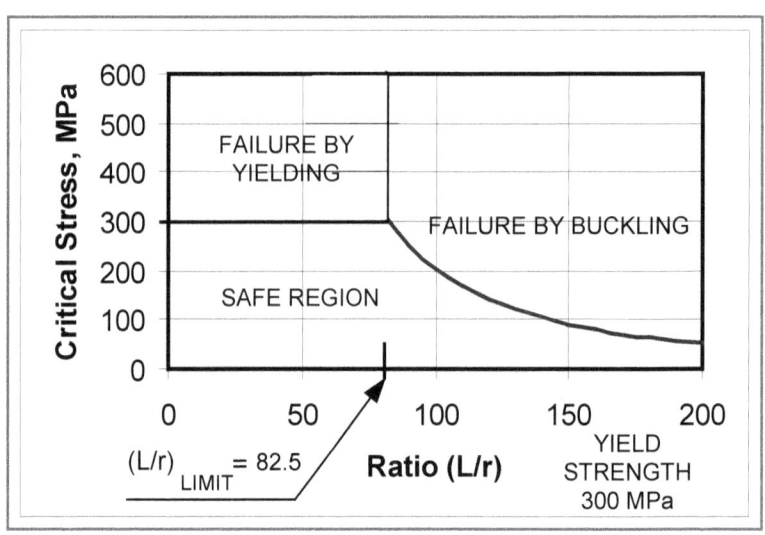

Fig. 11.9 Safety and failure regions for a centric loaded, pin-pin ended column.

11.5 ECCENTRICALLY LOADED COLUMNS

The developments in the previous sections of this chapter assumed that the axial compressive load was applied to the column through the centroid of its cross section. In practice this centric loading is nearly impossible to achieve. Even if the load is placed at the centroid of the cross section, columns are never perfectly straight. As a consequence, the loading always exhibits some small eccentricity e. Let's explore the effect of the eccentricity of the load on the buckling relations by considering the pin-pin ended column, shown in Fig. 11.10.

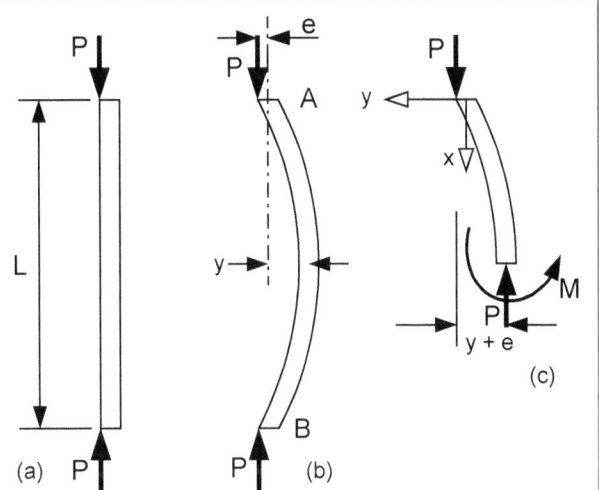

Fig. 11.10 A column with an eccentric axial load P.

The analysis follows the same procedure as described previously. We deform the column as a sine wave, because the ends are pinned. The column is cut at a location x from the top end and a FBD of the end is made as shown in Fig. 11.10c. Writing the equation for the moment M gives:

$$M = -P(y + e) \tag{11.44}$$

Substituting Eq. (11.44) into Eq. (8.4) yields:

$$\frac{d^2y}{dx^2} = \frac{M(x)}{EI} = -\frac{1}{EI}(Py + Pe) \tag{11.45}$$

Rearranging Eq. (11.45) to write the differential equation in standard format gives:

$$\frac{d^2y}{dx^2} + p^2 y = -p^2 e \tag{11.46}$$

where p is given in Eq. (11.3)

Solving Eq. (11.46) yields:

$$y = y_h + y_p = C_1 \sin(px) + C_2 \cos(px) - e \tag{11.47}$$

We follow the usual procedure to solve for the constants of integration using boundary conditions. Because both ends of the column are pinned, we write:

$$y(0) = 0 \qquad y(L) = 0 \tag{11.48}$$

From the first boundary condition (y = 0 at x = 0), it is clear that C_2 = e and from the second boundary condition (y = 0 at x = L), we can show that:

$$C_1 = e[1 - \cos(pL)]/\sin(pL) \tag{11.49}$$

Using the following trigometric identities:

$$1 - \cos(pL) = 2\sin^2(pL/2) \qquad \sin(pL) = 2\sin(pL/2)\cos(pL/2)$$

the expression for C_1 can be rewritten as:

$$C_1 = e\tan(pL/2) \tag{11.50}$$

Substituting the values for C_1 and C_2 into Eq. (11.47) yields:

$$y = = e[\tan(pL/2)\sin(px) + \cos(px) - 1] \tag{11.51}$$

This relation shows that the amplitude of the deflection of the column prior to the initiation of buckling depends upon the eccentricity e. We will show a graph illustrating this fact later in this section. The maximum value of y occurs at the mid point of the column where $x = L/2$. At that location, we can write the expression for y_{Max} as:

$$y_{Max} = e[\sec(pL/2) - 1] \tag{11.52}$$

The column will buckle when y_{Max} becomes large. For small eccentricity, y_{Max} becomes large when:

$$\sec(pL/2) \to \infty \quad \Rightarrow \quad pL/2 = \pi/2$$

From this result and Eq. (11.3), it is clear that:

$$P_{CR} = \frac{\pi^2 EI}{L^2} \tag{11.53}$$

The buckling load is identical with that derived for centrically loaded columns when $e = 0$. While the end result is the same, the bucking process differs. With centric loading the column remained straight as the load P was increased until P_{CR} was achieved. At that point the column became unstable and suddenly buckled (collapsed). With eccentricity the column deflects laterally with increasing magnitude until the critical load is achieved. We will illustrate this behavior in the next example.

EXAMPLE E11.7

Consider a column with both ends pin loaded with an axial force P. The force P is applied with an eccentricity e. Determine y_{Max}, as a function of P, when the quantity $pL/2$ varies from 0 to $\pi/2$. Let $e = 1$ and 2 in. The column is 30 ft long and fabricated from a square steel tube depicted in Fig. E11.7.

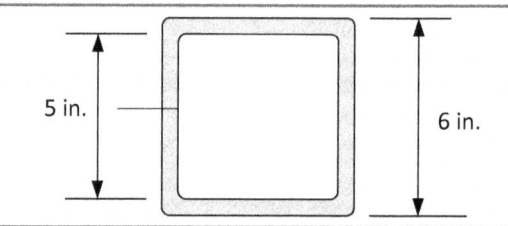

Fig. E11.7

Solution: Before determining y_{Max}, establish the critical buckling load from Eq. (11.53) as:

$$P_{CR} = \frac{\pi^2 EI}{L^2} = \frac{\pi^2 (30 \times 10^6)[(6)^4 - (5)^4]}{(12)(30)^2(12)^2} = 127.8 \text{ kip} \tag{a}$$

This result enables us to bound the calculations needed in preparing a graph of y_{Max}, as a function of P. Recall Eq. (11.52) and rewrite it as:

$$y_{Max} = e\left(\sec\left(\frac{pL}{2}\right) - 1\right) = e\left(\sec\sqrt{\frac{PL^2}{4EI}} - 1\right) \quad \text{(b)}$$

Substituting numerical parameters into Eq. (b) gives:

$$y_{Max} = e\left(\sec\sqrt{\frac{PL^2}{4EI}} - 1\right) = e\left(\sec\sqrt{\frac{P(360)^2}{4(30\times 10^6)(55.92)}} - 1\right) = e\left(\sec\sqrt{19.31\times 10^{-6}P} - 1\right) \quad \text{(c)}$$

Next, employ a spreadsheet to evaluate Eq. (c) for P varying from 0 to 130 kip to obtain the graph shown in Fig. E11.7a.

Fig. E11.7a

Examination of the results in Fig. E11.7a shows that the deflection is not a linear function of the axial load P. Instead the deflection increases exponentially, as the critical buckling load is approached. This non-linear behavior is due to the moment $M = P(e + y)$. As P increases, the deflection y also increases; hence, the moment increases in a non-linear manner.

11.6 STRESSES IN COLUMNS WITH ECCENTRIC LOADING

In some structural applications, it is necessary to apply loads to columns with significant eccentricity. While this practice does not affect the critical buckling load, as determined by the Euler theory, it may result in failure of the column due to excessive compressive stress at loads less than P_{CR}. Let's consider the free-ended column, shown in Fig. 11.11, with an axial force P that is applied with an eccentricity e.

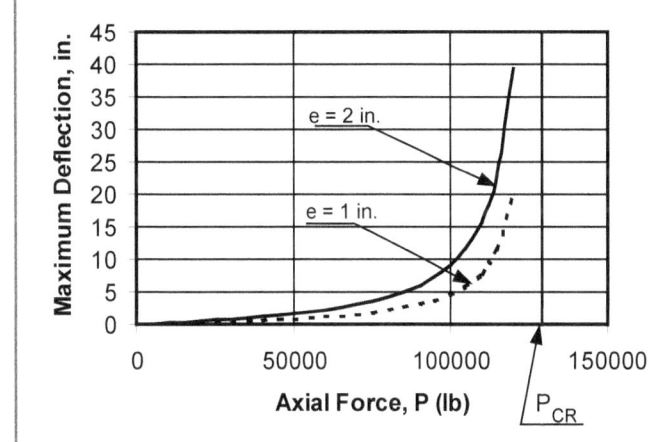

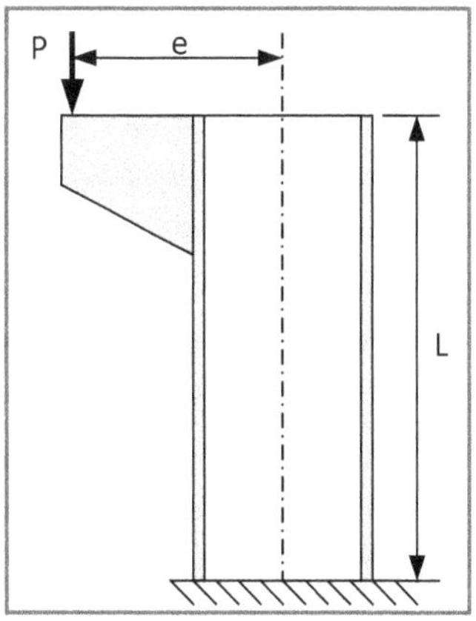

Fig. 11.11 A free-ended column with an eccentric load P.

The load P generates an internal force and a bending moment, both of which are constant over the length of the column. A FBD of the top portion of the column, presented in Fig. 11.12, shows this force and moment.

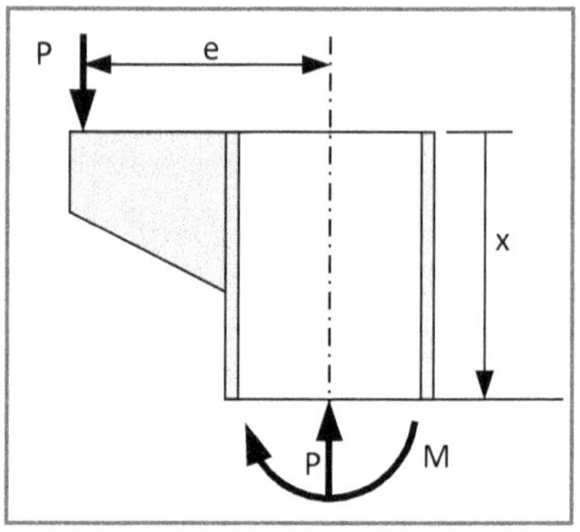

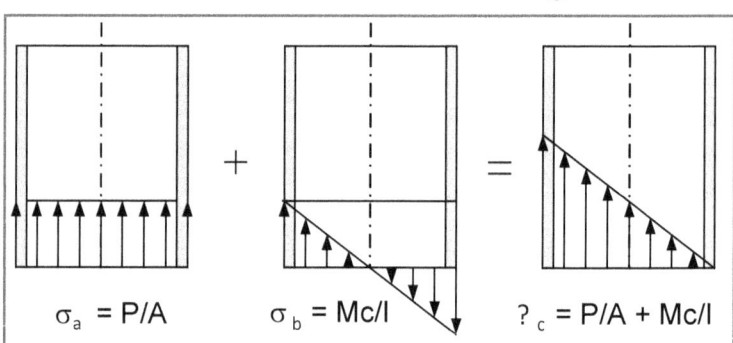

Fig. 11.12 A FBD showing an internal force P and a bending moment M.

Considering equilibrium permits us to write:

$$M = Pe \qquad (11.54)$$

The axial stresses produced in the column due to P and M are:

$$\sigma_c = \sigma_a + \sigma_b = -P/A \pm Mc/I \qquad (11.55)$$

It is evident from Eq. (11.55) that the axial stresses due to the axial force and the bending moment combine. The combined distribution of axial stresses across the column is illustrated in Fig. 11.13.

Fig. 11.13 Combined stresses due to eccentric loading of a column.

$\sigma_a = P/A$ $\sigma_b = Mc/I$ $\sigma_c = P/A + Mc/I$

For an axial compressive force applied to a column, the maximum combined stress produced is a compressive stress given by:

$$\sigma_c = \sigma_a + \sigma_b = -(P/A + Mc/I) \qquad (11.56)$$

Substituting Eq. (11.54) into Eq. (11.56) gives:

$$\sigma_{Max} = -[P/A + P(y_{Max} + e)c/I] \qquad (11.57)$$

Using the value for y_{Max} from Eq. (11.52), we can write:

$$\sigma_{Max} = -P\left(\frac{1}{A} + \frac{ce}{I}\sec\left(\frac{pL}{2}\right)\right) \qquad (11.58)$$

An alternate formula for σ_{Max} can be obtained by substituting Eqs. (11.3) and (11.53) into Eq. (11.58). After simplification:

$$\sigma_{Max} = -P\left(\frac{1}{A} + \frac{ce}{I}\sec\left(\frac{\pi}{2}\sqrt{\frac{P}{P_{CR}}}\right)\right) \qquad (11.59)$$

This equation is the well-known secant formula for eccentrically loaded columns. The equation may be used for columns with any end conditions, providing the appropriate formula for P_{CR} is selected. Also observe that Eq. (11.59) is non-linear in terms of the load P. Thus, it is often necessary to solve the equation by trial and error or by using advanced methods (i.e. programmable calculator, computer software, etc.). For the same reason, the safety factor pertains only to the critical load and not to the stresses.

In the design of an eccentrically loaded column, the procedure is to determine the maximum stress using Eq. (11.59) and the critical buckling stress using the appropriate relation that depends on the end conditions of the column. We then compare these two values to determine the mode of failure for the column. We will illustrate this procedure in the next example.

EXAMPLE E11.8

Consider a column with both ends pin loaded with an axial force P, as shown in Fig. E11.8. The force P is applied with an eccentricity e = 350 mm. Determine the maximum load P that can be applied if a safety factor SF = 2.5 is to be maintained against failure by excessive compressive stress or by buckling. The steel column is 8.0 m long and is fabricated from an American Standard section with a designation of S127 × 22. The steel is 1020 HR with S_y = 290 MPa.

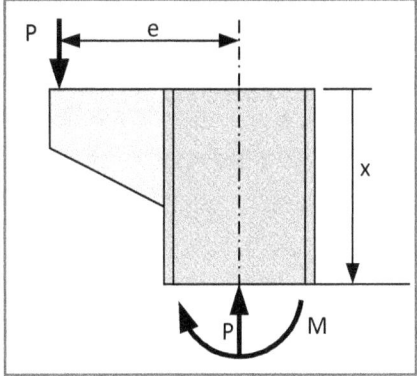

Fig. E11.8

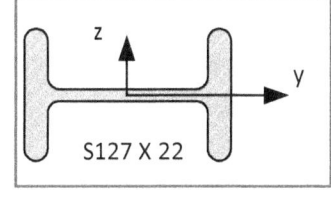

S127 X 22

Solution: We employ Eq. (11.8) to write the expression for the critical buckling force for this pinned ended column as:

$$P_{CR} = \frac{\pi^2 EI}{L^2} \qquad (a)$$

The modulus of elasticity E = 207 GPa for the steel column is from Appendix B-1.

The moment of inertia is a function of the orientation of the pin. If the pin is inserted along the y axis (see Fig. E11.8), then the end of the column will rotate about this axis. However, if the pin is inserted in the z direction, the column will rotate about this axis. Both I_y and I_z are given in Appendix D as:

$$I_y = 0.695 \times 10^6 \text{ mm}^4 \qquad (b)$$

$$I_z = 6.33 \times 10^6 \text{ mm}^4 \tag{c}$$

Substituting the minimum value I_y into Eq. (a) yields:

$$(P_{CR})_y = \frac{\pi^2 E I_y}{L^2} = \frac{\pi^2 (207 \times 10^3)(0.695 \times 10^6)}{(8,000)^2} = 22.19 \text{ kN} \tag{d}$$

Examination of the results indicates the importance of the direction of the pin, when the cross section is not circular or square. In this case, the difference in the moment of inertia I is in excess of a factor of nine. Clearly, for the section shown in Fig. E11.8, the pin should be inserted in the z direction. Many conservative designers will base their recommendations on the minimum critical load, even with the column end free to rotate about the z-axis. Let's follow this conservative approach. Then the allowable load for buckling is given by:

$$(P_{Allowable})_{CR} = P_{CR}/SF = 22.19/2.5 = 8.876 \text{ kN} \tag{e}$$

The maximum stress is given by Eq. (11.59) as:

$$\sigma_{Max} = -S_y = -P_y \left(\frac{1}{A} + \frac{ce}{I_y} \sec\left(\frac{\pi}{2} \sqrt{\frac{P_y}{P_{CR}}} \right) \right) \tag{f}$$

Additional properties for the American Standard section are obtained from Appendix C as:

$$A = 2,800 \text{ mm}^2 \tag{g}$$

$$c = w_f/2 = (83.4)/2 = 41.7 \text{ mm} \tag{h}$$

Substituting all numerical parameters for the S127 × 22 section and the yield strength of the 1020 HR steel into Eq. (f) gives:

$$290 = P_y \left(\frac{1}{2,800} + \frac{(41.7)(350)}{0.695 \times 10^6} \sec\left(\frac{\pi}{2} \sqrt{\frac{P_y}{22,190}} \right) \right) \tag{i}$$

Solving Eq. (i) by trial and error:

$$P_y = 8,018 \text{ N} = 8.018 \text{ kN} \tag{j}$$

Then the allowable load for yielding is:

$$(P_{Allowable})_y = P_y/SF = 8.018/2.5 = 3.207 \text{ kN} \tag{k}$$

Compare both allowable loads from Eqs. (e) and (k) and choose the smallest one, which gives:

$$P_{Allowable} = 3.207 \text{ kN} \tag{l}$$

In this case, the column will fail by yielding rather than buckling.

11.7 DESIGN CODES

Design codes have been developed by several industrial associations to facilitate the design of columns fabricated from steel, aluminum and wood. These codes accommodate both long slender columns and intermediate length columns. Local governments controlling building permits have adopted these codes and structural engineers employ the formulas contained in the codes when sizing columns.

11.7.1 Steel Columns

The American Institute for Steel Construction (AISI) has adopted two different equations for the design of steel columns. These equations permit a designer to determine the maximum allowable stress for centric loaded columns for specified slenderness ratios. The first of these two equations for long slender columns is based on Euler's theory, but with a safety factor SF = 1.92.

$$\sigma_{Allowable} = \frac{\pi^2 E}{1.92(kL/r)^2} \quad (11.60)$$

where k depends on the column's end conditions and is given in Table 11.1.

Equation (11.60) is applied for specified values of (kL/r) as indicated below:

$$SR_C = \sqrt{\frac{2\pi^2 E}{S_y}} \leq \left(\frac{kL}{r}\right) \leq 200 \quad (11.61)$$

The limitations Eq. (11.60) imposed by Eq. (11.61) insure that the column will remain elastic and that $\sigma_{Allowable} \leq S_y$, even when the residual stresses in the rolled steel section approach one half of the yield strength of the steel.

For columns with slenderness ratios less than $SR_C = [(2\pi^2 E)/S_y]^{1/2}$, a more conservative equation is employed to determine the allowable stress. Because of plastic behavior of the steel with these lower slenderness ratios, a larger safety factor is incorporated in the design equation. The safety factor specified in the design code is a variable depending on (kL/r).

$$SF = 1.667 + 0.375 \,[(kL/r)/SR_c] - 0.125 \,[(kL/r)^3/SR_C^3] \quad (11.62)$$

The allowable stress in this range of (kL/r) is given by:

$$\sigma_{Allowable} = \left(1 - \frac{(kL/r)^2}{2(SR_C)^2}\right)\left(\frac{S_y}{SF}\right) \quad (11.63)$$

Equation (11.63) is employed for columns with (kL/r) < SR_c. A graph showing the allowable stress as a function of the slenderness ratio (kL/r) is presented in Fig. 11.14.

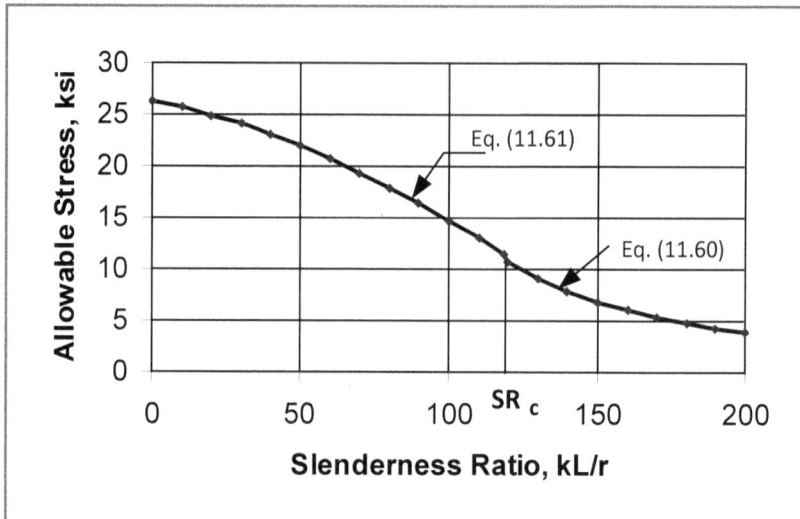

Fig. 11.14 Design curve for pin ended column fabricated from 1020 HR steel with $S_y = 42$ ksi.

A smooth transition from one equation to the other is clear at the critical slenderness ratio that occurs at $SR_c = 118.7$ for a steel with $S_y = 30 \times 10^3$ ksi.

EXAMPLE E11.9

A column fabricated from a steel wide flange section W 12 × 50 is pinned at both ends. Use the curve resulting from the code equations that is shown in Fig. 11.14, to determine the allowable centric load applied to the column. The column is 24 ft long.

Solution: First reference Appendix D and note for the W 12 × 50 section that the minimum value of r = 1.96 in. and A = 14.7 in^2. Also note k = 1 for a column with pinned ends. Determine the slenderness ratio (kL/r):

$$(kL/r) = (1.0)(24)(12)/1.96 = 146.9 \qquad (a)$$

Employing the graph in Fig. 11.14, we locate the point (kL/r) = 146.9 and draw a line upward to intersect the curve, as shown in Fig. E11.9. We then construct a horizontal line from the intersection point to the ordinate to obtain the value of $\sigma_{Allowable} = 7.15$ ksi.

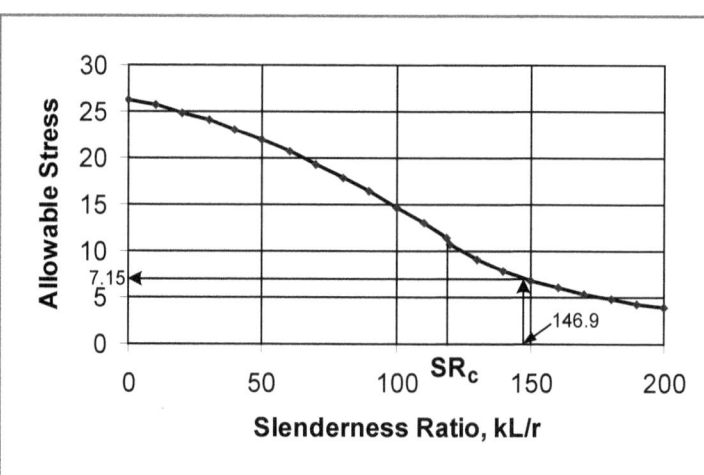

Fig. E11.9

The maximum allowable load on the column is then given by:

$$P_{Allowable} = \sigma_{Allowable} A = (7.15)(14.7) = 105.1 \text{ kip} \qquad (b)$$

Clearly, the use of the design curve and the geometric properties of the wide flange section from Appendix D enable a rapid determination of the maximum allowable centric force applied to the column.

11.7.2 Aluminum Columns

The Aluminum Association has also developed a design code for aluminum columns. It accounts for the elastic and plastic behavior of aluminum by using three different equations to determine the allowable stress. Because the physical properties of aluminum alloys vary considerably from one alloy to another, three equations are provided by the Aluminum Association for each alloy commonly employed in construction. The relations presented below are for the alloy 2014 T-6.

$$\sigma_{Allowable} = 28 \text{ ksi} \qquad 0 \leq (kL/r) \leq 12 \qquad (11.64)$$

$$\sigma_{Allowable} = [30.7 - 0.23(kL/r)] \text{ ksi} \qquad 12 \leq (kL/r) \leq 55 \qquad (11.65)$$

$$\sigma_{Allowable} = [54{,}000/(kL/r)^2] \text{ ksi} \qquad 55 \leq (kL/r) \qquad (11.66)$$

The allowable stress $\sigma_{Allowable}$ as a function of (kL/r) was determined from these three equations to develop the design curve presented in Fig. 11.15. Recall that this curve is valid only for the aluminum alloy 2014 T-6.

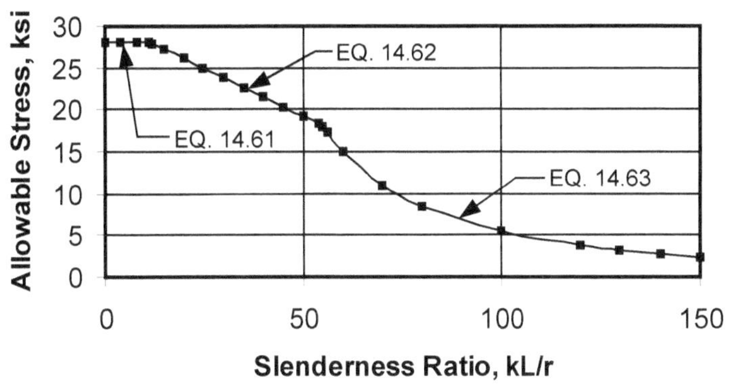

Fig. 11.15 Design curve for a column fabricated from aluminum alloy 2014 T-6 ($S_y = 41$ ksi).

EXAMPLE E11.10

A column fabricated from an aluminum (2014 T-6) structural T section WT 6 × 48 is pinned at one end and built-in at the other. Use the curve resulting from the code equations that is shown in Fig. 11.15, to determine the allowable centric load applied to the column. The column is 18 ft long.

Solution: First reference Appendix D and note for the WT 6 × 48 section that the minimum value of r = 1.51 in. Also the area A = 14.1 in². Note k = 0.7 for a column pinned at one end and built-in at the other. Determine the slenderness ratio (kL/r):

$$(kL/r) = (0.7)(18)(12)/(1.51) = 100.1 \qquad (a)$$

Employing the graph in Fig. 11.15, we locate the point (kL/r) = 101.1 and draw a line upward to intersect the curve, as shown in Fig. E11.10. We then construct a horizontal line from the intersection point to the ordinate, to obtain the value of $\sigma_{Allowable}$ = 5.40 ksi.

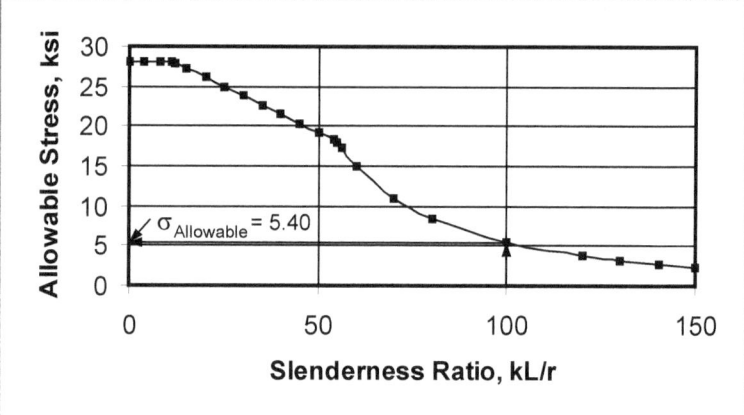

Fig. E11.10

The maximum allowable load on the column is then given by:

$$P_{Allowable} = \sigma_{Allowable} A = (5.40)(14.1) = 76.14 \text{ kip} \qquad (b)$$

Clearly, the use of the design curve and the geometric properties of the structural tee section from Appendix D enable a rapid determination of the maximum allowable centric force applied to the column.

11.7.3 Wood Columns

Design of both columns and beams fabricated from wood is difficult, because the physical properties of wood depend on the type of tree used to produce the structural members and even the part of the country where the tree was grown. Even the moisture content of the wood affects its properties. To deal with these many variables, the American Forest and Paper Association has published the National Design Specification for Wood Construction. The Western Woods Products Association has incorporated these design specifications in an excellent reference for designing structural members fabricated from many types of woods in their Western Woods Use Book[3].

A computer program is provided with the Western Woods Use Book that has merged the physical properties of several types of woods with the design relations adopted by the wood products and construction industries. This program enables the user to determine allowable column loading and beam loading while taking into account several factors affecting the physical properties of the wood, such as section size, duration of the load, temperature and moisture content.

11.8 SUMMARY

The concept of elastic instability was introduced. Columns can fail by buckling at stress levels lower than the yield strength of the materials from which they are fabricated. Failure by buckling is catastrophic as the column collapses suddenly. The critical load P_{CR} depends on the end conditions of the column with the critical load increasing as the constraint at the ends increase.

Equations for P_{CR} were derived for four different end conditions. The results are given by:

---

[3] Available from Western Woods Products Association, 522 SW 5th Avenue, Suite 400, Portland, OR 97204-3930

$$P_{CR} = \frac{\pi^2 EI}{(kL)^2} \quad (11.24)$$

where k, given in the table below, accounts for the constraint provided by the column supports.

Case No.	End Constraints	k
1	Pinned—Pinned	1
2	Pinned—Built-in	0.7
3	Free—Built-in	2
4	Built-in—Built-in	0.5

The inertia I of the cross section of the column is almost always taken as the minimum value.

The critical stress σ_{CR} in centric loaded columns was defined as:

$$\sigma_{CR} = P_{CR}/A \quad (11.42)$$

The critical stress may be written as:

$$\sigma_{CR} = \frac{\pi^2 EI}{L^2 A} = \frac{\pi^2 E A r^2}{L^2 A} = \frac{\pi^2 E}{(L/r)^2} \quad (11.43)$$

where A is the cross sectional area of the column and r is the radius of gyration of the area.

The term (L/r) is known as the slenderness ratio of the column. As (L/r) increases, the column will buckle at lower and lower stress levels.

When the slenderness ratio is large, columns buckle in accordance with Eq. (11.24); however, for shorter and stiffer columns $\sigma_{CR} > S_y$. In these cases the column fails by yielding. The transition from the Euler theory of buckling to yielding occurs at a limit value of (L/r), which depends on the modulus of elasticity and yield strength of the column material, as indicated in Fig. E11.6. Safety and failure regions for a typical column are illustrated in Fig. 11.9.

When the load applied to the column exhibits some eccentricity, a bending moment is produced that affects the stresses in the column. However, the eccentricity does not affect the critical bucking load. While the result for P_{CR} is the same, the bucking process differs. With centric loading, the column remained straight, as the load P was increased until P_{CR} was achieved. At that point the column became unstable and suddenly buckled (collapsed). With eccentricity, the column deflects laterally with increasing magnitude until the critical load is achieved.

The stress in an eccentrically loaded column is due to a combination of a compressive stress σ_a due to P and a bending stress σ_b due to the moment M = Pe. The two stresses superimpose as indicated by:

$$\sigma_c = \sigma_a + \sigma_b = -(P/A + Mc/I) \quad (11.56)$$

An alternate formula for σ_{Max} was derived as:

$$\sigma_{Max} = -P\left(\frac{1}{A} + \frac{ce}{I}\sec\left(\frac{\pi}{2}\sqrt{\frac{P}{P_{CR}}}\right)\right) \qquad (11.59)$$

This equation, known as the secant formula, may be used for columns with any end conditions, by calculating P_{CR} from Eq. (11.24). Because the stress is not linearly related to load, it is often necessary to solve the equation by trial and error or by using advanced methods, such as a programmable calculator or computer software. Also, any safety factor should only be applied to the loads and not to the stresses. For eccentrically loaded columns, both the maximum stress and the critical buckling stress must be computed and compared to determine the actual mode of failure.

Finally, a sampling of the design codes for columns fabricated from steel and aluminum were described. These codes are extensive and in practice control the design of columns placed in most major structures. Reference was also given for design codes for columns and beams fabricated from timber.

PROBLEMS

11.1. Write an engineering brief explaining why understanding buckling is important in ensuring public safety.

11.2. A column L long is fabricated from a metal tube, with an outside diameter D_o and an inside diameter D_i. Determine the critical buckling force P_{CR} if the tube is pinned at both ends.

Prob. No.	L	D_o	D_i	Material
11.2a	18 ft.	8 in.	7.6 in.	Aluminum
11.2b	6 m	150 mm	125 mm	Aluminum
11.2c	21 ft	6.0 in.	5.5 in.	Steel
11.2d	8 m	220 mm	200 mm	Steel

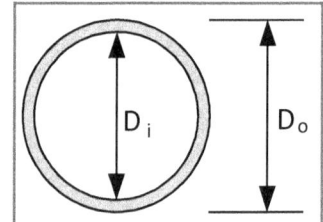

11.3. A column L long is fabricated from a square tube. Determine the critical buckling force P_{CR}, if the tube is pinned at both ends. Also determine the compressive stress in the tube at the critical load and comment on its magnitude.

Prob. No.	L	S_o	S_i	Material
11.3a	18 ft.	8 in.	7.6 in.	Aluminum
11.3b	6 m	150 mm	125 mm	Aluminum
11.3c	21 ft	6.0 in.	5.5 in.	Steel
11.3d	8 m	220 mm	200 mm	Steel

11.4. A column L long is fabricated from a tube with the rectangular cross section. The column is pinned at one end and built-in at the other. Determine the critical buckling force P_{CR}.

Prob. No.	L	w_o	w_i	d_o	d_i	Matrl
11.4a	16 ft.	8 in.	7.6 in.	5.0 in.	4.6 in.	Alum.
11.4b	5.0 m	150 mm	125 mm	100 mm	86 mm	Alum.
11.4c	24 ft	6.0 in.	5.5 in.	4.0 in.	3.8 in.	Steel
11.4d	7.3 m	220 mm	200 mm	125 mm	110 mm	Steel

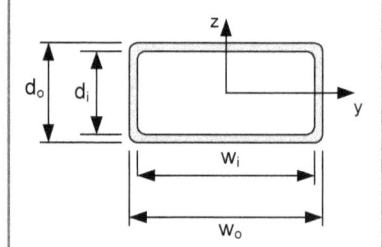

11.5. Design a series of round tubular columns for a construction firm that are intended to meet the specifications, shown in the table below.

Prob. No.	L	P_{CR}	D_i/D_o	End Support	Material
11.5a	20 ft.	80.0 kip	0.9	Free/Built-in	Aluminum
11.5b	6.0 m	300 kN	0.85	Pinned/Pinned	Aluminum
11.5c	19 ft	120 kip	0.80	Pinned/Built-in	Steel
11.5d	7.0 m	500 kN	0.88	Built-in/Built-in	Steel

11.6. Design a series of square tubular columns for a construction firm that are intended to meet the specifications, shown in the table below.

Prob. No.	L	P_{CR}	S_i/S_o	End Support	Material
11.6a	24 ft.	100 kip	0.9	Free/Built-in	Aluminum
11.6b	8.0 m	360 kN	0.85	Pinned/Pinned	Aluminum
11.6c	28 ft	150 kip	0.80	Pinned/Built-in	Steel
11.6d	7.4 m	600 kN	0.88	Built-in/Built-in	Steel

11.7. Design a series of rectangular tubular columns for a construction firm, which are intended to meet the specifications, shown in the table below. Note $w_i/w_o = d_i/d_o$.

Prob. No.	L	P_{CR}	w_i/w_o	End Support	Material
11.7a	22 ft.	140 kip	0.9	Free/Built-in	Aluminum
11.7b	8.4 m	420 kN	0.85	Pinned/Pinned	Aluminum
11.7c	26 ft	175 kip	0.80	Pinned/Built-in	Steel
11.7d	7.1 m	500 kN	0.88	Built-in/Built-in	Steel

11.8. Select the most suitable wide flange section for the design of a column to meet the following specifications.

Prob. No.	L	P_{CR}	End Support	Material
11.8a	22 ft.	140 kip	Free/Built-in	Aluminum
11.8b	8.4 m	420 kN	Pinned/Pinned	Aluminum
11.8c	26 ft	175 kip	Pinned/Built-in	Steel
11.8d	7.1 m	500 kN	Built-in/Built-in	Steel

11.9. Select the most suitable American Standard section for the design of a column to meet the following specifications.

Prob. No.	L	P_{CR}	End Support	Material
11.9a	16 ft.	100 kip	Free/Built-in	Aluminum
11.9b	5.4 m	320 kN	Pinned/Pinned	Aluminum
11.9c	14 ft	115 kip	Pinned/Built-in	Steel
11.9d	6.3 m	200 kN	Built-in/Built-in	Steel

11.10. Select the most suitable structural tee section for the design of a column to meet the following specifications.

Prob. No.	L	P_{CR}	End Support	Material
11.10a	15 ft.	75 kip	Free/Built-in	Aluminum
11.10b	4.9 m	300 kN	Pinned/Pinned	Aluminum
11.10c	12 ft	85 kip	Pinned/Built-in	Steel
11.10d	5.3 m	260 kN	Built-in/Built-in	Steel

11.11. A control rod in a vehicle is subjected to axial forces of ± 2,800 lb. If it is fabricated from stainless steel determine its diameter, if the design incorporates a safety factor of 2.5. Consider both ends pinned.

11.12. A lever is used to provide an axial force P to the vertical member, shown in the figure below. Determine the maximum force P that may be exerted if the vertical link is fabricated from a rectangular bar with dimensions 15 mm by 40 mm. Assume pinned ends for the link.

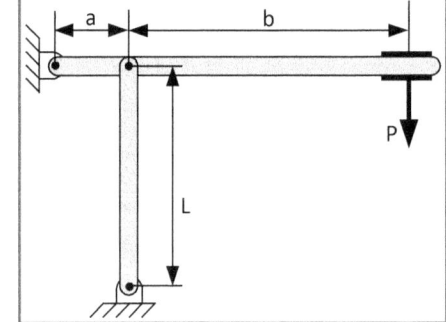

Prob. No.	L	a	b	Material
11.12a	30 in.	6.0 in.	30 in.	Aluminum
11.12b	1.0 m	100 mm	1.5 m	Aluminum
11.12c	42 in.	7.5 in.	52 in.	Steel
11.12d	0.8 m	200 mm	2.0 m	Steel

11.13. The truss, shown in the figure below, is fabricated from round tubular rods. If the rods all have the same outside diameter of 80 mm, determine the inside diameter of the compression members required to prevent buckling. Employ a safety factor of SF and specify steel tubes. Numerical parameters are given in the table below.

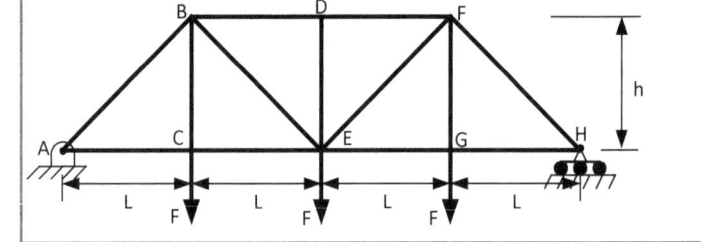

Prob. No.	L	h	F	SF
11.13a	20 ft.	20 ft.	75 kip	3.0
11.13b	6 m	6 m	200 kN	2.5
11.13c	24 ft	20 ft.	60 kip.	3.4
11.13d	8.0 m	7.0 m	250 kN	4.0

11.14. The truss, shown in the figure below, is fabricated from square tubular rods. If the rods all have the same outside dimension of s_o = 4.0 in., determine the inside dimension s_i of the compression members required to prevent buckling. Employ a safety factor of SF and specify steel tubes. Numerical parameters are given in the table below.

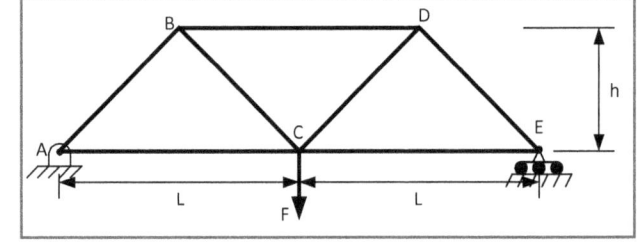

Prob. No.	L	h	F	SF
11.14a	20 ft.	14 ft	100 kip	2.5
11.14b	6 m	4 m	300 kN	3.5
11.14c	24 ft	18 ft.	90 kip.	4.0
11.14d	8.0 m	6.5 m	400 kN	3.0

11.15. A force F acts on a simple truss with a horizontal tie rod and an inclined strut, as shown in the figure below. The strut, fabricated from an aluminum alloy, is solid rectangular with dimensions w and d. Determine those dimensions if w = 2d for the numerical parameters given in the table below.

Prob. No.	a	b	F	SF
11.15a	10 ft.	14 ft	50 kip	2.5
11.15b	3 m	4 m	30 kN	3.5
11.15c	8.0 ft	16 ft.	25 kip.	4.0
11.15d	2.5 m	4.5 m	40 kN	3.0

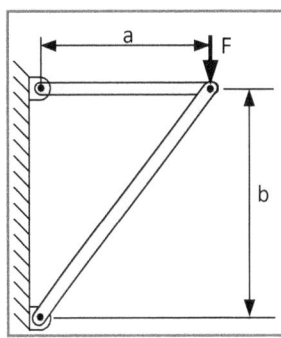

11.16. For the frame, shown in the figure below, determine the maximum force that can be supported prior to the inclined member buckling. The inclined member is fabricated from a rectangular steel bar with dimensions w and d with w = 1.5d.

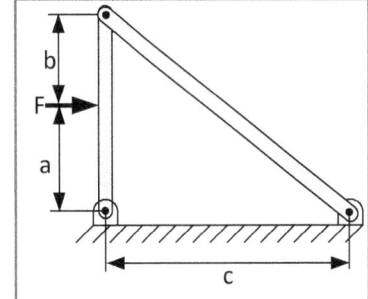

Prob. No.	a	b	c	w
11.16a	2.0 ft.	2.0 ft	5.0 ft	1.5 in.
11.16b	1.0 m	1.2 m	3.0 m	40 mm
11.16c	1.6 ft	1.2 ft.	2.5 ft	2.0 in.
11.16d	0.75 m	1.0 m	2.6 m	50 mm

11.17. The truss, shown in the figure below, is fabricated from rectangular tubular rods. If the rods all have the same outside dimension of w_o = 4.0 in. and d_o = 2.0 in. determine the wall thickness of the compression members required to prevent buckling. Employ a safety factor of SF and specify steel tubes. Numerical parameters are given in the table below.

Prob. No.	L	h	F	SF
11.17a	20 ft.	20 ft	50 kip	2.5
11.17b	5.0 m	4.0 m	100 kN	4.0
11.17c	16 ft	15 ft.	75 kip	2.0
11.17d	7.5 m	6.0 m	120 kN	3.5

11.18. A wooden column with a square cross section is subjected to an eccentric load, as shown in Fig. P11.18. The column is embedded in a concrete footer at one end and free at the other. Determine the load that can be applied to the column without causing it to fail by either buckling or yielding. The modulus of elasticity of the wood is 10 GPa and its yield strength S_y = 50 MPa.

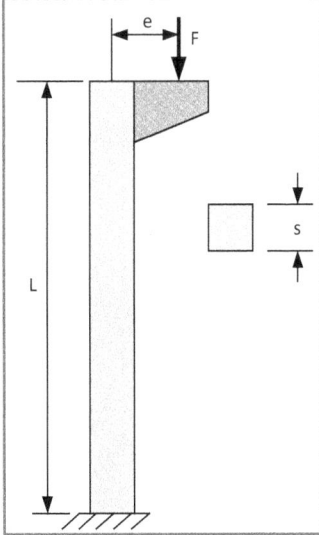

Prob. No.	L	e	s
11.18a	20 ft.	1.2 ft	5.5 in.
11.18b	5.0 m	0.40 m	100 mm
11.18c	16 ft	1.6 ft	7.5 in.
11.18d	7.5 m	0.60 m	150 mm

11.19. A column fabricated from a wide flange section is loaded eccentrically, as indicated in the figure below. Determine the force F required for the column to fail by either buckling or yielding. The wide flange section is fabricated from 1020 HR steel. The column is braced so that it does not buckle about the minimum inertia axis.

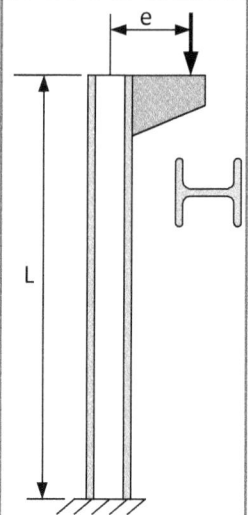

Prob. No.	L	e	Section
11.19a	20 ft.	1.2 ft	W 12 × 50
11.19b	5.0 m	0.40 m	W 305 × 74
11.19c	16 ft	1.6 ft	W 8 × 31
11.19d	7.5 m	0.60 m	W 254 × 89

11.20. Determine the deflection y_{Max} at the top of the column, shown in the figure above at 80% of the load producing failure.

11.21. A column fabricated from an American Standard section is loaded eccentrically, as indicated in the figure below. Determine the force F required for the column to fail by either buckling or yielding. The wide flange section is fabricated from 1020 HR steel. Assume both ends of the column are built-in, and that bracing prevents buckling about the minimum inertia axis.

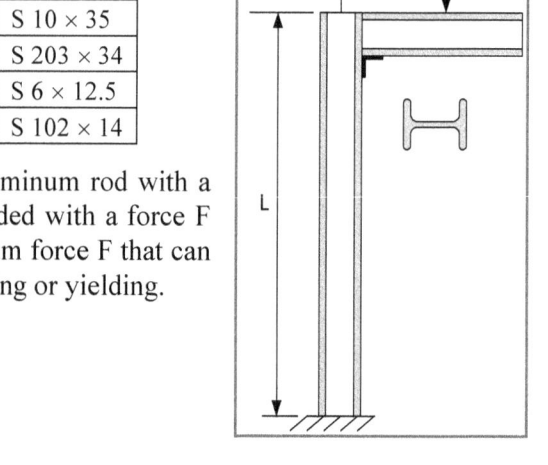

Prob. No.	L	e	Section
11.21a	18 ft.	1.8 ft	S 10 × 35
11.21b	4.5 m	0.75 m	S 203 × 34
11.21c	15 ft	2.6 ft	S 6 × 12.5
11.21d	6.5 m	0.50 m	S 102 × 14

11.22. A column of length L is fabricated from a solid aluminum rod with a diameter D, as shown in the figure below. It is loaded with a force F with an eccentricity e = D/2. Determine the maximum force F that can be applied if the column is not to fail by either buckling or yielding.

Prob. No.	L	D	Material
11.22a	8 ft	3.00 in.	2024 T-4
11.22b	1.5 m	60 mm	7075 T-6
11.22c	6.5 ft	2.5 in.	2024 T-4
11.22d	2.5 m	50 mm	7075 T-6

11.23. Develop a design curve similar to the one shown in Fig. 11.14 for steel columns with specified ends and with a yield strength of S_y. Also determine the limiting value of the slenderness ratio SR_c for this case.

Prob. No.	S_y	End Support
11.23a	50 ksi	Pinned/Pinned
11.23b	52 ksi	Pinned/Built-in
11.23c	38 ksi	Free/Built-in
11.23d	42 ksi	Built-in/Built-in

11.24. Determine the safety factor as a function of (kL/r) that is employed using the design code for steel columns. Consider yield strengths of 35, 40, 45, and 50 ksi.

11.25. A column of length L is fabricated from a steel wide flange section and specified in the table below. Use the curve resulting from the code equations that is shown in Fig. 11.14 to determine the allowable centric load applied to the column.

Prob. No.	L	Section	End Support
11.25a	22 ft.	W12 × 50	Free/Built-in
11.25b	8.4 m	W254 × 45	Pinned/Pinned
11.25c	26 ft	W18 × 60	Pinned/Built-in
11.25d	7.1 m	W356 × 64	Built-in/Built-in

11.26. A column of length L is fabricated from a steel American Standard section and specified in the table below. Use the curve resulting from the code equations that is shown in Fig. 11.14 to determine the allowable centric load applied to the column.

Prob. No.	L	Section	End Support
11.26a	20 ft.	S8 × 23	Free/Built-in
11.26b	8.0 m	S178 × 23	Pinned/Pinned
11.26c	30 ft	S5 × 10	Pinned/Built-in
11.26d	8.5 m	S102 × 14	Built-in/Built-in

11.27. A column fabricated from aluminum (2014 T-6) in the form of a structural tee section and is specified in the table below. Use the curve resulting from the code equations that is shown in Fig. 11.15 to determine the allowable centric load applied to the column.

Prob. No.	L	Section	End Support
11.27a	16 ft.	WT5 × 56	Free/Built-in
11.27b	6.4 m	WT127 × 45	Pinned/Pinned
11.27c	12 ft	WT8 × 20	Pinned/Built-in
11.27d	4.3 m	WT178 × 36	Built-in/Built-in

11.28. An aluminum (2014-T6) tube with a square cross section is to be employed in a building lobby, because of its appearance. Determine the largest axial (centric) load that it can carry.

Prob. No.	L	S_o	S_i	End Support
11.28a	18 ft.	8 in.	7.5 in.	Free/Built-in
11.28b	6.0 m	150 mm	130 mm	Pinned/Pinned
11.28c	15 ft	6.0 in.	5.75 in.	Pinned/Built-in
11.28d	5.6 m	250 mm	230 mm	Built-in/Built-in

CHAPTER 12

ENERGY METHODS

12.1 INTRODUCTION

To introduce the concept of energy in the analysis of structural elements, let's consider the tie bar subjected to an axial tension force F. The tie bar extends an amount δ as the force is applied. Because we have moved a force F through a distance δ, work **W** has been performed. The amount of work is established from the force-displacement graph presented in Fig. 12.1.

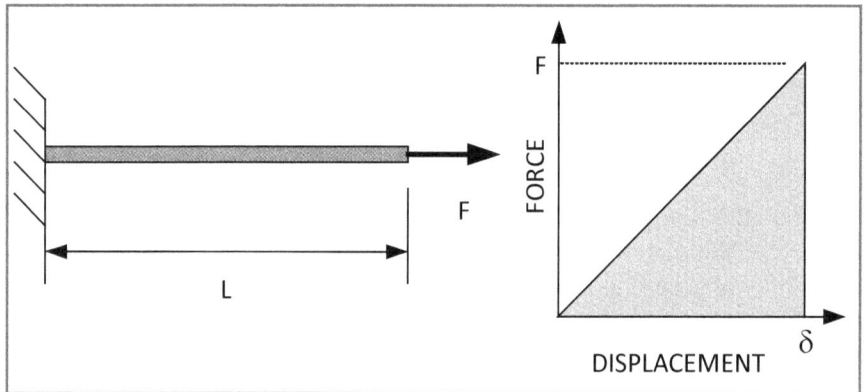

Fig. 12.1 Force-displacement graph for a tie rod with an applied force F.

The work performed on the tie rod is the area under the F-δ curve in Fig. 12.1. Hence:

$$\mathbf{W} = F\delta/2 \quad (12.1)$$

The tie bar may be considered as a system in which energy is conserved. This fact implies that work performed on the bar is conserved as strain energy $\mathbf{E}_\varepsilon$. Accordingly, we equate the work and strain energy and express the strain energy stored in the tie rod as:

$$\mathbf{E}_\varepsilon = F\delta/2 \quad (12.2)$$

In later sections of this chapter, we will derive the equations for the strain energy stored in beams and torsion bars. The ability to compute the strain energy stored in structural members is important, because three different energy theorems can be employed to extend our methods of analysis including:

1. Theorem of virtual work.
2. Theorem of Castigliano.
3. Theorem of least work (minimum potential energy).

12.2 THE ENERGY THEOREMS

12.2.1 The Theorem of Virtual Work

Consider a structural member subjected to a number of forces that maintain the member in equilibrium, as illustrated in Fig. 12.2.

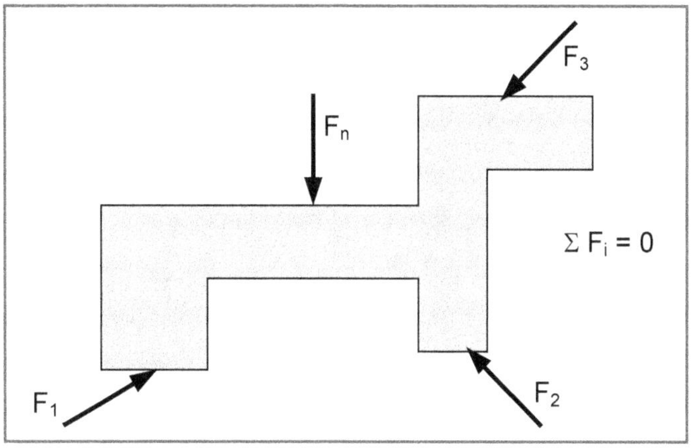

Fig. 12.2 A structural member in equilibrium subjected to forces F_1, F_2, F_3 and F_n.

Suppose we apply a virtual[1] displacement $\Delta\delta_n$ at the location and in the direction of F_n. Then the strain energy is increased by a small amount given by:

$$\Delta E_\varepsilon = F_n \Delta \delta_n \qquad (12.3)$$

In writing Eq. (12.3), we have assumed that the other external forces F_1, F_2 and F_3 do not move as the virtual displacement is applied. With this assumption, we can rewrite Eq. (12.3) as:

$$\frac{\partial E_e}{\partial \delta_n} = F_n \qquad (12.4)$$

In applying Eq. (12.4), E_ε is expressed in terms of displacements so that the applied loads F_1, F_2 and F_3 do not appear and are not an issue. The application is in the determination of the force F_n, when the loading of a structural element is accomplished by applying displacements at specified locations.

12.2.2 The Theorem of Castigliano

Consider again the structural member in Fig. 12.2, which is in equilibrium when subjected to forces F_1, F_2, F_3 and F_n. Let's increase the force F_n by a virtual amount ΔF_n. Accordingly, the increase in the strain energy is given by:

$$\Delta E_\varepsilon = \delta_n \Delta F_n \qquad (12.5)$$

Again we have assumed that ΔF_n is sufficiently small that the external forces F_1, F_2 and F_3 do not move as ΔF_n is applied. We can then rewrite Eq. (12.5) as:

$$\frac{\partial E_e}{\partial F_n} = \delta_n \qquad (12.6)$$

---

[1] The word virtual means the displacement $\Delta\delta_n$ is small when compared to δ_n.

In applying Eq. (12.6), we write the expression for $\mathbf{E}_\varepsilon$ so that the displacements $\delta_1, \delta_2, \delta_3$ and δ_n do not appear. Castigliano's theorem is useful in determining displacements at selected locations on structural members or in solving for reactions in statically indeterminate structures.

12.2.3 The Theorem of Least Work[2]

Consider a number of different stress distributions that are possible for some structural member subjected to external loading. Moreover, assume that all of these different stress distributions yield internal forces that satisfy the equations of equilibrium. Only one of these stress distributions is correct and the others violate the compatibility conditions[3].

The correct (compatible) stress distribution results in the least amount of strain energy stored in the structural member. We can express this fact by writing:

$$\partial \mathbf{E}_e = 0 \qquad (12.7)$$

All three of the theorems are useful in the analysis of structures. Castigliano's theorem is often employed to determine the reaction at the redundant support in a statically indeterminate structure. The theorem of virtual work is helpful in solving for the critical load in the buckling of structural members. The theorem of least work is used with finite elements to determine displacement fields that satisfy the compatibility conditions. Energy methods can also be used to solve for forces generated during impact events. We will demonstrate many of these applications later in this chapter. The theorems of virtual work and least work are deferred until a later course.

12.3 STRAIN ENERGY DENSITY

Consider an arbitrary element removed from the tie rod, shown previously in Fig. 12.1. When it is subjected to an axial force F, a normal stress σ_{xx} results on opposites sides of the element, as shown in Fig. 12.3a.

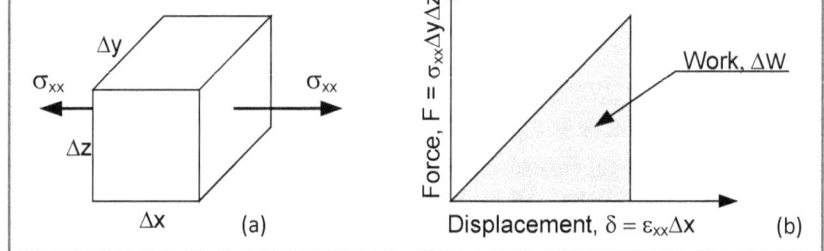

Fig. 12.3 An element removed from the tie rod is subjected to a uniaxial stress σ_{xx}.

The work performed on the element, illustrated in Fig. 12.3b, is the shaded area under the force-displacement line. We equate this work to strain energy and write:

$$\Delta \mathbf{W} = \Delta \mathbf{E}_\varepsilon = (1/2)(\sigma_{xx}\,\varepsilon_{xx})\Delta x \Delta y \Delta z \qquad (12.8)$$

Let's rewrite Eq. (12.8) as:

$$\mathbf{e}_\varepsilon = \Delta \mathbf{E}_\varepsilon / \mathbf{V} = (1/2)\sigma_{xx}\,\varepsilon_{xx} \qquad (12.9)$$

where $\mathbf{e}_\varepsilon$ is the strain energy density, which is the amount of strain energy stored per unit volume of the member, and $\mathbf{V} = \Delta x \Delta y \Delta z$ is the volume of the element.

---

[2] The theorem of least work is also referred to as the theorem for minimum potential energy.
[3] Many different stress distributions can be devised to satisfy equilibrium; however, only the true (correct) distribution will yield the correct displacement field for the member. This requirement is called the compatibility condition.

Equation (12.9) is valid for a uniaxial state of stress where $\sigma_{yy} = \sigma_{zz} = \tau_{xy} = \tau_{yz} = \tau_{zx} = 0$. However, if the member is subjected to a general three-dimensional state of stress, we write the expression for $\mathbf{e}_\varepsilon$ as:

$$\mathbf{e}_\varepsilon = (1/2)(\sigma_{xx}\varepsilon_{xx} + \sigma_{yy}\varepsilon_{yy} + \sigma_{zz}\varepsilon_{zz} + \tau_{xy}\gamma_{xy} + \tau_{yz}\gamma_{yz} + \tau_{zx}\gamma_{zx}) \qquad (12.10)$$

If the state of stress in the structural member is a principal state, then x, y and z are principal axes and $\gamma_{xy} = \gamma_{yz} = \gamma_{zx} = 0$. Then Eq. (12.10) reduces to:

$$\mathbf{e}_\varepsilon = (1/2)(\sigma_1\varepsilon_1 + \sigma_2\varepsilon_2 + \sigma_3\varepsilon_3) \qquad (12.11)$$

where $\sigma_{xx} = \sigma_1$, $\varepsilon_{xx} = \varepsilon_1$, etc.

Let's consider two examples to illustrate the methods for computing strain energy density.

EXAMPLE E12.1

A point in a machine component is subjected to a three-dimensional state of stress that is described in terms of three principal stresses given by $\sigma_1 = 16,200$ psi, $\sigma_2 = 9,300$ psi and $\sigma_3 = -12,500$ psi. Determine the strain energy density for this point, if the machine component is fabricated from steel.

Solution: Let's begin by writing the expression of the strain energy density in terms of the principal stresses and strains, which is given by Eq. (12.11) as:

$$\mathbf{e}_\varepsilon = (1/2)(\sigma_1\varepsilon_1 + \sigma_2\varepsilon_2 + \sigma_3\varepsilon_3) \qquad (a)$$

Recall the stress-strain relations as:

$$\varepsilon_1 = (1/E)[\sigma_1 - \nu(\sigma_2 + \sigma_3)]$$

$$\varepsilon_2 = (1/E)[\sigma_2 - \nu(\sigma_1 + \sigma_3)] \qquad (b)$$

$$\varepsilon_3 = (1/E)[\sigma_3 - \nu(\sigma_1 + \sigma_2)]$$

Substituting Eq. (b) into Eq. (a) gives:

$$\mathbf{e}_\varepsilon = (1/2E)[\sigma_1^2 + \sigma_2^2 + \sigma_3^2 - 2\nu(\sigma_1\sigma_2 + \sigma_2\sigma_3 + \sigma_3\sigma_1)] \qquad (12.12)$$

Substituting numerical parameters from the problem statement into Eq. (12.12) gives:

$$\mathbf{e}_e = \frac{1}{2(30 \times 10^6)} \{(16.2)^2 + (9.3)^2 + (-12.5)^2 - (2)(0.30)[(16.2)(9.3) + (9.3)(-12.5) + (-12.5)(16.2)]\}(10^6)$$

$$\mathbf{e}_e = 10.11 \text{ in.-lb/in}^3$$

We have used the units in.-lb/in.^{3} instead if lb/in^2 for the strain energy density, because the former conveys the concept of energy per unit volume. Actually both representations of the units are correct.

EXAMPLE E12.2

A point in a structural member is subjected to a three dimensional state of stress described by the six Cartesian stress components σ_{xx} = 185 MPa, σ_{yy} = – 114 MPa, σ_{zz} = 75.6 MPa, τ_{xy} = 105 MPa, τ_{yz} = 82.6 MPa, τ_{zx} = 67.9 MPa. Determine the strain energy density, if the member is fabricated from aluminum alloy 2024 T-4.

Solution: Let's begin by writing the expression for the strain energy density in terms of the Cartesian components of stress, which is given by Eq. (12.10) as:

$$\mathbf{e}_\varepsilon = (1/2)(\sigma_{xx}\varepsilon_{xx} + \sigma_{yy}\varepsilon_{yy} + \sigma_{zz}\varepsilon_{zz} + \tau_{xy}\gamma_{xy} + \tau_{yz}\gamma_{yz} + \tau_{zx}\gamma_{zx}) \qquad (a)$$

Recall the stress strain relations as:

$$\varepsilon_{xx} = (1/E)[\sigma_{xx} - \nu(\sigma_{yy} + \sigma_{zz})] \qquad \gamma_{xy} = \tau_{xy}/G$$

$$\varepsilon_{yy} = (1/E)[\sigma_{yy} - \nu(\sigma_{xx} + \sigma_{zz})] \qquad \gamma_{yz} = \tau_{yz}/G \qquad (b)$$

$$\varepsilon_{zz} = (1/E)[\sigma_{zz} - \nu(\sigma_{xx} + \sigma_{yy})] \qquad \gamma_{zx} = \tau_{zx}/G$$

Substituting Eq. (b) into Eq. (a) gives:

$$\mathbf{e}_\varepsilon = (1/2E)[\sigma_{xx}^2 + \sigma_{yy}^2 + \sigma_{zz}^2 - 2\nu(\sigma_{xx}\sigma_{yy} + \sigma_{yy}\sigma_{zz} + \sigma_{zz}\sigma_{xx})] + (1/2G)[\tau_{xy}^2 + \tau_{yz}^2 + \tau_{zx}^2] \quad (12.13)$$

Substituting numerical parameters from the problem statement into Eq. (12.13) gives:

$$\mathbf{e}_e = \frac{10^{12}}{2(72 \times 10^9)}\left[(185)^2 + (-114)^2 + (75.6)^2 - (2)(0.32)[(185)(-114) + (-114)(75.6) + (75.6)(185)]\right]$$

$$+ \frac{10^{12}}{2(27 \times 10^9)}\left[(105)^2 + (82.6)^2 + (67.9)^2\right]$$

$$\mathbf{e}_e = 853.4 \times 10^3 \text{ N-m/m}^3$$

12.4 STRAIN ENERGY IN STRUCTURAL ELEMENTS

This section describes methods employed to determine the strain energy stored in common structural elements, such as tie rods, beams in bending and torsion bars. First, consider the tie rod from Fig. 12.1.

12.4.1 Strain Energy in Tie Rods

Previously we described a technique for determining the strain energy stored in an axially loaded rod in Section 12.1. Let's consider a second approach based on strain energy density. Recall Eq. (12.9) and substitute Hooke's law into this relation to obtain:

$$\mathbf{e}_\varepsilon = (1/2)(\sigma_{xx}\varepsilon_{xx}) = (1/2)E\varepsilon_{xx}^2 = [1/(2E)]\sigma_{xx}^2 \qquad (12.14)$$

Integrate Eq. (12.14) over the volume **V** of the tie rod to obtain:

$$E_e = \frac{1}{2E}\int \sigma_{xx}^2 dV \qquad (12.15)$$

Because $\sigma_{xx} = F/A$ over each volume element in the tie rod, we can rewrite Eq. (12.15) as:

$$E_e = \frac{1}{2E}\frac{F^2}{A^2}AL = \frac{F^2 L}{2AE} \qquad (12.16)$$

Let's consider two examples to demonstrate this approach for determining the strain energy in a tie bar.

EXAMPLE E12.3

A tie rod, 16 ft long fabricated from steel with a yield strength $S_y = 42$ ksi is subjected to an axial force of 9,500 lb. If the bar diameter is 0.75 in., determine the strain energy stored in the tie rod and the strain energy density.

Solution: We employ Eq. (12.16) to determine the strain energy E_ε as:

$$E_e = \frac{F^2 L}{2AE} = \frac{(9500)^2 (16)(12)}{2(\pi)(0.375)^2 (30\times 10^6)} = 653.7 \text{ in.-lb} \qquad (a)$$

The strain energy density is given by Eq. (12.14) as:

$$e_e = \frac{\sigma_{xx}^2}{2E} = \left(\frac{9,500}{\pi (0.375)^2}\right)^2 \left(\frac{1}{(2)(30\times 10^6)}\right) = 7.707 \text{ in.-lb/in.}^3 \qquad (b)$$

It is difficult to physically interpret the significance of these results, because we do not have a basis for comparison. What is a large strain energy or strain energy density? The total strain energy increases with the volume of the member. Consequently, the strain energy stored in a member depends upon both the square of the stress and the rod's volume. To assess the results of this analysis, it is advantageous to examine the strain energy density, which is independent of the volume. Hence we seek a maximum value of the strain energy density to be used for comparison purposes. From Eq. (12.14) it is evident that:

$$(e_\varepsilon)_{Max} = [1/(2E)]S_y^2 \qquad (12.17)$$

$$(e_\varepsilon)_{Max} = (42,000)^2/[(2)(30\times 10^6)] = 29.40 \text{ in.-lb/in.}^3 \qquad (c)$$

Comparing the results of Eq. (b) with Eq. (c) indicates that the tie rod in this example was loaded to a strain energy density of (7.707/29.40) = 0.2621, or 26.21% of the maximum value possible was achieved.

EXAMPLE E12.4

A tie rod, fabricated from stainless steel is shown in Fig. E12.4. If the axial loading is F = 450 kN, determine the strain energy stored in the tie rod and the strain energy density.

Fig. E12.4

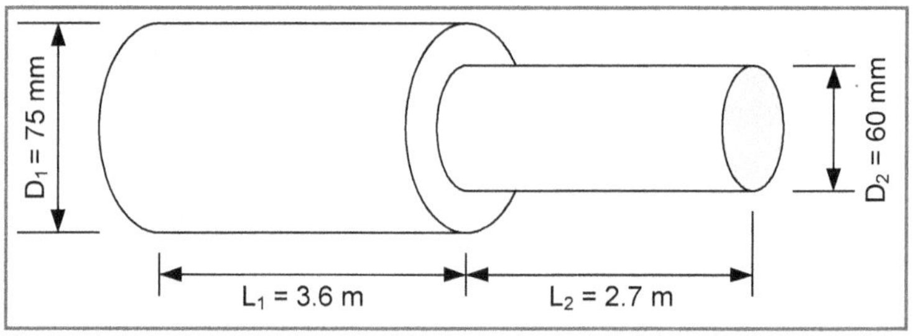

Solution:

Let's begin by computing the stress in each segment of the tie rod as:

$$\sigma_1 = \frac{4F}{\pi D_1^2} = \frac{(4)(450 \times 10^3)}{\pi (75)^2} = 101.9 \text{ MPa}$$

(a)

$$\sigma_2 = \frac{4F}{\pi D_2^2} = \frac{(4)(450 \times 10^3)}{\pi (60)^2} = 159.2 \text{ MPa}$$

Using Eq. (12.14) and Eq. (a), we determine the strain energy density for each segment of the bar as:

$$(e_\varepsilon)_1 = \frac{\sigma_1^2}{2E} = \frac{(101.9)^2 \times 10^{12}}{2(190 \times 10^9)} = 27.33 \times 10^3 \text{ N-m/m}^3$$

(b)

$$(e_\varepsilon)_2 = \frac{\sigma_2^2}{2E} = \frac{(159.2)^2 \times 10^{12}}{2(190 \times 10^9)} = 66.70 \times 10^3 \text{ N-m/m}^3$$

The total strain energy stored in the tie rod is determined from Eq. (12.15) as:

$$E_\varepsilon = (e_\varepsilon)_1 V_1 + (e_\varepsilon)_2 V_2 = (27.33 \times 10^3)(.0375)^2 \pi (3.6)$$

$$+ (66.70 \times 10^3)(.03)^2 \pi (2.7) = 943.9 \text{ N-m}$$

Because the bar was not of constant diameter, it was necessary to divide it into two segments and determine the strain energy density in each segment. The total strain energy is then computed by multiplying each value of e_ε by the appropriate volume and summing the two quantities.

12.4.2 Strain Energy for a Beam in Bending

For a beam subjected to transverse loading, as shown in Fig. 12.4, the flexural stress σ_{xx} can be determined from:

$$\sigma_{xx} = My/I_z \tag{12.18}$$

where M is the bending moment at any location defined by x.

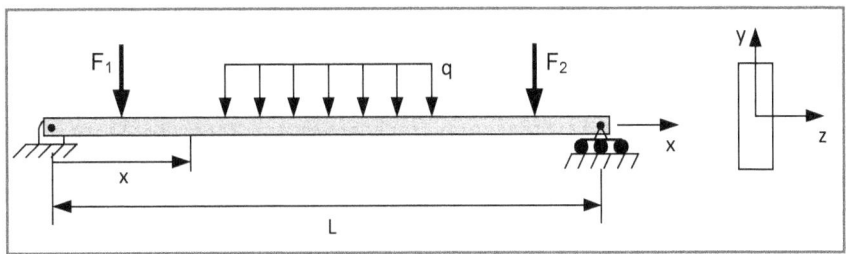

Fig. 12.4 A beam subjected to transverse forces.

To determine the strain energy density, we use Eq. (12.14), which is valid for uniaxial states of stress. Combing Eq. (12.14) and Eq. (12.18), we obtain:

$$\mathbf{e_e} = \left(\frac{1}{2E}\right)\left(\frac{My}{I_z}\right)^2 \quad (12.19)$$

An examination of Eq. (12.19) shows that the strain energy density varies with the location x over the length of the beam because M is a function of x. The strain energy density also depends on the position y, measured from the neutral axis of the beam.

We integrate Eq. (12.19) over the volume to obtain the strain energy stored in the beam as:

$$\mathbf{E_e} = \int \frac{1}{2E}\left(\frac{My}{I_z}\right)^2 dAdx = \int \frac{1}{2E}\left(\frac{M}{I_z}\right)^2 \left[\int y^2 dA\right] dx \quad (12.20)$$

In Eq. (12.20), we recognize $\int y^2 dA = I_z$. Hence, we rewrite Eq. (12.20) as:

$$\mathbf{E_e} = \left(\frac{1}{2EI_z}\right)\int_0^L M^2 dx \quad (12.21)$$

We have assumed that both E and I_z are constant over the volume of the beam in the derivation of Eq. (12.21). Let's consider two examples to demonstrate the use of Eq. (12.21) in determining the strain energy stored in a beam subjected to transverse loading.

EXAMPLE E12.5

For the beam shown in Fig. E12.5, determine the expression for the strain energy due to the normal stress σ_{xx}.

Fig. E12.5

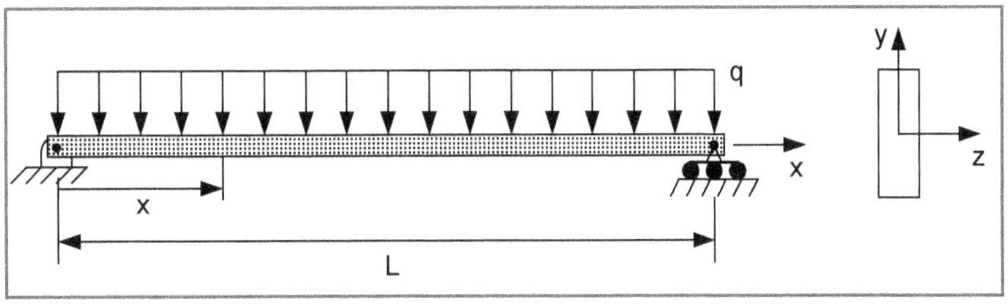

Solution:

The reactions at the simple supports are determined from an appropriate FBD and the application of the equilibrium equations as:

$$\Sigma F_x = 0 \quad \Rightarrow \quad R_{Lx} = 0$$

$$\Sigma M_L = 0 \quad \Rightarrow \quad R_R = qL/2 \quad \text{(a)}$$

$$\Sigma F_y = 0 \quad \Rightarrow \quad R_{Ly} = qL/2$$

Next prepare a FBD of the left-hand portion of the beam, as shown in Fig. E12.5a.

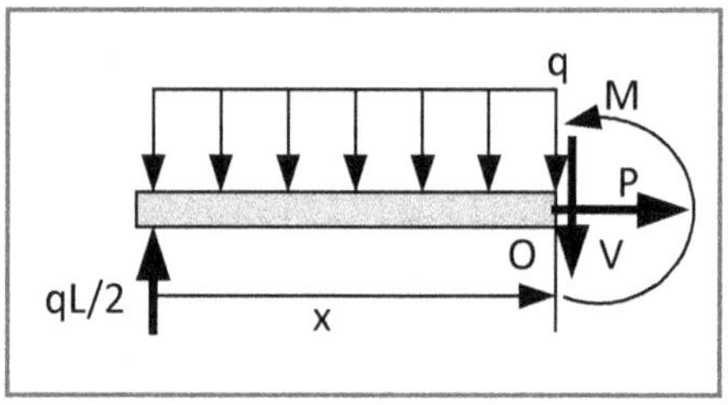

Fig. E12.5a

Applying the equilibrium equation yields:

$$\Sigma F_x = 0 \quad \Rightarrow \quad P = 0$$

$$\Sigma F_y = 0 \quad \Rightarrow \quad qL/2 - qx - V = 0 \quad \text{(b)}$$

$$\Sigma M_O = 0 \quad \Rightarrow \quad (qL/2)x - (qx)(x/2) + M = 0$$

Solving Eqs. (b) for V and M yields:

$$V = (q/2)(L - 2x) \quad \text{(c)}$$

$$M = (qx/2)(x - L) \quad \text{(d)}$$

Substituting Eq. (d) into Eq. (12.21) gives:

$$E_e = \frac{1}{2EI_z} \int_0^L \left[\left(\frac{qx}{2} \right)(x - L) \right]^2 dx \quad \text{(e)}$$

Integrating Eq. (e) and simplifying gives:

$$E_e = \frac{q^2 L^5}{240 EI_z} \quad (12.22)$$

It is important to recognize that the strain energy represented in Eq. (12.22) is due to the normal stress σ_{xx}, which in turn is produced by the bending moment M. However, a shear stress τ_{xy} due to the shear force V exists. This shear stress τ_{xy} produces additional strain energy so that the total strain energy includes contributions from both σ_{xx} and τ_{xy}. We will describe methods to determine the strain energy due to shearing stresses later in this chapter.

EXAMPLE E12.6

For the beam shown in Fig. E12.6, determine the expression for the strain energy due to the normal stress σ_{xx}.

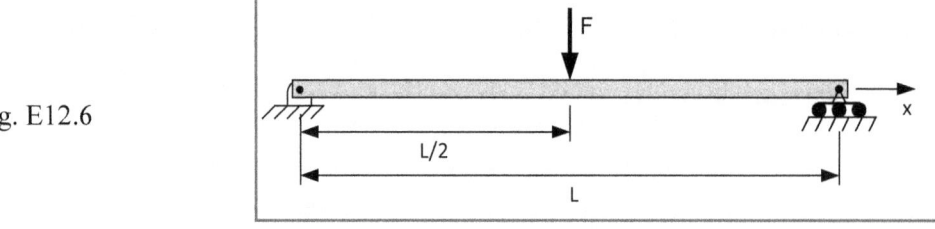

Fig. E12.6

Solution: The reactions at the simple supports are determined from an appropriate FBD and the application of the equilibrium equations to obtain:

$$R_{Lx} = 0 \qquad R_{Ly} = F/2 \qquad R_R = F/2 \qquad (a)$$

Next prepare a FBD of the left-hand portion of the beam as shown in Fig. E12.6a.

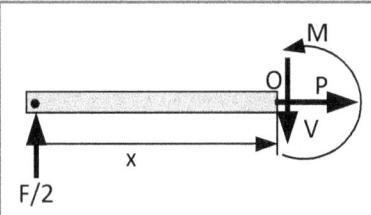

Fig. E12.6a

Applying the equilibrium equations yield:

$$\Sigma F_x = 0 \quad \Rightarrow \quad P = 0 \qquad 0 \leq x \leq (L/2)$$

$$\Sigma F_y = 0 \quad \Rightarrow \quad V = F/2 \qquad 0 \leq x \leq (L/2) \qquad (b)$$

$$\Sigma M_O = 0 \quad \Rightarrow \quad M = (F/2)x \qquad 0 \leq x \leq (L/2)$$

Substituting the expression for the moment M in Eq. (b) into Eq. (12.21) gives a relation that is valid for the left half of the beam.

$$E_e = \frac{1}{2EI_z} \int_0^{L/2} \left(\frac{Fx}{2}\right)^2 dx \qquad (c)$$

Integrating Eq. (c) gives:

$$E_e = \frac{F^2 L^3}{192 EI_z} \qquad (d)$$

The result in Eq. (d) represents the strain energy due to σ_{xx} in the left half of the beam. Because the loading on the beam is symmetric, we multiply this result by two to obtain the strain energy over the entire volume of the beam as:

$$E_e = \frac{F^2 L^3}{96 EI_z} \qquad (12.23)$$

Again, it is important to recognize that the strain energy represented in Eq. (12.23) is due to the normal stress σ_{xx}; however, a shear stress τ_{xy} due to the shear force V exists. This shear stress τ_{xy} produces additional strain energy so that the total strain energy includes contributions from both σ_{xx} and τ_{xy}.

12.4.3 Strain Energy Due to Shearing Stresses

The most general expression for strain energy density is given in Eq. (12.10), where the contributions due to all six Cartesian components of stress are included. To develop a simpler form of this expression, consider the state of stress where $\sigma_{xx} = \sigma_{yy} = \sigma_{zz} = \tau_{yz} = \tau_{zx} = 0$. Then we can rewrite Eq. (12.10) as:

$$\mathbf{e}_\varepsilon = \tau_{xy}\gamma_{xy}/2 = \tau_{xy}^2/(2G) \tag{12.24}$$

Integrate Eq. (12.24) over the volume $\mathbf{V}$ of a structural member to obtain the relation for the strain energy due to τ_{xy} as:

$$\mathbf{E}_e = \frac{1}{2G}\int \tau_{xy}^2 \, dV \tag{12.25}$$

Let's consider a circular shaft subjected to a torsional load to demonstrate the application of Eq. (12.25) in the determination of strain energy due to shear stress.

EXAMPLE E12.7

A 55 in. long steel shaft is subjected to a torque T = 20,000 in.-lb. If the shaft is 1.5 in. in diameter, compute the strain energy stored in the shaft.

Solution: Let's begin by writing the relation for the shear stress τ_{xy}, as a function of the radial position ρ.

$$\tau_{xy} = T\rho/J \tag{a}$$

Substituting Eq. (a) into Eq. (12.25) gives:

$$\mathbf{E}_e = \frac{1}{2G}\int \left(\frac{T\rho}{J}\right)^2 dA\,dx = \frac{1}{2G}\int \left(\frac{T}{J}\right)^2 \left[\int \rho^2 dA\right] dx \tag{b}$$

Recognizing that $\int \rho^2 dA = J$, we rewrite Eq. (b) as:

$$\mathbf{E}_e = \frac{1}{2GJ}\int_0^L T^2 dx = \frac{T^2 L}{2GJ} \tag{12.26}$$

Note that T and J are considered constant over the length of the shaft in Eq. (12.26). Substituting numerical parameters into this relation yields:

$$\mathbf{E}_e = \frac{T^2 L}{2GJ} = \frac{(20\times 10^3)^2 (55)(32)}{(2)(11.5\times 10^6)\pi(1.5)^4} = 1925 \text{ in.-lb} \tag{c}$$

EXAMPLE E12.8

Let's again consider the beam with the uniformly distributed load q that we studied in Example 12.5. In this re-examination, determine the relation for the strain energy stored in the beam due to the shear stress τ_{xy}. The beam has a rectangular cross section as shown in the figure to the right.

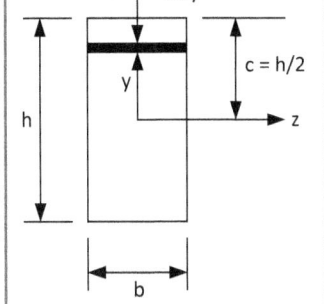

Solution:

For a beam with a rectangular cross section, it can be shown that:

$$\tau_{xy} = \frac{3V}{2A}\left(1 - \frac{y^2}{c^2}\right) \qquad (a)$$

Recall from Example 12.5 that the shear force V was given by:

$$V = qL/2 - qx \qquad (b)$$

Substitute Eq. (b) into Eq. (a) to obtain:

$$\tau_{xy} = \frac{3q}{2A}\left(1 - \frac{y^2}{c^2}\right)\left(\frac{L}{2} - x\right) \qquad (c)$$

Substitute Eq. (c) into Eq. (12.25) and note that $d\mathbf{V} = bdydx$ to obtain:

$$(\mathbf{E}_e)_\tau = \frac{1}{2G}\left(\frac{3q}{2A}\right)^2 \int \left(1 - \frac{y^2}{c^2}\right)^2 \left(\frac{L}{2} - x\right)^2 bdydx \qquad (d)$$

Expanding Eq. (d) gives:

$$(\mathbf{E}_e)_\tau = \frac{1}{8G}\left(\frac{9q^2}{bh^2}\right) \int_{-c}^{+c}\left(1 - \frac{2y^2}{c^2} + \frac{y^4}{c^4}\right)dy \int_0^L \left(\frac{L^2}{4} - Lx + x^2\right)dx \qquad (e)$$

Integrating and substituting the upper and lower limits into the results yields:

$$(\mathbf{E}_e)_\tau = \frac{q^2 L^3}{20Gbh} \qquad (12.27)$$

A comparison of the strain energies stored in the beam due to shear and normal stress is obtained from the ratio of the strain energies given by Eqs. (12.22) and (12.27) as:

$$\frac{(\mathbf{E}_e)_\tau}{(\mathbf{E}_e)_\sigma} = \frac{240 q^2 L^3 EI_z}{20 bhG q^2 L^5} \qquad (a)$$

Noting that $I_z = bh^3/12$ and $G = E/[2(1 + \nu)]$, we can reduce Eq. (a) to:

$$\frac{(\mathbf{E_e})_\tau}{(\mathbf{E_e})_\sigma} = 2(1+\nu)\left(\frac{h}{L}\right)^2 \tag{12.28}$$

Evaluating Eq. (12.28) for a series of beams with different ratios of h/L, gives the strain energy ratio for uniformly distributed loaded beams, shown in Table 12.1.

Table 12.1
Strain Energy Ratio for Steel Beams with Different Ratios of h/L.

h/L	$(E_\varepsilon)_\tau/(E_\varepsilon)_\sigma$	$(E_\varepsilon)_\tau/(E_\varepsilon)_\sigma$	Type of Beam
0.2	0.104	10.4%	Deep
0.1	0.026	2.6%	Long and Slender
0.05	0.0065	0.65%	Long and Slender
0.02	0.0010	0.10%	Long and Slender

Examination of the results of Table 12.1 shows that the fraction of strain energy due to shear stresses is very small if the beam is long and slender. The shear stress becomes an important contributor to strain energy only for short, deep beams. For this reason, we neglect the contribution of the shear stresses in beams to the total strain energy stored when h/L < 1/10.

12.5 DYNAMIC (IMPACT) LOADING

All of the methods introduced thus far in this textbook pertain to structures with loadings, which are gradually applied. Typical force-time and force-displacement curves for gradual application of loads to structures are presented in Fig. 12.5.

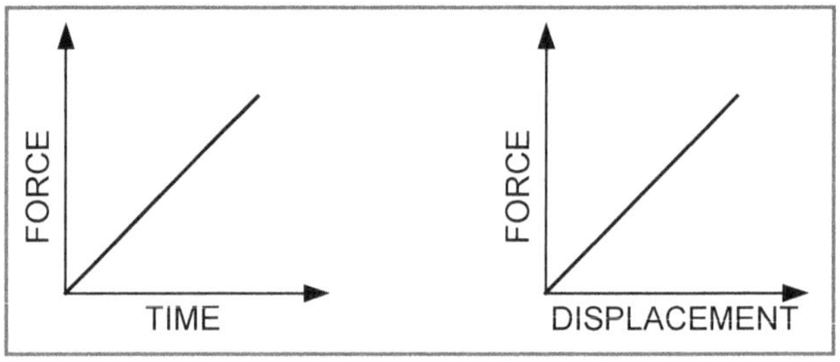

Fig. 12.5 Force-time and force-displacement functions associated with gradual (static) loading.

For dynamically applied forces, sometimes called impact forces, we treat the force as if it were applied instantaneously, with the force-time trace as illustrated in Fig. 12.6. The instantaneous application of load on a structural element has a dramatic effect on the magnitude of the stresses generated.

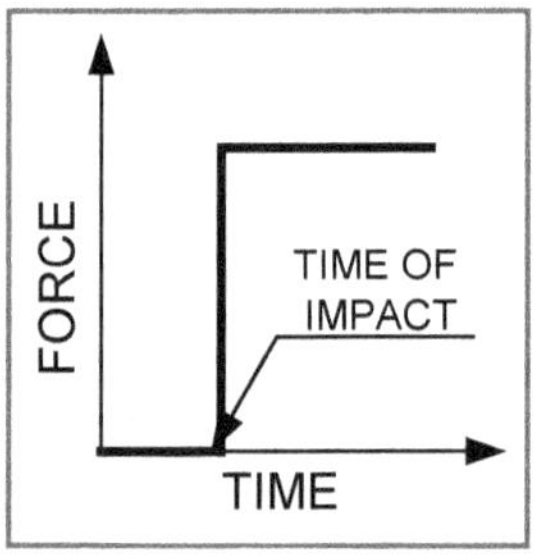

Fig. 12.6 Force-time trace for impact loading.

We are able to use energy methods to solve for the stresses in rods and beams when they are subjected to impact loading. With this approach, two assumptions are made:

1. Energy is conserved, which implies that the impact is elastic and that no work is converted to heat or to produce plastic deformation.

2. The impacting body remains in contact with the structure during the impact period and transfers all of its energy to the structure.

Next, we will describe the application of energy methods for determining the maximum stresses developed due to impact of both rods and beams.

12.5.1 Impact Loading of Rods

Consider the axial rod shown in Fig. 12.7 that is impacted by a ball traveling with a velocity v at the instant it strikes the rod. If the ball remains in contact with the rod during the entire impact period, the kinetic energy in the ball is converted into strain energy stored in the rod. To determine the stresses produced by the impact of the ball, recall the expression for the kinetic energy E_k of the ball is given by:

$$E_k = mv^2/2 \qquad (12.29)$$

where: m is the mass of the ball and v is the velocity of the ball at the instant of impact.

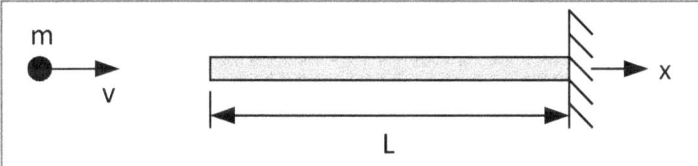

Fig. 12.7 Ball impacting rod with kinetic energy $(½)mv^2$.

If energy is conserved, the kinetic energy is converted to strain energy and we can write:

$$E_k = E_\varepsilon \qquad (12.30)$$

Substituting Eq. (12.16) into Eq. (12.30) yields:

$$(F^2L)/(2AE) = mv^2/2 \qquad (12.31)$$

Solving Eq. (12.31) for the maximum impact force gives:

$$F_{Max} = \sqrt{\frac{mv^2 AE}{L}} \qquad (12.32)$$

The axial stress developed during impact is then given by:

$$\sigma_{Max} = \frac{F_{Max}}{A} = \sqrt{\frac{mv^2 E}{AL}} = \sqrt{\frac{mv^2 E}{V}} \qquad (12.33)$$

Let's consider an example to demonstrate the application of this approach to the solution of impact problems involving rods.

EXAMPLE E12.9

A hammer strikes a steel rod with an axial impact. Determine the maximum force and the maximum stress produced if the hammer weighs 1.6 lb. The rod is 32 in. long with a rectangular cross section having dimensions of 1.25 by 1.5 in. The velocity of the head of the hammer at impact was 15 ft/s.

Solution: Substitute the numerical parameters from the problem statement into Eq. (12.32) to obtain the impact force as:

$$F_{Max} = \sqrt{\frac{mv^2 AE}{L}} = \sqrt{\frac{1.6(15 \times 12)^2(1.25)(1.5)(30 \times 10^6)}{(386)(32)}} = 15,360 \text{ lb} \qquad (a)$$

Note that $g = 386$ in./s^2 must be used in Eq. (a) to convert the 1.6 lb weight into mass.

The axial stress developed during impact is determined from Eq. (12.33) as:

$$\sigma_{Max} = \frac{F_{Max}}{A} = \frac{15,360}{(1.25)(1.5)} = 8,195 \text{ psi} \qquad (b)$$

The result, of this very simple example, clearly demonstrates that very large impact forces can be generated with small masses during an impact event. In this example, a dynamic force in excess of seven tons was produced with a 1.6 lb hammer traveling with a velocity of 10.23 MPH.

12.5.2 Impact Loading of Beams

Consider a simply supported beam that is subjected to impact by a falling weight W, as illustrated in Fig. 12.8.

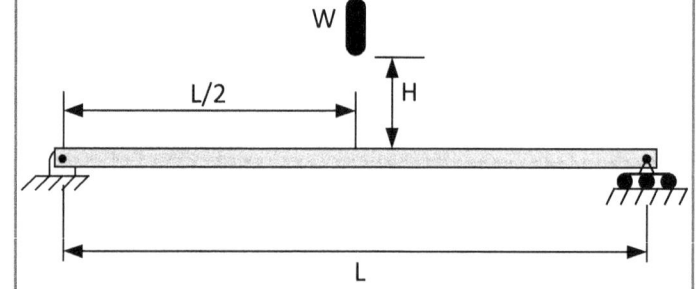

Fig. 12.8 A simply supported beam subjected to impact by a falling weight W.

We approach this problem by recognizing that the potential energy of the weight E_p is converted to kinetic energy prior to impact. Upon impact the kinetic energy E_k is converted to strain energy E_ε stored in the beam. Accordingly, we can write:

$$E_p = E_k = E_\varepsilon = W(H + \delta) \qquad (12.34)$$

where δ is the deflection of the beam at the point of impact.

The strain energy stored in the beam at impact is determined from Eq. (12.21) as:

$$E_e = \left(\frac{1}{2EI_z}\right)\int_0^L M^2 dx \qquad (a)$$

For the simply supported beam with a force F applied at the beam's midpoint, we can write the moment equation as:

$$M = Fx/2 \qquad 0 \leq x \leq (L/2) \qquad (b)$$

This equation is identical to the moment relation written for the beam in Example 12.6. Using the results from Example 12.6, we recall:

$$E_e = \frac{F^2 L^3}{96 EI_z} \qquad (c)$$

Let's first assume that the deflection of the beam is small with respect to the drop height H. Then the potential energy is reduced to $E_p = WH$. Equating Eq. (c) and Eq. (12.34) gives:

$$WH = \frac{F^2 L^3}{96 EI_z} \qquad (d)$$

Solving Eq. (d) for the maximum force gives:

$$F_{Max} = \sqrt{\frac{96 EI_z WH}{L^3}} \qquad (12.35)$$

The maximum bending stress occurs at the midpoint of the beam under the impact point. The bending moment at this location is given by:

$$M_{Max} = \frac{F_{Max} L}{4} = \sqrt{\frac{6 EI_z WH}{L}} \qquad (12.36)$$

The stress at the midpoint of the beam's span is given by:

$$(\sigma_{xx})_{Max} = M_{Max} c / I_z = M_{Max}/Z \qquad (e)$$

where $Z = I_z/c$ is the section modulus.

If the beam's cross section is rectangular with $c = h/2$ and $I_z = bh^3/12$, then Eq. (12.36) can be rewritten as:

$$(\sigma_{xx})_{Max} = \sqrt{\frac{18 EWH}{Lbh}} = \sqrt{\frac{18 EWH}{V}} \qquad (12.37)$$

where V is the volume of the beam.

EXAMPLE E12.10

An experimental impact machine is to be employed in testing the integrity of the weld made in steel beams. A schematic diagram of the impact machine is shown in Fig. E12.10. The cross section of the test beam is b = 40 mm, h = 60 mm and its length L = 400 mm. For a hammer height H set at 1.5 m, determine the effective weight of the hammer necessary to break the weld, if its estimated strength is 500 MPa.

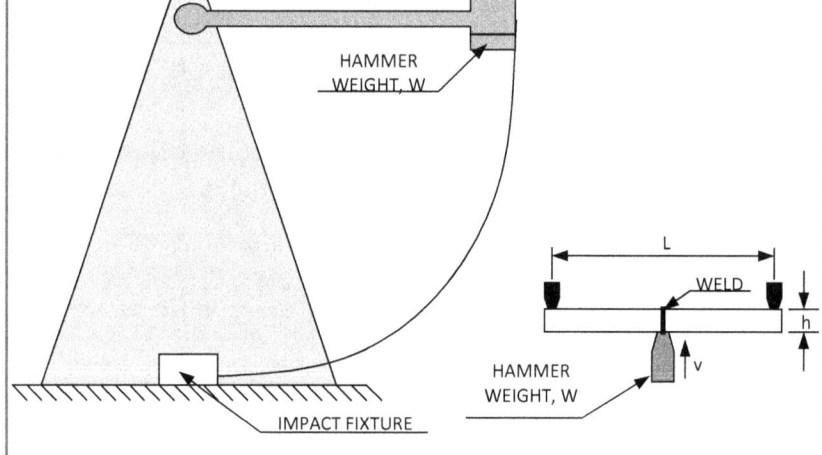

Fig. E12.10

Solution: While the hammer swings through an arc, the equivalent drop of 1.5 meters must be sufficient to generate a maximum stress that is equal to or greater than the strength of the weld material. Accordingly, we can write:

$$(\sigma_{xx})_{Max} = (S_u)_{Weld} \qquad (a)$$

Substituting Eq. (a) into Eq. (12.37) and squaring both sides yields:

$$\frac{18EWH}{Lbh} = S_u^2 \qquad (b)$$

Solving Eq. (b) for the hammer weight W using the numerical parameters in the problem statement gives:

$$W = \frac{S_u^2 bhL}{18EH} = \frac{(500 \times 10^6)^2 (0.04)(0.06)(0.4)}{(18)(207 \times 10^9)(1.5)} = 42.94 \text{ N} \qquad (c)$$

This solution shows that an impact machine incorporating a relatively small weight (43 N ≈ 10 lb) dropped through a short distance (1.5 m) can develop sufficiently high stress to produce failure of a relatively large steel test specimen. Of course, we are assuming an elastic impact. If plastic bending occurs prior to the failure of the weld, the amount of energy necessary to deform the beam increases dramatically.

The maximum force developed during impact is given by Eq. (12.35) as:

$$F_{Max} = \sqrt{\frac{96EI_z WH}{L^3}} = \sqrt{\frac{8EWHbh^3}{L^3}} = \sqrt{\frac{8(207 \times 10^9)(42.94)(1.5)(0.04)(0.06)^3}{(0.4)^3}} = 120.0 \text{ kN}$$

In this example, we observe that the impact force is more than 2,500 times larger than the weight attached to the impact machine. Clearly, dynamic loading is a very important consideration in the design of structures and machine components.

EXAMPLE E12.11

A weight W is positioned at the midpoint on a simply supported beam as shown in Fig. E12.11. While the weight is in contact with the beam (H = 0), its gravitational force has not been applied to the beam. The weight is suddenly released to produce an impact event, although the height of the drop is zero. Determine the equations for the maximum stress, the displacement of the load point and the maximum force developed during the impact.

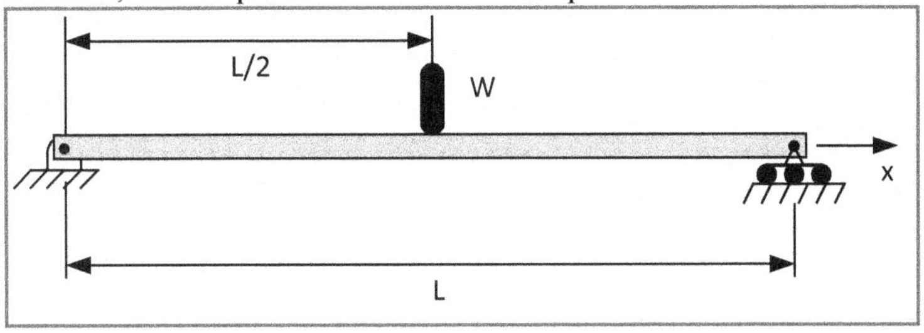

Fig. E12.11

Solution: The potential energy of the weight in contact with the beam before it is released is given by:

$$E_p = W\delta \qquad (a)$$

where δ is the beam's deflection at midpoint due to the sudden application of the weight W.

Let's assume that the elastic curve for the beam during impact is identical to that produced by static loading. Then reference to Appendix E indicates that the force F required to produce the deflection during impact is given by:

$$F = \left(\frac{48EI_z}{L^3}\right)\delta \qquad (b)$$

The strain energy stored in the beam is equal to the work performed on the beam by this force and is given by:

$$E_\varepsilon = W = F\delta/2 \qquad (c)$$

Substituting Eq. (b) into Eq. (c) yields:

$$E_e = \left(\frac{24EI_z}{L^3}\right)\delta^2 \qquad (d)$$

Noting that $E_p = E_e$, we can write:

$$W\delta = \left(\frac{24EI_z}{L^3}\right)\delta^2 \qquad (e)$$

Solving Eq. (e) for δ gives:

$$\delta = \frac{WL^3}{24EI_z} = 2\delta_{st} \qquad (f)$$

where δ_{st} is the static deflection of the beam as given in Appendix E.

> The maximum force F_{Max} during this impact event is given by substituting Eq. (f) into Eq. (b) to obtain:
>
> $$F_{Max} = \left(\frac{48EI_z}{L^3}\right)\left(\frac{WL^3}{24EI_z}\right) = 2W \quad (g)$$
>
> The results from Eqs. (f) and (g) show the effect of suddenly applying a weight W to a simply supported beam doubles both the maximum force and the beam deflection when compared to the case where the weight W is slowly applied.

Now that we have established the importance of the potential energy due to the deflection of the beam during an impact event, let's reconsider the assumption made in the derivation of Eq. (12.36). In this derivation, we assumed $\delta \ll H$ and dropped δ in writing the energy balance. Let's repeat this derivation retaining δ in the energy balance. Begin by recalling Eq. (12.34) as:

$$E_p = E_k = E_\varepsilon = W(H + \delta) \quad (a)$$

For the simply supported beam impacted by a weight W at its midpoint, we recall Eq. (12.23) and write:

$$E_e = \frac{F^2 L^3}{96 EI_z} \quad (b)$$

Again, let's assume that the elastic curve for the beam during impact is identical to that produced by static loading. Then the force F required to produce the deflection during impact is given by:

$$F = \left(\frac{48EI_z}{L^3}\right)\delta \quad (c)$$

Substituting Eq. (c) into (b) and equating the strain energy with the potential energy gives:

$$\left(\frac{48EI_z}{2L^3}\right)\delta^2 - W\delta - WH = 0 \quad (d)$$

Recall from Appendix E that the static deflection δ_{st} is given by:

$$\delta_{st} = \frac{WL^3}{48EI_z} \quad (e)$$

Combining Eqs. (d) and (e) and rearranging yields a quadratic equation in terms of δ:

$$\delta^2 - 2\delta_{st}\delta - 2\delta_{st}H = 0 \quad (f)$$

Solving Eq. (f) gives:

$$\delta = \delta_{st}\left(1 + \sqrt{1 + 2\left(\frac{H}{\delta_{st}}\right)}\right) \quad (12.38)$$

$$\delta = M\delta_{st} \quad (12.38a)$$

where the term $M = \left(1 + \sqrt{1 + 2\left(\frac{H}{\delta_{st}}\right)}\right)$ is a multiplier on the static deflection.

We examine the magnitude of this multiplier as a function of the ratio H/δ_{st} in Table 12.2.

Table 12.2
Multiplier for beam deflection for different ratios of H/δ_{st}.

H/δ_{st}	M	H/δ_{st}	M
0	2.00	50	11.05
1	2.73	100	15.17
2	3.24	200	21.02
5	4.32	500	32.64
10	5.58	1000	45.73
20	7.40	2000	64.25

This multiplier is the same for the maximum force developed during impact. Because δ_{st} is usually a small quantity, the ratio of H/δ_{st} becomes large even for relatively small values of the drop height H.

12.6 CASTIGLIANO'S THEOREM

In a previous section, we have developed relations for the strain energy for rods in tension or compression, beams in bending, and shafts in torsion. These expressions are useful, because they may be employed in the application of Castigliano's theorem. Castigliano's theorem enables us to determine the expressions for the displacement of a structural member, at any point where a force is applied. To illustrate this concept consider the strain energy stored in a tie rod, which is given by Eq. (12.16) as:

$$E_e = \frac{F^2 L}{2AE} \quad (a)$$

If we differentiate this relation with respect to the force F, in accordance with Castigliano's theorem and Eq. (12.6), we obtain:

$$\frac{\partial E_e}{\partial F} = \frac{FL}{AE} = \delta \quad (b)$$

Differentiating the expression for the strain energy with respect to a force, gives the displacement at the point of application of the force. Moreover, the displacement is in the direction of the applied force.

The general procedure for applying Castigliano's theorem is to write an expression for the strain energy of a structural member in terms of the applied loads F_1, F_2, F_3 and F_n. The expression for the strain energy is then differentiated with respect to one of the forces to obtain the displacement at the point of application of that force. The differentiation is repeated for other forces, to obtain additional solutions for displacements.

Several examples will be described to demonstrate the application of Castigliano's theorem.

EXAMPLE E12.12

For the simply supported beam shown in Fig. E12.12, determine the deflection at midspan using Castigliano's theorem.

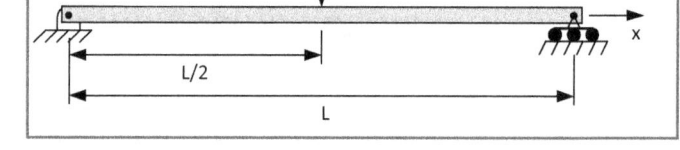

Fig. E12.12

Solution: We make use of Eq. (12.23) for the strain energy stored in this beam, which was derived in Example 12.6 as:

$$\mathbf{E_e} = \frac{F^2L^3}{96EI_z} \qquad \text{(a)}$$

Employ Eq. (12.6) with Eq. (a) to obtain the deflection at the midspan of the beam as:

$$\delta = \frac{\partial \mathbf{E_e}}{\partial F} = \frac{FL^3}{48EI_z} \qquad \text{(b)}$$

This result matches the answer given in Appendix E for the identical case.

It is clear that Castigliano's method provides a useful approach to determine the deflection at a point on a structure, where a concentrated load is applied. But what if we find it necessary to determine the displacement at some other point on the structure? The next example describes a clever approach to provide a solution for this class of problems.

EXAMPLE E12.13

For the cantilever beam, shown in Fig. E12.13, determine the deflection at its free end.

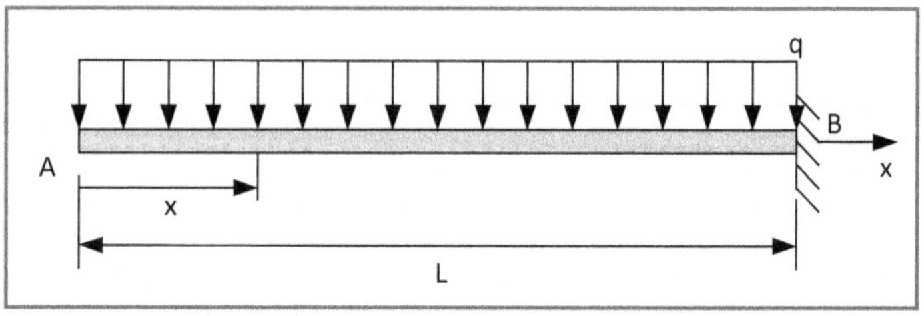

Fig. E12.13

Solution:

The application of Castigliano's theorem to the solution of this problem requires a concentrated force to be located at the point, where the deflection is to be determined. However, in this case the cantilever beam is subjected to a uniformly distributed load. To circumvent this difficulty, we apply a fictitious concentrated force f at the free end of the beam, as shown in Fig. E12.13a.

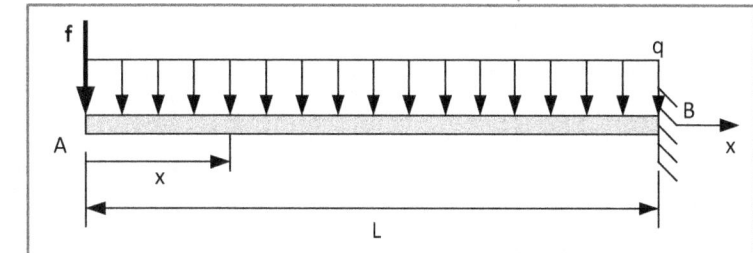

Fig. E12.13a

To apply Castigliano's theorem it is necessary to determine the strain energy stored in the beam, shown in Fig. E12.13a. We begin this determination by preparing a FBD of a segment of the beam, as shown in Fig. E12.13b.

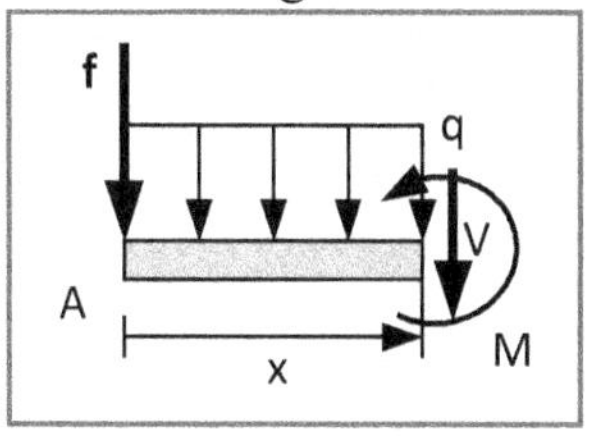

Fig. 12.13b

Writing the moment equation for this segment, we obtain:

$$M = -fx - qx^2/2 \qquad (a)$$

Next, use Eq. (12.21) to express the relation for the strain energy stored in the beam as:

$$E_e = \left(\frac{1}{2EI_z}\right) \int_0^L M^2 dx \qquad (b)$$

If we differentiate Eq. (b) with respect to f and employ Eq. (12.6), it is possible to write the expression for the deflection at the free end of the beam as:

$$\delta = \frac{\partial E_e}{\partial f} = \left(\frac{1}{EI_z}\right) \int_0^L M \frac{\partial M}{\partial f} dx \qquad (12.39)$$

From Eq. (a), it is evident that:

$$\frac{\partial M}{\partial f} = -x \qquad (c)$$

Substituting Eqs. (a) and (c) into Eq. (12.39) gives:

$$\delta = \left(\frac{1}{EI_z}\right) \int_0^L \left(fx^2 + \frac{qx^3}{2}\right) dx \qquad (d)$$

Integrating Eq. (d) yields:

$$\delta = \left(\frac{1}{EI_z}\right)\left[\frac{fx^3}{3} + \frac{qx^4}{8}\right]_0^L = \left(\frac{1}{EI_z}\right)\left[\frac{fL^3}{3} + \frac{qL^4}{8}\right] \qquad (e)$$

Let the fictitious force $f \Rightarrow 0$ and Eq. (e) becomes:

$$\delta = \left(\frac{qL^4}{8EI_z}\right) \qquad (12.40)$$

Again, this result agrees with the answer given in Appendix E for the identical case.

EXAMPLE E12.14

For the simply supported beam shown in Fig. E12.14, use Castigliano's theorem to determine the deflection at the midpoint of the span. Note that $b = L - 2a$.

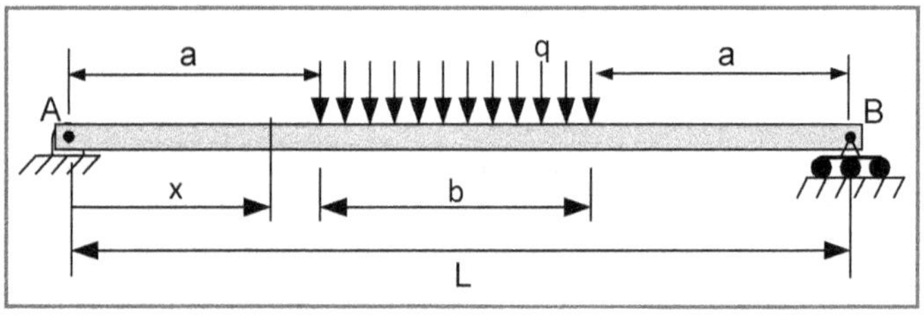

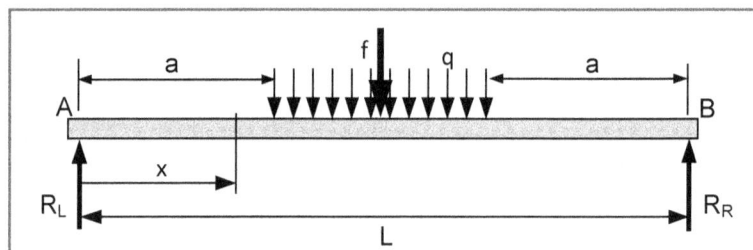

Fig. E12.14

Solution: We first add the fictitious force f at the midpoint of the span and prepare a FBD of the entire beam as shown in Fig. E12.14a.

Fig. E12.14a

From the equilibrium relations and the statement that $b = L - 2a$, it is clear that:

$$R = R_L = R_R = (1/2)(qb + f) \qquad (a)$$

An FBD of the beam segment for $x \leq a$ is presented in Fig. E12.14b. Writing the equation for the moment M from this FBD yields:

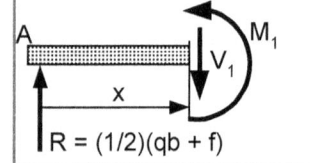

Fig. E12.14b

$$M_1 = \frac{x}{2}(qb + f) \qquad \text{for } x \leq a \qquad (b)$$

$$\frac{\partial M_1}{\partial f} = \frac{x}{2}$$

The FBD presented in Fig. E12.14c is helpful in writing the equation for the moment M which is valid from $a \leq x \leq L/2$.

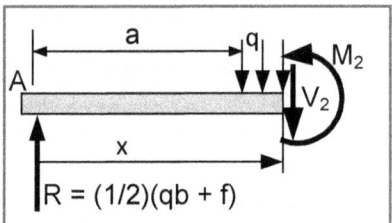

Fig. E12.14c

$$M_2 = \frac{x}{2}(qb+f) - \frac{q}{2}(x-a)^2 \quad \text{for } a \leq x \leq L/2 \qquad (c)$$

$$\frac{\partial M_2}{\partial f} = \frac{x}{2}$$

Note that the loading of the beam is symmetric with respect to x, even after the addition of the fictitious force f. We will use this symmetry in simplifying the calculations needed to determine the midspan deflection. Let's write Eq. (12.39) and adapt it to account for the symmetry of loading by setting the limits of integration from zero to L/2 and multiplying the resulting relation by two.

$$\delta = \left(\frac{1}{EI_z}\right)\left\{\int_0^a M_1 \frac{\partial M_1}{\partial f} dx + \int_a^{L/2} M_2 \frac{\partial M_2}{\partial f} dx\right\} 2 \qquad (d)$$

Substituting Eqs. (b) and (c) into Eq. (d) yields:

$$\delta = \left(\frac{2}{EI_z}\right)\left\{\int_0^a \left(\frac{x}{2}\right)(qb+f)\left(\frac{x}{2}\right)dx + \int_a^{L/2}\left(\left(\frac{x}{2}\right)(qb+f) - \frac{q}{2}(x-a)^2\right)\left(\frac{x}{2}\right)dx\right\} \qquad (e)$$

Let's simplify Eq. (e) by letting $f \Rightarrow 0$ to obtain:

$$\delta = \left(\frac{2}{EI_z}\right)\left\{\int_0^a \left(\frac{qbx^2}{4}\right)dx + \int_a^{L/2}\left(\frac{qbx^2}{4} - \frac{qx(x-a)^2}{4}\right)dx\right\}$$

$$= \left(\frac{q}{2EI_z}\right)\left\{\int_0^a bx^2 dx + \int_a^{L/2}\left[bx^2 - x(x-a)^2\right]dx\right\} \qquad (f)$$

Integrating Eq. (f) gives:

$$\delta = \left(\frac{q}{2EI_z}\right)\left\{\left[\frac{bx^3}{3}\right]_0^a + \left[\frac{bx^3}{3}\right]_a^{L/2} - \left[\frac{x^4}{4} - \frac{2ax^3}{3} + \frac{a^2 x^2}{2}\right]_a^{L/2}\right\} \qquad (g)$$

Substitute the upper and lower limits in Eq. (g), replace a with (1/2)(L − b), and simplify to obtain:

$$\delta = \left(\frac{qb}{384EI_z}\right)\left(8L^3 + b^3 - 4Lb^2\right) \qquad (h)$$

The result is the same as the deflection determined by the double integration method in Chapter 8 [see Eq. (8.19)]. In this case, the sign of the deflection is positive, because the fictitious force was directed downward. The positive sign indicates that the deflection is in the same direction of the fictitious force.

12.6.1 Castigliano's Method Applied to Statically Indeterminate Beams

In Chapter 8, we described techniques for solving for the reaction forces in statically indeterminate beams, by employing either the superposition or the integration method. We can also solve for the reaction forces by using Castigliano's theorem. This approach utilizes Eq. (12.6) applied at one of the structure's supports. We rewrite Eq. (12.6) as:

$$\frac{\partial E_e}{\partial F_n} = \frac{\partial E_e}{\partial R} = \delta_R = 0 \tag{12.41}$$

where $\delta_R = 0$ is the displacement of the reactive force R at the support.

In applying Eq. (12.41), we differentiate the expression for the strain energy in the beam with respect to an appropriate reaction force R. The result of the differentiation gives the displacement at the support; however, because the support is fixed this displacement is zero. Thus Eq. (12.41) provides the additional relation that can be used together with the usual equilibrium equations to solve for unknown reaction forces in statically indeterminate beams. Let's consider two examples to demonstrate this method of analysis.

EXAMPLE 12.15

For the uniformly loaded cantilever beam, shown in Fig. E12.15, use Castigliano's theorem to determine the reaction force at the prop.

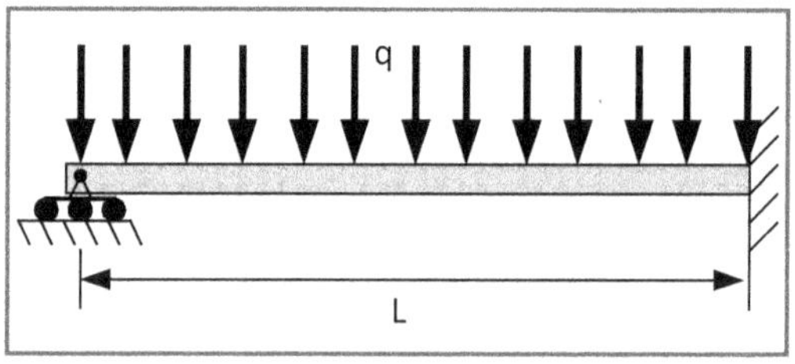

Fig. E12.15

Solution: The propped cantilever beam is statically indeterminate, because we have three unknown reactive forces—the reaction R at the prop, the moment and the reaction force at the built-in end of the beam. Let's begin the solution by drawing a FBD of a portion of the left end of the beam that includes the reaction force R at the prop, as shown in Fig. E12.15a.

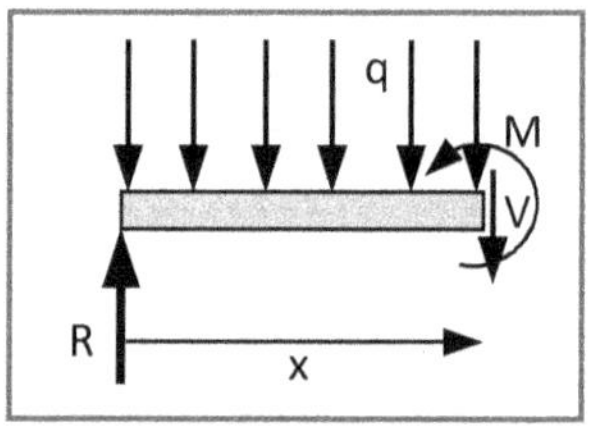

Fig. E12.15a

Writing the equation for the moment M from this FBD yields:

$$M = Rx - \frac{qx^2}{2}$$

$$\frac{\partial M}{\partial R} = x \tag{a}$$

Let's write Eq. (12.41) and use Eq. (12.39) to obtain:

$$\delta = \frac{\partial E_e}{\partial R} = \left(\frac{1}{EI_z}\right)\int_0^L M\frac{\partial M}{\partial R}dx = 0 \tag{b}$$

Substituting Eq. (a) into Eq. (b) and integrating gives:

$$\int_0^L M \frac{\partial M}{\partial R} dx = \int_0^L \left(Rx - \frac{qx^2}{2} \right) x \, dx = 0 \qquad (c)$$

$$\left[\frac{Rx^3}{3} - \frac{qx^4}{8} \right]_0^L = 0$$

Substituting the limits into Eq. (c) and simplifying yields:

$$R = 3qL/8 \qquad (d)$$

This example illustrates another useful approach for determining unknown reactive forces acting on statically indeterminate structures.

EXAMPLE E12.16

For the uniformly loaded continuous beam, shown in Fig. E12.16, use Castigliano's theorem to determine the reaction force at the center support.

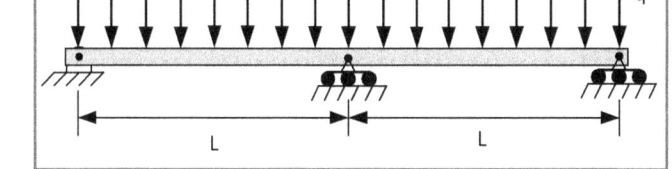

Fig. E12.16

Solution: The propped continuous beam is statically indeterminate, because we have three unknown reactive forces—the three reactions at the simple supports—and only two useful equations of equilibrium. Let's begin the solution by writing the equilibrium relation $\Sigma F_y = 0$ for the entire beam to obtain:

$$R_L + R_C + R_R = 2qL \qquad (a)$$

Because of symmetry $R_L = R_R$, and we can rewrite Eq. (a) as:

$$R_C = 2qL - 2R_L \qquad (b)$$

Now draw a FBD of a portion of the left end of the beam that includes the reaction force R_L at the left hand end, as shown in Fig. E12.16a.

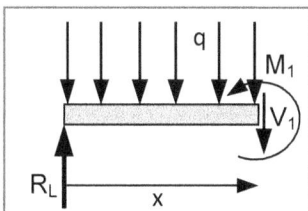

Fig. E12.16a

Writing the equation for the moment from this FBD yields:

$$M_1 = R_L x - \frac{qx^2}{2} \qquad \text{for } 0 \le x \le L \qquad (c)$$

$$\frac{\partial M_1}{\partial R_L} = x$$

Next, we prepare another FBD that includes the reaction R_C at the center support, as shown in Fig. E12.16b.

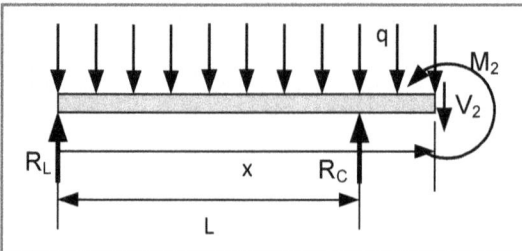

Fig. E12.16b

Writing the equation for the moment from this FBD yields:

$$M_2 = R_L x - \frac{qx^2}{2} + R_C(x-L) \qquad \text{for } L \le x \le 2L \qquad (d)$$

Substituting Eq. (b) into Eq. (d) and simplifying gives:

$$M_2 = R_L(2L-x) - \frac{qx^2}{2} + 2qL(x-L) \qquad \text{for } L \le x \le 2L \qquad (e)$$

$$\frac{\partial M_2}{\partial R_L} = 2L - x$$

Let's write Eq. (12.41) and adopt the form of the strain energy, shown in Eq. (12.39), to obtain:

$$\delta_L = \frac{\partial E_\varepsilon}{\partial R_L} = \left(\frac{1}{EI_z}\right)\left\{\int_0^L M_1 \frac{\partial M_1}{\partial R_L}dx + \int_L^{2L} M_2 \frac{\partial M_2}{\partial R_L}dx\right\} = 0 \qquad (f)$$

Substituting Eqs. (c) and (e) into Eq. (f) and integrating gives:

$$\delta_L = \int_0^L \left(R_L x - \frac{qx^2}{2}\right)(x)dx + \int_L^{2L}\left(R_L(2L-x) - \frac{qx^2}{2} + 2qL(x-L)\right)(2L-x)dx = 0$$

$$\left[\frac{R_L x^3}{3} - \frac{qx^4}{8}\right]_0^L + \left[R_L\left(4L^2 x - 2Lx^2 + \frac{x^3}{3}\right) + q\left(\frac{x^4}{8} - Lx^3 + 3L^2 x^2 - 4L^3 x\right)\right]_L^{2L} = 0 \qquad (g)$$

Substituting the limits into Eq. (g) and simplifying yields:

$$(2L^3/3)R_L - qL^4/4 = 0 \qquad (h)$$

Solving for R_L yields:

$$R_L = R_R = 3qL/8 \qquad (i)$$

Finally from Eq. (i) and Eq. (b), it is clear that:

$$R_C = 5qL/4 \qquad (j)$$

This example illustrates that the center support in a continuous beam carries a large portion of the loading. In this case, the center support carries over three times as much load as each end support. Note that these same results can be obtained by using the methods (i.e. superposition and singularity functions) presented in Chapter 8.

12.6.2 Castigliano's Method Applied to Determining Rotation

In previous sections, we considered a structural element subjected to forces F_1, F_2, F_3 and F_n to derive the relation:

$$\frac{\partial E_e}{\partial F_n} = \delta_n \qquad (12.6)$$

In the derivation of Eq. (12.6), we equated the incremental changes in the work $\Delta W = \delta_n \Delta F_n$ with the incremental changes in the strain energy ΔE_ε. However, Castigliano's method is more general and can also accommodate incremental changes in the work due either to the application of a concentrated moment M or a torque T. When a concentrated moment is applied to a structural member, the work is given by:

$$W = (1/2)M\theta \qquad (12.42)$$

where θ is the angle through which the moment rotates.

Similarly when applying a torque T to a shaft, the resulting work is:

$$W = (1/2)T\phi \qquad (12.43)$$

where ϕ is the angle of twist of the shaft.

Recognizing these alternative forms for expressing work enables us to write expressions for the angular displacements (rotations) as:

$$\theta_n = \partial E_\varepsilon / \partial M_n \qquad (12.44)$$

$$\phi_n = \partial E_\varepsilon / \partial T_n \qquad (12.45)$$

Let's demonstrate the use of these relations in the solution of two example problems.

EXAMPLE E12.17

A cantilever beam is formed in the shape of a quarter circle, as illustrated in Fig. E12.17. If the cantilever beam is loaded with a concentrated moment M_o and a concentrated force F at its free end, determine the angular displacement and the beam's deflection in the direction of the force F.

Fig. E12.17

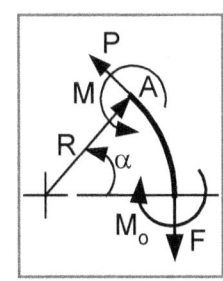

Solution: Let's begin by preparing a FBD of a segment of the circular arc as shown in to the right.

Writing the equation for the moment M at some arbitrary position defined by the angle α as shown in the FBD gives:

$$M = M_0 + FR(1 - \cos\alpha) \qquad (a)$$

The cantilever beam, in this example, is in the form of a circular arc; hence, it is necessary to modify Eq. (12.21) to read as:

$$E_e = \left(\frac{1}{2EI_z}\right)\int_0^{\pi/2} M^2 R d\alpha \qquad (12.46)$$

where $Rd\alpha = dx$

In this case, the upper limit on the integral is $\pi/2$, because the cantilever is formed as a quarter circle. The upper limit is adjusted to accommodate the shape of the circular arc. Let's determine the deflection downward δ_y in the direction of the force F, by using Eq. (12.6) to obtain:

$$\delta_y = \frac{\partial E_e}{\partial F} = \left(\frac{1}{EI_z}\right)\int_0^{\pi/2} M\frac{\partial M}{\partial F} R d\alpha \qquad (b)$$

Substitute Eq. (a) into Eq. (b) and integrate to obtain:

$$\delta_y = \left(\frac{1}{EI_z}\right)\int_0^{\pi/2}[M_0 + FR(1-\cos\alpha)][R(1-\cos\alpha)]Rd\alpha$$

$$\delta_y = \left(\frac{R^2}{EI_z}\right)\int_0^{\pi/2}\left[M_0(1-\cos\alpha) + FR(1-\cos\alpha)^2\right]d\alpha \qquad (c)$$

$$\delta_y = \left(\frac{R^2}{EI_z}\right)\left[M_0\alpha - M_0\sin\alpha + FR\left(\alpha - 2\sin\alpha + \frac{\alpha}{2} + \frac{\sin 2\alpha}{4}\right)\right]_0^{\pi/2}$$

Substituting the limits into Eq. (c) and simplifying yields:

$$\delta_y = \frac{R^2}{EI_z}\left\{\frac{M_0}{2}(\pi-2) + \frac{FR}{4}(3\pi-8)\right\} \qquad (d)$$

To determine the rotation at the free end of the cantilever beam, substitute Eq. (a) into Eq. (12.44) to obtain:

$$\theta = \frac{\partial E_e}{\partial M_0} = \left(\frac{1}{EI_z}\right)\int_0^{\pi/2} M\frac{\partial M}{\partial M_0} R d\alpha \qquad (e)$$

Integrating Eq. (e) gives:

$$\theta = \frac{\partial E_e}{\partial M_0} = \left(\frac{1}{EI_z}\right)\int_0^{\pi/2}[M_0 + FR(1-\cos\alpha)][1]Rd\alpha$$

$$\theta = \left(\frac{R}{EI_z}\right)[M_0\alpha + FR\alpha - FR\sin\alpha]_0^{\pi/2}$$

$$\theta = \left(\frac{R}{EI_z}\right)\left[M_0\frac{\pi}{2} + FR\left(\frac{\pi}{2}-1\right)\right] \qquad (f)$$

Castigliano's theorem has provided an approach for solving for the deflection and rotation at select points on **thin** curved beams. Additional exercises of this type will be given in the problems section of this chapter.

EXAMPLE E12.18

A shaft is fabricated by brazing together two sections of round tubing, as shown in Fig. E12.18. The shaft is fixed at one end and a torque T is applied at its free end. Determine the equation for the angle of twist ϕ. Assume the thickness of the braze metal is negligible.

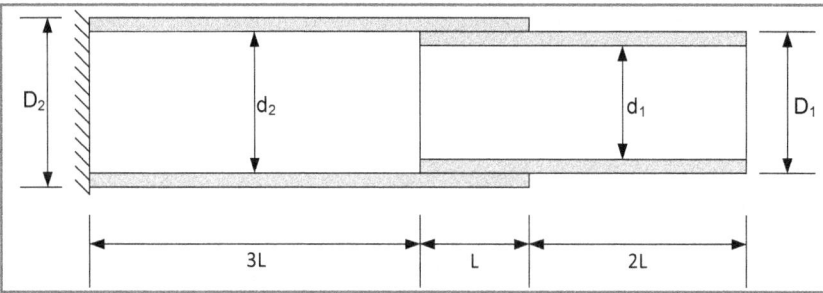

Fig. E12.18

Solution: The strain energy for a shaft subjected to torsion loading is given by Eq. (12.26) as:

$$E_e = \frac{1}{2GJ}\int_0^L T^2 dx = \frac{T^2 L}{2GJ} \tag{a}$$

We can apply this relation if it is adjusted to account for the variation of the polar moment of inertia J along the length of the shaft. Accordingly we divide the shaft into three parts—the left side, the center portion, and the right side and rewrite Eq. (a) as:

$$E_e = \frac{T^2 L}{2G}\left[\frac{3}{(\pi/32)(D_2^4 - d_2^4)} + \frac{1}{(\pi/32)(D_2^4 - d_1^4)} + \frac{2}{(\pi/32)(D_1^4 - d_1^4)}\right] \tag{b}$$

Recalling Eq. (12.45), substitute Eq. (b) into this relation and simplify to obtain:

$$\phi = \frac{\partial E_e}{\partial T} = \frac{32TL}{\pi G}\left[\frac{3}{(D_2^4 - d_2^4)} + \frac{1}{(D_2^4 - d_1^4)} + \frac{2}{(D_1^4 - d_1^4)}\right] \tag{c}$$

Again Castigliano's method provides another independent approach to determining displacements (angular or linear) at one point on the structural member.

12.6.3 Castigliano's Method Applied to Truss Deflections

In our previous discussion of trusses, we were concerned with determining forces, stresses and safety factors for the individual members. However, in many instances, it is important to determine the displacement of the joints in the truss. There are two approaches for determining the displacement. The first is to calculate the extension or contraction of the individual members, and then to solve a complex geometry problem to establish the movement of each of the joints. The second approach, which is described in this section, is the application of Castigliano's method for this determination.

We have derived the relation for the strain energy stored in a tie rod in Eq. (12.16). This equation is also valid for truss members, because the forces are always applied along the axis of the individual members. Let's rewrite this relation as:

$$\mathbf{E}_e = \frac{P_i^2 L_i}{2 A_i E} \tag{12.47}$$

where P_i is the internal force and A_i is the cross sectional area in truss member i and L_i is the length of member i.

The total strain energy in the truss system is then obtained by summing the strain energy stored in the individual members. Hence, we can write:

$$\mathbf{E}_e = \sum_{i=1}^{n} \frac{P_i^2 L_i}{2 A_i E} \tag{12.48}$$

where n is the number of members in the truss.

Substituting Eq. (12.48) into Eq. (12.6) yields:

$$\delta_k = \frac{\partial \mathbf{E}_e}{\partial F_k} = \frac{\partial}{\partial F_k} \sum_{i=1}^{n} \frac{P_i^2 L_i}{2 A_i E} \tag{12.49}$$

Let's differentiate Eq. (12.49) to obtain:

$$\delta_k = \sum_{i=1}^{n} \frac{P_i L_i}{A_i E} \left(\frac{\partial P_i}{\partial F_k} \right) \tag{12.50}$$

Care must be exercised in the application of Eq. (12.50). To determine the ($\partial P_i/\partial F_k$), it is necessary to write equations for the internal forces P_i in each member of the truss in terms of the external force F_k applied at a specified joint in the truss. Numbers must not be substituted into the equations, until after Eq. (12.50) has been executed in symbolic format.

Let's consider three examples to demonstrate this approach for determining the displacement of a joint in a truss.

EXAMPLE E12.19

Determine the displacement of joint B in the vertical direction for the two-member truss, depicted in Fig. E12.19 using Castigliano's method. Truss members AB and BC are fabricated from the same material, with identical length L and area A.

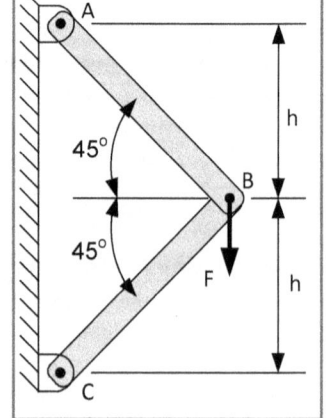

Fig. E12.19

Solution: We begin by writing the equations for the internal forces P_{AB} and P_{BC} in terms of the force F and the geometric parameters describing the truss. From the FBD of the pin at joint B, we write:

$$\Sigma F_x = -P_{BA} \cos 45° - P_{BC} \cos 45° = 0$$

$$P_{BA} = -P_{BC} \qquad (a)$$

$$\Sigma F_y = P_{BA} \sin 45° - P_{BC} \sin 45° - F = 0$$

$$P_{BA} = F/\sqrt{2} \qquad P_{BC} = -F/\sqrt{2} \qquad (b)$$

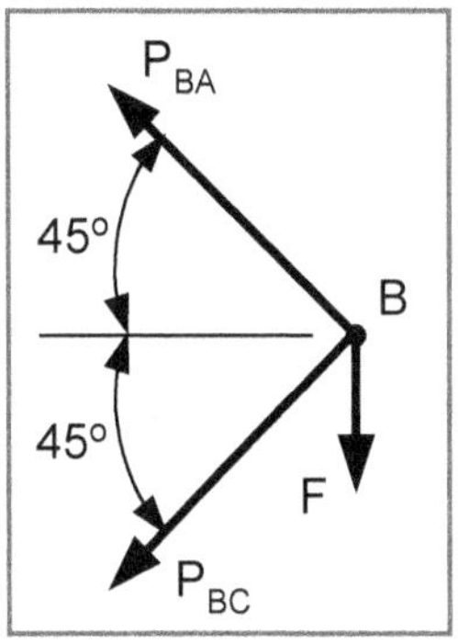

Fig. E12.19a

Next let's take the partial derivatives of the two internal forces to obtain:

$$\partial P_{BA}/\partial F = 1/\sqrt{2} \qquad \partial P_{BC}/\partial F = -1/\sqrt{2} \qquad (c)$$

Substituting the results of Eqs. (b) and (c) into Eq. (12.50) yields:

$$\delta_B = \frac{L}{AE} \sum_1^2 \left(\frac{\partial P_i}{\partial F_B}\right) P_i = \frac{L}{AE} \left[\left(\frac{F}{\sqrt{2}}\right)\left(\frac{1}{\sqrt{2}}\right) + \left(\frac{-F}{\sqrt{2}}\right)\left(\frac{-1}{\sqrt{2}}\right) \right] \qquad (d)$$

Reducing Eq. (d) gives:

$$\delta_B = FL/AE \qquad (e)$$

This result gives the deflection of the pin at joint B in the downward direction. The question of whether the pin at joint B moves in the horizontal direction is given as a problem at the end of the chapter.

EXAMPLE E12.20

Determine the displacement in the vertical direction at joint C for the seven-member truss, depicted in Fig. E12.20 using Castigliano's method. All of the truss members are fabricated from the same material. The cross sectional area of members BC, AE, BD and DE is A; the area of AB is 2A, and the area of members BE and CD is $(2/\sqrt{2})A$.

Fig. E12.20

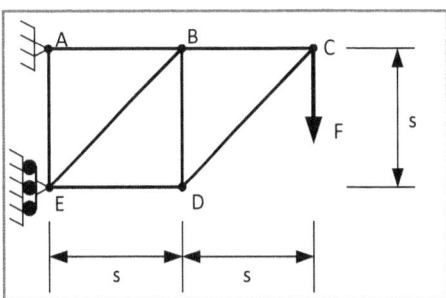

Solution:

Prepare FBDs for the entire truss and for the individual pins A, B, C and E and write the equilibrium relations, to determine the reaction forces and the forces in the seven truss members. The FBDs are shown below, and the forces determined by applying the equilibrium relations are listed in Table E12.20.

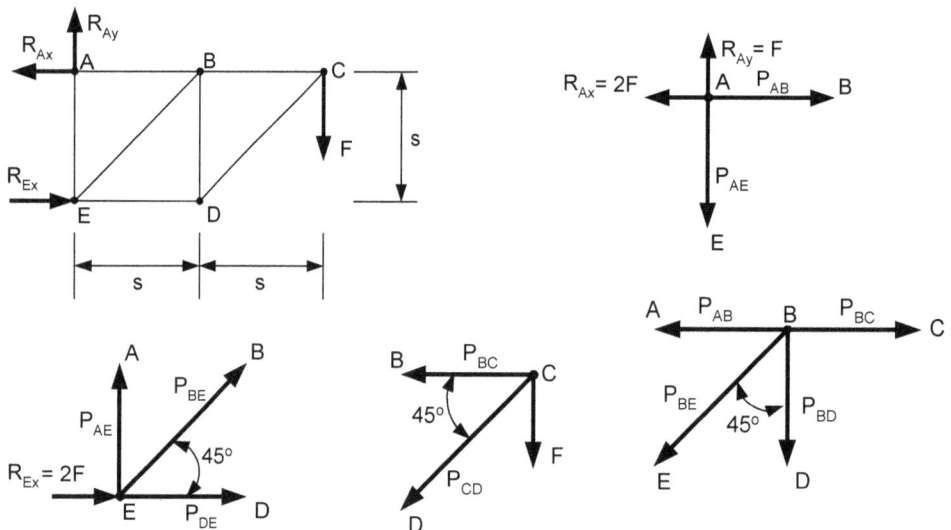

Table E12.20
Arrangement of results for internal forces, length and areas of truss members.

Member	P_i	$\partial P_i/\partial F$	Length. L_i	Area, A_i	$(P_iL_i/A_i)(\partial P_i/\partial F)$
AB	$2F$	2	s	2A	$2Fs/A$
BC	F	1	s	A	Fs/A
AE	F	1	s	A	Fs/A
BE	$-2F/\sqrt{2}$	$-2/\sqrt{2}$	$\sqrt{2}\,s$	$(2/\sqrt{2})A$	$2Fs/A$
BD	F	1	s	A	Fs/A
DE	$-F$	-1	s	A	Fs/A
CD	$-2F/\sqrt{2}$	$-2/\sqrt{2}$	$\sqrt{2}\,s$	$(2/\sqrt{2})A$	$2Fs/A$
					$\Sigma = 10(Fs/A)$

In the last row in Table E12.20, we have determined the sum of $(P_iL_i/A_i)(\partial P_i/\partial F)$ for the seven different truss members. Then it is evident from Eq. (12.50) that the downward deflection of the joint at C is given by:

$$\delta_y = (10Fs)/(AE)$$

The question remains regarding the horizontal deflection of the joint at C. Let's determine this deflection in the next example.

EXAMPLE E12.21

Determine the displacement in the horizontal direction at joint C for the seven-member truss, depicted in Fig. E12.20 using Castigliano's method. All of the truss members are fabricated from the same material. The cross sectional area of members BC, AE, BD and DE is A; the area of AB is 2A, and the area of members BE and CD is $(2/\sqrt{2})A$

Solution:

To determine the horizontal displacement of the pin at joint C, it is necessary to apply a fictitious force f acting in the horizontal direction at that location. The approach is the same as with Example 12.20. We prepare FBDs for the entire truss and for the individual pins A, B, C and E and write the equilibrium relations, to determine the reaction forces and the forces in the seven truss members. Because the reactions and the internal forces due to the force F have already been determined, we will consider only the fictitious force f. We will then superimpose the result due to f on the previous results for F in Table E12.21.

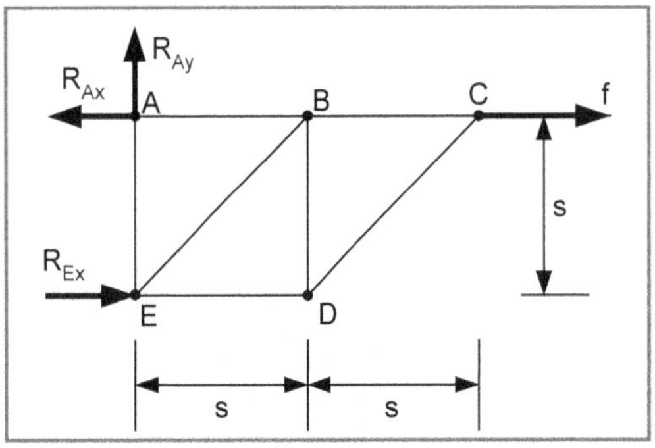

Fig. E12.21

$R_{Ax} = f \quad R_{Ay} = 0 \quad R_{Ex} = 0$

Considering the pin at joint A, we establish that $P_{AB} = f$ and $P_{AE} = 0$.

Considering the pin at joint C, we establish that $P_{BC} = f$ and $P_{CD} = 0$.

Considering the pin at joint B, we establish that $P_{BE} = 0$ and $P_{BD} = 0$.

Finally, considering the pin at joint E, we establish that $P_{DE} = 0$.

We superimpose these results on those obtained in Example 12.20 to give:

Table E12.21
Revision of Table E12.20 showing results for internal forces, length and areas of truss members.

Member	P_i	$\partial P_i / \partial f$	Length. L_i	Area, A_i	$(P_i L_i / A_i)(\partial P_i / \partial f)$
AB	$2F + f$	1	s	2A	$(2F + f)(s/2A)$
BC	$F + f$	1	s	A	$(F + f)(s/A)$
AE	F	0	s	A	0
BE	$-2F/\sqrt{2}$	0	$\sqrt{2}$ s	$(2/\sqrt{2})A$	0
BD	F	0	s	A	0
DE	$-F$	0	s	A	0
CD	$-2F/\sqrt{2}$	0	$\sqrt{2}$ s	$(2/\sqrt{2})A$	0
					$\Sigma = (2Fs/A)+(3fs/2A)$

> In the last row in Table E12.21, we have determined the sum of $(P_iL_i/A_i)(\partial P_i/\partial f)$ for the seven different truss members. Then it is evident from Eq. (12.50) that the deflection in the positive x direction of the joint at C is given by:
>
> $$\delta_y = (2Fs)/(AE) \qquad (a)$$
>
> Of course, in writing Eq. (a), we permitted the fictitious force f to go to zero. In this solution, we have only considered the displacement of one joint in the truss. We could determine the displacement of any joint by applying fictitious forces at the joint in question and repeating the analysis described in Examples 12.20 and 12.21.

12.7 SUMMARY

When a force F is applied to a structural member, it undergoes a displacement δ and work is performed. This work is conserved and stored as strain energy $\mathbf{E}_\varepsilon$.

$$\mathbf{E}_\varepsilon = F\delta/2 \qquad (12.2)$$

The ability to compute the strain energy stored in structural members is important, because three different energy theorems can be employed to extend our methods of analysis. These include: the theorem of virtual work, the theorem of Castigliano and the theorem of least work (minimum potential energy). We described several applications of Castigliano's theorem in this chapter.

The application of Castigliano's theorem involves the partial differentiation of the strain energy with respect to an applied force F to give the displacement δ as:

$$\frac{\partial \mathbf{E}_e}{\partial F_n} = \delta_n \qquad (12.6)$$

The concept of strain energy density was introduced and methods for computing this quantity were demonstrated based on the following equations:

$$\mathbf{e}_\varepsilon = \Delta \mathbf{E}_\varepsilon / \mathbf{V} = (1/2)\sigma_{xx}\varepsilon_{xx} \qquad (12.9)$$

$$\mathbf{e}_\varepsilon = (1/2)(\sigma_{xx}\varepsilon_{xx} + \sigma_{yy}\varepsilon_{yy} + \sigma_{zz}\varepsilon_{zz} + \tau_{xy}\gamma_{xy} + \tau_{yz}\gamma_{yz} + \tau_{zx}\gamma_{zx}) \qquad (12.10)$$

$$\mathbf{e}_\varepsilon = (1/2)(\sigma_1\varepsilon_1 + \sigma_2\varepsilon_2 + \sigma_3\varepsilon_3) \qquad (12.11)$$

The strain energy density served a useful function because it could be integrated over the volume of the structural member to establish the strain energy stored in the element.

$$\mathbf{E}_e = \frac{1}{2E}\int \sigma_{xx}^2 dV \qquad (12.15)$$

We showed that the strain energy stored in a tie rod is given by:

$$\boxed{E_e = \frac{1}{2E}\frac{F^2}{A^2}AL = \frac{F^2L}{2AE}} \quad (12.16)$$

We also showed the strain energy stored in a beam is given by:

$$\boxed{E_e = \left(\frac{1}{2EI_z}\right)\int_0^L M^2 dx} \quad (12.21)$$

And the strain energy stored in a shaft subjected to a torque T is given by:

$$\boxed{E_e = \frac{1}{2GJ}\int_0^L T^2 dx = \frac{T^2L}{2GJ}} \quad (12.26)$$

After describing and demonstrating techniques for determining the strain energy in structural members, we used energy concepts to write relations for impact stresses developed in rods and beams. For a rod subjected to axial impact by an object with a mass m traveling with a velocity v, we derived the relation for the stress as:

$$\boxed{\sigma_{Max} = \sqrt{\frac{mv^2 E}{AL}} = \sqrt{\frac{mv^2 E}{V}}} \quad (12.33)$$

We also considered impact of a beam by a weight W falling through a distance H. If the deflection of the beam was small relative to H, then the maximum stress is given by:

$$\boxed{(\sigma_{xx})_{Max} = \sqrt{\frac{18EWH}{Lbh}} = \sqrt{\frac{18EWH}{V}}} \quad (12.37)$$

When the displacement of the beam is not negligible relative to the height H of the falling weight, both the stress and the displacement is increased. The relation for the displacement is given by:

$$\boxed{\delta = \delta_{st}\left(1 + \sqrt{1 + 2\left(\frac{H}{\delta_{st}}\right)}\right)} \quad (12.38)$$

where $\delta_{st} = \dfrac{WL^3}{48EI_z}$ and $\left(1 + \sqrt{1 + 2\left(\dfrac{H}{\delta_{st}}\right)}\right)$ is a multiplier **M** on the static deflection.

Castigliano's theorem and different variations of Eq. (12.6) were applied to four different types of problems, which included determining the displacement of beams under an applied load, solving for unknown reactions in statically indeterminate beams, determining rotations in curved beams and shafts, and determining displacements of select pins in trusses. These applications were demonstrated with several example problems.

PROBLEMS

12.1 A rod of area A and length L is subjected to an axial force F, as shown in the figure below. Determine the work **W** performed on the rod and the strain energy stored in the rod. Numerical parameters pertaining to the problem are given in the table below:

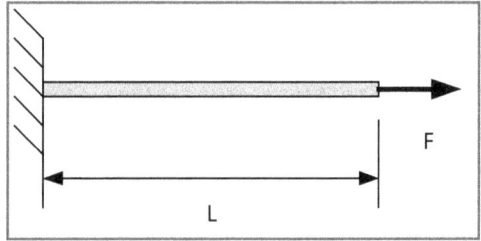

Prob. No.	Force, F	Area, A	Length, L	Material
12.1a	4.6 kip	1.25 in^2	60 in.	Aluminum
12.1b	32.5 kN	400 mm^2	1.0 m	Steel
12.1c	6.25 kip	1.35 in^2	36 in.	Titanium
12.1d	55.0 kN	300 mm^2	1.75 m	Brass
12.1e	3.35 kip	0.75 in^2	48 in.	Iron

12.2 A point in a machine component is subjected to a three-dimensional state of stress that is described in terms of three principal stresses σ_1, σ_2, and σ_3. Determine the strain energy density at this point, if the machine component is fabricated from the material described in the table below.

Prob. No.	σ_1	σ_2	σ_3	Material
12.2a	18.5 ksi	12.3 ksi	$-$ 11.6 ksi.	Aluminum
12.2b	120 MPa	85 MPa	$-$ 99.4 MPa	Steel
12.2c	12.3 ksi	10.3 ksi	$-$ 14.6 ksi.	Titanium
12.2d	80.5 MPa	37.8 MPa	$-$ 22.4 MPa	Brass
12.2e	$-$ 3.5 ksi	$-$ 4.3 ksi	$-$ 6.6 ksi.	Iron

12.3 A point in a structural member is subjected to a three dimensional state of stress described by the six Cartesian stress components σ_{xx}, σ_{yy}, σ_{zz}, τ_{xy}, τ_{yz}, and τ_{zx}. Determine the strain energy density at this point, if the member is fabricated from the material described in the table below.

Prob. No.	σ_{xx}	σ_{yy}	σ_{zz}	τ_{xy}	τ_{yz}	τ_{zx}	Material
12.3a	12.5 ksi	11.6 ksi	$-$ 10.3 ksi	$-$ 9.6 ksi.	11.3 ksi	7.5 ksi	Aluminum
12.3b	114 MPa	$-$ 91.4 MPa	65.0 MPa	$-$ 62.4 MPa	75.3 MPa	81.3 MPa	Steel
12.3c	10.6 ksi	13.3 ksi	$-$ 14.7 ksi	$-$ 10.7 ksi.	2.3 ksi	7.7 ksi	Titanium
12.3d	60.2 MPa	51.1 MPa	34.2 MPa	$-$ 25.3 MPa	48.7 MPa	38.8 MPa	Brass
12.3e	$-$ 6.4 ksi	$-$ 4.7 ksi	$-$ 5.6 ksi	$-$ 7.2 ksi.	$-$ 2.9 ksi	$-$ 4.6 ksi	Iron

12.4 A tie rod, with a length L and diameter D is fabricated from the material described in the table below. If the tie rod is subjected to an axial force F, determine the strain energy stored in the tie rod and the strain energy density.

Prob. No.	Length, L	Diameter D	Force, F	Material
12.4a	25 ft	10 in.	1570 ton	Steel
12.4b	4.0 m	120 mm	1100 kN	Aluminum
12.4c	6.0 ft	3.0 in.	150 ton	Titanium
12.4d	3.0 m	75 mm	200 kN	Brass
12.4e	20 ft	4.0 in.	160 ton	Stainless Steel

12.5 For the beam shown in the figure below, determine the strain energy stored in the beam due to normal stresses for the conditions listed in the table below: The beam is fabricated from steel.

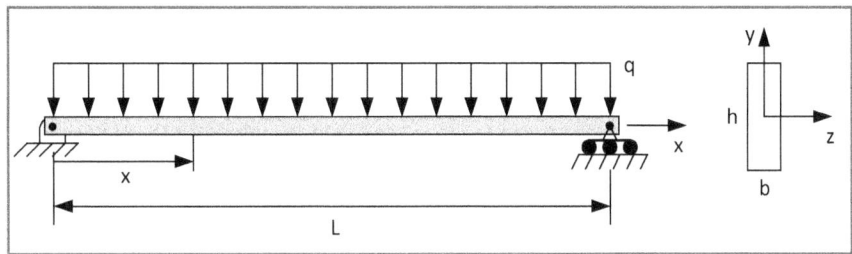

Prob. No.	Height, h	Width, b	Length, L	Load, q
12.5a	4.0 in.	1.20 in.	12.0 ft	180 lb/ft
12.5b	75 mm	20 mm	3.5 m	500 N/m
12.5c	5.0 in.	1.4 in.	10.0 ft	250 lb/ft
12.5d	100 mm	25 mm	4.0 m	900 N/m
12.5e	6.0 in.	1.6 in.	14.0 ft	225 lb/ft

12.6 For the beam shown in the figure below, determine the expression for the strain energy due to the normal stress σ_{xx}. Assume that EI_z is a constant.

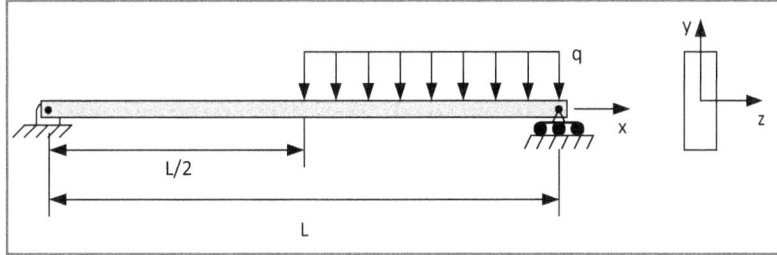

12.7 For the beam shown in the figure below, determine the expression for the strain energy due to the normal stress σ_{xx}. Assume that EI_z is a constant.

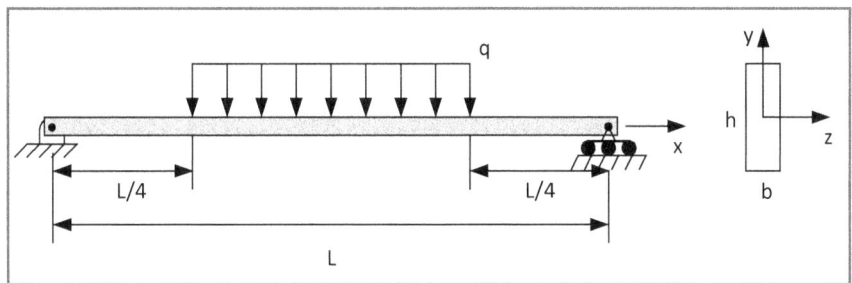

12.8 For the beam shown in the figure below, determine the expression for the strain energy due to the normal stress σ_{xx}. Assume that EI_z is a constant.

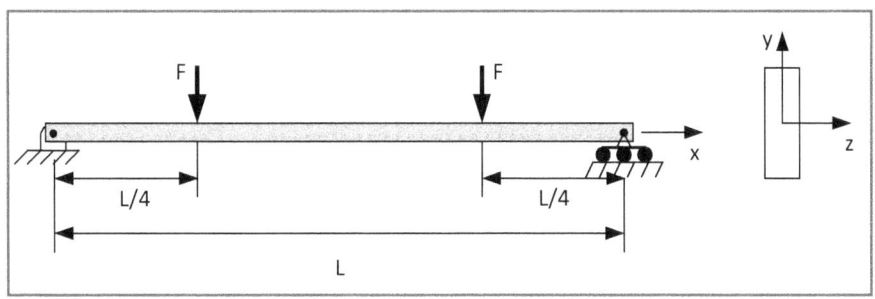

12.9 For the beam shown in the figure below, determine the expression for the strain energy due to the normal stress σ_{xx}. Assume that EI_z is a constant.

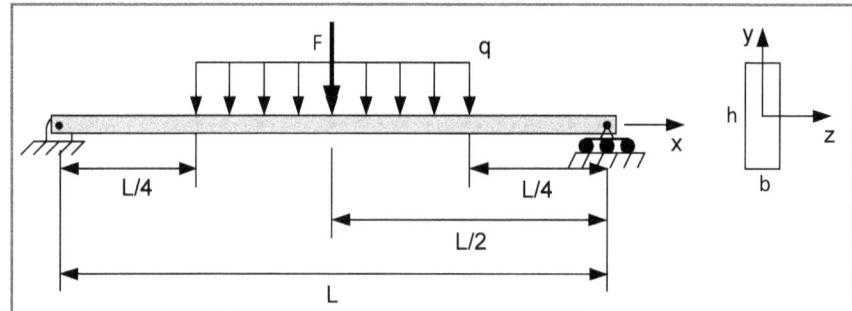

12.10 For the beam shown in Fig. P12.10, determine the expression for the strain energy due to the normal stress σ_{xx}. Assume that EI_z is a constant.

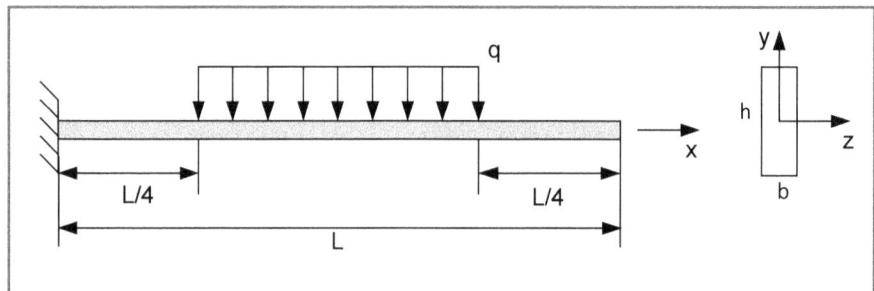

12.11 For the beam shown in the figure below, determine the expression for the strain energy due to the normal stress σ_{xx}. Assume that EI_z is a constant.

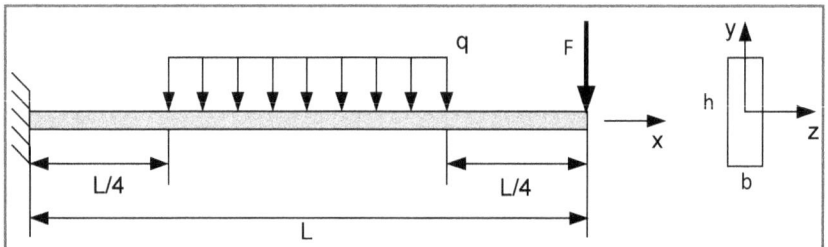

12.12 For the beam in Problem 12.6, determine the relation for the strain energy stored due to the shear stress τ_{xy}. The beam has a rectangular cross section with dimensions h and b.

12.13 For the beam in Problem 12.8, determine the relation for the strain energy stored due to the shear stress τ_{xy}. The beam has a rectangular cross section with dimensions h and b.

12.14 For the beam in Problem 12.9, determine the relation for the strain energy stored due to the shear stress τ_{xy}. The beam has a rectangular cross section with dimensions h and b.

12.15 For the beam in Problem 12.11, determine the relation for the strain energy stored due to the shear stress τ_{xy}. The beam has a rectangular cross section with dimensions h and b.

12.16 A shaft of length L and diameter D, shown in the figure below, is subjected to a torque T. Compute the strain energy stored in the shaft.

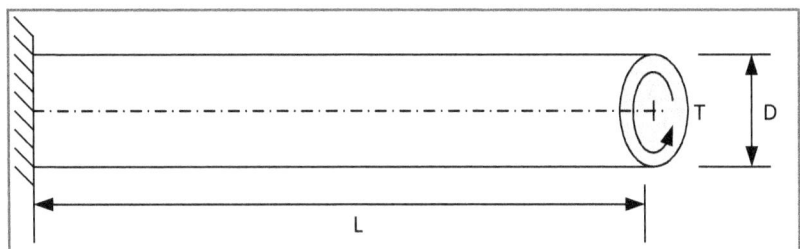

Prob. No.	Length, L	Diameter D	Torque, T	Material
12.16a	2.5 ft	2.0 in.	150 ft-lb	Steel
12.16b	800 mm	50 mm	200 N-m	Aluminum
12.16c	3.0 ft	1.75 in.	125 ft-lb	Titanium
12.16d	3.0 m	75 mm	450 N-m	Brass
12.16e	1.5 ft	1.0 in.	110 ft-lb	Stainless Steel

12.17 A stepped shaft, shown in the figure below, with lengths L_1 and L_2 and diameters D_1 and D_2 is subjected to a torque T. Compute the strain energy stored in the shaft.

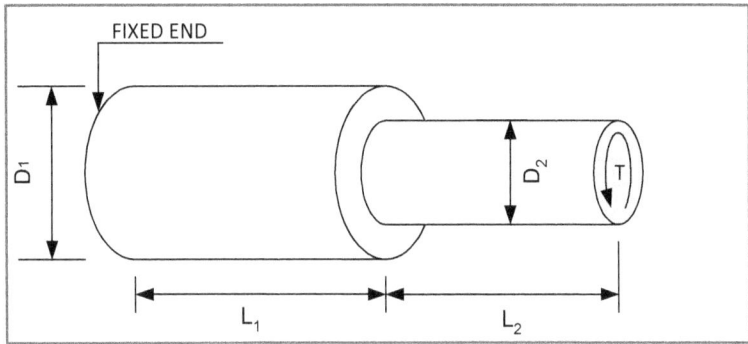

Prob. No.	L_1	L_2	D_1	D_2	Torque, T	Material
12.17a	2.5 ft	1.8 ft	2.0 in.	1.25 in.	150 ft-lb	Steel
12.17b	800 mm	650 mm	50 mm	30 mm	200 N-m	Aluminum
12.17c	3.0 ft	2.5 ft	1.75 in.	1.15 in.	125 ft-lb	Titanium
12.17d	3.0 m	2.2 m	75 mm	60 mm	450 N-m	Brass
12.17e	1.5 ft	1.25 ft	1.0 in.	0.75 in.	110 ft-lb	Stainless Steel

12.18 A sphere, shown in the figure below, fired from an air gun strikes a rod with an axial impact. Determine the maximum force and the maximum stress produced, if the sphere weighs W. The rod is L long with a diameter D. The velocity of the sphere at the instant of impact is v.

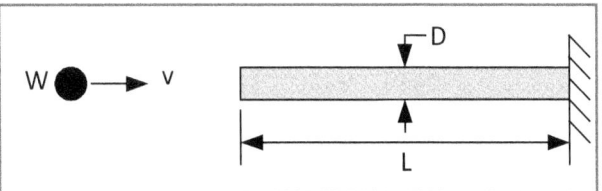

Prob. No.	Length, L	Diameter D	Velocity, v	Weight, W	Material
12.18a	2.5 ft	1.5 in.	500 ft/s	6.0 oz	Steel
12.18b	800 mm	30 mm	200 m/s	150 gm	Aluminum
12.18c	3.0 ft	1.25 in.	250 ft/s	4.5 oz	Titanium
12.18d	2.0 m	25 mm	375 m/s	120 gm	Brass
12.18e	1.5 ft	1.0 in.	1000 ft/s	5.0 oz	Stainless Steel

12.19 A steel cylinder, shown in the figure below, with a diameter d = 0.750 in. = 19 mm and length L is fired from an air gun and strikes a rod with an axial impact. Determine the maximum force and the maximum stress produced, if the cylinder has a density of γ. The rod is L long with a diameter D. The velocity of the cylinder at the instant of impact is v.

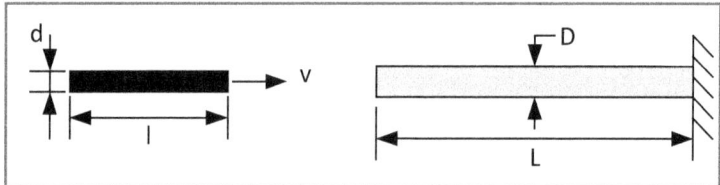

Prob. No.	Length, L	Diameter D	Velocity, v	Density, γ	l/d	Material
12.19a	2.5 ft	1.5 in.	500 ft/s	0.284 lb/in.^{3}	3.0	Steel
12.19b	800 mm	30 mm	150 m/s	2.77 Mg/m^3	4.0	Aluminum
12.19c	3.0 ft	1.25 in.	250 ft/s	0.167 lb/in.^{3}	6.0	Titanium
12.19d	2.0 m	25 mm	175 m/s	8.75 Mg/m^3	8.0	Brass
12.19e	1.5 ft	1.0 in.	300 ft/s	0.286 lb/in.^{3}	10.0	Stainless Steel

12.20 Consider a simply supported beam that is subjected to impact by a falling weight W, as illustrated in the figure below. Derive the relation for the maximum force and the maximum bending stress. Assume that the deflection of the beam is small relative to the height H.

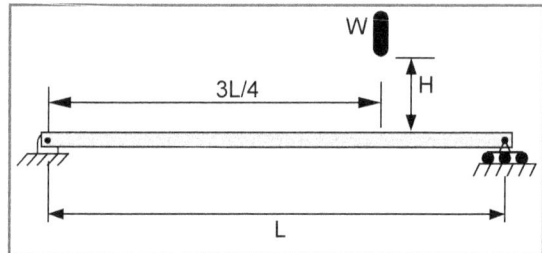

12.21 A weight W is positioned at the free end of a cantilever beam, as shown in the figure below. While the weight is in contact with the beam (H = 0), its gravitational force has not been released. The weight is suddenly released to produce an impact event, although the height of the drop is zero. Determine the equations for the maximum stress, the displacement of the load point and the maximum force developed during the impact.

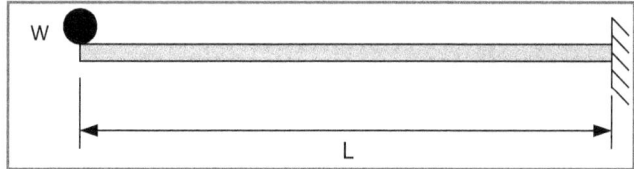

12.22 For the simply supported steel beam, shown in the figure below, determine the maximum force and maximum stress for the numerical parameters described in the table below.

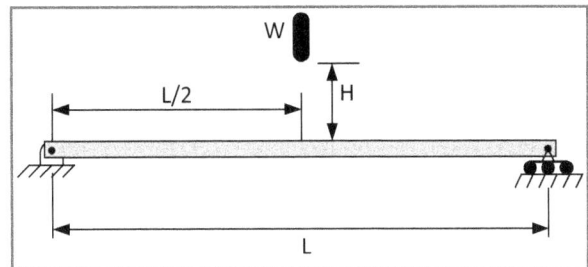

Prob. No.	Length, L	Weight, W	Height, H	Section
12.22a	14.5 ft	15 lb	5.0 ft	W12 × 50
12.22b	5 m	50 N	2.0 m	W203 × 60
12.22c	13.0 ft	10 lb	7.5 ft	S7 × 20
12.22d	4.5 m	25 N	3.2 m	S127 × 22
12.22e	10.5 ft	8 lb	10.0 ft	WT8 × 50

12.23 For the aluminum alloy cantilever beam, shown in the figure below, determine the maximum force and maximum stress for the numerical parameters described in the table below.

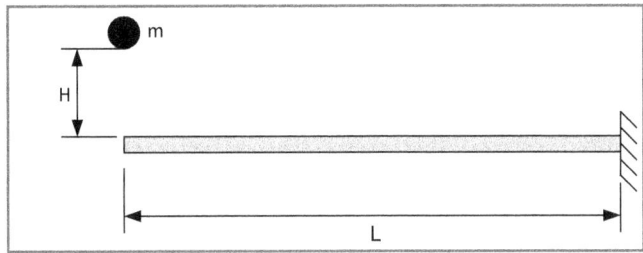

Prob. No.	Length, L	Mass, m	Height, H	Section
12.23a	10.0 ft	0.42 slug	5.0 ft	W12 × 50
12.23b	4.0 m	2.4 kg	1.2 m	W203 × 60
12.23c	12.0 ft	0.61 slug	4.5 ft	S7 × 20
12.23d	3.5 m	3.2 kg	2.6 m	S127 × 22
12.23e	8.5 ft	0.5 slug	6.8 ft	WT8 × 50

12.24 For the simply supported beam, shown in the figure below, determine the deflection at the quarter span ($x = L/4$) using Castigliano's theorem.

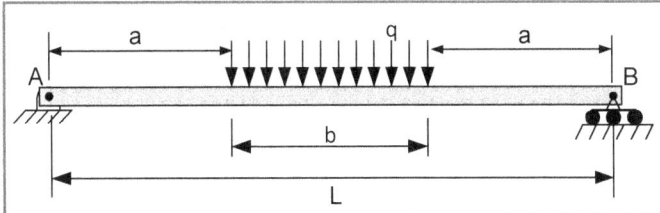

12.25 For the cantilever beam, shown in the figure below, determine the deflection at its free end using Castigliano's theorem.

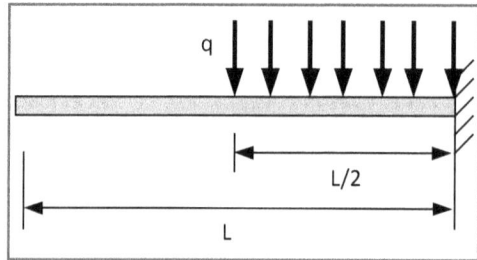

12.26 For the cantilever beam, shown in the figure below, determine the deflection at midspan and its free end using Castigliano's theorem.

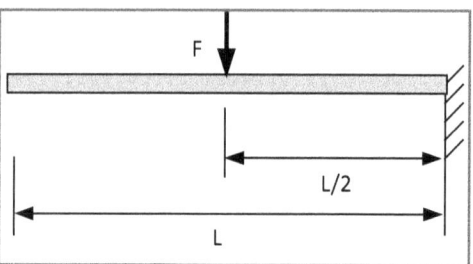

12.27 For the simply supported beam, shown in the figure below, use Castigliano's theorem to determine the deflection at the midpoint of the span.

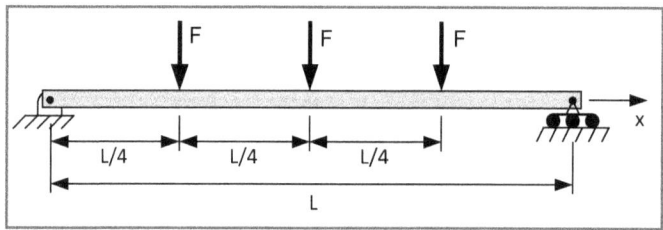

12.28 For the simply supported beam, in the figure below, determine the deflection at midspan and both quarter points using Castigliano's theorem.

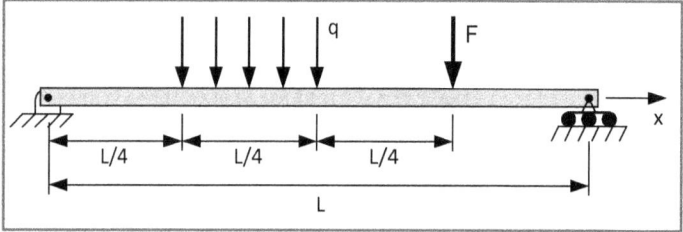

12.29 For the cantilever beam, shown in in the figure below, determine the deflection at midspan and its free end using Castigliano's theorem.

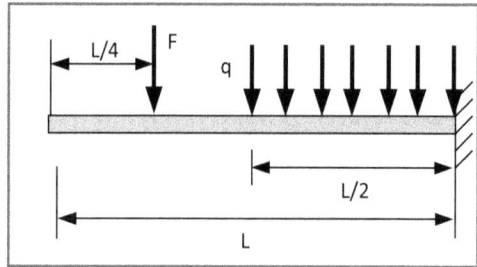

12.30 For the cantilever beam, shown in the figure below, use Castigliano's theorem to determine the reaction force at the prop.

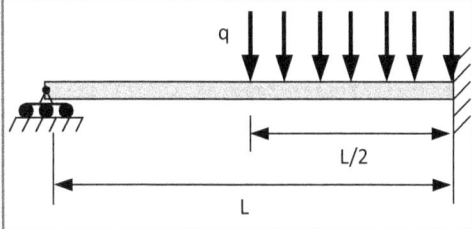

12.31 For the cantilever beam, shown in the figure below, use Castigliano's theorem to determine the reaction force at the prop.

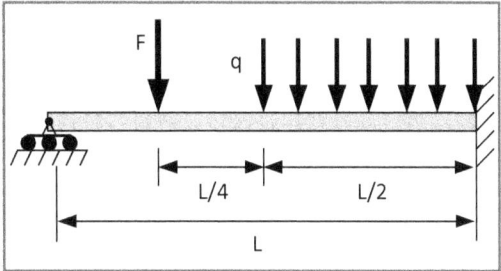

12.32 For the continuous beam, shown in the figure below, use Castigliano's theorem to determine the reaction force at the center support.

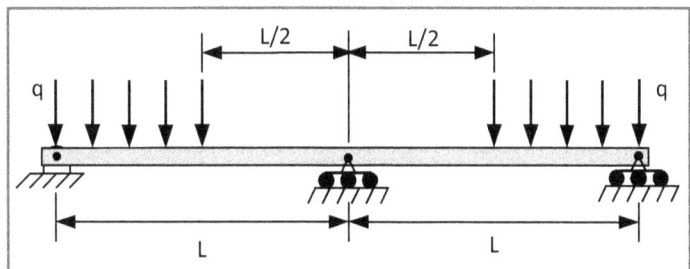

12.33 For the continuous beam, shown in the figure below and to the left, use Castigliano's theorem to determine the reaction force at the center support.

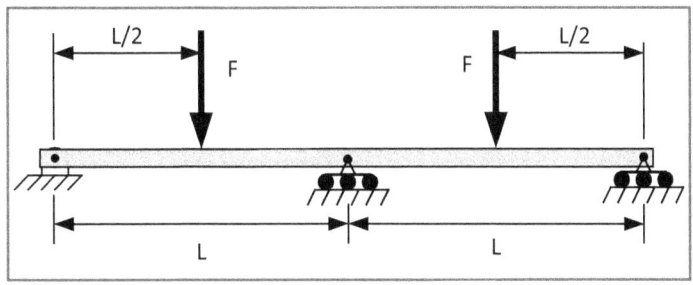
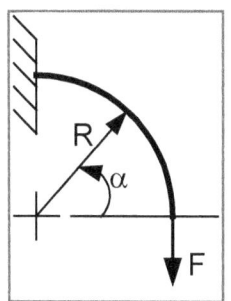

Fig. P12.34

12.34 A cantilever beam is formed in the shape of a quarter circle, as illustrated in Fig. P12.34. If the cantilever beam is loaded with a vertical force F at its free end, determine the beam's deflection in the direction of the applied force.

12.35 A cantilever beam is formed in the shape of a quarter circle, as illustrated in Fig. P12.35. If the cantilever beam is loaded with a horizontal force F at its free end, determine the beam's deflection in the direction of the applied force.

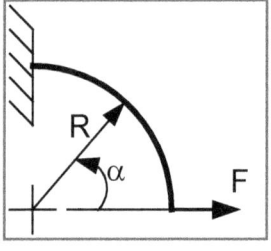

Fig. P12.35

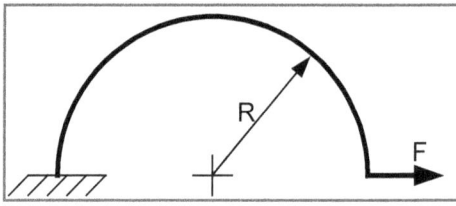

Fig. P12.36

12.36 A cantilever beam is formed in the shape of a semicircle, as illustrated in Fig. P12.36. If the cantilever beam is loaded with a horizontal force F at its free end, determine the beam's deflection in the direction of the applied force.

12.37 A cantilever beam is formed in the shape of a semicircle, as illustrated in Fig. P12.36. If the cantilever beam is stretched by an amount δ in the horizontal direction, determine the force that must be applied to produce this deflection.

12.38 A shaft is fabricated by brazing together two sections of round tubing, as shown in the figure below. The shaft is fixed at one end and a torque T is applied to it at its free end. Use Castigliano's theorem to determine the equation for the angle of twist ϕ. Assume the thickness of the braze metal is negligible.

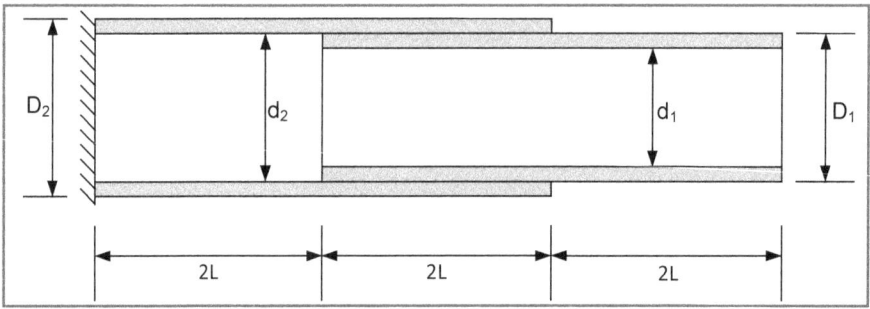

12.39 A shaft is fabricated by brazing together two sections of round tubing, as shown in in the figure below. The shaft is fixed at one end and a torque T is applied to it at its free end. Use Castigliano's theorem to determine the equation for the angle of twist ϕ. Assume the thickness of the braze metal is negligible.

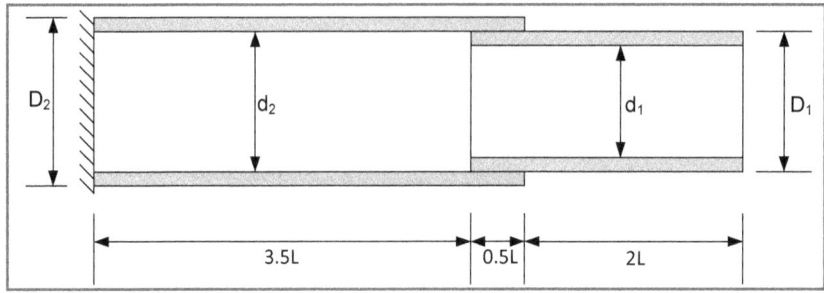

12.40 Determine the displacement in the vertical direction at joint B for the two-member truss, depicted in the figure below, using Castigliano's method. Truss members AB and BC are identical of the same material, length L, and area A. The ratio of h/w is given in the table below.

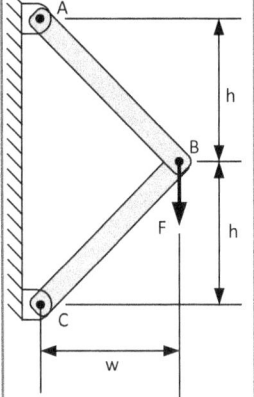

Prob. No.	h/w
12.40a	1.50
12.40b	1.25
12.40c	1.00
12.40d	0.75
12.40e	0.50

12.41 Determine the displacement in the horizontal direction at joint B for the two-member truss, depicted in the figure above, using Castigliano's method. Truss members AB and BC are identical of the same material, length L, and area A. See the table above for the ratio h/w.

12.42 Determine the displacement in the vertical direction at joint C for the seven-member truss, depicted in Fig. P12.42 using Castigliano's method. All of the truss members are fabricated from the same material. The cross sectional area of members BC, AE, BD and DE is 2A; the area of AB is A; and the area of members BE and CD is 3A.

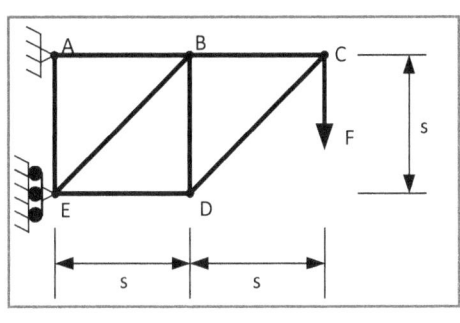
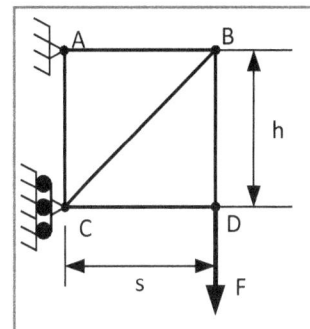

Fig. P12.42   Fig. P12.43

12.43 Determine the displacement in the vertical direction at joint D for the five-member truss, depicted in Fig. P12.43 using Castigliano's method. All of the truss members are fabricated from the same material with the same cross sectional area of A.

12.44 Determine the displacement in the vertical direction at joint C for the seven-member truss, depicted in Fig. P12.44 using Castigliano's method. All of the truss members are fabricated from the same material with the same cross sectional area of A.

Prob. No.	h/s
12.44a	1.30
12.44b	1.15
12.44c	1.00
12.44d	0.85
12.44e	0.70

12.45 Determine the displacement in the vertical direction at joint B for the seven-member truss, depicted in Fig. P12.44 using Castigliano's method. All of the truss members are fabricated from the same material with the same cross sectional area of A. Use the same ratio of h/s as listed in the table associated with Problem 12.44.

12.46 Determine the displacement in the vertical direction at joint D for the five-member truss, depicted in Fig. P12.46 using Castigliano's method. All of the truss members are fabricated from the same material with the same cross sectional area of A.

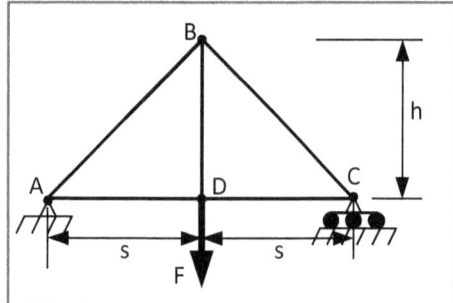

Prob. No.	h/s
12.46a	1.40
12.46b	1.20
12.46c	1.00
12.46d	0.80
12.46e	0.60

Fig. P12.46

12.47 Determine the displacement in the horizontal direction at joint C for the five-member truss, depicted in Fig. P12.46 using Castigliano's method. All of the truss members are fabricated from the same material with the same cross sectional area of A. Use the same ratio of h/s as listed in the table associated with Problem 12.46.

APPENDICES

APPENDIX A

WIRE AND SHEET METAL GAGES
DIMENSIONS ARE SHOWN IN INCHES

Gage No. of Wire	B & S Non Ferrous Metals	American S. & W Steel Wire	American S. & W Music Wire	Steel Manufactures' Sheet	Gage No. of Wire
7-0s	0.651354	0.4900	......	......	7-0s
6-0s	0.580049	0.4615	0.004	......	6-0s
5-0s	0.516549	0.4305	0.005	......	5-0s
4-0s	0.460	0.3938	0.006	......	4-0s
000s	0.40964	0.3625	0.007	......	000s
00	0.3648	0.3310	0.008	......	00
0	0.32486	0.3065	0.009	......	0
1	0.2893	0.2830	0.010	......	1
2	0.25763	0.2625	0.011	......	2
3	0.22942	0.2437	0.012	0.2391	3
4	0.20431	0.2253	0.013	0.2242	4
5	0.18194	0.2070	0.014	0.2092	5
6	0.16202	0.1920	0.016	0.1943	6
7	0.14428	0.1770	0.018	0.1793	7
8	0.12849	0.1620	0.020	0.1644	8
9	0.11443	0.1483	0.022	0.1495	9
10	0.10189	0.1350	0.024	0.1345	10
11	0.090742	0.1205	0.026	0.1196	11
12	0.080808	0.1055	0.029	0.1046	12
13	0.071961	0.0915	0.031	0.0897	13
14	0.064084	0.0800	0.033	0.0747	14
15	0.057068	0.0720	0.035	0.0763	15
16	0.05082	0.0625	0.037	0.0598	16
17	0.045257	0.0540	0.039	0.0538	17
18	0.040303	0.0475	0.041	0.0478	18
19	0.03589	0.0410	0.043	0.0418	19
20	0.031961	0.0348	0.045	0.0359	20
21	0.028462	0.0317	0.047	0.0329	21
22	0.025347	0.0286	0.049	0.0299	22
23	0.022571	0.0258	0.051	0.0269	23
24	0.0201	0.0230	0.055	0.0239	24
25	0.0179	0.0204	0.059	0.0209	25
26	0.01594	0.0181	0.063	0.0179	26
27	0.014195	0.0173	0.067	0.0164	27
28	0.012641	0.0162	0.071	0.0149	28
29	0.011257	0.0150	0.075	0.0135	29
30	0.010025	0.0140	0.080	0.0120	30
31	0.008928	0.0132	0.085	0.0105	31
32	0.00795	0.0128	0.090	0.0097	32
33	0.00708	0.0118	0.095	0.0090	33
34	0.006304	0.0104	......	0.0082	34
35	0.005614	0.0095	......	0.0075	35
36	0.005	0.0090	......	0.0067	36
37	0.004453	0.0085	......	0.0064	37
38	0.003965	0.0080	......	0.0060	38
39	0.003531	0.0075	......	......	39
40	0.003144	0.0070	......	......	40

The designation B & S is the Courtesy of Brown and Sharpe Manufacturing Company.

APPENDIX B-1

PHYSICAL PROPERTIES OF COMMON STRUCTURAL MATERIALS

Material	Elastic Modulus, E		Shear Modulus, G		Poisson's Ratio, ν	Coefficient Thermal Expansion, α	
	Mpsi	GPa	Mpsi	GPa	—	$\times 10^{-6}/°F$	$\times 10^{-6}/°C$
METAL							
Aluminum Alloy	10.4	72	3.9	27	0.32	12.9	23.2
Brass, Bronze	16	110	6.0	41	0.33	11.1	20.0
Copper	17.5	121	6.6	46	0.33	9.4	16.9
Cast Iron - Gray	15	103	6.0	41	0.26	6.7	12.1
Cast Iron - Malleable	25	170	9.9	68	0.26	6.7	12.1
Magnesium Alloy	6.5	45	2.4	17	0.35	14.4	25.9
Nickel Alloy	30	207	11.5	79	0.30	7.8	14.0
Steel	30	207	11.5	79	0.30	6.3	11.3
Stainless Steel	27.5	190	10.6	73	0.30	9.6	17.3
Titanium Alloy	16.5	114	6.2	43	0.33	4.9	8.8
WOOD[1]							
Douglas Fir	1.9	13	0.1	0.7	—	2.1	3.8
Sitka Spruce	1.5	10	0.07	0.5	—	2.1	3.8
Western White Pine	1.5	10	—	—	—	2.1	3.8
White Oak	1.8	12	—	—	—	2.1	3.8
Red Oak	1.8	12	—	—	—	2.1	3.8
Redwood	1.3	9	—	—	—	2.1	3.8
CONCRETE							
Medium Strength	3.6	25	—	—	—	5.5	9.9
High Strength	4.5	31	—	—	—	5.5	9.9
PLASTIC							
Nylon Type 6/6	0.4	2.8	—	—	—	80	144
Polycarbonate	0.35	2.4	—	—	—	68	122
Polyester, PBT	0.35	2.4	—	—	—	75	135
Polystyrene	0.45	3.1	—	—	—	70	125
Vinyl, Rigid PVC	0.45	3.1	—	—	—	75	135
STONE							
Granite	10	70	4	28	0.25	4	7.2
Marble	8	55	3	21	0.33	6	10.8
Sandstone	6	40	2	14	0.50	5	9.0
GLASS	9.6	65	4.1	28	0.17	44	80
RUBBER	0.22[2]	0.0015	0.073[2]	0.0005	0.50	125	225

The values for the properties given above are representative. Because processing methods and exact composition of the material influence the properties to some degree, the exact values may differ from those presented here.

1. Wood is an orthotropic material with different properties in different directions. The values given here are parallel to the grain.
2. The modulus for rubber is given in ksi.

APPENDIX B-2

TENSILE PROPERTIES OF COMMON STRUCTURAL MATERIALS

Material	Ultimate Tensile Strength, S_u		Yield Strength, S_y		Density, ρ	
	ksi	MPa	ksi	MPa	lb/in.^{3}	Mg/m^3
CARBON & ALLOY STEELS						
1010 A	44	303	29	200	0.284	7.87
1018 A	49.5	341	32	221	0.284	7.87
1020 HR	66	455	42	290	0.284	7.87
1045 HR	92.5	638	60	414	0.284	7.87
1212 HR	61.5	424	28	193	0.284	7.87
4340 HR	151	1041	132	910	0.283	7.84
52100 A	167	1151	131	903	0.284	7.87
STAINLESS STEELS						
302 A	92	634	34	234	0.286	7.92
303 A	87	600	35	241	0.286	7.92
304 A	83	572	40	276	0.286	7.92
440C A	117	807	67	462	0.286	7.92
CAST IRON						
Gray	25	170	—	—	0.260	7.20
Malleable	50	340	32	220	0.266	7.37
ALUMINUM ALLOYS						
1100-0	12	83	4.5	31	0.098	2.71
2024-T4	65	448	43	296	0.100	2.77
6061-T6	38	260	35	240	0.098	2.71
7075-0	34	234	14.3	99	0.100	2.77
7075 T6	86	593	78	538	0.100	2.77
MAGNESIUM ALLOYS						
HK31XA-0	25.5	176	19	131	0.066	1.83
HK31XA-H24	36.2	250	31	214	0.066	1.83
NICKEL ALLOYS						
Monel 400 A	80	550	32	220	0.319	8.83
Cupronickel A	53	365	16	110	0.323	8.94
COPPER ALLOYS						
Oxygen-free (99.9%) A	32	220	10	70	0.322	8.91
90-10 Brass A	36.4	251	8.4	58	0.316	8.75
80-20 Brass A	35.8	247	7.2	50	0.316	8.75
70-30 Brass A	44.0	303	10.5	72	0.316	8.75
Naval Brass	54.5	376	17	117	0.316	8.75
Tin Bronze	45	310	21	145	0.318	8.80
Aluminum Bronze	90	620	40	275	0.301	8.33
TITANIUM ALLOY						
Annealed	155	1070	135	930	0.167	4.63

A = Annealed and HR = Hot Rolled

APPENDIX B-3

TENSILE PROPERTIES OF NON-METALLIC MATERIALS

Materials	Ultimate Tensile Strength, S_u		Yield Strength, S_y		Density, ρ	
	ksi	MPa	ksi	MPa	lb/in.^{3}	Mg/m^3
WOOD						
Douglas Fir	15	100	—	—	0.017	0.470
Sitka Spruce	8.6	60	—	—	0.015	0.415
Western White Pine	5.0	34	—	—	0.014	0.390
White Oak	7.4	51	—	—	0.025	0.690
Red Oak	6.8	47	—	—	0.024	0.660
Redwood	9.4	65	—	—	0.015	0.415
CONCRETE[1]						
Medium Strength	4.0	28	—	—	0.084	2.32
High Strength	6.0	40	—	—	0.084	2.32
PLASTIC						
Nylon Type 6/6	11	75	6.5	45	0.0412	1.14
Polycarbonate	9.5	65	9	62	0.0433	1.20
Polyester, PBT	8	55	8	55	0.0484	1.34
Polystyrene	8	55	8	55	0.0374	1.03
Vinyl, Rigid PVC	6.5	45	6	40	0.0520	1.44
STONE[1]						
Granite	35	240	—	—	0.100	2.77
Marble	18	125	—	—	0.100	2.77
Sandstone	12	85	—	—	0.083	2.30
GLASS[1]						
98% Silica	7	50	—	—	0.079	2.19
RUBBER						
Natural, Vulcanized	4	28	—	—	0.034	0.95

1. The tensile strength of concrete, stone and bulk glass is negligible. Only the compressive strength of these materials is reported in this table.

APPENDIX C

PROPERTIES OF AREAS

C.1 AREA

The cross sectional area, A of structural members plays an extremely important role in the efficiency and the adequacy of any member to safely carry its load. For example, the stress in a uniaxial tension or compression member is given by:

$$\sigma = P/A$$

The force P is in the numerator and the area A of the cross section is in the denominator of this relation for the normal stresses. To lower the stresses we have only two options — decrease the force P or increase the area A. The area A of the cross section depends on its shape as indicated in Fig. C.1:

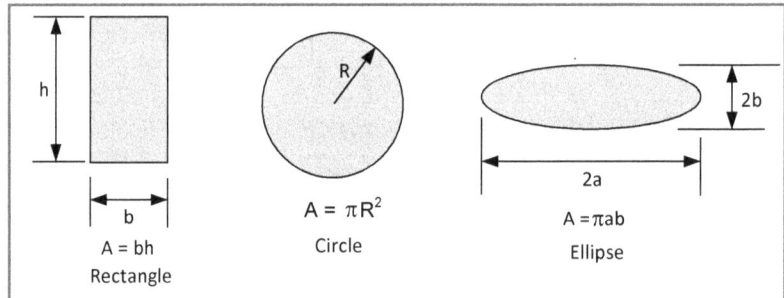

Fig. C.1 Dimensions and equations for areas of common cross-sections.

For areas of arbitrary shape, we determine their area by integration, as indicated in Fig. C.2:

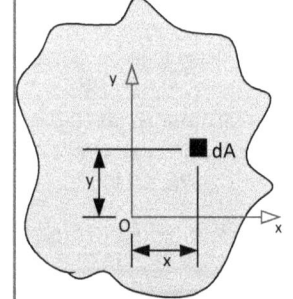

Fig. C.2 An arbitrary area A.

The area, A is determined by summing the incremental area dA in an integration process:

$$A = \int_A dA \qquad (C.1)$$

If the boundaries of the area are not known in terms of well-defined mathematical expressions, it is not possible to integrate to determine the area A. However, it is always possible to divide the area into many small squares or rectangles each with an area ΔA. If these squares or rectangles closely follow the boundary of the shape in question and completely fill the interior region, the area is then given by:

$$A = \sum_{i=1}^{N} \Delta A_i = N\Delta A \qquad (C.2)$$

where N is the number of small squares or rectangles.

C.2 FIRST MOMENT OF AN AREA

The first moment of an area is important, because it is useful in locating the position of the centroid of a given cross sectional area. Let's consider an arbitrary area with the coordinate system Oxy, as shown in Fig. C.2. The first moment of the area A about the x-axis is defined as:

$$Q_x = \int_A y \, dA \qquad (C.3)$$

Also, the first moment of the area A with respect to the y-axis is defined as:

$$Q_y = \int_A x \, dA \qquad (C.4)$$

Depending on the location of the coordinate system relative to the area, the numerical values obtained for the first moments Q_x and Q_y may be either positive or negative. The units for Q_x and Q_y are mm^3 in the SI system or in^3 in the U. S. Customary system.

C.3 CENTROID OF AN AREA

The centroid of an area A is defined by point C located relative to an arbitrary coordinate system Oxy in Fig. C.3. A centroid is defined as the point that locates the center of gravity of a line, an area or a volume.

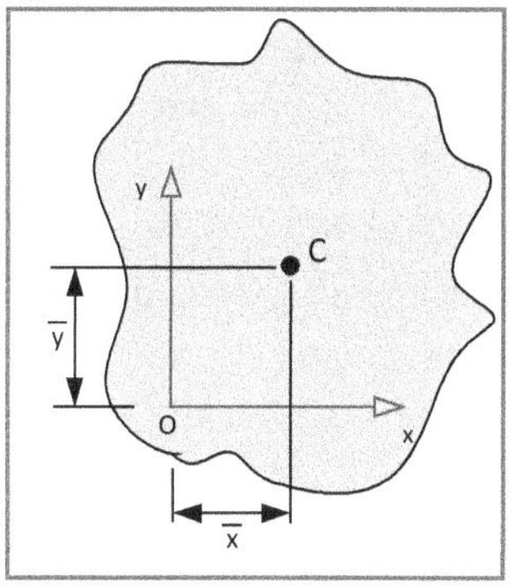

Fig. C.3 The centroid C of an area A is located with coordinates $\bar{x}$ and $\bar{y}$.

The coordinates $\bar{x}$ and $\bar{y}$ locating the centroid of an area are determined from the first moments as:

$$Q_y = \int_A x \, dA = A\bar{x}$$

$$Q_x = \int_A y \, dA = A\bar{y} \qquad (C.5)$$

where $\bar{x}$ and $\bar{y}$ are dimensions locating the centroid as shown in Fig. C.3.

Let's illustrate the method for determining the first moment of the area and the location of a centroid by considering a few elementary shapes, in the examples presented below.

EXAMPLE C1

Consider the rectangular area illustrated in Fig C.4, with the origin of the Oxy coordinate system positioned at its lower left-hand corner. Determine the first moments of the rectangular area and the location of its centroid relative to the Oxy coordinate system.

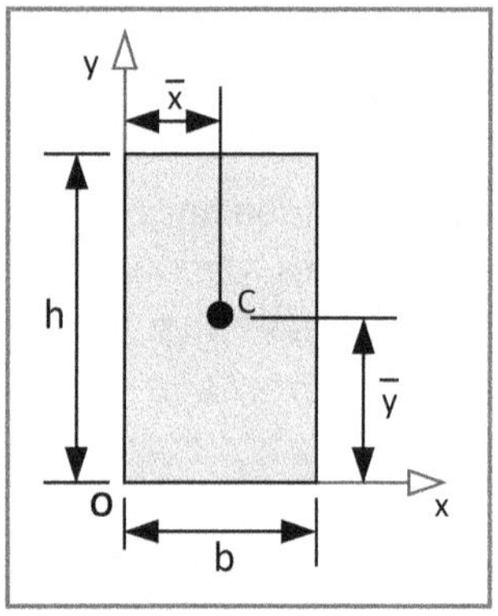

Fig. C.4 A rectangle with the coordinate system located along its edges.

Solution: For the rectangle area presented in Fig. C.4, the first moments of the area about the x and y axes are given by Eq. (C.5) as:

$$Q_x = A\bar{y} = (bh)\frac{h}{2} = \frac{bh^2}{2}$$

$$Q_y = A\bar{x} = (bh)\frac{b}{2} = \frac{b^2h}{2}$$

(C.6)

It was possible to quickly solve for the first moments, Q_x and Q_y, because we recognized the location of the centroid for the rectangular area. When an axis of symmetry exists for a given area, the centroid is located somewhere on this axis of symmetry. With the rectangle, two axes of symmetry exist; hence, the location of the centroid is at the intersection of its two symmetric axes.

For the circular cross section shown in Fig. C.5, the center of the circle clearly locates the centroid. The center also serves as the origin C for a special set of axes known as the centroidal axes x_c and y_c.

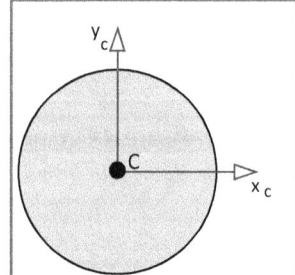

Fig. C.5 The centroid serves as the origin for the centroidal axes x_c, y_c.

For cross sectional shapes such as ellipses, circles, squares and rectangles, the center may be located by inspection, because these geometries have at least two axes of symmetry. However, for non-symmetric figures, such as triangles, portions of circles, parabolic areas, etc., locating the center of the area is not obvious. We will demonstrate a method for determining the centroid's location for an area that does not exhibit two or more axes of symmetry.

With respect to a centroidal coordinate system, the first moment of the area must vanish for both axes. Therefore:

$$Q_{\bar{x}} = \int_A y dA = 0 \qquad Q_{\bar{y}} = \int_A x dA = 0 \qquad (C.7)$$

These relations are employed to locate the centroid of an area of any shape, providing its boundary can be defined with some mathematical function.

EXAMPLE C2

For a right triangle, determine the first moment of the area about its base and vertical side, and the position of its centroid relative to these two sides. The right triangle with a base b and a height h is illustrated in Fig. C.6:

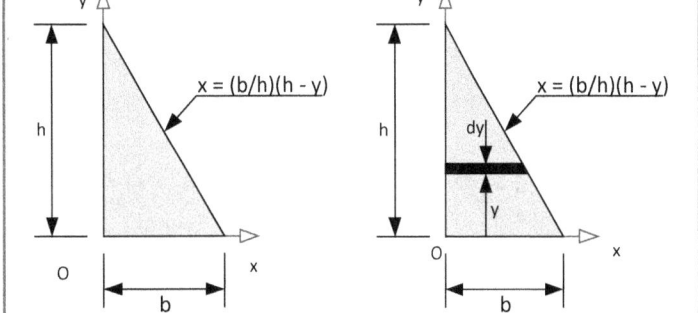

Fig C.6 A right triangle with a coordinate system coincident with its base and vertical side.

Solution: To begin, let's determine the first moment of the area of the right triangle relative to the x-axis (its base side). Writing Eq. C.3 gives:

$$Q_x = \int_A y dA = \int_y y(xdy) \qquad (a)$$

where $dA = x\,dy$ is located a constant distance y from the x axis.

Note, the equation for the inclined boundary of the triangle is given by:

$$x = (b/h)(h - y) \qquad (b)$$

The limits on the integral go from 0 to h to encompass the area of the triangle. We substitute Eq. (b) as the limits on y for the integral into Eq. (a), and write:

$$Q_x = \frac{b}{h} \int_0^h (h-y) y\, dy \qquad (c)$$

Integrating Eq. (c) gives:

$$Q_x = \frac{b}{h}\left[\frac{hy^2}{2} - \frac{y^3}{3}\right]_0^h \qquad (d)$$

Evaluating Eq. (d) gives:

$$Q_x = bh^2/6 \qquad (C.8)$$

By using Eq. C.4 and following the same procedure, we find the first moment about the vertical side of the triangle is given by:

$$Q_y = b^2h/6 \qquad (C.9)$$

Equation (C.8) gives the first moment of the area of a right triangle about its base. This is an interesting exercise in calculus, but what does it have to do with determining the location of the centroid of the right triangle? The result presented in Eq. (C.8) is an intermediate step. We continue the solution, by combining the results of Eqs. (C.8) and (C.9) with Eqs. (C.5) to obtain:

$$Q_x = bh^2/6 = A\,\overline{y} = (bh/2)\,\overline{y} \qquad (e)$$

$$Q_y = b^2h/6 = A\,\overline{x} = (bh/2)\,\overline{x} \qquad (f)$$

where $\overline{x}$ and $\overline{y}$ locate the C, x_c, y_c coordinates relative to the Oxy coordinates (see Fig. C.7).

Fig. C.7 A right triangular area with a base axes Oxy and centroidal axes C, x_c, y_c.

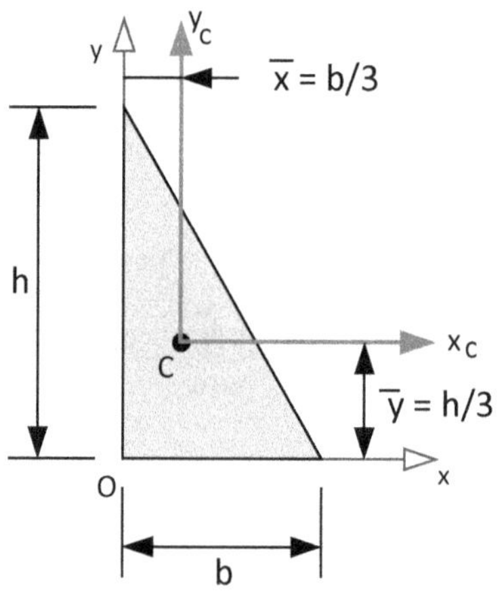

To determine the position of the centroid, let's solve Eqs. (e) and (f) for $\overline{x}$ and $\overline{y}$ to obtain:

$$\overline{y} = h/3 \qquad \overline{x} = b/3 \qquad (C.10)$$

We employ Eqs. (C.3) and (C.4) to determine the first moment of the area Q relative to either the x or y axes. The location of the centroid is then established from Eq. (C.5). The location of the centroid for common shapes has been determined, and the results are presented together with the definition of the shape of an area in Fig. C.8.

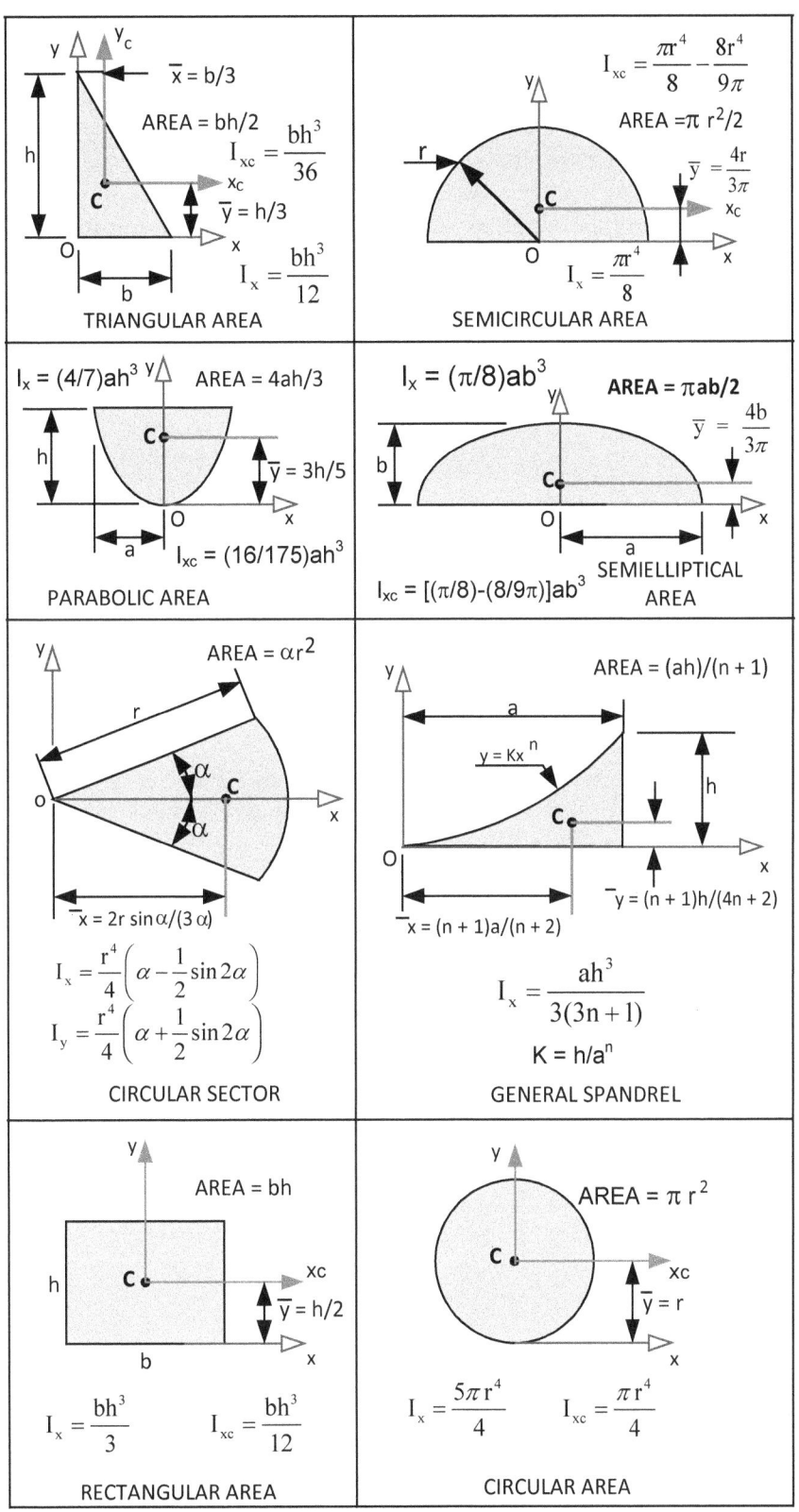

Fig. C.8 Area properties of some common shapes.

C.4 LOCATING THE CENTROID OF A COMPOSITE AREA

In many cases, the shape of a cross section is unusual and differs from the common geometries described in Fig. C.8. To locate the centroid of areas with irregular shapes, we divide its area into several different common shapes, for which we have solutions for the location of the centroid. Then we combine the product of these individual areas and their centroid locations to give the location of the centroid of the composite area. Let's consider the irregular shape defined in Fig. C.9, and locate the position of its centroid by employing the composite area technique.

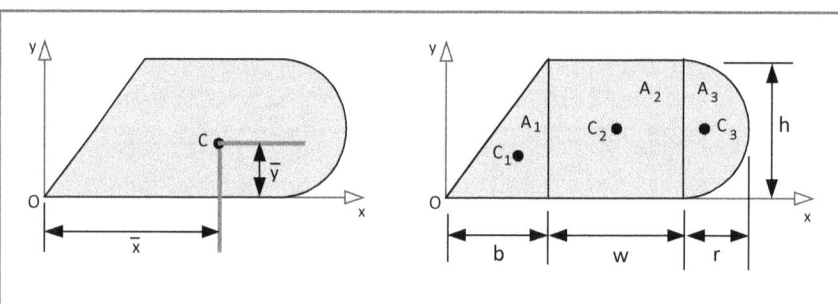

Fig. C.9 The unusual shaped area on the left is divided into common shaped areas on the right.

EXAMPLE C3

Determine the location of the centroid of the irregular shape defined in Fig. C.9.

Solution: To begin the solution, we divide the area into three sub-regions—A_1, A_2 and A_3 as shown in Fig. C.9. Note the sub-regions are a right triangle, rectangle and semicircle. The dimensions of the three different shapes are:

- The triangle — base b = 8 units and height h = 12 units.
- The rectangle — width w = 16 units and height h = 12 units.
- The semicircle — radius r = 6 units.

We apply Eq. (C.5) to the composite area, and write:

$$\Sigma Q_x \quad \Rightarrow \quad \overline{Y} A_t = \Sigma \overline{y}_n A_n \qquad (C.11)$$

$$\Sigma Q_y \quad \Rightarrow \quad \overline{X} A_t = \Sigma \overline{x}_n A_n \qquad (C.12)$$

where A_t is the total area of the composite area: $\overline{X}$ and $\overline{Y}$ are the coordinates of the centroid of the irregular (composite) area..

Let's first determine $\overline{Y}$ from Eq. (C.11):

$$\overline{Y}(A_1 + A_2 + A_3) = \overline{y}_1 A_1 + \overline{y}_2 A_2 + \overline{y}_3 A_3 \qquad (a)$$

Solving for $\overline{Y}$ yields:

$$\overline{Y} = (\overline{y}_1 A_1 + \overline{y}_2 A_2 + \overline{y}_3 A_3)/(A_1 + A_2 + A_3) \qquad (b)$$

Substituting results from Fig. C.8 into Eq. (b) gives:

$$\overline{Y} = \frac{\left(\dfrac{h}{3}\dfrac{bh}{2} + \dfrac{h}{2}wh + r\dfrac{\pi r^2}{2}\right)}{\left(\dfrac{bh}{2} + wh + \dfrac{\pi r^2}{2}\right)} \qquad \text{(c)}$$

Substituting b = 8, h = 12, w = 16 and r = 6 units into Eq. (c) yields:

$$\overline{Y} = 5.676 \text{ units}$$

This result is slightly less than h/2, as we would anticipate. The presence of the triangle shifts the location of the centroid downward from the centerline of the rectangle. Next, let's determine the position of the centroid in the direction of the x-axis. We begin by using Eq. (C.12), and write:

$$\overline{X}(A_1 + A_2 + A_3) = \overline{x}_1 A_1 + \overline{x}_2 A_2 + \overline{x}_3 A_3 \qquad \text{(d)}$$

Solving for $\overline{X}$ yields:

$$\overline{X} = (\overline{x}_1 A_1 + \overline{x}_2 A_2 + \overline{x}_3 A_3)/(A_1 + A_2 + A_3) \qquad \text{(e)}$$

From the information listed in Fig. C.8, we determine the centroid location for each of the shapes in the composite area as indicated below:

$$\overline{X} = \frac{\dfrac{2b}{3}\dfrac{bh}{2} + \left(b + \dfrac{w}{2}\right)wh + \left(b + w + \dfrac{4r}{3\pi}\right)\dfrac{\pi r^2}{2}}{\dfrac{bh}{2} + wh + \dfrac{\pi r^2}{2}} \qquad \text{(f)}$$

Substituting b = 8, h = 12, w = 16 and r = 6 units into Eq. (f) yields:

$$\overline{X} = 16.28 \text{ units} \qquad \text{(g)}$$

We note that the location of the centroid of the irregular area is slightly to the right of the center of the rectangular area. This position is to be expected, because the orientation of the right triangle with its area concentrated toward the right side tends to shift the centroid to the right.

C.5 SECOND MOMENT OF THE AREA

The second moment of the area is also known as the area moment of inertia. You will encounter the second moment of the area in your study of Mechanics of Materials, when determining the stresses produced in beams by an internal moment, and when determining the load required to buckle long, slender columns. Two different second moments of the area A, illustrated in Fig. C.10, are defined in accordance to the axis referenced.

The second moment of the area, is referenced to one or both of its coordinate axes. The moment of inertia relative to the y and z-axes is defined by:

$$I_z = \int_A y^2 dA \qquad I_y = \int_A z^2 dA \qquad \text{(C.13)}$$

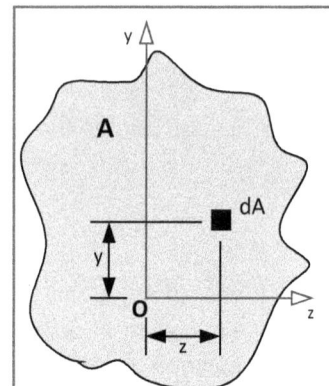

Fig. C.10 An elemental area dA is used when integrating to determine the second moment of the area A.

We also define a polar moment of inertia of the area A relative to the origin O of the y – z coordinate system, as indicated in Fig. C.11 as:

$$J_O = \int_A r^2 dA = \int_A (z^2 + y^2) dA = \int_A z^2 dA + \int_A y^2 dA \qquad (C.14)$$

and from Eq. (C.13) it is evident that:

$$J_0 = I_y + I_z \qquad (C.15)$$

The units for the second moments of the area are in^4 in the U. S. Customary system and m^4 or mm^4 in the SI system.

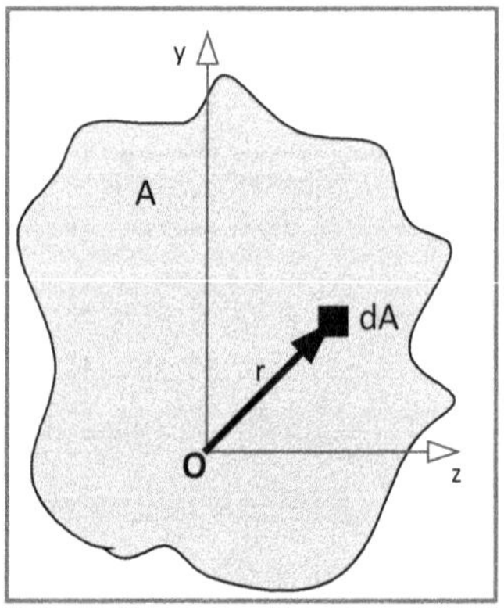

Fig. C.11 Coordinate system for the polar moment of inertia J_O.

Finally, the radius of gyration of an area A with respect to its axes is defined by:

$$I_z = r_z^2 A \qquad I_y = r_y^2 A \qquad J_O = r_O^2 A \qquad (C.16)$$

The symbols r_z and r_y reference the radii of gyration relative to the z and y-axes, respectively. The radius of gyration for the polar moment of inertia is r_O. The radius of gyration is simply a number when squared and multiplied by the area gives the area moment of inertia.

Let's consider a few examples to demonstrate the use of Eq. (C.13) in determining the second moment of the area.

EXAMPLE C4

For the rectangular cross sectional area and coordinate system, as defined in Fig C.12, determine the equations for I_z and I_y.

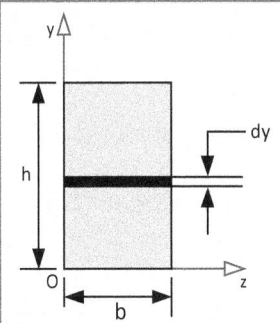

Fig. C.12 A rectangular area with a coordinate system along its edges.

Solution: To determine I_z, we write Eq. (C.13) and observe that $dA = b\,dy$. Then:

$$I_z = \int_A y^2 dA = b\int_0^h y^2 dy = b\left[\frac{y^3}{3}\right]_0^h = \frac{bh^3}{3} \tag{C.17}$$

Similarly for I_y, we write Eq. (C.13) and observe that $dA = h\,dz$. Then:

$$I_y = \int_A z^2 dA = h\int_0^b z^2 dz = h\left[\frac{z^3}{3}\right]_0^b = \frac{hb^3}{3} \tag{C.18}$$

To show the importance of the location of the coordinate system in determining the second moment of the area, let's shift the origin of the coordinate system to the center of the rectangle as shown in Fig C.13.

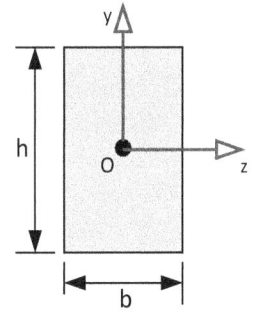

Fig. C.13 A rectangular area with a centroidal coordinate system.

EXAMPLE C5

Determine the moment of inertia of a rectangular area relative to its centroidal axes.

Solution: To determine the moment of inertia I_z, we write Eq. (C.13) and note that $dA = b\,dy$.

$$I_z = \int_A y^2 dA = b\int_{-h/2}^{h/2} y^2 dy = b\left[\frac{y^3}{3}\right]_{-h/2}^{h/2} = \frac{bh^3}{12} \tag{C.19}$$

To determine the moment of inertia I_y, we write Eq. (C.13) and note that $dA = h\,dz$.

$$I_y = \int_A z^2 dA = h\int_{-b/2}^{b/2} z^2 dz = h\left[\frac{z^3}{3}\right]_{-b/2}^{b/2} = \frac{hb^3}{12} \qquad (C.20)$$

A comparison of the results for the moments of inertia for the rectangle, clearly indicates the importance of the location of the coordinate system relative to the area in question. As we may be required to determine the moment of inertia about axes with arbitrary locations, a useful method for accounting for shifting the position of axes is presented in Section C.6.

C.6 <u>THE PARALLEL AXIS THEOREM</u>

Let's again consider an arbitrary area A positioned some distance from the z-axis, as shown in Fig C.14. The centroidal axis of the area A is known and is identified with the z' axis. Assume that we have determined the second moment of the area with respect to the centroidal axis z'. Let's compute the moment of inertia I_z with respect to an axis z that is parallel to the centroidal axis, but located some distance d below it. We begin again with Eq. C.13 and write:

$$I_z = \int_A y^2 dA = \int_A (y_1 + d)^2 dA = \int_A y_1^2 dA + 2d\int_A y_1 dA + d^2 \int_A dA \qquad (C.21)$$

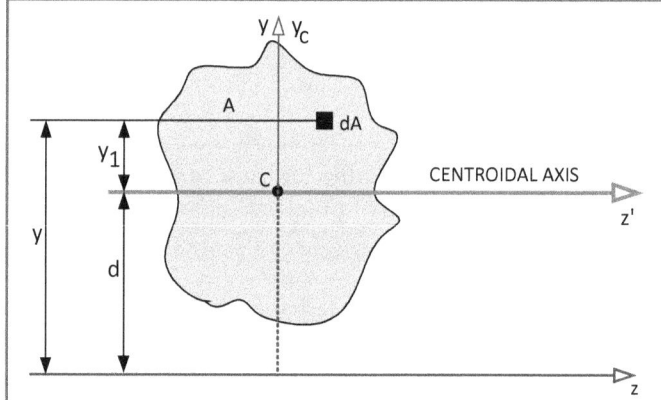

Fig. C.14 An arbitrary area A positioned a distance d from the z-axis.

The first term on the right hand side of Eq. (C.21) is the moment inertia $I_{z'}$ of the area about the centroidal axis. The second term $\int y_1 dA$ is the first moment of the area about its centroidal axis, which is zero by definition of the centroid. The final term is simply Ad^2. Hence, the parallel axis theorem may be written as:

$$I_z = I_{z'} + Ad^2 \qquad (C.22)$$

where $I_{z'}$ is the second moment of the area about the centroidal axis.

EXAMPLE C6

To demonstrate the use of Eq. (C.22), let's determine the moment of inertia I_z of the rectangle shown in Fig C.15.

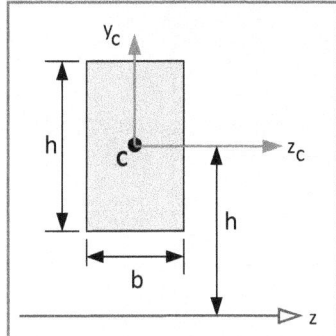

Fig. C.15 A rectangular area shifted by an amount d = h relative to the z-axis.

Solution: To determine the moment of inertia I_z, we recall Eq. (C.22) and write:

$$I_z = I_{z'} + Ad^2$$

From Eq. (C.19) it is clear that $I_{z'} = bh^3/12$. Then from Eq. (C.22) we obtain:

$$I_z = bh^3/12 + (bh)(h^2) = (13/12)\,bh^3$$

This example illustrates three points. First, the parallel axis theorem is very helpful in determining the increase in the moment of inertia, when the reference axis is some distance removed from the centroidal axis. Second, the moment of inertia is very sensitive to the movement of the reference axis relative to the centroidal axis. In this example, we moved the reference axis by an amount equal to the height h of the section and increased the inertia by a factor of 12. Third, the moment of inertia is a minimum about the centroidal axis.

C.7 MOMENTS OF INERTIA OF COMPOSITE AREAS

To determine the moment of inertia of areas with complex shapes, we divide the complex area into subsections. Each subsection is a simple shape, such as a square, rectangle, circle or semi-circle with known properties. Let's consider a structural tee with a web and a flange, as illustrated in Fig. C.16, and demonstrate the procedure for determining its moment of inertia.

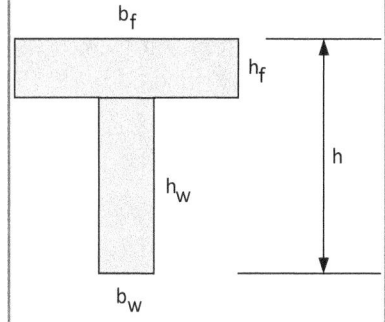

Fig. C.16 A structural tee is divided into two rectangular areas A_{web} and A_{flange}.

The procedure for determining the properties of the composite area representing the structural tee involves three steps:

- Determine the location of the centroidal axis of the composite area.
- Determine the moment of inertia of each area of the composite section about its centroidal axis.
- Employ the parallel axis theorem to determine the moment of inertia of the total section relative to its centroidal axis.

EXAMPLE C7

Determine the moment of inertia I_z of the structural tee, shown in Fig. C.16, relative to its centroidal axis. The dimensions of the structural tee are given by:

$$h_f = 0.15\, h,\ h_w = 0.85\, h,\ b_w = 0.15\, b_f,\ \text{and}\ b_f = 170\text{ mm and } h = 250\text{ mm}.$$

Solution: To determine the location of the centroidal axis of the composite area, subdivide the structural tee into two rectangular areas A_1 and A_2, as shown in Fig. C.17. Then apply Eq. (C.11) to determine the location of the centroid relative to the z (reference) axis.

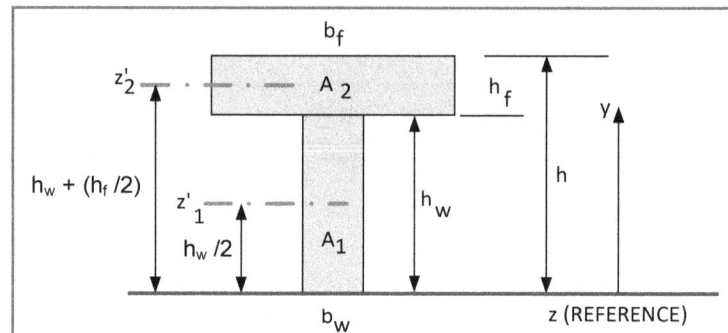

Fig. C.17 Subdivide the structural tee to form two rectangular areas and establish a convenient reference axis z.

From Eq. (C.11), we write:

$$\Sigma Q_{z(\text{REFERENCE})} = \overline{Y} A_t = \Sigma \overline{y}_n A_n = \overline{y}_1 A_1 + \overline{y}_2 A_2 \qquad (a)$$

Substituting the dimensional quantities from Fig. C.17 into Eq. (a) yields:

$$\overline{Y} = \frac{A_1 \dfrac{h_w}{2} + A_2(h_w + \dfrac{h_f}{2})}{A_1 + A_2} = \frac{\dfrac{b_w h_w^2}{2} + b_f h_f h_w + \dfrac{b_f h_f^2}{2}}{b_w h_w + b_f h_f} \qquad (b)$$

Recall $h_f = 0.15\, h$, $h_w = 0.85\, h$, $b_w = 0.15\, b_f$, $b_f = 170$ mm, and $h = 250$ mm. Substitute these quantities into Eq. (b) to obtain:

$$\overline{Y} = 0.69528\, h = 0.69528 \times 250 = 173.8 \text{ mm} \qquad (c)$$

Next determine the moment for inertia of each area of the composite section, about its own centroidal axis. For A_1 (the web) with a width $b_w = 0.15\, b_f = 25.5$ mm, and a height $h_w = 0.85\, h = 212.5$ mm, we use Eq. (C.19) and write:

$$I_{z'1} = b_w h_w^3 / 12 = (25.5)(212.5)^3 / 12 = 20.39 \times 10^6 \text{ mm}^4 \qquad (d)$$

For A_2 (the flange) with a width $b_f = 170$ mm, and a height $h_f = 0.15\, h = 37.5$ mm, we write:

$$I_{z'2} = b_f h_f^3 / 12 = (170)(37.5)^3 / 12 = 0.7471 \times 10^6 \text{ mm}^4 \qquad (e)$$

Finally, employ the parallel axis theorem to determine the moment of inertia of the total area relative to the centroidal axis. Note for the composite area, we express the moment of inertia due to the two areas as:

$$I_z = I_{z1} + I_{z2} \tag{f}$$

We use Eq. (C.22) to expand Eq. (f) as:

$$I_z = I_{z1'} + A_1 d_1^2 + I_{z2'} + A_2 d_2^2 \tag{g}$$

The dimensions d_1 and d_2 are given by:

$$d_1 = \overline{Y} - h_w/2 = 173.8 - 106.25 = 67.55 \text{ mm} \tag{h}$$

$$d_2 = h_w + (h_f/2) - \overline{Y} = 212.5 + 18.75 - 173.8 = 57.45 \text{ mm} \tag{i}$$

Substituting numerical values for the terms in Eq. (g) yields the final result:

$$I_z = (20.39 \times 10^6) + (25.5)(212.5)(67.55)^2 + (0.7471 \times 10^6) + (170)(37.5)(57.45)^2 = 66.96 \times 10^6 \text{ mm}^4$$

The determination of the moment of inertia for complex shapes is simple but tedious. Care must be exercised to avoid numerical errors in computing each of the quantities shown in Eq. (g). The moment of inertia is important in determining the bending stresses in beams, because I_z occurs in the denominator of the well know flexural formula $\sigma = -My/I_z$. The procedure for establishing the position of the centroidal axis is also important, because it locates the neutral axis about which bending occurs. When the position of the neutral axis is known, we can establish y_{max}, I_z and the maximum bending stress σ. The minimum moment of inertia is also used to determine the buckling load for long slender columns.

EXAMPLE C8

Determine the quantities for the unsymmetrical wide flanged section, shown in Fig. C.18, which are listed below:

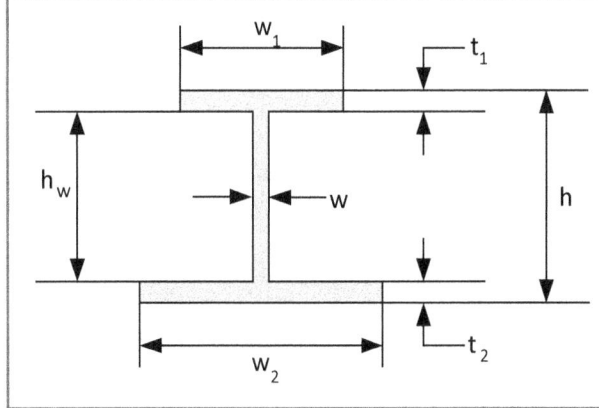

Fig. C.18 Dimensions of the unsymmetrical wide flanged section.

1. The location of the centroid relative to a defined reference axis.
2. The moment of inertia of the web, top flange, and bottom flange relative to the centroidal axis.
3. The moment of inertia of the complete section relative to the centroidal axis.

Solution: To determine the location of the centroidal axis of the composite area, divide the unsymmetrical wide flanged section into three different areas consisting of the top flange, web and bottom flange, as shown in Fig. C.19. Also establish a reference axis, z that is used as the datum for dimensioning the location of the centroid.

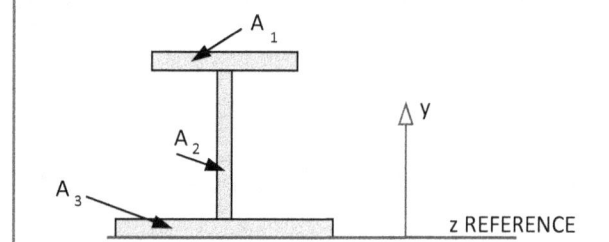

Fig. C.19 Divide the unsymmetrical wide flanged section into three parts.

Let's apply Eq. (C.11) to determine the location of the centroid of the unsymmetrical wide-flange section relative to the z (reference) axis.

$$\Sigma Q_{z(REFERENCE)} = \overline{Y} A_t = \Sigma \overline{y}_n A_n = \overline{y}_1 A_1 + \overline{y}_2 A_2 + \overline{y}_3 A_3 \tag{a}$$

Substituting the symbols for the dimensions from Fig. C.18 into Eq. (a) yields:

$$\overline{Y} = \frac{A_1(h_w + t_2 + t_1/2) + A_2(t_2 + h_w/2) + A_3(t_2/2)}{A_1 + A_2 + A_3} \tag{b}$$

$$\overline{Y} = \frac{1}{2}\left[\frac{t_1 w_1(2h_w + 2t_2 + t_1) + h_w w(2t_2 + h_w) + w_2 t_2^2}{w_1 t_1 + h_w w + w_2 t_2}\right]$$

Suppose we assume the proportions of the cross section as:

$$w_1 = 12\ w;\ w_2 = 16\ w;\ t_1 = 0.15\ h;\ t_2 = 0.15\ h;\ \text{and}\ h_w = 0.70\ h \tag{c}$$

Then Eq. (b) reduces to:

$$\overline{Y} = 0.4480\ h \tag{d}$$

Finally if the height h = 400 mm and web thickness w = 20 mm, the location of the centroidal axis is given by:

$$\overline{Y} = 179.2\ \text{mm} \tag{e}$$

To determine the moment of inertia of each area of the composite section about its centroidal axis, apply Eq. (C.19) to each of the three rectangular areas to obtain:

$$I_{z'1} = (1/12)w_1 t_1^3 = (1/12)(240)(60)^3 = 4.320 \times 10^6\ \text{mm}^4 \tag{f}$$

$$I_{z'2} = (1/12)w h_w^3 = (1/12)(20)(280)^3 = 36.59 \times 10^6\ \text{mm}^4 \tag{g}$$

$$I_{z'3} = (1/12) w_2 t_2^3 = (1/12)(320)(60)^3 = 5.760 \times 10^6\ \text{mm}^4 \tag{h}$$

Next use the parallel axis theorem [Eq. (C.22)] to determine the moment of inertia of each of the three areas relative to the centroidal axis. For the top flange (Area 1):

$$I_{z1} = I_{z'1} + A_1 d_1^2 = I_{z'1} + (w_1 t_1/4)(2h - t_1 - 2\overline{Y})^2$$

$$I_{z1} = 4.320 \times 10^6 + (240)(15)(800 - 60 - 358.4)^2 \qquad (i)$$

$$I_{z1} = 4.320 \times 10^6 + 524.2 \times 10^6 = 528.5 \times 10^6 \text{ mm}^4$$

For the web (Area 2):

$$I_{z2} = I_{z'2} + A_2 d_2^2 = I_{z'2} + (wh_w/4)(2t_2 + h_w - 2\overline{Y})^2$$

$$I_{z2} = 36.59 \times 10^6 + (20)(70)(120 + 280 - 358.4)^2 \qquad (j)$$

$$I_{z2} = 36.59 \times 10^6 + 2.42 \times 10^6 = 39.01 \times 10^6 \text{ mm}^4$$

For the bottom flange (Area 3):

$$I_{z3} = I_{z'3} + A_3 d_3^2 = I_{z'3} + (w_2 t_2/4)(2\overline{Y} - t_2)^2$$

$$I_{z3} = 5.760 \times 10^6 + (320)(15)(358.4 - 60)^2 \qquad (k)$$

$$I_{z3} = 5.760 \times 10^6 + 427.4 \times 10^6 = 433.2 \times 10^6 \text{ mm}^4$$

Before we complete the computation, examine the results listed in Eqs. (i – k). For both of the flanges, we note that the contribution of the $I_{z'}$ term was negligible compared to the Ad^2 term. For the web the $I_{z'}$ term was dominant and the $A_2 d_2^2$ term was negligible. To complete the solution for the moment of inertia of this composite area, we sum the moments of inertia due to the three areas as:

$$I_z = I_{z1} + I_{z2} + I_{z3} \qquad (l)$$

Substituting the results for Eqs. (i – k) into Eq. (l) yields:

$$I_z = (528.5 + 39.01 + 433.2) \times 10^6 = 1000.7 \times 10^6 \text{ mm}^4 \qquad (m)$$

The contribution of the web to the moment of inertia about the centroidal axis of this unsymmetrical wide flange section was a small part of the total (less than 4%). This example shows that the main purpose of the web is to move the flanges outward from the neutral (centroidal) axis. The height of the web is limited by buckling, where it becomes elastically unstable.

C.8 SUMMARY

Properties of areas is an important topic in mechanics, because stresses produced in structures and machine components depend upon the cross sectional area of the member, its centroidal location and the first and second moments of the area. Most of the equations required to compute the quantities found in the study of Statics, Mechanics of Materials and Dynamics are introduced in this chapter. Many of the topics described here should be familiar to the reader from his or her studies of Calculus and Physics.

The area of a shape that can be defined by a mathematical function is determined by integration according to:

$$A = \int_A dA \quad (C.1)$$

If the shape cannot be expressed with a mathematical function, numerical methods are used to sum small areas ΔA to obtain the total area A.

$$A = \sum_{i=1}^{N} \Delta A_i = N \Delta A \quad (C.2)$$

The first moment of the area about the x or y axes is defined as:

$$Q_x = \int_A y\, dA \quad (C.3)$$
$$Q_y = \int_A x\, dA \quad (C.4)$$

The first moment of the area provides a method for determining the centroid locations as:

$$Q_y = \int_A x\, dA = A\bar{x}$$
$$Q_x = \int_A y\, dA = A\bar{y} \quad (C.5)$$

The first moment of the area with respect to the centroidal axes vanishes; thus, enabling a technique for determining the location of the centroid of an arbitrary area.

$$Q_{\bar{x}} = \int_A y\, dA = 0 \qquad Q_{\bar{y}} = \int_A x\, dA = 0 \quad (C.7)$$

For complex shapes, the location of the centroid can be determined by dividing the complex shape into a composite of common shapes and summing first moments as indicated below:

$$\Sigma Q_x \Rightarrow \bar{Y} A_t = \Sigma \bar{y}_n A_n \quad (C.11)$$
$$\Sigma Q_y \Rightarrow \bar{X} A_t = \Sigma \bar{x}_n A_n \quad (C.12)$$

The second moment of the area also called the moment of inertia is determined from:

$$I_z = \int_A y^2 dA \qquad I_y = \int_A z^2 dA \qquad (C.13)$$

The polar moment of inertia is determined from:

$$J_O = \int_A r^2 dA = \int_A (z^2 + y^2) dA = \int_A z^2 dA + \int_A y^2 dA \qquad (C.14)$$

and

$$J_0 = I_y + I_z \qquad (C.15)$$

The radius of gyration is related to the area and the moment of inertia by:

$$I_z = r_z^2 A \qquad I_y = r_y^2 A \qquad J_O = r_O^2 A \qquad (C.16)$$

The parallel axis theorem, used to determine the moment of inertia about a parallel axis some distance removed from a centroidal axis, is derived from the following relation:

$$I_z = \int_A y^2 dA = \int_A (y_1 + d)^2 dA = \int_A y_1^2 dA + 2d \int_A y_1 dA + d^2 \int_A dA \qquad (C.21)$$

Equation (C.21) reduces to:

$$I_z = I_{z'} + Ad^2 \qquad (C.22)$$

APPENDIX D

GEOMETRIC PROPERTIES OF ROLLED STEEL SHAPES

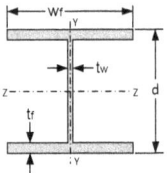

Wide-Flange Beams (U.S. Customary Units)

Designation*	Area (in.^{2})	Depth (in.)	Flange Width (in.)	Flange Thickness (in.)	Web Thickness (in.)	Axis Z-Z I (in.^{4})	Axis Z-Z Z (in.^{3})	Axis Z-Z r (in.)	Axis Y-Y I (in.^{4})	Axis Y-Y Z (in.^{3})	Axis Y-Y r (in.)
W36 × 230	67.6	35.90	16.470	1.260	0.760	15000	837	14.9	940	114	3.73
× 160	47.0	36.01	12.000	1.020	0.650	9750	542	14.4	295	49.1	2.50
W33 × 201	59.1	33.68	15.745	1.150	0.715	11500	684	14.0	749	95.2	3.56
× 152	44.7	33.49	11.565	1.055	0.635	8160	487	13.5	273	47.2	2.47
× 130	38.3	33.09	11.510	0.855	0.580	6710	406	13.2	218	37.9	2.39
W30 × 132	38.9	30.31	10.545	1.000	0.615	5770	380	12.2	196	37.2	2.25
× 108	31.7	29.83	10.475	0.760	0.545	4470	299	11.9	146	27.9	2.15
W27 × 146	42.9	27.38	13.965	0.975	0.605	5630	411	11.4	443	63.5	3.21
× 94	27.7	26.92	9.990	0.745	0.490	3270	243	10.9	124	24.8	2.12
W24 × 104	30.6	24.06	12.750	0.750	0.500	3100	258	10.1	259	40.7	2.91
× 84	24.7	24.10	9.020	0.770	0.470	2370	196	9.79	94.4	20.9	1.95
× 62	18.2	23.74	7.040	0.590	0.430	1550	131	9.23	34.5	9.80	1.38
W21 × 101	29.8	21.36	12.290	0.800	0.500	2420	227	9.02	248	40.3	2.89
× 83	24.3	21.43	8.355	0.835	0.515	1830	171	8.67	81.4	19.5	1.83
× 62	18.3	20.99	8.240	0.615	0.400	1330	127	8.54	57.5	13.9	1.77
W18 × 97	28.5	18.59	11.145	0.870	0.535	1750	188	7.82	201	36.1	2.65
× 76	22.3	18.21	11.035	0.680	0.425	1330	146	7.73	152	27.6	2.61
× 60	17.6	18.24	7.555	0.695	0.415	984	108	7.47	50.1	13.3	1.69
W16 × 100	29.4	16.97	10.425	0.985	0.585	1490	175	7.10	186	35.7	2.52
× 67	19.7	16.33	10.235	0.665	0.395	954	117	6.96	119	23.2	2.46
× 40	11.8	16.01	6.995	0.505	0.305	518	64.7	6.63	28.9	8.25	1.57
× 26	7.68	15.69	5.500	0.345	0.250	301	38.4	6.26	9.59	3.49	1.12
W14 × 120	35.3	14.48	14.670	0.940	0.590	1380	190	6.24	495	67.5	3.74
× 82	24.1	14.31	10.130	0.855	0.510	882	123	6.05	148	29.3	2.48
× 43	12.6	13.66	7.995	0.530	0.305	428	62.7	5.82	45.2	11.3	1.89
× 30	8.85	13.84	6.730	0.385	0.270	291	42.0	5.73	19.6	5.82	1.49
W12 × 96	28.2	12.71	12.160	0.900	0.550	833	131	5.44	270	44.4	3.09
× 65	19.1	12.12	12.000	0.605	0.390	533	87.9	5.28	174	29.1	3.02
× 50	14.7	12.19	8.080	0.640	0.370	394	64.2	5.18	56.3	13.9	1.96
× 30	8.79	12.34	6.520	0.440	0.260	238	38.6	5.21	20.3	6.24	1.52
W10 × 60	17.6	10.22	10.080	0.680	0.420	341	66.7	4.39	116	23.0	2.57
× 45	13.3	10.10	8.020	0.620	0.350	248	49.1	4.33	53.4	13.3	2.01
× 30	8.84	10.47	5.810	0.510	0.300	170	32.4	4.38	16.7	5.75	1.37
× 22	6.49	10.17	5.750	0.360	0.240	118	23.2	4.27	11.4	3.97	1.33
W8 × 40	11.7	8.25	8.070	0.560	0.360	146	35.5	3.53	49.1	12.2	2.04
× 31	9.13	8.00	7.995	0.435	0.285	110	27.5	3.47	37.1	9.27	2.02
× 24	7.08	7.93	6.495	0.400	0.245	82.8	20.9	3.42	18.3	5.63	1.61
× 15	4.44	8.11	4.015	0.315	0.245	48.0	11.8	3.29	3.41	1.70	0.876
W6 × 25	7.34	6.38	6.080	0.455	0.320	53.4	16.7	2.70	17.1	5.61	1.52
× 16	4.74	6.28	4.030	0.405	0.260	32.1	10.2	2.60	4.43	2.20	0.967
W5 × 16	4.68	5.01	5.000	0.360	0.240	21.3	8.51	2.13	7.51	3.00	1.27
W4 × 13	3.83	4.16	4.060	0.345	0.280	11.3	5.46	1.72	3.86	1.90	1.00

Courtesy of the American Institute of Steel Construction.
*W is the symbol for a wide-flange beam, followed by the nominal depth in inches, and weight in pounds per foot of length.

GEOMETRIC PROPERTIES OF ROLLED STEEL SHAPES

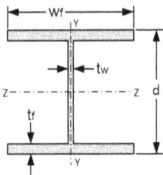

Wide-Flange Beams (SI Units)

Designation*	Area (mm²)	Depth (mm)	Width (mm)	Flange Thickness (mm)	Web Thickness (mm)	Axis Z-Z I (10^6 mm⁴)	Axis Z-Z Z (10^3 mm³)	Axis Z-Z r (mm)	Axis Y-Y I (10^6 mm⁴)	Axis Y-Y Z (10^3 mm³)	Axis Y-Y r (mm)
W914 × 342	43610	912	418	32.0	19.3	6245	13715	378	391	1870	94.7
× 238	30325	915	305	25.9	16.5	4060	8880	366	123	805	63.5
W838 × 299	38130	855	400	29.2	18.2	4785	11210	356	312	1560	90.4
× 226	28850	851	294	26.8	16.1	3395	7980	343	114	775	62.7
× 193	24710	840	292	21.7	14.7	2795	6655	335	90.7	620	60.7
W762 × 196	25100	770	268	25.4	15.6	2400	6225	310	81.6	610	57.2
× 161	20450	758	266	19.3	13.8	1860	4900	302	60.8	457	54.6
W686 × 217	27675	695	355	24.8	15.4	2345	6735	290	184	1040	81.5
× 140	17870	684	254	18.9	12.4	1360	3980	277	51.6	406	53.8
W610 × 155	19740	611	324	19.1	12.7	1290	4230	257	108	667	73.9
× 125	15935	612	229	19.6	11.9	985	3210	249	39.3	342	49.5
× 92	11750	603	179	15.0	10.9	645	2145	234	14.4	161	35.1
W533 × 150	19225	543	312	20.3	12.7	1005	3720	229	103	660	73.4
× 124	15675	544	212	21.2	13.1	762	2800	220	33.9	320	46.5
× 92	11805	533	209	15.6	10.2	554	2080	217	23.9	228	45.0
W457 × 144	18365	472	283	22.1	13.6	728	3080	199	83.7	592	67.3
× 113	14385	463	280	17.3	10.8	554	2395	196	63.3	452	66.3
× 89	11355	463	192	17.7	10.5	410	1770	190	20.9	218	42.9
W406 × 149	18970	431	265	25.0	14.9	620	2870	180	77.4	585	64.0
× 100	12710	415	260	16.9	10.0	397	1915	177	49.5	380	62.5
× 60	7615	407	178	12.8	7.7	216	1060	168	12.0	135	39.9
× 39	4950	399	140	8.8	6.4	125	629	159	3.99	57.2	28.4
W356 × 179	22775	368	373	23.9	15.0	574	3115	158	206	1105	95.0
× 122	15550	363	257	21.7	13.0	367	2015	154	61.6	480	63.0
× 64	8130	347	203	13.5	7.7	178	1025	148	18.8	185	48.0
× 45	5710	352	171	9.8	6.9	121	688	146	8.16	95.4	37.8
W305 × 143	18195	323	309	22.9	14.0	347	2145	138	112	728	78.5
× 97	12325	308	305	15.4	9.9	222	1440	134	72.4	477	76.7
× 74	9485	310	205	16.3	9.4	164	1060	132	23.4	228	49.8
× 45	5670	313	166	11.2	6.6	99.1	633	132	8.45	102	38.6
W254 × 89	11355	260	256	17.3	10.7	142	1095	112	48.3	377	65.3
× 67	8580	257	204	15.7	8.9	103	805	110	22.2	218	51.1
× 45	5705	266	148	13.0	7.6	70.8	531	111	6.95	94.2	34.8
× 33	4185	258	146	9.1	6.1	49.1	380	108	4.75	65.1	33.8
W203 × 60	7550	210	205	14.2	9.1	60.8	582	89.7	20.4	200	51.8
× 46	5890	203	203	11.0	7.2	45.8	451	88.1	15.4	152	51.3
× 36	4570	201	165	10.2	6.2	34.5	342	86.7	7.61	92.3	40.9
× 22	2865	206	102	8.0	6.2	20.0	193	83.6	1.42	27.9	22.3
W152 × 37	4735	162	154	11.6	8.1	22.2	274	68.6	7.12	91.9	38.6
× 24	3060	160	102	10.3	6.6	13.4	167	66.0	1.84	36.1	24.6
W127 × 24	3020	127	127	9.1	6.1	8.87	139	54.1	3.13	49.2	32.3
W102 × 19	2470	106	103	8.8	7.1	4.70	89.5	43.7	1.61	31.1	25.4

Courtesy of The American Institute of Steel Construction.
*W is the symbol for a wide-flange beam, followed by the nominal depth in mm, and the mass in kg per meter of length.

GEOMETRIC PROPERTIES OF ROLLED STEEL SHAPES

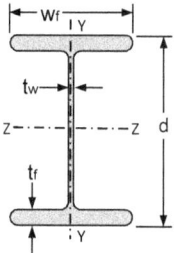

American Standard Beams (U.S. Customary Units)

Designation*	Area (in^2)	Depth (in.)	Width (in.)	Flange Thickness (in.)	Web Thickness (in.)	Axis Z-Z I (in.^{4})	Axis Z-Z Z (in.^{3})	Axis Z-Z r (in.)	Axis Y-Y I (in.^{4})	Axis Y-Y Z (in.^{3})	Axis Y-Y r (in.)
S24 × 121	35.6	24.50	8.050	1.090	0.800	3160	258	9.43	83.3	20.7	1.53
× 106	31.2	24.50	7.870	1.090	0.620	2940	240	9.71	77.1	19.6	1.57
× 100	29.3	24.00	7.245	0.870	0.745	2390	199	9.02	47.7	13.2	1.27
× 90	26.5	24.00	7.125	0.870	0.625	2250	187	9.21	44.9	12.6	1.30
× 80	23.5	24.00	7.000	0.870	0.500	2100	175	9.47	42.2	12.1	1.34
S20 × 96	28.2	20.30	7.200	0.920	0.800	1670	165	7.71	50.2	13.9	1.33
× 86	25.3	20.30	7.060	0.920	0.660	1580	155	7.89	46.8	13.3	1.36
× 75	22.0	20.00	6.385	0.795	0.635	1280	128	7.62	29.8	9.32	1.16
× 66	19.4	20.00	6.255	0.795	0.505	1190	119	7.83	27.7	8.85	1.19
S18 × 70	20.6	18.00	6.251	0.691	0.711	926	103	6.71	24.1	7.72	1.08
× 54.7	16.1	18.00	6.001	0.691	0.461	804	89.4	7.07	20.8	6.94	1.14
S15 × 50	14.7	15.00	5.640	0.622	0.550	486	64.8	5.75	15.7	5.57	1.03
× 42.9	12.6	15.00	5.501	0.622	0.411	447	59.6	5.95	14.4	5.23	1.07
S12 × 50	14.7	12.00	5.477	0.659	0.687	305	50.8	4.55	15.7	5.74	1.03
× 40.8	12.0	12.00	5.252	0.659	0.462	272	45.4	4.77	13.6	5.16	1.06
× 35	10.3	12.00	5.078	0.544	0.428	229	38.2	4.72	9.87	3.89	0.980
× 31.8	9.35	12.00	5.000	0.544	0.350	218	36.4	4.83	9.36	3.74	1.000
S10 × 35	10.3	10.00	4.944	0.491	0.594	147	29.4	3.78	8.36	3.38	0.901
× 25.4	7.46	10.00	4.661	0.491	0.311	124	24.7	4.07	6.79	2.91	0.954
S8 × 23	6.77	8.00	4.171	0.426	0.441	64.9	16.2	3.10	4.31	2.07	0.798
× 18.4	5.41	8.00	4.001	0.426	0.271	57.6	14.4	3.26	3.73	1.86	0.831
S7 × 20	5.88	7.00	3.860	0.392	0.450	42.4	12.1	2.69	3.17	1.64	0.734
× 15.3	4.50	7.00	3.662	0.392	0.252	36.7	10.5	2.86	2.64	1.44	0.766
S6 × 17.25	5.07	6.00	3.565	0.359	0.465	26.3	8.77	2.28	2.31	1.30	0.675
× 12.5	3.67	6.00	3.332	0.359	0.232	22.1	7.37	2.45	1.82	1.09	0.705
S5 × 14.75	4.34	5.00	3.284	0.326	0.494	15.2	6.09	1.87	1.67	1.01	0.620
× 10	2.94	5.00	3.004	0.326	0.214	12.3	4.92	2.05	1.22	0.809	0.643
S4 × 9.5	2.79	4.00	2.796	0.293	0.326	6.79	3.39	1.56	0.903	0.646	0.569
× 7.7	2.26	4.00	2.663	0.293	0.193	6.08	3.04	1.64	0.764	0.574	0.581
S3 × 7.5	2.21	3.00	2.509	0.260	0.349	2.93	1.95	1.15	0.586	0.468	0.516
× 5.7	1.67	3.00	2.330	0.260	0.170	2.52	1.68	1.23	0.455	0.390	0.522

Courtesy of The American Institute of Steel Construction.
*S is the symbol for a standard beam, followed by the nominal depth in inches, then the weight in pounds per foot of length.

GEOMETRIC PROPERTIES OF ROLLED STEEL SHAPES

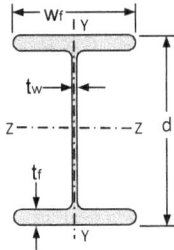

American Standard Beams (SI Units)

Designation*	Area (mm²)	Depth (mm)	Flange Width (mm)	Flange Thickness (mm)	Web Thickness (mm)	Axis Z-Z I (10^6 mm⁴)	Axis Z-Z Z (10^3 mm³)	r (mm)	Axis Y-Y I (10^6 mm⁴)	Axis Y-Y Z (10^3 mm³)	r (mm)
S610 × 180	22970	622.3	204.5	27.7	20.3	1315	4225	240	34.7	339	38.9
× 158	20130	622.3	199.9	27.7	15.7	1225	3935	247	32.1	321	39.9
× 149	18900	609.6	184.0	22.1	18.9	995	3260	229	19.9	216	32.3
× 134	17100	609.6	181.0	22.1	15.9	937	3065	234	18.7	206	33.0
× 119	15160	609.6	177.8	22.1	12.7	874	2870	241	17.6	198	34.0
S508 × 143	18190	515.6	182.9	23.4	20.3	695	2705	196	20.9	228	33.8
× 128	16320	515.6	179.3	23.4	16.8	658	2540	200	19.5	218	34.5
× 112	14190	508.0	162.2	20.2	16.1	533	2100	194	12.4	153	29.5
× 98	12520	508.0	158.9	20.2	12.8	495	1950	199	11.5	145	30.2
S457 × 104	13290	457.2	158.8	17.6	18.1	358	1690	170	10.0	127	27.4
× 81	10390	457.2	152.4	17.6	11.7	335	1465	180	8.66	114	29.0
S381 × 74	9485	381.0	143.3	15.8	14.0	202	1060	146	6.53	91.3	26.2
× 64	8130	381.0	139.7	15.8	10.4	186	977	151	5.99	85.7	27.2
S305 × 74	9485	304.8	139.1	16.7	17.4	127	832	116	6.53	94.1	26.2
× 61	7740	304.8	133.4	16.7	11.7	113	744	121	5.66	84.6	26.9
× 52	6645	304.8	129.0	13.8	10.9	95.3	626	120	4.11	63.7	24.1
× 47	6030	304.8	127.0	13.8	8.9	90.7	596	123	3.90	61.3	25.4
S254 × 52	6645	254.0	125.6	12.5	15.1	61.2	482	96.0	3.48	55.4	22.9
× 38	4815	254.0	118.4	12.5	7.9	51.6	408	103	2.83	47.7	24.2
S203 × 34	4370	203.2	105.9	10.8	11.2	27.0	265	78.7	1.79	33.9	20.3
× 27	3490	203.2	101.6	10.8	6.9	24.0	236	82.8	1.55	30.5	21.1
S178 × 30	3795	177.8	98.0	10.0	11.4	17.6	198	68.3	1.32	26.9	18.6
× 23	2905	177.8	93.0	10.0	6.4	15.3	172	72.6	1.10	23.6	19.5
S152 × 26	3270	152.4	90.6	9.1	11.8	10.9	144	57.9	0.961	21.3	17.1
× 19	2370	152.4	84.6	9.1	5.9	9.20	121	62.2	0.758	17.9	17.9
S127 × 22	2800	127.0	83.4	8.3	12.5	6.33	99.8	47.5	0.695	16.6	15.7
× 15	1895	127.0	76.3	8.3	5.4	5.12	80.6	52.1	0.508	13.3	16.3
S102 × 14	1800	101.6	71.0	7.4	8.3	2.83	55.6	39.6	0.376	10.6	14.5
× 11	1460	101.6	67.6	7.4	4.9	2.53	49.8	41.7	0.318	9.41	14.8
S76 × 11	1425	76.2	63.7	6.6	8.9	1.22	32.0	29.2	0.244	7.67	13.1
× 8.5	1075	76.2	59.2	6.6	4.3	1.05	27.5	31.2	0.189	6.39	13.3

Courtesy of The American Institute of Steel Construction.
*S is the symbol for a standard beam, followed by the nominal depth in mm, then the mass in kg per meter of length.

GEOMETRIC PROPERTIES OF ROLLED STEEL SHAPES

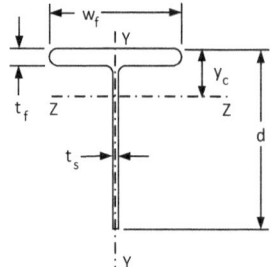

Structural Tees (U.S. Customary Units)

Designation*	Area (in.^{2})	Depth of Tee (in.)	Flange Width (in.)	Flange Thickness (in.)	Stem Thickness (in.)	Axis Z-Z I (in.^{4})	Axis Z-Z Z (in.^{3})	Axis Z-Z r (in.)	y_c (in.)	Axis Y-Y I (in.^{4})	Axis Y-Y Z (in.^{3})	Axis Y-Y r (in.)
WT18 × 115	33.8	17.950	16.470	1.260	0.760	934	67.0	5.25	4.01	470	57.1	3.73
× 80	23.5	18.005	12.000	1.020	0.650	740	55.8	5.61	4.74	147	24.6	2.50
WT15 × 66	19.4	15.155	10.545	1.000	0.615	421	37.4	4.66	3.90	98.0	18.6	2.25
× 54	15.9	14.915	10.475	0.760	0.545	349	32.0	4.69	4.01	73.0	13.9	2.15
WT12 × 52	15.3	12.030	12.750	0.750	0.500	189	20.0	3.51	2.59	130	20.3	2.91
× 47	13.8	12.155	9.065	0.875	0.515	186	20.3	3.67	2.99	54.5	12.0	1.98
× 42	12.4	12.050	9.020	0.770	0.470	166	18.3	3.67	2.97	47.2	10.5	1.95
× 31	9.11	11.870	7.040	0.590	0.430	131	15.6	3.79	3.46	17.2	4.90	1.38
WT9 × 38	11.2	9.105	11.035	0.680	0.425	71.8	9.83	2.54	1.80	76.2	13.8	2.61
× 30	8.82	9.120	7.555	0.695	0.415	64.7	9.29	2.71	2.16	25.0	6.63	1.69
× 25	7.33	8.995	7.495	0.570	0.355	53.5	7.79	2.70	2.12	20.0	5.35	1.65
× 20	5.88	8.950	6.015	0.525	0.315	44.8	6.73	2.76	2.29	9.55	3.17	1.27
WT8 × 50	14.7	8.485	10.425	0.985	0.585	76.8	11.4	2.28	1.76	93.1	17.9	2.51
× 25	7.37	8.130	7.070	0.630	0.380	42.3	6.78	2.40	1.89	18.6	5.26	1.59
× 20	5.89	8.005	6.995	0.505	0.305	33.1	5.35	2.37	1.81	14.4	4.12	1.57
× 13	3.84	7.845	5.500	0.345	0.250	23.5	4.09	2.47	2.09	4.80	1.74	1.12
WT7 × 60	17.7	7.240	14.670	0.940	0.590	51.7	8.61	1.71	1.24	247	33.7	3.74
× 41	12.0	7.155	10.130	0.855	0.510	41.2	7.14	1.85	1.39	74.2	14.6	2.48
× 34	9.99	7.020	10.035	0.720	0.415	32.6	5.69	1.81	1.29	60.7	12.1	2.46
× 24	7.07	6.985	8.030	0.595	0.340	24.9	4.48	1.87	1.35	25.7	6.40	1.91
× 15	4.42	6.920	6.730	0.385	0.270	19.0	3.55	2.07	1.58	9.79	2.91	1.49
× 11	3.25	6.870	5.000	0.335	0.230	14.8	2.91	2.14	1.76	3.50	1.40	1.04
WT6 × 60	17.6	6.560	12.320	1.105	0.710	43.4	8.22	1.57	1.28	172	28.0	3.13
× 48	14.1	6.355	12.160	0.900	0.550	32.0	6.12	1.51	1.13	135	22.2	3.09
× 36	10.6	6.125	12.040	0.670	0.430	23.2	4.54	1.48	1.02	97.5	16.2	3.04
× 25	7.34	6.095	8.080	0.640	0.370	18.7	3.79	1.60	1.17	28.2	6.97	1.96
× 15	4.40	6.170	6.520	0.440	0.260	13.5	2.75	1.75	1.27	10.2	3.12	1.52
× 8	2.36	5.995	3.990	0.265	0.220	8.70	2.04	1.92	1.74	1.41	0.706	0.773
WT5 × 56	16.5	5.680	10.415	1.250	0.755	28.6	6.40	1.32	1.21	118	22.6	2.68
× 44	12.9	5.420	10.265	0.990	0.605	20.8	4.77	1.27	1.06	89.3	17.4	2.63
× 30	8.82	5.110	10.080	0.680	0.420	12.9	3.04	1.21	0.884	58.1	11.5	2.57
× 15	4.42	5.235	5.810	0.510	0.300	9.28	2.24	1.45	1.10	8.35	2.87	1.37
× 6	1.77	4.935	3.960	0.210	0.190	4.35	1.22	1.57	1.36	1.09	0.551	0.785
WT4 × 29	8.55	4.375	8.220	0.810	0.510	9.12	2.61	1.03	0.874	37.5	9.13	2.10
× 20	5.87	4.125	8.070	0.560	0.360	5.73	1.69	0.988	0.735	24.5	6.08	2.04
× 12	3.54	3.965	6.495	0.400	0.245	3.53	1.08	0.999	0.695	9.14	2.81	1.61
× 9	2.63	4.070	5.250	0.330	0.230	3.41	1.05	1.14	0.834	3.98	1.52	1.23
× 5	1.48	3.945	3.940	0.205	0.170	2.15	0.717	1.20	0.953	1.05	0.532	0.841
WT3 × 10	2.94	3.100	6.020	0.365	0.260	1.76	0.693	0.774	0.560	6.64	2.21	1.50
× 6	1.78	3.015	4.000	0.280	0.230	1.32	0.564	0.861	0.677	1.50	0.748	0.918
WT2 × 6.5	1.91	2.080	4.060	0.345	0.280	0.526	0.321	0.524	0.440	1.93	0.950	1.00

Courtesy of The American Institute of Steel Construction.
*WT is the symbol for a structural T-section (cut from a W-section), followed by the nominal depth in inches, and the weight in pounds per foot of length.

GEOMETRIC PROPERTIES OF ROLLED STEEL SHAPES

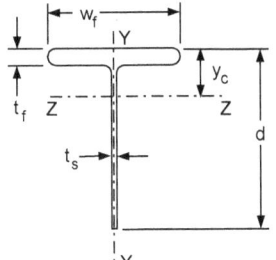

Structural Tees (SI Units)

Designation*	Area (mm^2)	Depth of Tee (mm)	Flange Width (mm)	Flange Thickness (mm)	Stem Thickness (mm)	Axis Z-Z I (10^6 mm^4)	Axis Z-Z Z (10^3 mm^3)	r (mm)	y_c (mm)	Axis Y-Y I (10^6 mm^4)	Axis Y-Y Z (10^3 mm^3)	r (mm)
WT457 × 171	21805	455.9	418.3	32.0	19.3	389	1098	133	102	196	936	94.7
× 119	15160	457.3	304.8	25.9	16.5	308	914	142	120	61.2	403	63.5
WT381 × 98	12515	384.9	267.8	25.4	15.6	175	613	118	99.1	40.8	305	57.2
× 80	10260	378.8	266.1	19.3	13.8	145	524	119	102	30.4	228	54.6
WT305 × 77	9870	305.6	323.9	19.1	12.7	78.7	328	89.2	65.8	54.1	333	73.9
× 70	8905	308.7	230.3	22.2	13.1	77.4	333	93.2	75.9	22.7	197	50.3
× 63	8000	306.1	229.1	19.6	11.9	69.1	300	93.2	75.4	19.6	172	49.5
× 46	5875	301.5	178.8	15.0	10.9	54.5	256	96.3	87.9	7.16	80.3	35.1
WT229 × 57	7225	231.3	280.3	17.3	10.8	29.9	161	64.5	45.7	31.7	226	66.3
× 45	5690	231.6	191.9	17.7	10.5	26.9	152	68.8	54.9	10.4	109	42.9
× 37	4730	228.5	190.4	14.5	9.0	22.3	128	68.6	53.8	8.32	87.7	41.9
× 30	3795	227.3	152.8	13.3	8.0	18.6	110	70.1	58.2	3.98	51.9	32.3
WT203 × 74	9485	215.5	264.8	25.0	14.9	32.0	187	57.9	44.7	38.8	293	63.8
× 37	4755	206.5	179.6	16.0	9.7	17.6	111	61.0	48.0	7.74	86.2	40.4
× 30	3800	203.3	177.7	12.8	7.7	13.8	87.7	60.2	46.0	5.99	67.5	39.9
× 19	2475	199.3	139.7	8.8	6.4	9.78	67.0	62.7	53.1	2.00	28.5	28.4
WT178 × 89	11420	183.9	372.6	23.9	15.0	21.5	141	43.4	31.5	103	552	95.0
× 61	7740	181.7	257.3	21.7	13.0	17.1	117	47.0	35.3	30.9	239	63.0
× 51	6445	178.3	254.9	18.3	10.5	13.6	93.2	46.0	32.8	25.3	198	62.5
× 36	4560	177.4	204.0	15.1	8.6	10.4	73.4	47.5	34.3	10.7	105	48.5
× 22	2850	175.8	170.9	9.8	6.9	7.91	58.2	52.6	40.1	4.07	47.7	37.8
× 16	2095	174.5	127.0	8.5	5.8	6.16	47.7	54.4	44.7	1.46	22.9	26.4
WT152 × 89	11355	166.6	312.9	28.1	18.0	18.1	135	39.9	32.5	71.6	459	79.5
× 71	9095	161.4	308.9	22.9	14.0	13.3	100	38.4	28.7	56.2	364	78.5
× 54	6840	155.6	305.8	17.0	10.9	9.66	74.4	37.6	25.9	40.6	265	77.2
× 37	4735	154.8	205.2	16.2	9.4	7.78	62.1	40.6	29.7	11.7	114	49.8
× 22	2840	156.7	165.6	11.2	6.6	5.62	45.1	44.5	32.3	4.25	51.1	38.6
× 12	1525	152.3	101.3	6.7	5.6	3.62	33.4	48.8	44.2	0.587	11.6	19.6
WT127 × 83	10645	144.3	264.5	31.8	19.2	11.9	105	33.5	30.7	49.1	370	68.1
× 65	8325	137.7	260.7	25.1	15.4	8.66	78.2	32.3	26.9	37.2	285	66.8
× 45	5690	129.8	256.0	17.3	10.7	5.37	49.8	30.7	22.5	24.2	188	65.3
× 22	2850	133.0	147.6	13.0	7.6	3.86	36.7	36.8	27.9	3.48	47.0	34.8
× 9	1140	125.3	100.6	5.3	4.8	1.81	20.0	39.9	34.5	0.454	9.03	19.9
WT102 × 43	5515	111.1	208.8	20.6	13.0	3.80	42.8	26.2	22.2	15.6	150	53.3
× 30	3785	104.8	205.0	14.2	9.1	2.39	27.7	25.1	18.7	10.2	99.6	51.8
× 18	2285	100.7	165.0	10.2	6.2	1.47	17.7	25.4	17.7	3.80	46.0	40.9
× 13	1695	103.4	133.4	8.4	5.8	1.42	17.2	29.0	21.2	1.66	24.9	31.2
× 7	955	100.2	100.1	5.2	4.3	0.895	11.7	30.5	24.2	0.437	8.72	21.4
WT76 × 15	1895	78.7	152.9	9.3	6.6	0.733	11.4	19.7	14.2	2.76	36.2	38.1
× 9	1150	76.6	101.6	7.1	5.8	0.549	9.24	21.9	17.2	0.624	12.3	23.3
WT51 × 10	1230	52.8	103.1	8.8	7.1	0.219	5.26	13.3	11.2	0.803	15.6	25.4

Courtesy of The American Institute of Steel Construction.
*WT is the symbol for a structural T-section (cut from a W-section), followed by the nominal depth in mm, and the mass in kg per meter of length.

APPENDIX E
EQUATIONS FOR DEFLECTIONS OF SIMPLY SUPPORTED BEAMS

SIMPLY SUPPORTED BEAMS LENGTH, L	SLOPE	DEFLECTION EQUATIONS
Center point load F at $L/2$	$\left(\dfrac{dy}{dx}\right)_{Max} = \dfrac{-FL^2}{16EI_z}$	$y = \dfrac{-Fx}{48EI_z}(3L^2 - 4x^2)$, $0 \le x \le L/2$ $y_{Max} = \dfrac{-FL^3}{48EI_z}$
Off-center point load F, distances a and b	$\left(\dfrac{dy}{dx}\right)_A = \dfrac{-Fab(L+b)}{6EI_z L}$ $\left(\dfrac{dy}{dx}\right)_B = \dfrac{Fab(L+a)}{6EI_z L}$	$y = \dfrac{-Fbx}{6EI_z L}(L^2 - b^2 - x^2)$, $0 \le x \le a$ $y_{x=a} = \dfrac{-Fba}{6EI_z L}(L^2 - b^2 - a^2)$
Moment M at A	$\left(\dfrac{dy}{dx}\right)_A = -\dfrac{ML}{3EI_z}$ $\left(\dfrac{dy}{dx}\right)_B = \dfrac{ML}{6EI_z}$	$y = \dfrac{-Mx}{6EI_z L}(x^2 - 3Lx + 2L^2)$ $y_{Max} = \dfrac{-ML^2}{\sqrt{243}EI_z}$ at $x = 0.4226L$
Uniform distributed load q_0	$\left(\dfrac{dy}{dx}\right)_{Max} = -\dfrac{q_0 L^3}{24EI_z}$	$y = \dfrac{-q_0 x}{24EI_z}(x^3 - 2Lx^2 + L^3)$ $y_{Max} = \dfrac{-5q_0 L^4}{384EI_z}$
Uniform distributed load q_0 over left half $L/2$	$\left(\dfrac{dy}{dx}\right)_A = -\dfrac{3q_0 L^3}{128EI_z}$ $\left(\dfrac{dy}{dx}\right)_B = \dfrac{7q_0 L^3}{384EI_z}$	$y = \dfrac{-q_0 x}{384EI_z}(16x^3 - 24Lx^2 + 9L^3)$, $0 \le x \le L/2$ $y = \dfrac{-q_0 L}{384EI_z}(8x^3 - 24Lx^2 + 17L^2 x - L^3)$, $L/2 \le x \le L$ $y_{Max} = \dfrac{-6.563 \times 10^{-3} q_0 L^4}{EI_z}$ at $x = 0.4598L$
Triangular load, q_0 at B	$\left(\dfrac{dy}{dx}\right)_A = \dfrac{-7q_0 L^3}{360EI_z}$ $\left(\dfrac{dy}{dx}\right)_B = \dfrac{q_0 L^3}{45EI_z}$	$y = \dfrac{-q_0 x}{360EI_z L}(3x^4 - 10L^2 x^2 + 7L^4)$ $y_{Max} = \dfrac{-6.52 \times 10^{-3} q_0 L^4}{EI_z}$ at $x = 0.5193$

EQUATIONS FOR DEFLECTIONS OF CANTILEVER BEAMS

CANTILEVER BEAMS LENGTH, L	SLOPE	DEFLECTION EQUATIONS
Cantilever with point load F at free end	$\left(\dfrac{dy}{dx}\right)_{Max} = \dfrac{-FL^2}{2EI_z}$	$y = \dfrac{-Fx^2}{6EI_z}(3L - x)$ $y_{Max} = \dfrac{-FL^3}{3EI_z}$
Cantilever with point load F at $L/2$	$\left(\dfrac{dy}{dx}\right)_{Max} = \dfrac{-FL^2}{8EI_z}$	$y = \dfrac{-Fx^2}{6EI_z}\left(\dfrac{3}{2}L - x\right)$ $0 \le x \le L/2$ $y = \dfrac{-FL^2}{24EI_z}\left(3x - \dfrac{1}{2}L\right)$ $L/2 \le x \le L$
Cantilever with uniform load q_0 over full length	$\left(\dfrac{dy}{dx}\right)_{Max} = \dfrac{-q_0 L^3}{6EI_z}$	$y = \dfrac{-q_0 x^2}{24EI_z}\left(x^2 - 4Lx + 6L^2\right)$ $y_{Max} = \dfrac{-q_0 L^4}{8EI_z}$
Cantilever with end moment M_0	$\left(\dfrac{dy}{dx}\right)_{Max} = \dfrac{ML}{EI_z}$	$y = \dfrac{Mx^2}{2EI_z}$ $y_{Max} = \dfrac{ML^2}{2EI_z}$
Cantilever with uniform load q_0 over $L/2$	$\left(\dfrac{dy}{dx}\right)_{Max} = \dfrac{-q_0 L^3}{48EI_z}$	$y = \dfrac{-q_0 x^2}{24EI_z}\left(x^2 - 2Lx + \dfrac{3}{2}L^2\right)$ $0 \le x \le L/2$ $y = \dfrac{-q_0 L^3}{192EI_z}\left(4x - \dfrac{L}{2}\right)$ $L/2 \le x \le L$
Cantilever with triangular load q_0	$\left(\dfrac{dy}{dx}\right)_{Max} = \dfrac{-q_0 L^3}{24EI_z}$	$y = \dfrac{-q_0 x^2}{120EI_z L}\left(10L^3 - 10L^2 x + 5Lx^2 - x^3\right)$ $y_{Max} = \dfrac{-q_0 L^4}{30EI_z}$

INDEX

American Standard Beams, 436
American Steel and Wire C.,
Angle of twist, 164
Area, 416
    first moment of, 417
    centroid, 417
    percent reduction, 78
    polar moment of, 423
    second moment of, 423
    table of properties, 425
Axial loading, 99

Bars,
    design analysis of, 100-101
    tapered, 120
    stepped, 123
Beams, 196
    cantilever, 214
    composite, 222
    deformation, 199, 254
    elastic curve, 254-255
    foam core, 222
    plastic bending, 230
    pure bending, 196
    rectangular cross section, 205
    reinforced concrete, 225
    shear forces, 217-219
    shear stresses, 217
    simply supported, 207
    singularity functions, 272
    statically indeterminate, 293
    strains, 200
    stresses, 201
    symmetric, 196-197
Bending,
    plastic, 230-234
    transverse forces, 206
bending moment,
    in beams, 197-198
Bending moment diagram, 210
Bending moment function, 205
Boundary conditions, 267
Brittle, 75
Buckling, 333
Buckling of columns, 333
    end conditions, 337
    fixed & fixed, 341
    free & fixed, 340
    pinned-ends, 334
    pinned & fixed, 337

Cables,
    Characteristics, 99

Cables,
    design analysis of, 100-101
    forces in, 100
Cantilever beam, 214
Castigliano's theorem, 383
    beam deflection, 384
    indeterminate beams, 388
    rotations, 391
    truss deflections, 393
Center of gravity, 417
Centroid, 417
    area, 111, 417
    chart of, 421
    composite area, 427
Coefficient thermal expansion,
    table of, 68
Columns, 333
    buckling, 333
    eccentric loading, 347
    stresses, 349
Combined loading, 313
Compatibility conditions, 267
Components, 3
    of force, 3, 7, 11
    of vector, 3
Composite area, 442
Composite beams, 222
Concrete,
    reinforced, 225
Concurrent forces, 12
Conversion factors, 6
Coplanar forces, 12
Critical force, 334
Cross section,
    properties of, 205
Curvature, 196-199, 254
Cyclic stresses, 84
    crack growth, 85

Deformation, 2
    equations for beams, 400-401
    of bars, 101, 114
    of beams, 199, 254
    of circular shafts, 164
    of tapered bars, 120-122
    of stepped bars, 123-124
    of wire, 101, 114
Design stress, 73-75
Diagram, 209-210
    bending moment, 209
    free body, 11-14
    S_f-N, 87
    shear, 209-210

    stress-strain, 76
Displacement, 50
Distributed load, 9
Ductile,
    failure, 77
Ductility,
    measures of, 78
Dynamics, 3
Dynamic loading, 366

Eccentricity, 347
Eccentric loading, 347
    stresses in columns, 349
Elastic constants, 80
Elastic curve, 254
    by integration, 255
Elastic instability,
Elastic limit, 76
Elastic region, 76-77
Elongation, percent, 78
Energy methods, 364-366
Energy theorems, 365
    Castigliano's theorem, 383
    least work, 366
    virtual work, 365
Equations for:
    deformation of beams, 440
Equations of equilibrium, 2
    coplanar, concurrent, 12
    coplanar, non–concurrent, 12
    non–coplanar, concurrent, 12
    non–coplanar, non–
      concurrent, 12
    scalar form, 12
    vector form, 12
Equations of transformation:
    strain, 54
    stress, 26
Equilibrium
    equations, 2, 3
    three-dimensional eqs. 11
Euler theory, 336
Extensometer, 75
External forces, 4, 8

Failure, 2
    types of, 4, 106
Failure theories, 42
    max distortion energy, 44
    max principal stress, 43
    max shear stress, 43

Fatigue, 84
    crack, 85
    corrosion, 94
    strength, 84
    S_f–N diagram, 87
    surface finish, 95
    testing, 85
Fixed support, 13
Force,
    axial, 8, 18
    Cartesian components, 12
    component, 12
    concentrated, 9
    concurrent, 12
    coplanar, 12
    distributed, 9
    external, 4, 8
    gravitational, 4
    internal, 4, 8, 104
    magnitude, 14
    non-concurrent, 11
    non-coplanar, 11, 12
    reactive, 9
    shear, 14
    tension, 9
    transverse, 45
    uniformly distributed, 8
    vectors, 8-10
Free body diagram, 11, 13
    construction of, 13
    partial bodies, 16

Geometric properties steel shapes, 434
    American standard beams, 436
    structural tees, 438
    wide flange beams, 434
Gerber, 93
Goodman diagram, 93
Goodman method, 93
Gravitation,
    force, 4
    constant, 4
    universal constant, 4

Hooke, Robert, 103
Hooke's law, 103
Horsepower, 149
Impact loading, 366, 376
    of beams, 378
    of rods, 377
Inertia, moments of, 423
Instability, 333
Integrating,
    load distributions, 268
    shear force, 268
    singularity functions, 269

Internal forces, 2, 4   solving for, 8

Key, 116-117

Law
    of gravitation, 4
    Newton's, 3
Load,
    critical, 334
Load cell, 75

Magnitude of a vector, 8
Material properties, 77
Mechanics of materials, 2
Method
    of integration, 268
    of singularity functions, 269
Models,
    scale, 129
Modeling
    loads, 14
    supports, 14-15
Modulus
    of elasticity, 80
    of a section, 203
Mohr's circle, 36, 42, 56
Moment, 10
    arm, 10
    diagram, 209
    direction, 10
    external, 197
    curvature relation, 196
Moment of an area, 416
    first moment, 417
    polar moment, 424
    second moment, 423
Moment of a force, 10-11, 197
    about a point, 10
    sense, 10
Moment of inertia,
    composite area, 427
    polar, 424
Momentum, 11

Neutral axis, 254
Newton, Sir Isaac, 1
Newton's laws, 3
    first law, 3
    of gravitational attraction, 4
    of motion, 1
Newton's laws, 3
    second law, 3
    third law, 3
Non–concurrent, 11
Non–coplanar, 11

Parallel axis theorem, 426
Percent elongation, 78
Percent reduction in area, 78
Physical properties materials, 413
Plane stress, 61
Plastic bending, 230
Plastic hinge, 236
Plastic regime, 76-77
Poisson's ratio, 73, 80
Polar moment of inertia, 424
Power, 149
    transmission shafts, 168
Pressure vessel stresses, 305
    spherical, 305
    cylindrical, 306
Principal stresses, 30
    from strain gages, 62
    in shafts, 163
Properties,
    of area, 17, 416
    of cross sections, 205
    physical, 413
    structural steels, 413-414
    rolled steel shapes, 434
    strength, 77
    tensile, 75
    yield, 77
Prototype, 130
Pure bending, 196

Radius of curvature, 199, 254
Radius of gyration, 345
Reactive force, 14
    solving for, 3, 13
Rebar, 225-227
Reinforced concrete, 225
Right hand rule, 10
Rigid body, 1
Rods,
    design analysis of, 100-101
Rosette, 64
    rectangular, 64-65

Safety factor, 107
Scalars, 7
    equations, 3
Scale factors, 129
    for displacements, 131
    for loads, 130
    for modulus, 131
    for stresses, 129
    geometric, 129
Scale models, 129
Section modulus,
    in bending, 203
    in torsion, 157-158

Sense, 10
Shafting,
  circular, 150-152
  hollow, 156
  power transmission, 168
Shear center, 174
Shear diagram, 209
Shear force,
  in beams, 209
Shear modulus, 83
Shear strain, 53
  circular shaft, 151
Shear stress, 34, 116
  in beams, 217
  in shafts, 152
  maximum, 34
  on oblique planes, 117, 161
  on orthogonal planes, 160
Sheet metal gages, 412
Significant figures, 7
Singularity functions, 269
  table of, 271
Slenderness ratio, 345
Springs,
  torsion bar, 167
Stability, 333
Statically determinate, 2
Statically indeterminate,
  axial members, 132
  beams, 288
  singularity functions, 290
  superposition, 288
Statics, 1
Steel
  properties, 414
  reinforcing bars,
  wire, 412
Strain, 50
  beams,
  displacement Eqs., 53
  energy, 346, 349
  engineering, 75, 77
  hardening, 77
  Mohr's circle, 56
  normal, 53
  plane, 61
  plastic, 75-77
  range, 78-79
  rosettes,
  shear, 98
  stress Eqs., 59
  transformation eqs. 54
  thermal, 66
  true, 75
  wire, 100

Strain energy,
  density, 366
  in beams in bending, 371
  in shear, 374
  in structural elements, 368
  in tie rods, 368
Strain gages, 62
  rosettes, 64
Strength, 2, 107
  fatigue, 84
  shear, 6,
  ultimate tensile, 75
  yield, 77
Stress, 2, 7
  alternating, 91
  bars, 8
  beams, 196
  bearing,
  Cartesian, 24-25
  columns, 333
  concentration factors, 125, 234
  cyclic, 84-85
  design, 73-75
  distributions, 190
  engineering, 75
  equations of transformation, 26
  internal, 2, 104
  mean, 90
  nominal, 125
  normal, 24
  oblique planes, 117
  plane, 54, 59, 61
  point, 23
  pressure vessels, 305
  principal, 30, 163
  resultant, 23
  rods, 113
  range, 78-79
  shafts in torsion, 149-152
  shear, 34
  states, 60
  stepped bars, 123
  strain Eqs., 59
Stress, 2, 7
  tangential, 24
  tapered bars, 120
  thermal, 66
  transformations, 26
  true, 75
  yield, 73, 78

Sheet metal gages, 412
Stress concentration factors, 125
  bar with circular hole, 125
  grooved shaft in bending, 236
  notched beam, 235

Stress concentration factors, 125
  shouldered bar, 127
  shouldered shaft torsion, 184
  stepped beam, 238
  stepped shaft in bending, 238
  tension members, 125
  transverse hole bending, 240
Stress–strain curves, 76-79
Stress–strain eqs. 59
Structural effectiveness, 160, 205
Structural tees, 438
Superposition, 281
Supports, 14-17
  built-in, 16
  simple, 13
Support reactions, 13-15
Symmetry,
  axis of, 196
  conditions, 197
System of units, 6

Tapered bars, 120
Tensile properties
  metals, 414
  non-metals, 415
Tensile test, 73
  specimen, 74
Tension,
  forces, 9
  rod, 99-101
Tensors, 7
Torque, 149
Torsion, 149
  bar, 167
  circular shafts, 151-154
  hollow circular shafts, 156
  non-circular shafts, 171
  thin-walled tube, 178

Ultimate tensile strength, 75
Units
  basic, 5
  international system, 6
  U. S. Customary, 6
Universal testing machine, 74
Unstable, 333

Vector, 7
  Cartesian, 3, 11
  components, 3, 7, 11
  concurrent, 12
  coplanar, 12
  direction, 8
  force, 8-10
  moment, 10
  non-coplanar, 12

Watt, James, 149
Weight, 4
Wide flange beam, 434-435
Wire,
    deformation, 101, 114
    gages, 412
    rope, 108
Work, 6,
Wrench, 10,

Yielding, 73-78
Yield strength, 73
Young's modulus, 80

Basic Quantities and Units

System of Units	Length	Time	Mass	Force	g
International System of Units (SI)	meter (m)	second (s)	kilogram (kg)	newton (N) (kg-m)/s^2	9.807 m/s^2
U. S. Customary (FPS)	foot (ft)	second (s)	slug (lb-s^2)/ft	pound (lb)	32.17 ft/s^2

Units of Other Frequently Used Quantities

System of Units	Moment M	Stress σ	Strain ε
SI	newton-meter (N-m)	pascal (Pa) = (N/m^2) mega pascal MPa = N/mm^2	dimensionless
U. S. Customary (FPS)	foot-pound (ft-lb)	pound/square foot (lb/ft^2) pound/square in. (lb/in.^{2}) = psi	dimensionless

SI Prefixes

Multiplication Factor	Prefix Name	Prefix Symbol
10^{18}	exa	E
10^{15}	peta	P
10^{12}	tera	T
10^{9}	giga	G
10^{6}	mega	M
10^{3}	kilo	k
10^{2}	hecto*	h
10^{1}	deka*	da
10^{-1}	deci*	d
10^{-2}	centi*	c
10^{-3}	milli	m
10^{-6}	micro	μ
10^{-9}	nano	n
10^{-12}	pico	p
10^{-15}	femto	f
10^{-18}	atto	a

*To be avoided when possible.

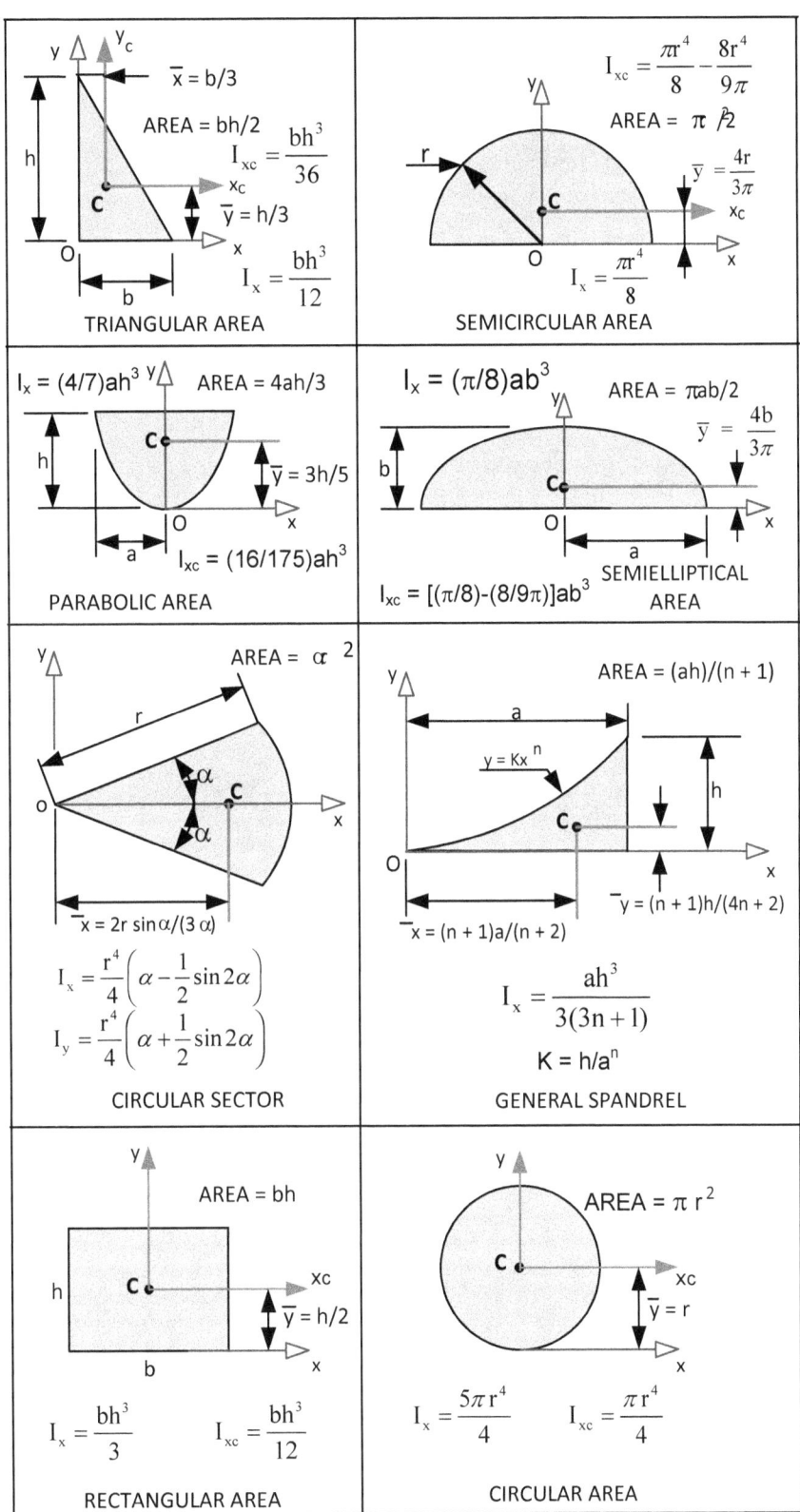

www.ingramcontent.com/pod-product-compliance
Lightning Source LLC
LaVergne TN
LVHW080134260326
834688LV00042B/1170